Extra High Voltage A.C. Transmission Engineering

Rakosh Das Begamudre

Indian Institute of Technology
Kanpur, India

A HALSTED PRESS BOOK

JOHN WILEY & SONS

New York   Chichester   Brisbane   Toronto   Singapore

Copyright © 1986, WILEY EASTERN LIMITED
New Delhi

Published in the Western Hemisphere by
Halsted Press, A Division of
John Wiley & Sons, Inc., New York

Library of Congress Cataloging in Publication Data

Begamudre, Rakosh Das.
    Extra high voltage A.C. transmission engineering.

    "A Halsted Press book".
    1. Electric power distribution—High tension.
I. Title.
TK3144.B37 1986 621.319'13 85-26303

ISBN 0-470-20267-X

Printed in India at Prabhat Press, Meerut.

TO
my mother and sister

both brave women who fought and won over their common enemy 'cancer' in their seventies and sixties to give hope to younger generation of their sex, this book is lovingly dedicated.

Foreword

It gives me pleasure to write the foreword to the book 'Extra High Voltage A.C. Transmission Engineering' authored by Rakosh Das Begamudre, Visiting Professor in Electrical Engineering at the Indian Institute of Technology, Kanpur. The field of e.h.v. is a very growing and dynamic one on which depends to a large measure the industrial growth of a developing country like ours. A course in this subject is offered for advanced undergraduate and postgraduate students, and the Institute also organized a short-term course for teachers in other learned institutions and practising engineers in India under the Quality Improvement Programme.

With a background of nearly 35 years in this area in India, Japan, U.S.A. and Canada in teaching, design, research, and development areas, I consider Dr. Begamudre one of the ablest persons to undertake the task of writing a book, placing his wide experience at the disposal of younger engineers. He has worked at notable institutions such as the National Research Council of Canada and the Central Power Research Institute, Bangalore, and several other places. His publications in the field of e.h.v. transmission are numerous and varied in extent. The I.I.T. Kanpur offered him Visiting Professorship and I am delighted to introduce the book by him to learned readers in this important field. It is not only a worthy addition to the technical literature in this topic but also to the list of text books published in India.

S. SAMPATH
Director
Indian Institute of Technology
Kanpur

Preface

Extra High Voltage (EHV) A.C. transmission may be considered to have come of age in 1952 when the first 380–400 kV line was put into service in Sweden. Since then, industrialized countries all over the world have adopted this and higher voltage levels. Very soon it was found that the impact of such voltage levels on the environment needed careful attention because of high surface voltage gradients on conductors which brought interference problems from power frequency to TV frequencies. Thus electrostatic fields in the line vicinity, corona effects, losses, audible noise, carrier interference, radio interference and TVI became recognized as steady-state problems governing the line conductor design, line height, and phase-spacing to keep the interfering fields within specified limits. The line-charging current is so high that providing synchronous condensers at load end only was impractical to control voltages at the sending-end and receiving-end buses. Shunt compensating reactors for voltage control at no load and switched capacitors at load conditions became necessary. The use of series capacitors to increase power-handling capacity has brought its own problems such as increased current density, temperature rise of conductors, increased short-circuit current and subsynchronous resonance. All these are still steady-state problems.

However, the single serious problem encountered with e.h.v. voltage levels is the overvoltages during switching operations, commonly called switching-surge overvoltages. Very soon it was found that a long airgap was weakest for positive polarity switching surges. The coordination of insulation must now be based on switching impulse levels (SIL) and not on lightning impulse levels.

From time to time, outdoor research projects have been established to investigate high-voltage effects from e.h.v. and u.h.v. lines to place line designs on a more scientific basis, although all variables in the problem are statistical in nature and require long-term observations to be carried out. Along with field data, analysis of various problems and calculations using the Digital Computer have advanced the state of the art of e.h.v. line designs to a high level of scientific attainment. Most basic mechanisms are now placed on a firm footing, although there are still an endless list of problems that require satisfactory solution.

During his lecturing career for undergraduate and postgraduate classes in High Voltage A.C. Transmission the author was unable to find a text book suitable for the courses. The existing text books are for first courses in High Voltage Engineering concentrating on breakdown phenomena of solid, liquid, and gaseous insulation, together with high-voltage laboratory and measurement techniques. On the other hand, reference books are very highly specialized which deal with results obtained from one of the outdoor projects mentioned earlier. To bridge the gap, this text-reference book for a course in EHV A.C. Transmission is presented. The material has been tried out on advanced undergraduate and post-graduate courses at the I.I.T. Kanpur, in special short-term courses offered to teachers in Universities and practising engineers through the Quality Improvement Programme, and during the course of his lectures offered at other Universities and Institutes. Some of the material is based on the author's own work at reputed research and development organizations such as the National Research Council of Canada, and similar organizations in India and at Universities and Institutes, over the past 25 years. But no one single person or organization can hope to deal with all problems so that over the years, the author's notes have grown through reference work of technical and scientific journals which have crystallized into the contents of the book. It is hoped that it will be useful also for engineers as well as scientists engaged in research, development, design, and decision making about e.h.v. a.c. transmission lines.

Acknowledgement: The preparation of such a work has depended on the influence, cooperation and courtesy of many organizations and individuals. To start with, I acknowledge the deep influence which three of my venerable teachers had on my career—Principal Manoranjan Sengupta at the Banaras Hindu University, Professor Dr. Shigenori Hayashi at the Kyoto University, Japan, and finally to Dean Loyal Vivian Bewley who exercised the greatest impact on me in the High Voltage field at the Lehigh University, Bethlehem, Pennsylvania, USA. To the Council and Director of the I.I.T. Kanpur, I am indebted for giving me a Visiting Professorship, to Dr. S.S. Prabhu, the Head of EE Department, for constant help and encouragement at all times. To the coordinator, Q.I.P. Programme at I.I.T., Dr. A. Ghosh, I owe the courtesy of defraying the expense for preparation of the manuscript. To the individuals who have done the typing and drafting, I owe my thanks. My special thanks are due to Mr. H.S. Poplai, Publishing Manager, Wiley Eastern Publishing Company, for his cooperation and tolerence of delays in preparing the manuscript. Thanks finally are due to my colleagues, both postgraduate students and professors, who have helped me at many stages of the work involved in preparing this book.

Electrical Engineering Department
Indian Institute of Technology
Kanpur, U.P., 208 016, India.

RAKOSH DAS BEGAMUDRE

Contents

CHAPTER 1

Introduction to EHV AC Transmission

1.1 Role of EHV AC Transmission

Industrial-minded countries of the world require a vast amount of energy of which electrical energy forms a major fraction. There are other types of energy such as oil for transportation and industry, natural gas for domestic and industrial consumption, which form a considerable proportion of the total energy consumption. Thus, electrical energy does not represent the only form in which energy is consumed but an important part nevertheless. It is only 150 years since the invention of the dynamo by Faraday and 120 years since the installation of the first central station by Edison using dc. But the world has already consumed major portion of its natural resources in this short period and is looking for sources of energy other than hydro and thermal to cater for the rapid rate of consumption which is outpacing the discovery of new resources. This will not slow down with time and therefore there exists a need to reduce the rate of annual increase in energy consumption by any intelligent society if resources have to be preserved for posterity. After the end of the Second World War, countries all over the world have become independent and are showing a tremendous rate of industrial development, mostly on the lines of North-American and European countries, the U.S.S.R. and Japan. Therefore the need for energy is very urgent in these developing countries, and national policies and their relation to other countries are sometimes based on energy requirements, chiefly nuclear. Hydro-electric and coal or oil-fired stations are located very far from load centres for various reasons which requires the transmission of the generated electric power over very long distances. This requires very high voltages for transmission. The very rapid strides taken by development of dc transmission since 1950 is playing a major role in extra-long-distance transmission, complementing or supplementing e.h.v. ac transmission. Both have their roles to play and a country must make intelligent assessment of both in order to decide which is best suited for the country's economy. This book concerns itself with problems of e.h.v. ac transmission only.

1.2 Brief Description of Energy Sources and Their Development

Any engineer interested in electrical power transmission must concern himself or herself with energy problems. Electrical energy sources for industrial and domestic use can be divided into two broad categories; (1) Transportable; and (2) Locally Usable.

Transportable type is obviously hydro-electric and conventional thermal power. But locally generated and usable power is by far more numerous and exotic. Several countries, including India, have adopted national policies to investigate and develop them, earmarking vast sums of money in their multi-year plans to accelerate the rate of development. These are also called 'Alternative Sources of Power'. Twelve such sources of electric power are listed here, but there are others also which the reader will do well to research.

Locally Usable Power
(1) Conventional thermal power in urban load centres;
(2) Micro-hydel power stations;
(3) Nuclear Thermal: Fission and Fusion;
(4) Wind Energy;
(5) Ocean Energy: (a) Tidal Power, (b) Wave Power, and (c) Ocean thermal gradient power;
(6) Solar thermal;
(7) Solar cells, or photo-voltaic power;
(8) Geo-thermal;
(9) Magneto hydro-dynamic or fluid dynamic;
(10) Coal gasification and liquefaction;
(11) Hydrogen power; and last but not least,
(12) Biomass Energy: (a) Forests; (b) Vegetation; and (c) animal refuse.

To these can also be added bacterial energy sources where bacteria are cultured to decompose forests and vegetation to evolve methane gas. The water hyacinth is a very rich source of this gas and grows wildly in water-logged ponds and lakes in India. A brief description of these energy sources and their limitation as far as India is concerned is given below, with some geographical points.

(1) *Hydro-Electric Power*: The known potential in India is 50,000 MW (50 GW) with 10 GW in Nepal and Bhutan and the rest within the borders of India. Of this potential, almost 30% or 12 GW lies in the north-eastern part in the Brahmaputra Valley which has not been tapped. When this power is developed it will necessitate transmission lines of 1000 to 1500 kilometres in length so that the obvious choice is extra high voltage, ac or dc. The hydel power in India can be categorized as (a) high-head (26% of total potential), (b) medium-head (47%), (c) low-head (7%, less than 30 metres head), and (d) run-of-the-river (20%). Thus, micro-hydel plants and run-of-the-river plants (using may be bulb turbines) have a great future for remote loads in hilly tracts.

(2) *Coal.* The five broad categories of coal available in India are Peat (4500 *BTU/Lb**), Lignite (6500), Sub-Bituminous (7000–1200), Bituminous (14,000), and Anthracite (15,500 *BTU/Lb*). Only non-cooking coal of the sub-bituminous type is available for electric power production whose deposit is estimated at 50 giga tonnes in the Central Indian coal fields. With 50% of this allocated for thermal stations, it is estimated that the life of coal deposits will be 140 years if the rate of annual increase in installed capacity is 5%. Thus, the country cannot rely on this source of power to be perennial. Nuclear thermal power must be developed rapidly to replace conventional thermal power.

(3) *Oil and Natural Gas.* At present, all oil is used for transportation and none is available for electric power generation. Natural gas deposits are very meager at the oil fields in the North-Eastern region and only a few gas-turbine stations are installed to provide the electric power for the oil operations.

(4) *Coal Liquefaction and Gasification.* Indian coal contains 45% ash and the efficiency of a conventional thermal station rarely exceeds 25 to 30%. Also transportation of coal from mines to urban load centres is impossible because of the 45% ash, pilferage of coal at stations where coal-hauling trains stop, and more importantly the lack of availability of railway wagons for coal transportation. These are needed for food transportation only. Therefore, the national policy is to generate electric power in super thermal stations of 2100 MW capacity located at the mine mouths and transmit the power by e.h.v. transmission lines. If coal is liquified and pumped to load centres, power up to 7 times its weight in coal can be generated in high efficiency internal cumbustion engines.

(5) *Nuclear Energy.* The recent advances made in Liquid Metal Fast Breeder Reactors (*LMFBR*) is helping many developing countries, including India, to install large nuclear thermal plants. Although India has very limited Uranium deposits, it does possess nearly 50% of the world's Thorium deposits. The use of this material for *LMFBR* is still in infant stages and is being developed rapidly.

(6) *Wind Energy.* It is estimated that 20% of India's power requirement can be met with development of wind energy. There are areas in the Deccan Plateau in South-Central India where winds of 30 km/hour blow nearly constantly. Wind power is intermittent and storage facilities are required which can take the form of storage batteries or compressed air. For an electrical engineer, the challenge lies in devising control circuitry to generate a constant magnitude constant-frequency voltage from the variable-speed generator and to make the generator

---

*1000 *BTU/Lb* = 555.5 k-cal/kg.

operate in synchronism with an existing grid system.

(7) *Solar-Cell Energy.* Photo-voltaic power is very expensive, being nearly the same as nuclear power costing U.S. $ 1000/KW of peak power. (At the time of writing, 1 U.S. $ = Rs. 12). Solar cells are being manufactured to some extent in India, but the U.S.A. is the largest supplier still. Indian insolation level is 600 calories/sq. cm./day on the average which will generate 1.5 KW, and solar energy is renewable as compared to some other sources of energy.

(8) *Magneto Hydro-Dynamic.* The largest MHD generator successfully completed in the world is a 500 KW unit of AVCO in the U.S.A. Thus, this type of generation of electric energy has very local applications.

(9) *Fuel-Cell Energy.* The fuel-cell uses H–O interaction through a Phosphoric Acid catalyzer to yield a flow of electrons in a load connected externally. The most recent installation is by the Consolidated Edison Co. of New York which uses a module operating at 190°C. Each cell develops 0.7 V and there are sufficient modules in series to yield an output voltage of 13.8 kV, the same as a conventional control-station generator. The power output is expected to reach 1 MW.

(10) *Ocean Energy.* Energy from the vast oceans of the earth can be developed in 3 different ways: (i) Tidal; (ii) Wave; and (iii) Thermal Gradient.

TIDAL POWER. The highest tides in the world occur at 40 to 50° latitudes with tides up to 12 m existing twice daily. Therefore, Indian tides are low being about 3.5 m in the Western Coast and Eastern rivers in estuaries. The U.K. development is expected to reach 8000 MW by the year 2000 A.D. (10% of total national power). France has successfully operated a 2400 MW station at the Rance-River estuary using bulb turbines. Several installations in the world have followed suit. The development of Indian tidal power at the Gujarat Coast in the West is very ambitious and is taking shape very well. Like wind power, tidal power is intermittent in nature.

The seawater during high tides is allowed to run in the same passage where the turbine-generators are located to fill a reservoir whose retaining walls may be up to 30 km long. At low-tide periods, the stored water flows back to the sea through the turbines and power is generated.

WAVE ENERGY. An average power of 25 to 75 kw can be developed per metre of wave length depending on the wave height. The scheme uses air turbines coupled to generators located in chambers open to the sea at the bottom and closed at the top. There may be as many as 200-300 such chambers connected together at the top through pipes. A wave crest underneath some chambers will compress the air which will flow into other chambers underneath which the wave-trough is passing resulting in lower pressure. This runs the air turbines and generates power.

OCEAN THERMAL POWER. This scheme utilizes the natural temperature difference between the warm surface water (25-30°C) and the cooler ocean-bed water at 5°C. The turbine uses NH_3 as the working fluid in one type of installation which is vaporized in a heat-exchanger by the warm water. The condenser uses the cooler ocean-bed water and the cycle is complete as in a conventional power station. The cost of such an installation is nearly the same as a nuclear power station.

This brief description of 'alternative' sources of electric power should provide the reader with an interest to delve deeper into modern energy sources and their development.

1.3 Description of Subject Matter of this Book

Extra High Voltage (EHV) ac transmission can be assumed to have seen its development since the end of the Second World War, with the installation of 345 kV in North America and 400 kV in Europe. The distance between generating stations and load centres as well as the amount of power to be handled increased to such an extent that 220 kV was inadequate to handle the problem. In these nearly 40 years, the highest commercial voltage has increased to 1150 kV (1200 kV maximum) and research is under way at 1500 kV by the AEP-ASEA group. In India, the highest voltage used is 400 kV ac, but will be increased after 1990 to higher levels. The problems posed in using such high voltages are different from those encountered at lower voltages. These are:

(a) Increased Current Density because of increase in line loading by using series capacitors.

(b) Use of bundled conductors.

(c) High surface voltage gradient on conductors.

(d) Corona problems: Audible Noise, Radio Interference, Corona Energy Loss, Carrier Interference, and TV Interference.

(e) High electrostatic field under the line.

(f) Switching Surge Overvoltages which cause more havoc to air-gap insulation than lightning or power frequency voltages.

(g) Increased Short-Circuit currents and possibility of ferro-resonance conditions.

(h) Use of gapless metal-oxide arresters replacing the conventional gap-type Silicon Carbide arresters, for both lightning and switching-surge duty.

(i) Shunt reactor compensation and use of series capacitors, resulting in possible sub-synchronous resonance conditions and high short-circuit currents.

(j) Insulation coordination based upon switching impulse levels.

(k) Single-pole reclosing to improve stability, but causing problems with arcing.

The subject is so vast that no one single book can cope to handle with a description, analysis, and discussion of all topics. The book has been limited to the transmission line only and has not dealt with transient and dynamic stability, load flow, and circuit breaking. Overvoltages and characteristics of long airgaps to withstand them have been discussed at length which can be classified as transient problems. Items (a) to (e) are steady-state problems and a line must be designed to stay within specified limits for interference problems, corona loss, electrostatic field, and voltages at the sending end and receiving end buses through proper reactive power compensation.

Chapter 2 is devoted to an introduction to the e.h.v. problem, such as choice of voltage for transmission, line losses and power-handling capacity for a given line length between source and load and bulk power required to be transmitted. The problem of vibration of bundled conductors is touched upon since this is the main mechanical problem in e.h.v. lines. Chapters 3 and 4 are basic to the remaining parts of the book and deal with calculation of line resistance, inductance, capacitance, and ground-return parameters, modes of propagation, electrostatics to understand charge distribution and the resulting surface voltage gradients. All these are directed towards an N-conductor bundle. Corona loss and Audible Noise from e.h.v. lines are consequences of high surface voltage gradient on conductors. This is dealt fully in Chapter 5. In several cases of line design, the audible noise has become a controlling factor with its attendant pollution of the environment of the line causing psycho-acoustics problems. The material on interference is continued in Chapter 6 where Radio Interference is discussed. Since this problem has occupied researchers for longer than AN, the available literature on RI investigation is more detailed than AN and a separate chapter is devoted to it. Commencing with corona pulses, their frequency spectrum, and the lateral profile of RI from lines, the reader is led into the modern concept of 'Excitation Function' and its utility in pre-determining the RI level of a line yet to be designed. For lines up to 750 kV, the C.I.G.R.E. formula applies. Its use in design is also discussed, and a relation between the excitation function and RI level calculated by the C.I.G.R.E. formula is given.

Chapter 7 relates to power frequency electrostatic field near an e.h.v. line which causes harmful effects to human beings, animals, vehicles, plant life, etc. The limits which a designer has to bear in mind in evolving a line design are discussed. Chapters 8-11 are devoted to the discussion of high transient overvoltages experienced by an e.h.v. line due to lightning and switching operations. Chapter 8 introduces the reader to the theoretical aspects of travelling waves caused by lightning and switching operations, and the method of standing waves which yields the same results as travelling waves but in many cases gives more convenient formulas. With the advent of the Digital Computers, the standing-wave method poses no problems for handling the calculation. The Laplace-Transform and Fourier-Transform Methods for handling transients on e.h.v. lines are described.

Chapter 9 deals with important aspects of lightning over-voltages and protection. The latest type of Metal Oxide Varistor known as gapless Zinc Oxide arrester is discussed as well as the conventional gap-type SiC arresters of both the non current-limiting and current-limiting types. The chapter commences with outage level aimed by a designer, and leads step by step in describing the factors affecting it, namely the isokeraunik level, probability of number of strokes to a tower or midspan, the tower-footing resistance, probability of lightning-stroke currents, and finally the insulator flash-over. Pre-discharge currents on towers and hardware are taken into account. Chapter 10 discusses all the possible conditions of internal overvoltages on e.h.v. lines commencing with circuit-breaker recovery voltage, terminal and short-line faults, interruption of low inductive current and overvoltages resulting from 'current chopping', line dropping and restrike in circuit breakers and ferroresonance conditions. The bulk of the chapter, however, is devoted to calculation of switching-surge overvoltages. Measures used for reduction of overvoltages are thoroughly discussed. Equations in matrix form dealing with the resulting transients are developed and examples using the Fourier Transform method for obtaining the switching overvoltages are worked out.

Having known the magnitude of overvoltages that can be expected on a system, the next aspect is to design air-gap clearances on a tower. This requires a thorough knowledge of the flashover and withstand characteristics of long air gaps. Chapter 11 is devoted to a description of these characteristics. Commencing with the basic mechanisms postulated by engineers and physicists for the breakdown of a long air gap, the reader is exposed to the statistical nature of insulation design. The work of the eminent Italian engineer, Dr. Luigi Paris [44], is described and examples of using his equations for insulation design are given.

Although transients caused by lightning and switching surges have been studied extensively by e.h.v. engineers, overvoltages caused under power-frequency are important for the design of line compensation. This is covered in chapter 12. The power-circle diagram and the geometrical relations resulting from it are used throughout for evaluating synchronous-condenser design, switched capacitors under load, shunt-reactor compensation including an intermediate station for a very long line, and finally a line with series-capacitor compensation is discussed. This problem leads logically to the problems of high short-circuit current and possible sub-synchronous resonance conditions. These are described from the point of view of the line. Countermeasures for SSR are described fully as used on the Navajo Project ([15], IEEE) and elsewhere. The chapter then describes Static Var compensating systems (SVS) of several types which are now finding more and more use instead of un-regulated or fixed shunt reactors. The chapter ends with a short description of high phase order transmission (6 phase) even though it does not yet belong to the e.h.v. class.

Chapter 13 deals with e.h.v. laboratories, equipment and testing. The design of impulse generators for lightning and switching impulses is fully worked out and waveshaping circuits are discussed. The effect of inductance in generator, h.v. lead, and the voltage divider are analyzed. Cascade-connected power-frequency transformers and the Greinacher chain for generation of high dc voltages are described. Measuring equipment such as the voltage divider, oscilloscope, peak volt-meter, and digital recording devices are covered and the use of fibre optics in large e.h.v. switchyards and laboratory measurements is discussed.

The last chapter, Chapter 14, uses the material of previous chapters to evolve methods for design of e.h.v. lines. Several examples are given from which the reader will be able to effect his or her own design of e.h.v. transmission lines in so far as steadystate and transient overvoltages are concerned.

Each chapter is provided with a large number of worked examples to illustrate all ideas in a step by step manner. The author feels that this will help to emphasize every formula or idea when the going is hot, and not give all theory in one place and provide examples at the end of each chapter. It is expected that the reader will work through these to be better able to apply the equations.

No references are provided at the end of each chapter since there are cases where one work can cover many aspects discussed in several chapters. Therefore, a consolidated bibliography is appended at the end after Chapter 14 which will help the reader who has access to a fine library or can get copies made from proper sources.

Review Questions and Problems

1. Give ten levels of transmission voltages that are used in the world.
2. Write an essay giving your ideas whether industrial progress is really a measure of human progress.
3. What is a micro-hydel station?
4. How can electric power be generated from run-of-the-river plants? Is this possible or impossible?
5. What is the fuel used in (a) Thermal reactors, and (b) LMFBR? Why is it called LMFBR? What is the liquid metal used? Is there a moderator in LMFBR? Why is it called a Breeder Reactor? Why is it termed Fast?
6. Draw sketches of a wind turbine with (a) horizontal axis, and (b) vertical axis. How can the efficiency of a conventional wind turbine be increased?

7. Give a schematic sketch of a tidal power development. Why is it called a 'bulb turbine'?
8. Give a schematic sketch of an ocean thermal gradient project showing a heat exchanger, turbine-generator, and condenser.
9. List at least ten important problems encountered in e.h.v. transmission which may or may not be important at voltages of 220 kV and lower.

CHAPTER 2

Transmission Line Trends and Preliminaries

2.1 Standard Transmission Voltages

Voltages adopted for transmission of bulk power have to conform to standard specifications formulated in all countries and internationally. They are necessary in view of import, export, and domestic manufacture and use. The following voltage levels are recognized in India as per IS-2026 for line-to-line voltages of 132 kV and higher.

Nominal System Voltage, kV	132	220	275	345	400	500	750
Maximum Operating Voltage, kV	145	245	300	362	420	525	765

There exist two further voltage classes which have found use in the world but have not been accepted as standard. They are: 1000 kV (1050 kV maximum) and 1150 kV (1200 kV maximum). The maximum operating voltages specified above should in no case be exceeded in any part of the system, since insulation levels of all equipment are based upon them. It is therefore the primary responsibility of a design engineer to provide sufficient and proper type of reactive power at suitable places in the system. For voltage rises, inductive compensation and for voltage drops, capacitive compensation must usually be provided. As example, consider the following cases.

Example 2.1: A single-circuit 3-phase 400 kV line has a series reactance per phase of 0.327 ohm/km. Neglect line resistance. The line is 400 km in length and the receiving-end load is 600 MW at 400 kV and 0.9 p.f. lag. Calculate the sending-end voltage in the absence of reactive compensation equipment. Work on bases of 400 kV and 1000 MVA.

Solution: Load voltage = 1.0 p.u., Load current = 0.6 $(1-j\,0.434)$ p.u. Base impedance $Z_B = 400^2/1000 = 160$ ohms.

Total series reactance of line, $X = j\,0.327 \times 400 = j\,130.8$ ohms

$$= j\,0.8175 \text{ p.u.}$$

Therefore, the sending-end voltage is, from Figure 2.1(a),

$$E_s = 1.0 + 0.6\,(1 - j\,0.434)\,j\,0.8175 = 1.308 \; \angle 22°.25, \text{ p.u.}$$

There is a 30.8% voltage rise at the generating station h.v. bus which exceeds the IS requirement of 420 kV.

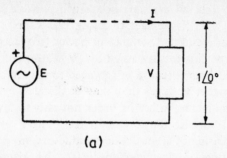

(a)

Fig. 2.1 (a).   400-kV line without compensation.

Example 2.2:   In the previous example, suggest suitable reactive-compensation equipment to be provided at the load end to maintain 400 kV at both ends of line.

Solution: Since the load is drawing lagging (inductive) current, obviously we have to provide capacitive compensating equipment across the load. Figure 2.1(b) shows the arrangement. Then, if I_c = compensation current,

$$E_s = 1 + (0.6 - j0.2604 + jI_c)\, j0.8175 = (1.213 - 0.8175\, I_c) + j\, 0.4905$$

Since $|\,E_s\,| = 1$, there results $I_c = 0.418$ p.u.

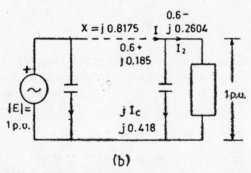

(b)

Fig. 2.1(b).   With capacitive compensation for voltage control.

The rating of the compensating equipment is 0.418 p.u. or 418 MVAR, capacitive. The load draws 600 MW and 290 MVAR inductive at 0.9 p.f. lag.

In a practical situation, the line capacitance will help to lower the rating of the compensating capacitor. The detailed discussion of power frequency voltage control on e.h.v. lines will be taken up in Chapter 12. We observe that the control of 50 Hz (or 60 Hz) voltage on a system is a very expensive proposition.

2.2   Average Values of Line Parameters

Detailed calculation of line parameters will be described in Chapter 3. In order to be able to estimate how much power a single-circuit at a given voltage can handle, we need to know the value of positive-sequence line inductance and its reactance at power frequency. Furthermore, in modern practice, line losses caused by $I^2 r$ heating of the conductors is gaining in importance because of the need to conserve energy. Therefore, the use of higher voltages than may be dictated by purely economic considerations might be used in order not only to lower the current I to be transmitted but also the conductor resistance r by using bundled conductors comprising of several sub-conductors in parallel. We will utilize average values of parameters for lines with horizontal configuration as shown in Table I for preliminary estimates.

Table I: Average Values of Line Parameters

System kV	400	750	1000	1200
Average Height, m	15	18	21	21
Conductor	2×32 mm	4×30 mm	6×46 mm	8×46 mm
Bundle Spacing, m	0.4572	0.4572	—	—
Bundle dia., m	—	—	1.2	1.2
r, ohm/km	0.031	0.0136	0.0036	0.0027
X, ohm/km (50 Hz)	0.327	0.272	0.231	0.231
X/r	10.55	20	64.2	85.6

When line resistance is neglected, the power that can be transmitted depends upon (a) the magnitudes of voltages at the ends (E_s, E_r), (b) their phase difference δ, and (c) the total positive-sequence reactance X per phase, when the shunt capacitive admittance is neglected.

Thus, $$P = E_s E_r \sin \delta / (L.X) \qquad (2.1)$$

where P = power in MW, 3-phase, E_s, E_r = voltages at the sending-end and receiving ends, respectively in kV line-line, δ = phase difference between E_s and E_r, X = positive-sequence reactance per phase, ohm/km, and L = line length, km.

From considerations of stability, the limiting value of δ is 30°, and for a preliminary estimate of P, we will take $E_s = E_r = E$.

2.3   Power-Handling Capacity and Line Loss

According to the above criteria, the power-handling capacity of a single circuit is $P = 0.5 E^2 / LX$. At unity power factor, at the load P, the current flowing is

$$I = 0.5 E/\sqrt{3} LX = 0.289 E/LX \qquad (2.2)$$

and the total power loss in the 3-phases will amount to

$$p = 3I^2rL = 0.5 \, E^2 \, r/LX^2 \tag{2.3}$$

Therefore, the percentage power loss is

$$\%p = 100 \, p/P = 50 \, r/X \tag{2.4}$$

Table II shows the percentage power loss and power-handling capacity of lines at various voltage levels shown in Table I.

Table II: Per Cent Power Loss and Power-Handling Capacity

System kV	400	750	1000	1200
Percentage Power Loss $P = 0.5E^2/LX$, MW Line Length, km	$\dfrac{50}{10.55} = 4.76$	$\dfrac{50}{20} = 2.5$	$\dfrac{50}{64.2} = 0.78$	$\dfrac{50}{85.6} = 0.584$
400	670	2860	6000	8625
600	450	1900	4000	5750
800	335	1430	3000	4310
1000	270	1140	2400	3450
1200	225	950	2000	2875

The following important and useful conclusions can be drawn for preliminary understanding of trends relating to power-handling capacity of a.c. transmission lines and line losses.

(1) One 750-kV line can normally carry as much power as four 400-kV circuits for equal distance of transmission.

(2) One 1200-kV circuit can carry the power of three 750-kV circuits and 12,400-kV circuits for the same transmission distance.

(3) Similar such relations can be found from the table.

(4) The power-handling capacity of line at a given voltage level decreases with line length, being inversely proportional to line length L.   From equation (2.2) the same holds for current to be carried.

(5) From the above property, we observe that if the conductor size is based on current rating, as line length increases, smaller sizes of conductor will be necessary.   This will increase the danger of high voltage effects caused by smaller diameter of conductor giving rise to corona on the conductors and intensifying radio interference levels and audible noise as well as corona loss.

(6) However, the *percentage* power loss in transmission remains independent of line length since it depends on the *ratio* of conductor resistance to the positive-sequence reactance per unit length.

(7) From the values of $\% \, p$ given in Table II, it is evident that it decreases as the system voltage is increased.   This is very strongly in favour of using higher voltages if energy is to be conserved. With the enormous increase in world oil prices and the need for conserving natural resources, this could sometimes become the

governing criterion for selection of voltage for transmission. The Bonneville Power Administration B.P.A. in the U.S.A. has based the choice of 1150 kV for transmission over only 280 km length of line since the power is enormous (10,000 MW over one circuit).

(8) In comparison to the % power loss at 400 kV, we observe that if the same power is transmitted at 750 kV, the line loss is reduced to $(2.5/4.76) = 0.525$, at 1000 kV it is $0.78/4.76 = 0.165$, and at 1200 kV it is reduced further to 0.124.

Some examples will serve to illustrate the benefits accrued by using very high transmission voltages.

Example 2.3: A power of 12,000 MW is required to be transmitted over a distance of 1000 km. At voltage levels of 400 kV, 750 kV, 1000 kV, and 1200 kV, determine

(a) possible number of circuits required with equal magnitudes for sending-and receiving-end voltages with 30° phase difference;

(b) the currents transmitted; and

(c) the total line losses.

Assume the values of X given in Table I.

Solution: This is carried out in tabular form.

	System, kV	400	750	1000	1200
	X, ohm/km	0.327	0.272	0.231	0.231
	$P=0.5\,E^2/LX$	268	1150	2400	3450
(a)	No. of circuits ($=12000/P$)	45	10–11	5	3–4
(b)	Current, kA	17.31	9.232	6.924	5.77
(c)	% power loss, p	4.76	2.5	0.78	0.584
	Total power loss, MW	571	300	93.6	70

The above situation might occur when the power potential of the Brahmaputra River in North-East India will be harnessed and the power transmitted to West Bengal and Bihar. Note that the total power loss incurred by using 1200 kV ac transmission is almost one-eighth that for 400 kV. The width of land required is far less while using higher voltages, as will be detailed later on.

Example 2.4: A power of 2000 MW is to be transmitted from a super thermal power station in Central India over 800 km to Delhi. Use 400 kV and 750 kV alternatives. Suggest the number of circuits required with 50% series capacitor compensation, and calculate the total power loss and loss per km.

Solution: With 50% of line reactance compensated, the total reactance will be half of the positive-sequence reactance of the 800-km line.

Therefore $P = 0.5 \times 400^2/400 \times 0.327 = 670$ MW/Circuit at 400 kV

and $P = 0.5 \times 750^2/400 \times 0.272 = 2860$ MW/Circuit at 750 kV

	400 kV	750 kV
No. of circuits required	3	1
Current per circuit, kA	$667/\sqrt{3} \times 400 = 0.963$	1.54
Resistance for 800 km, ohms	$0.031 \times 800 = 24.8$	$0.0136 \times 800 = 10.88$
Loss per circuit, MW	$3 \times 24.8 \times 0.963^2 = 69$ MW	$3 \times 10.88 \times 1.54^2$
		$= 77.4$ MW
Total power loss, MW	$3 \times 69 = 207$	77.4
Loss/km, kW	86.25 kW/km	97 kW/km

During normal operation, lines will be loaded less than what are indi-cated above because the phase difference will be lower than 30°. This might necessitate in certain situations, a slightly larger number of lines. During line outages the maximum power transmitted will be governed by the above criteria.

2.4 Examples of Giant Power Pools and Number of Lines

From the discussion of the previous section it becomes apparent that the choice of transmission voltage depends upon (a) the total power transmitted, (b) the distance of transmission, (c) the % power loss allowed, and (d) the number of circuits permissible from the point of view of land acquisition for the line corridor. For example, a single circuit 1200 kV line requires a width of 56 m while 6 single-circuit 500 kV lines for transmitting the same power require 220 m of right-of-way (R-O-W). An additional factor is the technological know-how in the country. Two examples of similar situations with regard to available hydro-electric power will be described in order to draw a parallel for deciding upon the transmission voltage selection. The first is from Canada and the second from India. These ideas will then be extended to thermal generation stations situated at mine mouths requiring long transmission lines for evacuating the bulk power to load centres.

2.4.1 Canadian Experience

The power situation in the province of Quebec comes closest to the power situation in India, in that nearly equal amounts of power will be developed eventually and transmitted over nearly the same distances. Hence the Canadian experience might prove of some use in making deci-sions in India also. The power to be developed from the La Grande River located in the James Bay area of Northern Quebec are as follows: Total 11,340 MW split into 4 stations [LG-1: 1140, LG-2: 5300, LG-3: 2300, and LG-4: 2600 MW]. The distance to load centres at Montreal and Quebec cities is 1100 km. The Hydro-Quebec company has vast experience with their existing 735 kV system from the earlier hydro-electric development at Manicouagan-Outardes Rivers so that the choice of transmission voltage fell between the existing 735 kV or a future 1200 kV. However, on account of the vast experience accumulated at the 735 kV level, this voltage was finally chosen. The number of circuits required from Table II can be seen to be 10–11 for 735 kV and 3–4 for 1200 kV. The lines run practically in wilderness and land acquisition is not as difficult a problem as in more thickly populated areas. Plans might however change as the development proceeds. The 1200 kV level is new to the industry and equipment manufacture is in the infant stages for this level. As an alternative, the company could have investigated the possibility of using e.h.v. dc transmission. But the final decision was to use 735 kV, ac.

2.4.2 Indian Requirement (as of 1985)

The giant hydro-electric power pools in India are located in the northern border of the country on the Himalayan Mountain valleys. These are in Kashmir, Upper Ganga on the Alakhananda and Bhagirathi Rivers, Nepal, Bhutan, and the Brahmaputra River. Power surveys indicate the following power generation and distances of transmission:

(1) 2500 MW, 250 km, (2) 3000 MW, 300 km, (3) 4000 MW, 400 km, (4) 5000 MW, 300 km, (5) 12000 MW over distances of (a) 250 km, (b) 450 km, and (c) 1000-1200 km.

Using the power-handling capacities given in Table II, we can construct a table showing the possible number of circuits required at different voltage levels (Table III).

Table III: Voltage Levels and Number of Circuits for Evacuating from Hydro-Electric Power Pools in India

Power, MW	2500	3000	4000	5000	12000		
Distance, km	250	300	400	300	250	450	1000
No. of Circuits/	3/400	4/400	6/400	6/400	12/400	20/400	48/400
Voltage Level		1/750	2/750	2/750	3/750	6/750	12/750
(AC only)			(70% loaded)	(75%)	1/1200	2/1200	6/1000
							4/1200

One can draw certain conclusions from the above table. For example, for powers up to 5000 MW, 400 KV transmission might be adequate. For 12000 MW, we observe that 750 kV level for distances up to 450 km and 1200 kV for 1000 km might be used, although even for this distance 750 kV might serve the purpose. It is the duty of a design engineer to work out such alternatives in order that final decisions might be taken. For the sake of reliability, it is usual to have at least 2 circuits.

While the previous discussion is limited to ac lines, the dc alternatives must also be worked out based upon 2000 Amperes per pole. The usual voltages used are $\pm$ 400 kV (1600 MW/bipole), $\pm$ 500 kV (2000 MW) and $\pm$ 600 kV (2400 MW). These power-handling capacities do not depend on distances of transmission. It is left as an exercise at the end of the chapter for the reader to work out the dc alternatives for powers given in Table III.

2.5 Costs of Transmission Lines and Equipment

It is universally accepted that cost of equipment all over the world is escalating every year. Therefore, a designer must ascertain current prices from manufacturers of equipment and line materials. These include conductors, hardware, towers, transformers, shunt reactors, capacitors, synchronous condensers, land for switchyards and line corridor, and so on. Generating station costs are not considered here, since we are only dealing with transmission in this book. In this section, some idea of costs of important equipment are given (which may be current in 1985) for comparison purposes only. These are not to be used for decision-making purpose.

(1 US $ = Rs. 12; 1 Lakh = 100,000; 1 Crore=100 Lakhs=10 Million)

(a) *High Voltage DC* $\pm$ 400 kV Bipole

Back-to-back terminals: Rs. 17 Lakhs/MVA for 150 MVA

Rs. 13 Lakhs/MVA for 300 MVA

Cost of 2 terminals: Rs. 12 Lakhs/MVA

Transmission line: Rs. 8.3 Lakhs/Circuit (cct) km

Switchyards: Rs. 1000 Lakhs/bay

(b) *400 kV AC*

Transformers: 400/220 kV Autotransformers

Rs. 1.25 Lakhs/MVA for 200 MVA 3-phase unit

to   Rs. 1.00 Lakh/MVA for 500 MVA 3-phase unit

400 kV/13.8 kV Generator Transformers

Rs. 0.75 Lakh/MVA for 250 MVA 3-phase unit

to   Rs. 0.5 Lakh/MVA for 550 MVA 3-phase unit

Shunt Reactors

Non-switchable   Rs. 0.86 Lakh/MVA for 50 MVA unit

to   Rs. 0.74 Lakh/MVA for 80 MVA unit

Switchable   Rs. 3 Lakhs to Rs. 2 Lakhs/MVA for 50 to 80 MVA units.

Shunt Capacitors Rs. 0.3 Lakh/MVA

Synchronous Condensers (including transformers):

Rs. 4.5 Lakhs/MVA for 70 MVA

to Rs. 2.5 Lakhs/MVA for 300 MVA

Transmission Line Cost: 400 kV Single Circuit: Rs. 9 Lakhs/cct km

220 kV: S/C: Rs. 4.5 Lakhs/cct km

D/C: Rs. 7.5 Lakhs/cct km

Example 2.5:   A power of 900 MW is to be transmitted over a length of 875 km (Singrauli-Delhi). Estimate the cost difference when using ± 400 kV dc line and 400 kV ac lines.

Solution: Power carried by a single circuit dc line = 1600 MW. Therefore 1 circuit is sufficient although it allows for future expansion.

Power carried by ac line = $0.5 \, E^2/XL = 0.5 \times 400^2/(0 \cdot 32 \times 875)$

= 285 MW/Circuit.

3 circuits will be necessary to carry 900 MW.

DC Alternative

Cost of

(a)	Terminal Stations	Rs. 10,800 Lakhs
(b)	Transmission Line	Rs. 7,200 Lakhs
(c)	2 switchyard bays	Rs. 2,000 Lakhs

Total   Rs. 20,000 Lakhs

= Rs. 200 Crores

(Computing at Rs. 10 per US $, total cost $\approx$ $ 200 Million).

AC Alternative

Cost of

(a)	6 switchyard bays	Rs. 6000 Lakhs
(b)	Shunt reactors 500 MVA	
	(80 MVA for each circuit at each end)	Rs. 375 Lakhs
(c)	Shunt capacitors 500 MVA	Rs. 150 Lakhs
(d)	Line cost: $(3 \times 875 \times 9$ Lakhs)	Rs. 23625 Lakhs

Total   Rs. 30,150 Lakhs

= Rs. 301.5 Crores

$\approx$ $ 301.5 Million

Difference in cost = Rs. 101 Crores, dc being lower than ac.

[*Note*.   Certain items common to both dc and ac transmission have been omitted.   Also, series capacitor compensation has not been considered].

Example 2.6:   Repeat the problem above if the transmission distance is 600 km.

Solution: For the dc alternative, the total cost is

Rs. [10,800 + 4,980 + 2,000] Lakhs = Rs. 17,780 Lakhs.

For the ac alternative, the power-handling capacity per circuit is increased to $285 \times 875/600 = 420$ MW. This requires 2 circuits for handling 900 MW.

The reactive powers will also be reduced to 120 MVA for each line in shunt reactors and switched capacitors.

Cost estimate will include:

(a)	4 switchyard bays	Rs. 4000 Lakhs
(b)	Shunt reactors 240 MVA	Rs. 190 Lakhs
(c)	Shunt capacitors 240 MVA	Rs. 72 Lakhs
(d)	Line cost: $2 \times 600 \times 9$ Lakhs	Rs. 10800 Lakhs

Total	Rs. 15,062 Lakhs
	= Rs. 150.62 Crores

The dc alternative has become more expensive by Rs. 2720 Lakhs or Rs. 27.2 Crores (U. S. $ 27.2 Million). In between line lengths of 600 km and 875 km for transmitting the same power, the two alternatives will cost nearly equal. This is called the "Break-Even Distance".

2.6 Mechanical Considerations in Line Performance

2.6.1 Types of Vibrations and Oscillations

In this section a brief description will be given of the enormous importance which designers place on the problems created by vibrations and oscillations of the very heavy conductor arrangements required for e.h.v. transmission lines. As the number of sub-conductors used in a bundle increases, these vibrations and countermeasures and spacings of sub-conductors will also affect the electrical design, particularly the surface voltage gradient. The mechanical designer will recommend the tower dimensions, phase spacings, conductor height, sub-conductor spacings, etc. from which the electrical designer has to commence his calculations of resistance, inductance, capacitance, electrostatic field, corona effects, and all other performance characteristics. Thus, the two go hand in hand.

The sub-conductors in a bundle are separated by spacers of suitable type, which bring their own problems such as fatigue to themselves and to the outer strands of the conductor during vibrations. The design of spacers will not be described here but manufacturers' catalogues should be consulted for a variety of spacers available. These spacers are provided at intervals ranging from 60 to 75 metres between each span which is in the neighbourhood of 300 metres for e.h.v. lines. Thus, there may be two end spans and two or three subspans in the middle. The spacers prevent conductors from rubbing or colliding with each other in wind and ice storms, if any. However, under less severe wind conditions the bundle spacer can damage itself or cause damage to the conductor under certain critical vibration conditions. Electrically speaking, since the charges on the sub-conductors are of the same polarity, there exists electrostatic repulsion among them. On the other hand, since they carry currents in the same direction, there is electromagnetic attraction. This force is especially severe during short-circuit currents so that the spacer has a force exerted on it during normal or abnormal electrical operation.

Three types of vibration are recognized as being important for e.h.v. conductors, their degree of severity depending on many factors, chief among which are: (a) conductor tension, (b) span length, (c) conductor size, (d) type of conductor, (e) terrain of line, (f) direction of prevailing winds, (g) type of supporting clamp of conductor insulator assemblies from the tower, (h) tower type, (i) height of tower, (j) type of spacers and dampers, and (k) the vegetation in the vicinity of line. In general, the most severe vibration conditions are created by winds without turbulence so that hills, buildings, and trees help in reducing the severity. The types of vibration are: (1) Aeolian Vibration, (2) Galloping, and (3) Wake-Induced Oscillations. The first two are present for both single- and multi-conductor bundles, while the wake-induced oscillation is confined to a bundle only. Standard forms of bundle conductors have sub-conductors ranging from 2.54 to 5 cm diameters with bundle spacing of 40 to 50 cm between adjacent conductors. For e.h.v. transmission, the number ranges from 2 to 8 sub-conductors for transmission voltages from 400 kV to 1200 kV, and up to 12 or even 18 for higher voltages which are not yet commercially in operation. We will briefly describe the mechanism causing these types of vibrations and the problems created by them.

2.6.2 Aeolian Vibration

When a conductor is under tension and a comparatively steady wind blows across it, small vortices are formed on the leeward side called Karman Vortices (which were first observed on aircraft wings). These vortices detach themselves and when they do alternately from the top and bottom they cause a minute vertical force on the conductor. The frequency of the forces is given by the accepted formula

$$f = 2.065 \ v/d, \ \text{Hz} \tag{2.5}$$

where v = component of wind velocity normal to the conductor in km/ hour, and d = diameter of conductor in centimetres. [The constant factor of equation (2.5) becomes 3.26 when v is in mph and d in inches.]

The resulting oscillation or vibrational forces cause fatigue of conductor and supporting structure and are known as aeolian vibrations. The frequency of detachment of the Karman vortices might correspond to one of the natural mechanical frequencies of the span, which if not damped properly, can build up and destroy individual strands of the conductor at points of restraint such as at supports or at bundle spacers. They also give rise to wave effects in which the vibration travels along the conductor suffering reflection at discontinuities at points of different mechanical characteristics. Thus, there is associated with them a mechanical impedance. Dampers are designed on this property and provide suitable points of negative reflection to reduce the wave amplitudes. Aeolian vibrations are not observed at wind velocities in excess of 25 km/hour. They occur principally in terrains which do not disturb the wind so that turbulence helps to reduce aeolian vibrations.

In a bundle of 2 conductors, the amplitude of vibration is less than for a single conductor due to some cancellation effect through the bundle spacer. This occurs when the conductors are not located in a vertical plane which is normally the case in practice. The conductors are located in nearly a horizontal plane. But with more than 2 conductors in a bundle, conductors are located in both planes. Dampers such as the Stockbridge type or other types help to damp the vibrations in the sub-spans connected to them, namely the end sub-spans, but there are usually two or three sub-spans in the middle of the span which are not protected by these dampers provided only at the towers. Flexible spacers are generally provided which may or may not be designed to offer damping. In cases where they are purposely designed to damp the sub-span oscillations, they are known as spacer-dampers.

Since the aeolian vibration depends upon the power imparted by the wind to the conductor, measurements under controlled conditions in the laboratory are carried out in wind tunnels. The frequency of vibration is usually limited to 20 Hz and the amplitudes less than 2.5 cm.

2.6.3 Galloping

Galloping of a conductor is a very high amplitude, low-frequency type of conductor motion and occurs mainly in areas of relatively flat terrain under freezing rain and icing of conductors. The flat terrain provides winds that are uniform and of a low turbulence. When a conductor is iced, it presents an unsymmetrical cross-section with the windward side having less ice accumulation than the leeward side of the conductor. When the wind blows across such a surface, there is an aerodynamic lift as well as a drag force due to the direct pressure of the wind. The two forces give rise to torsional modes of oscillation and they combine to oscillate the conductor with very large amplitudes sufficient to cause contact of two adjacent phases, which may be 10 to 15 metres apart in the rest position. Galloping is induced by winds ranging from 15 to 50 km/hour, which may normally be higher than that required for aeolian vibrations but there could be an overlap. The conductor oscillates at frequencies between 0.1 and 1 Hz. Galloping is controlled by using "detuning pendulums" which take the form of weights applied at different locations on the span.

Galloping may not be a problem in a hot country like India where temperatures are normally above freezing in winter. But in hilly tracts in the North, the temperatures may dip to below the freezing point. When the ice loosens from the conductor, it brings another oscillatory motion called Whipping but is not present like galloping during only winds.

2.6.4 Wake-Induced Oscillation

The wake-induced oscillation is peculiar to a bundle conductor, and similar to aeolian vibration and galloping occurring principally in flat terrain with winds of steady velocity and low turbulence. The frequency of the oscillation does not exceed 3 Hz but may be of sufficient amplitude to cause clashing of adjacent sub-conductors, which are separated by about 50 cm. Wind speeds for causing wake-induced oscillation must be normally in the range 25 to 65 km/hour. As compared to this, aeolian vibration occurs at wind speeds less than 25 km/hour, has frequencies less than 20 Hz and amplitudes less than 2.5 cm. Galloping occurs at wind speeds between 15 and 50 km/hour, has a low frequency of less than 1 Hz, but amplitudes exceeding 10 metres. Fatigue failure to spacers is one of the chief causes for damage to insulators and conductors.

Wake-induced oscillations, also called "flutter instability", is caused when one conductor on the windward side aerodynamically shields the leeward conductor. To cause this type of oscillation, the leeward conductor must be positioned at rest toward the limits of the wake or wind-shadow of the windward conductor. The oscillation occurs when the bundle tilts 5 to 15° with respect to a flat ground surface. Therefore, a gently sloping ground with this angle can create conditions favourable to wake-induced oscillations. The conductor spacing to diameter ratio in the bundle is also critical. If the spacing B is less than $15d$, d being the conductor diameter, a tendency to oscillate is created while for $B/d > 15$ the bundle is found to be more stable. As mentioned earlier, the electrical design, such as calculating the surface voltage gradient on the conductors, will depend upon these mechanical considerations.

2.6.5 Dampers and Spacers

When the wind energy imparted to the conductor achieves a balance with the energy dissipated by the vibrating conductor, steady amplitudes for the oscillations occur. A damping device helps to achieve this balance at smaller amplitudes of aeolian vibrations than an undamped conductor. The damper controls the intensity of the wave-like properties of travel of the oscillation and provides an equivalent heavy mass which absorbs the energy in the wave. A sketch of a Stockbridge damper is shown in Fig. 2.2.

A simpler form of damper is called the Armour Rod which is a set of wires twisted around the line conductor, at the insulator supporting conductor and hardware, and extending for about 5 metres on either side. This is used for small conductors to provide a change in mechanical impedance. But for heavier conductors, weights must be used, such as the Stockbridge, which range from 5 kg for conductors of 2.5 cm diameter to 14 kg for 4.5 cm. Because of the steel strands inside them ACSR conductors have better built-in property against oscillations than ACAR conductors.

There are a large number of types of spacers which keep the conduc-

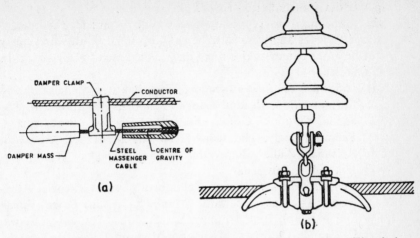

Fig. 2.2. (a) Stockbridge Damper; (b) Suspension Clamp (Courtesy: Electrical Manufacturing Co., Calcutta).

tors apart. Most modern spacers have some flexibility built into them to allow rotation of the conductor inside them such as lining the clamps with high-strength plastic or rubber washers. Some spacers are specially designed to act as dampers and may also take the form of heavy springs. The selection of the spacers is also determined by the wind speed in the locality.   Figure 2.3 shows a spacer used for a bundle conductor.

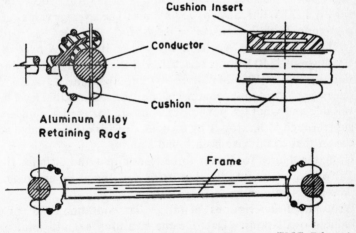

Fig. 2.3. Spacer for two-conductor bundle (Courtesy: EMC Calcutta).

Review Questions and Problems

1. Of the following transmission voltages (given in kV) used in the world, which ones are used in India at present: 66, 132, 169, 220, 275, 345, 400, 500–525, 735–765, 1000, 1150.
2. The geography and civilization of any country can best be understood by a knowledge of location of its rivers. In modern days,

power development also depends on the rivers. On an atlas, locate the following rivers in the countries listed. Note: Due to international disputes and wars between nations, some of the rivers may not be in the countries given, but only in the general geographic area in an atlas.

(a) *Indian sub-continent* (including Tibet, Pakistan, Bangladesh and Burma)—Sind, Jhelum, Ganga, Yamuna, Bhagirathi, Alakhnanda, Gandak, Gomti, Tsang-Po, Dehang, Brahmaputra, Padma, Hoogly, Narmada, Damodar, Mahanadi, Godavari, Krishna, Kali, Sharavathi, Kaveri, Vaigai, Tamraparni, Irrawaddy, Salween.

(b) *Canada and U.S.*—Fraser, Columbia, Slave, Mackenzie, Athabasca, The Saskatchewans, Winnipeg, Nelson, Peace (Finlay and Parsnip), Red, St. Lawrence, St. John, Churchill, Ottawa, La Grande, Colorado, Mississippi-Missouri, Ohio, Rio Grande, Delaware, Hudson, Mohawk, Niagara.

(c) *Europe, including the U.S.S.R. and Siberia*—Severn, Clyde, Thames, Danube, Rhone, Rhine, Elb, Po, Seine, Ob, Volga, Dneiper, Lena, Yenesei.

(d) *Other parts of the world*—Yangtze, Yellow, Senju, Marude, Yalu, Nile, Zambezi, Congo (Zaire), Amazon, Itiapu, Orinaco, Makong, La Plata, Sikiang, Volta.

3. Using equations (2.1) and (2.2), draw on a graph the variation of P and I as the distance of transmission is varied from 200 to 800/km for (a) 400 kV line, and (b) 750 kV line. Use average values of r and x given in Table I. Take $P = 0.5 \, E^2/Lx$.

4. In the U.S.A., for transmitting a power of 10,000 MW over 285 km, a voltage of 1150 kV was selected. In the U.S.S.R., for transmitting a power of 5000 MW over 800 km, the same voltage level was selected. Give your comments on the reasons this level is most suitable and what the possible reasons are for such a choice. Discuss through % line losses by comparing with other suitable voltage classes that could have been found suitable.

5. Using the figures for power to be transmitted and distance given in Table III, work out dc alternative for India to evacuate these powers to load centres.

6. Write brief descriptions of (a) aeolian vibration, and (b) wake-induced oscillations. Describe the measures taken to minimize the damage due to them.

CHAPTER 3

Calculation of Line and Ground Parameters

3.1 Resistance of Conductors

Conductors used for e.h.v. transmission lines are always stranded. Most common conductors use a steel core for reinforcement of the strength of aluminium, but recently high tensile strength aluminium is being increasingly used, replacing the steel. The former is known as ACSR (Aluminium Conductor Steel Reinforced) and the latter ACAR (Aluminium Conductor Aluminium Reinforced). When a steel core is used, because of its high permeability and inductance, power-frequency current flows only in the aluminium strands. In ACAR conductor, the cross-section is better utilized. Figure 3.1 shows an example of a stranded conductor.

Fig. 3.1. Cross-section of typical ACSR conductor.

If n_s = number of strands of aluminium; d_s = diameter of each strand in metre and ρ_a = specific resistance of Al, ohm-m, at temperature t, the resistance of the stranded conductor per km is

$$R = \rho_a \, 1.05 \times 10^3 / (\pi \, d_s^2 \, n_s / 4) = 1333 \, \rho_a / d_s^2 \, n_s \qquad (3.1)$$

The factor 1.05 accounts for the twist or lay whereby the strand length is increased by 5%.

Example 3.1: A Drake conductor of North-American manufacture has an outer diameter of 1.108 inches having an Al cross-sectional area of 795,000 circular mils. The stranding is 26 Al/7 Fe. Its resistance is given as 0.0215 ohm/1000' at 20°C under dc, and 0.1284 ohm/mile at 50°C and 50/60 Hz. Calculate

   (a) diameter of each strand of Al and Fe in mils, inch, and metre units;
   (b) check the values of resistances given above taking $\rho_a = 2.7 \times 10^{-8}$ ohm-metre at 20°C and temperature-resistance coefficient $\sigma = 4.46 \times 10^{-3}/°C$ at 20°C;

(c) find increase in resistance due to skin effect.

Note. 1 inch = 1000 mils; 1 in^2 = $\dfrac{4}{\pi} \times 10^6$ cir-mils;

10^6 cir-mils = $\dfrac{\pi}{4}$ sq. cm. = 5.067 sq. cm.

Solution: Area of each strand of Al = 795,000/26 = 30,577 c-m.

(a) Diameter of each strand, d_s = $\sqrt{30,577}$ = 175 mils = 0.175 inch
$$= 0.4444 \text{ cm} = 0.00444 \text{ m}$$

Referring to Fig. 3.1, there are 4 strands of Al along any diameter occupying 700 mils. The 3 diameters of Fe occupy 1108 − 700 = 408 mils since the overall dia. of the conductor is 1.108″ = 1108 mils.
Therefore,

diameter of each steel strand = 408/3 = 136 mils = 0.136″ = 0.3454 cm

(b) Because of the high permeability of steel, the steel strands do not carry current. Then, for 1000 feet,

$$R_a = 2.7 \times 10^{-8} \ (1000 \times 1.05/3.28)/ \left[\frac{\pi}{4} \times (4.444 \times 10^{-3})^2 \times 26 \right]$$
$$= 0.02144 \text{ ohm.}$$

This is close to 0.0215 ohm/1000′

At 50°C, $\rho_{50} = \dfrac{1 + 4.46 \times 10^{-3} \times 50}{1 + 4.46 \times 10^{-3} \times 20} \rho_{20} = 1.123 \ \rho_{20}$

∴ R_{50} = 1.123 × 0.0215 × 5.28 = 0.1275 ohm/mile.

(c) Increase in resistance due to skin effect at 50/60 Hz is
$$0.1284 - 0.1275 = 0.0009 \text{ ohm} = 0.706\%.$$

3.1.1 Effect of Resistance of Conductor

The effect of conductor resistance of e.h.v. lines is manifested in the following forms:

(1) Power loss in transmission caused by I^2R heating;
(2) Reduced current-carrying capacity of conductor in high ambient-temperature regions. This problem is particularly severe in Northern India where summer temperatures in the plains reach 50°C. The combination of intense solar irradiation of conductor combined with the I^2R heating raise the temperature of Aluminium beyond the maximum allowable temperature which stands at 65°C as per Indian Standards. At an ambient of 48°C, even the solar irradiation is sufficient to raise the temperature to 65°C for 400 KV line, so that no current can be carried. If there is improvement in material and the maximum temperature raised to 75°C, it is estimated that a current of 600 amperes can be transmitted for the same ambient temperature of 48°C.
(3) The conductor resistance affects the attenuation of travelling wavse

due to lightning and switching operations, as well as radio-frequency energy generated by corona. In these cases, the resistance is computed at the following range of frequencies: Lightning—100 to 200 KHz; Switching—4000-5000 Hz; Radio frequency—0.5 to 2 MHz.

We shall consider the high-frequency resistance later on.

3.1.2 Power Loss in Transmission

In Chapter 2, average resistance values were given in Table I. For various amounts of power transmitted at e.h.v. voltage levels, the I^2R heating loss in KW/km are shown in Table 3.1 below. The power factor is taken as unity. In every case the phase angle difference $\delta = 30°$ between E_s and E_r.

Table 3.1: I^2R Loss in KW/km of E.H.V. Lines

System KV	400	750	1000	1200
Resistance, Ohm/km	0.031	0.0136	0.0036	0.0027
Power Transmitted	I^2R Loss, KW/Km			
1000 MW	195	24.2	3.6	2.04
2000	780	97	14.4	8.16
5000	4,875	605	90	51
10,000	19,500	2420	360	204
20,000	78,000	9680	1440	816

We notice the vast reduction in KW/km loss occurring with increase in transmission voltage for transmitting the same power. The above calculations are based on the following equations:

(1) Current : $I = P/\sqrt{3}V$ $\qquad$ (3.2)

(2) Loss : $P = 3I^2R = P^2R/V^2$ $\qquad$ (3.3)

When P is in MW, V in KV, and R in ohm/km, the power loss in KW/km is $(P^2R/V^2)\,10^3$. $\qquad$ (3.4)

(3) If the line length is L km, the total power loss in transmission will be P^2RL/V^2, MW.

Consequently, efficiency of transmission is

$$\eta = 100\,P/(P + P^2RL/V^2) = 100/(1 + PRL/V^2)\% \qquad (3.5)$$

3.1.3 Skin Effect Resistance in Round Conductors

It was mentioned earlier that the resistance of overhead line conductors must be evaluated at frequencies ranging from power frequency (50/60 Hz) to radio frequencies up to 2 MHz or more. With increase in frequency, the current tends to flow nearer the surface resulting in a decrease in area for current conduction. This gives rise to increase in effective resistance due to the 'Skin Effect'. The physical mechanism for this effect is based on the fact that the inner filaments of the conductor link larger amounts of flux as the centre is approached which causes an increase in reactance. The reactance is proportional to frequency so that the impedance to current flow is larger in the inside, thus preventing flow of current easily. The result is a crowding of current at the outer filaments of the conductor. The increase in resistance of a stranded conductor is more difficult to calculate than that of a single round solid conductor because of the close proximity of the strands which distort the magnetic field still further. It is easier to determine the resistance of a stranded conductor by experiment at the manufacturer's premises for all conductor sizes manufactured and at various frequencies.

In this section, a method of estimating the ratio $R_{ac}(f)/R_{dc}$ will be described. The rigorous formulas involve the use of Bessel Functions and the resistance ratio has been tabulated or given in the form of curves by the National Bureau of Standards, Washington, several decades ago. Figure 3.2(a) shows some results where the ordinate is R_{ac}/R_{dc} at any frequency f and the abscissa is $X = mr = 0.0636\sqrt{f/R_0}$, where R_0 is the dc resistance of conductor in ohms/mile. When using SI units, $X = 1.59 \times 10^{-3}\sqrt{f/R_m}$, where $R_m = $ dc resistance in ohm/metre.

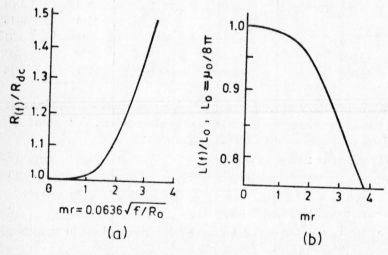

Fig. 3.2. Variation with frequency parameter (mr) of (a) skin effect resistance ratio $R_{ac}(f)/R_{dc}$ and (b) skin effect inductance $L(f)/L_0$, with $L_0 = \mu_0/8\pi$, the inductance with uniform current distribution in round conductor.

Example 3.2: A Moose conductor has a resistance of 62 milli-ohm/km. Using Fig. 3.2(a), determine the highest frequency for which the graph is applicable for a round conductor.

Solution: Maximum value of $X = 4 = 0.0636\sqrt{f/R_0}$.

Now,　　　　　$R_0 = 62 \times 10^{-3} \times 1.609 = 0.1$

Therefore　　　$f = (X/0.0636)^2 R_0 \approx 400$ Hz.

For other frequencies the functional relationship between $R_{ac}(f)/R_{dc}$ is as follows:

Let $\text{Ber}(X) = 1 - \dfrac{X^4}{2^2 . 4^2} + \dfrac{X^8}{2^2 . 4^2 . 6^2 . 8^2} \cdots$

$\text{Bei}(X) = \dfrac{X^2}{2^2} - \dfrac{X^6}{2^2 . 4^2 . 6^2} + \dfrac{X^{10}}{2^2 . 4^2 . 6^2 . 8^2 . 10^2} \cdots \qquad (3.6)$

$\text{B}'\text{er}(X) = d\,\text{Ber}(X)/dX, \quad \text{B}'\text{ei}(X) = d\,\text{Bei}(X)/dX$

Then, $\quad \dfrac{R_{ac}(f)}{R_{dc}} = \left(\dfrac{X}{2}\right) \dfrac{\text{Ber}(X) . \text{B}'\text{ei}(X) - \text{Bei}(X) . \text{B}'\text{er}(X)}{[\text{B}'\text{er}(X)]^2 + [\text{B}'\text{ei}(X)]^2} \qquad (3.7)$

The Bessel Functions are tabulated and values from there must be used [see H.B. Dwight: Mathematical Tables (Dover Publications) pages 194 onwards]. The following example will illustrate the increase in resistance of a round copper conductor up to a frequency of 100 KHz.

Example 3.3: A round $7/0$ copper conductor $0.5''$ (12.7 mm) in diameter has $\rho = 1.7 \times 10^{-8}$ ohm-m at 20°C. Calculate the variation of R_{ac}/R_{dc} as a function of frequency up to 10^5 Hz.

Solution: $R_0 = 1.7 \times 10^{-8} \times 1609/(\pi/4 \times 12.7^2 \times 10^{-6}) = 0.216$ ohm/mile

$\therefore \quad 0.0636/\sqrt{R_0} = 0.0137.$

We will use a logarithmic increase for frequency.

f	100	300	600	1000	3000	6000	10^4	3×10^4	6×10^4	10^5
$X = 0.0137\sqrt{f}$	.137	.237	.335	.4326	.749	1.06	1.37	2.37	3.35	4.326
R_{ac}/R_{dc}	1+	1+	1+	1+						
	0.37	0.4	.8	0.8	1.0017	1.0066	1.0182	1.148	1.35	1.8
	$\times 10^{-5}$	$\times 10^{-4}$	$\times 10^{-4}$	$\times 10^{-3}$						

1.0000037, 1.00004, 1.00008, 1.0008

[These values are taken from N.B.S. Tables and T. and D. Reference Book (Westinghouse)].

3.2 Temperature Rise of Conductors and Current-Carrying Capacity

When a conductor is carrying current and its temperature has reached a steady value, heat balance requires

$$\left(\begin{array}{c}\text{Internal Heat}\\ \text{Developed by } I^2R\end{array}\right) + \left(\begin{array}{c}\text{External Heat Supplied}\\ \text{by Solar Irradiation}\end{array}\right)$$

$$= \left(\begin{array}{c}\text{Heat Lost by}\\ \text{Convection to Air}\end{array}\right) + \left(\begin{array}{c}\text{Heat Lost by}\\ \text{Radiation}\end{array}\right) \qquad (3.8)$$

Let $\quad W_i = I^2R$ heating in watts/metre length of conductor

$\quad W_s =$ solar irradiation „ „ „ „ „

W_c = convection loss in watts/metre length of conductor

and　　　W_r = radiation loss　　　,,　　,,　　,,　　,,　　,,

Then the heat balance equation becomes

$$W_i + W_s = W_c + W_r \tag{3.9}$$

Each of these four terms depends upon several factors which must be written out in terms of temperature, conductor dimensions, wind velocity, atmospheric pressure, current, resistance, conductor surface condition, etc. It will then be possible to find a relation between the temperature rise and current. The maximum allowable temperature of an Al conductor is 65°C at present, but will be increased to 75°C. Many countries in the world have already specified the limit as 75°C above which the metal loses its tensile strength. The four quantities given above are as follows:

(1) *I^2R heating.* $W_i = I^2 R_m$ watts/metre where R_m = resistance of conductor per metre length at the maximum temperature.

$$R_m = \frac{1 + \alpha t}{1 + 20\alpha} R_{20},$$

with α = temperature resistance coefficient in ohm/°C and R_{20} = conductor resistance at 20°C.

(2) *Solar irradiation.*

$$W_s = s_a . I_s . d_m \text{ watts/metre}$$

where d_m = diameter of conductor in metre, s_a = solar absorption coefficient = 1 for black body or well-weathered conductor = 0.6 for new conductor, and I_s = solar irradiation intensity in watts/m².

At New Delhi in a summer's day at noon, I_s has a value of approximately 1000-1500 W/m².

[*Note.* 10⁴ calories/sq. cm/day = 4860 watts/m²]

(3) *Convection loss.*

$$w_c = 5.73\sqrt{p\ v_m/d_m} . \Delta t, \text{ watts/m}^2$$

where p=pressure of air in atmospheres, v_m=wind velocity in metres/sec., and Δt = temperature rise in °C above ambient = $t - t_a$.

Since 1 metre length of conductor has an area of πd_m sq. m., the convection loss is

$$W_c = 18. \Delta t.\sqrt{p.v_m.d_m}, \text{ watts/metre}$$

(4) *Radiation loss.* This is given by Stefan-Boltzmann Law

$$w_r = 5.702 \times 10^{-8}e\ (T^4 - T_a^4), \text{ watts/m}^2$$

where e = relative emissivity of conductor-surface = 1 for black body and 0.5 for oxidized Al or Cu, T=conductor temperature in °K=273+t and T_a = ambient temperature in °K = 273 + t_a.

The radiation loss per metre length of conductor is

$$W_r = 17.9 \times 10^{-8}\ e\ (T^4 - T_a^4)\ d_m, \text{ watts/m}.$$

Equation (3.9) for the heat balance then becomes

$$I^2R_m + s_a I_s d_m = 18\ \Delta t\sqrt{p.v_m.d_m} + 17.9.\ d_m\left[\left(\frac{T}{100}\right)^4 - \left(\frac{T_a}{100}\right)^4\right] \tag{3.10}$$

Example 3.4:   A 400-kV line in India uses a 2-conductor bundle with $d_m = 0.0318$m for each conductor. The phase current is 1000 Amps (500 Amps per conductor). The area of each conductor is 515.7 mm², $\rho_a = 2.7 \times 10^{-8}$ ohm-m at 20°C, $\alpha = 0.0045$ ohm/°C at 20°. Take the ambient temperature $t_a = 40$°C, atmospheric pressure $p = 1$, wind velocity $v_m = 1$ m/s, $e = 0.5$ and neglect solar irradiation. Calculate the final temperature of conductor due only to I^2R heating.

Solution: Let the final temperature $= t$ °C.

Then, $R_m = 2.7 \times 10^{-8} \dfrac{1 + 0.0045 \times t}{1 + 0.0045 \times 20} \dfrac{1.05}{515.7 \times 10^{-6}}$

$\qquad\qquad = 0.5 \times 10^{-4} (1 + 0.0045t)$, ohm/m

Therefore $W_i = I^2 R_m = 12.5 (1 + 0.0045t)$, watts/m

$\qquad W_c = 18\sqrt{1 \times 0.0318}.(t - 40) = 3.21 (t - 40)$, watts/m

$\qquad W_r = 17.9 \times 0.5 \times 0.0318 \left[\left(\dfrac{273 + t}{100} \right)^4 - \left(\dfrac{273 + 40}{100} \right)^4 \right]$

$\qquad\qquad = 0.2845 \{[(273 + t)/100]^4 - 95.95\}$.

Using equation (3.9), the equation for t comes out as

$\qquad 12.5 (1 + 0.0045t) = 3.21t + 0.2845 \times 10^{-8} (273 + t)^4 - 155.7$

or, $\qquad (273 + t)^4 = (590 - 11.28t) \, 10^8$.

A trial and error solution yields $t \approx 44$°C. (At this final temperature, we can calculate the values of the three heats which are $I^2R_m = 14.38$, $W_c = 12.84$, and $W_r = 1.54$, watts/m.).

Example 3.5:   In the previous example, calculate the final temperature (or temperature rise) if the solar irradiation adds (a) 10 watts/m, and (b) 1160 W/m² giving a contribution of 37 watts/m to the conductor.

Solution: By going through similar procedure, the answers turn out to be

(a) $t = 45.5$°C, $\Delta t = 5.5$°C;

(b) $t = 54.1$°C, $\Delta t = 14.1$°C.

We observe that had the ambient temperature been 50°C, the temperature rise would reach nearly the maximum. This is left as an exercise at the end of the chapter.

3.3   Properties of Bundled Conductors

Bundled conductors are exclusively used for e.h.v. transmission lines. Only one line in the world, that of the Bonneville Power Administration in the U.S.A., has used a special expanded ACSR conductor of 2.5 inch diameter for their 525 KV line. Figure 3.3 shows examples of conductor configurations used for each phase of ac lines or each pole of a dc line.

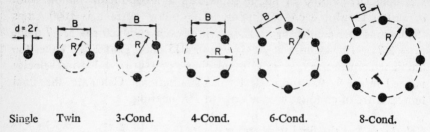

Single   Twin   3-Cond.   4-Cond.   6-Cond.   8-Cond.

Fig. 3.3.   Conductor configurations used for bundles in e.h.v. lines.

As of now a maximum of 18 sub-conductors have been tried on experimental lines but for commercial lines the largest number is 8 for 1150-1200 KV lines.

3.3.1   Bundle Spacing and Bundle Radius (or Diameter)

In almost all cases, the sub-conductors of a bundle are uniformly distributed on a circle of radius R. There are proposals to space them non-uniformly to lower the audible noise generated by the bundle conductor, but we will develop the relevant geometrical properties of an N-conductor bundle on the assumption of uniform spacing of the sub-conductors (Fig. 3.4).

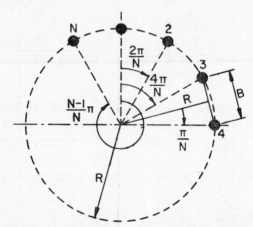

Fig. 3.4.   Bundle spacing B, and bundle radius R.

The spacing between adjacent sub-conductors is termed 'Bundle Spacing' and denoted by B. The radius of the pitch circle on which the sub-conductors are located will be called the 'Bundle Radius', denoted as R. The radius of each sub-conductor is r with diameter d. The angle subtended at the centre by adjacent sub-conductor is $(2\pi/N)$ radians, and it is readily seen that

$$\frac{B}{2} = R \sin (\pi/N) \text{ giving } R = B/2 \sin (\pi/N) \tag{3.11}$$

For $N = 1$ to 18, the following table gives (R/B) and (B/R).

$N = 2$	3	4	6	8	12	18
$R/B = 0.5$	0.578	0.7071	1	1.308	1.874	2.884
$B/R = 2$	$\sqrt{3}$	$\sqrt{2}$	1	0.7654	0.5344	0.3472

3.3.2 Geometric Mean Radius of Bundle (Equivalent Radius)

Except for calculating the surface voltage gradient from the charge of each sub-conductor, for most other calculations the bundle of N-sub-conductors can be replaced by a single conductor having an equivalent radius. This is called the 'Geometric Mean Radius' or simply the 'Equivalent Radius'. It will be shown below that its value is

$$r_{eq} = (N.r.R^{N-1})^{1/N} = r[N.(R/r)^{N-1}]^{1/N} = R(N.r/R)^{1/N} \qquad (3.12)$$

It is the Nth root of the product of the sub-conductor radius r, and the distance of this sub-conductor from all the other $(N-1)$ companions in the bundle. Equation (3.12) is derived as follows:

Referring to Fig. 3.4, the product of $(N-1)$ mutual distances is

$$\left(2R\sin\frac{\pi}{N}\right)\left(2R\sin\frac{2\pi}{N}\right)\left(2R\sin\frac{3\pi}{N}\right)\ldots\left(2R\sin\frac{N-1}{N}\pi\right)$$

$$= (2R)^{N-1}\left(\sin\frac{\pi}{N}\right)\left(\sin\frac{2\pi}{N}\right)\ldots\left(\sin\frac{N-1}{N}\pi\right).$$

$$r_{eq} = \left[r.(2R)^{N-1}\sin\frac{\pi}{N}\cdot\sin\frac{2\pi}{N}\ldots\sin\frac{N-1}{N}\pi\right]^{1/N} \qquad (3.13)$$

For $N = 2$, $r_{eq} = (2rR)^{1/2}$

For $N = 3$, $r_{eq} = \left(2^2.R^2.r.\sin\frac{\pi}{3}\cdot\sin\frac{2\pi}{3}\right)^{1/3} = (3rR^2)^{1/3}$

For $N = 4$, $r_{eq} = \left(2^3.R^3.r.\sin\frac{\pi}{4}\cdot\sin\frac{2\pi}{4}\cdot\sin\frac{3\pi}{4}\right)^{1/4} = (4rR^3)^{1/4}$

For $N = 6$, $r_{eq} = \left(2^5.R^5.r.\sin\frac{\pi}{6}\cdot\sin\frac{2\pi}{6}\ldots\sin\frac{5\pi}{6}\right)^{1/6} = (6.r.R^5)^{1/6}$

This is equation (3.12) where the general formula is $r_{eq}=(N.r.R^{N-1})^{1/N}$. The reader should verify the result for $N = 8, 12, 18$.

Example 3.6: The configurations of some e.h.v. lines for 400 KV to 1200 KV are given. Calculate r_{eq} for each.

(a) 400 KV: $N = 2$, $d = 2r = 3.18$ cm, $B = 45$ cm
(b) 750 KV: $N = 4$, $d = 3.46$ cm, $B = 45$ cm
(c) 1000 KV: $N = 6$, $d = 4.6$ cm, $B = 12\ d$
(d) 1200 KV: $N = 8$, $d = 4.6$ cm, $R = 0.6$ m

Solution: The problem will be solved in different ways.

(a) $r_{eq} = \sqrt{r.B} = (1.59 \times 45)^{1/2} = 8.46$ cm $= 0.0846$ m

(b) $r_{eq} = [4 \times 1.73 \times (45/\sqrt{2})^3]^{1/4} = 21.73$ cm $= 0.2173$ m

(c) $r_{eq} = [6 \times 2.3 \times 55.2^5]^{1/6} = 43.81$ cm $= 0.4381$ m

Also, $r_{eq} = 55.2 (6 \times 2.3/55.2)^{1/6} = 43.81$ cm

(d) $r_{eq} = 60 (8 \times 2.3/60)^{1/8} = 51.74$ cm $= 0.5174$ m

We observe that as the number of sub-conductor increases, the equivalent radius of bundle is approaching the bundle radius. The ratio r_{eq}/R is $(Nr/R)^{1/N}$. The concept of equivalent bundle radius will be utilized for calculation of inductance, capacitance, charge, and several other line parameters in the sections to follow.

3.4 Inductance of e.h.v. Line Configurations

Figure 3.5 shows several examples of line configuration used in various

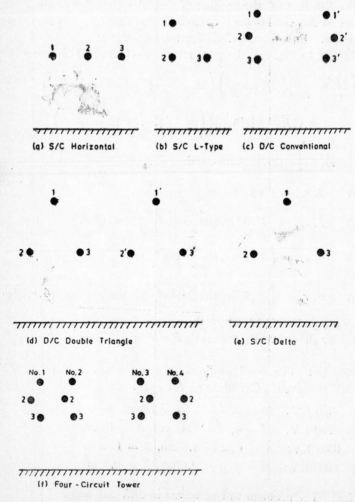

(a) S/C Horizontal   (b) S/C L-Type   (c) D/C Conventional

(d) D/C Double Triangle   (e) S/C Delta

(f) Four-Circuit Tower

Fig. 3.5. e.h.v. line configurations used.

parts of the world. They range from single-circuit (S/C) 400 KV lines to proposed 1200 KV lines. Double-circuit (D/C) lines are not very common, but will come into practice to save land for the line corridor. As pointed out in Chapter 2, one 750 KV circuit can transmit as much power as 4–400 KV circuits and in those countries where technology for 400 KV level exists there is a tendency to favour the four-circuit 400 KV line instead of using the higher voltage level. This will save on import of equipment from other countries and utilize the know-how of one's own country. This is a National Policy and will not be discussed further.

3.4.1 Inductance of Two Conductors

We shall very quickly consider the method of handling the calculation of inductance of two conductors each of external radius r and separated by a distance D which forms the basis for the calculation of the matrix of inductance of multi-conductor configurations.

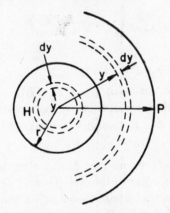

Fig. 3.6. Round conductor with internal and external flux linkages.

Figure 3.6 shows a round conductor carrying a current I. We first investigate the flux linkage experienced by it due, up to to a distance X, to its own current, and then extend it to two conductors. The conductor for the present is assumed round and solid, and the current is also assumed to be uniformly distributed with a constant value for current density $J = I/\pi r^2$. There are two components to the flux linkage: (1) flux internal to the conductor up to r; and (2) flux external to the conductor from r up to x.

Inductance Due to Internal Flux

At a radius y inside the conductor, Ampere's circuital law gives $\oint H.$

dl = current enclosed. With a uniform current density J, the current enclosed up to radius y is $I_y = y^2 I/r^2$. This gives,

$$H_y.2\pi y = Iy^2/r^2 \text{ or, } H_y = \frac{I}{2\pi r^2}.y \tag{3.14}$$

Now, the energy stored in a magnetic field per unit volume is

$$w_y = \tfrac{1}{2}\,\mu_0\mu_r H_y^2 = \frac{I^2\mu_0\mu_r}{8\pi^2 r^2}\,y^2, \text{ Joules/m}^3 \tag{3.15}$$

Consider an annular volume at y, thickness dy, and one metre length of conductor. Its volume is $(2\pi y.dy.l)$ and the energy stored is

$$dW = 2\pi\,y.w_y.dy = \frac{I^2\mu_0\mu_r}{4\pi r^4}\,y^3.dy$$

Consequently, the total energy stored up to radius r in the conductor can be calculated. But this is equal to $\tfrac{1}{2}L_i I^2$, where L_i = inductance of the conductor per metre due to the internal flux linkage.
Therefore

$$\tfrac{1}{2}L_i I^2 = \int_0^r dW = \frac{I^2\mu_0\mu_r}{4\pi r^4}\int_0^r y^3.dy = \frac{\mu_0\mu_r}{16\pi}\,I^2 \tag{3.16}$$

Consequently,

$$L_i = \mu_0\mu_r/8\pi, \text{ Henry/metre} \tag{3.17}$$

For a non-magnetic material, $\mu_r = 1$. With $\mu_0 = 4\pi \times 10^{-7}$ H/m, we obtain the interesting result that irrespective of the size of the conductor, the inductance due to internal flux linkage is

$$L_i = 0.05\,\mu \text{ Henry/metre for } \mu_r = 1$$

The effect of non-uniform current distribution at high frequencies is handled in a manner similar to the resistance. Due to skin effect, the internal flux linkage decreases with frequency, contrary to the behaviour of resistance. The equation for the inductive reactance is (W. D. Stevenson, 2nd Ed.)

$$X_i(f) = R_0.(X/2).\frac{\text{Ber}(X).\text{B}'\text{er}(X) + \text{Bei}(X).\text{B}'\text{ei}(X)}{[\text{B}'\text{er}(X)]^2 + [\text{B}'\text{ei}(X)]^2} \tag{3.18}$$

where $X_i(f)$ = reactance due to internal flux linkage at any frequency f, R_0=dc resistance of conductor per mile in ohms, and $X=0.0636\sqrt{f/R_0}$. [If R_m = resistance per metre, then $X = 1.59\times 10^{-3}\sqrt{f/R_m}$.]

Fig. 3.2(b) shows the ratio L_i/L_0 plotted against X, where $L_i = X_i/2\pi f$ and $L_0 = \mu_0/8\pi$ derived before.

Inductance Due to External Flux

Referring to Fig. 3.6 and applying Ampere's circuital law around a circle of radius y_e on which the field strength H is same everywhere, the magnetic field strength is given as

$$H = I/2\pi y_e \text{ giving } B = \mu_0\mu_r\,I/2\pi y_e$$

Since e.h.v. line conductors are always located in air, $\mu_r = 1$. In a differential distance dy_e, the magnetic flux is $d\phi = B_y.dy_e$ per metre length of conductor. Consequently, the flux linkage of conductor due to external flux up to a distance x is

$$\psi_e = \int_r^x B_y.dy_e = \frac{\mu_0\mu_r}{2\pi} I.\ln(x/r) \tag{3.19}$$

The inductance is $L_e = \psi_e/I = \frac{\mu_0\mu_r}{2\pi} \ln(x/r)$ (3.20)

In air, and with

$$\mu_0 = 4\pi \times 10^{-7} \text{ H/m}, L_e = 0.2 \ln(x/r) \text{ }\mu\text{H/m}.$$

For a round conductor with uniform current density, the combined inductance due to internal and external flux linkage up to distance x from the centre of conductor is

$$L = 0.2[0.25 + \ln(x/r)] = 0.2[\ln 1.284 + \ln(x/r)]$$
$$= 0.2 \ln(x/0.7788r), \text{ }\mu\text{H/m or mH/km}$$

This expression can be interpreted as though the effective radius of conductor becomes $r_e = 0.7788 \times$ actual radius. We emphasize here that this applies only to a solid round conductor with uniform current density distribution inside. It does not apply to stranded conductors nor at alternating currents where the current density is not uniformly distributed. For stranded conductors and at power frequency, conductor manufacturers provide data of the effective radius to be used for inductance calculation. This is known as the 'Geometric Mean Radius' and the reader should consult catalogues of conductor details. Its average value lies between 0.8 and 0.85.

The Two-Conductor Line

Figure 3.7 depicts two conductors each of radius r, separated by a centre-to-centre distance D, and carrying currents I and $-I$. We will derive expression for the flux linkage and inductance of 1 metre length of the 2-conductor system which will enable us to translate the result to the case of a single conductor located at a height $H = D/2$ above a ground plane.

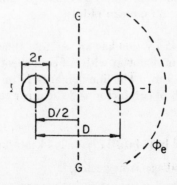

Fig. 3.7. Flux linkage calculation of 2-conductor line.

First consider a flux line ϕ_e flowing external to both conductors. It is clear that this line links zero current and so the magnetic field strength is zero. Therefore, all the flux must flow in between the conductors from r to $(D-r)$. The flux linkage of conductor 1 shown on the left has two parts:

(i) due to its own current I, and (ii) due to the current $-I$ in conductor 2. Neglecting internal flux linkage, the flux linkage due to its own current in the absence of current in conductor 2 up to the distance $(D-r)$ is

$$\psi_{11} = \int_r^{D-r} d\psi_{11} = \frac{\mu_0 \mu_r}{2\pi} I \int_r^{D-r} dx/x = \frac{\mu_0 \mu_r}{2\pi} I \ln \frac{D-r}{r} \qquad (3.21)$$

Consider the effect of current in conductor 2. Fleming's rule shows that the flux is in the same direction as that produced by current in conductor 1. Therefore, the flux linkage of conductor 1 due to current in conductor 2 is

$$\psi_{12} = \int_r^{D-r} d\psi_{12} = \frac{\mu_0 \mu_r}{2\pi} I \ln \frac{D-r}{r} \qquad (3.22)$$

Hence, the total flux linkage of conductor 1 due to both currents is

$$\psi_1 = \psi_{11} + \psi_{12} = \frac{\mu_0 \mu_r}{\pi} I . \ln \frac{D-r}{r} \approx \frac{\mu_0 \mu_r}{\pi} I . \ln(D/r) \qquad (3.23)$$

(when $D \gg r$). The centre line $G-G$ between the two conductors is a flux line in the field of two equal but opposite currents. The inductance of any one of the conductors due to flux flowing up to the plane $G-G$ will be one-half that obtained from equation (3.23). This is

$$L = \frac{\mu_0 \mu_r}{2\pi} \ln (D/r) \qquad (3.24)$$

Using $\mu_0 = 4\pi \times 10^{-7}$, $\mu_r = 1$, and $D = 2H$, the inductance of a single overhead-line conductor above a ground plane can be written as

$$L = 0.2 \ln (2H/r), \mu \text{ Henry/metre (milli Henry/km)} \qquad (3.25)$$

To this can be added the internal flux linkage and the resulting inductance using the geometric mean radius.

Example 3.7: A 345-kV line has an ACSR Bluebird conductor 1.762 inches (0.04477 m) in diameter with an equivalent radius for inductance calculation of 0.0179 m. The line height is 12 m. Calculate the inductance per km length of conductor and the error caused by neglecting the internal flux linkage.

Solution: $L = 0.2 \ln (24/0.0179) = 1.44$ mH/km.

If internal flux linkage is neglected,

$$L = 0.2 \ln (24/0.02238) = 1.3955 \text{ mH/km}$$

$$\text{Error} = (1.44 - 1.3955) 100/1.44 = 3.09\%.$$

We also note that GMR/outer radius $= 0.0179/0.02238 = 0.8$. For a round solid conductor, GMR $= 0.7788 \times$ outer radius.

3.4.2   Inductance of Multi-Conductor Lines—Maxwell's Coefficients

In the expression for the inductance $L = 0.2 \ln (2H/r)$ of a single conductor located above a ground plane, the factor $P = \ln (2H/r)$ is called Maxwell's coefficient. When several conductors are present above a ground at different heights each with its own current, the system of n-conductors can be assumed to consist of the actual conductors in air and their images below ground carrying equal currents but in the opposite direction which will preserve the ground plane as a flux line. This is shown in Fig. 3.8.

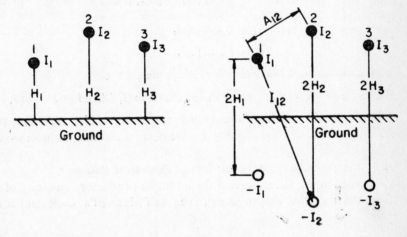

Fig. 3.8.   Multi-conductor line above ground with image conductors below ground.

The flux linkage of any conductor, say 1, consists of 3 parts in a 3-phase line, due its own current and the contribution from other conductors. The self flux linkage is $\psi_{11} = (\mu_0/2\pi) I_1 \ln (2H/r)$. We may use the geometric mean radius instead of r to account for internal flux linkage so that we write $\psi_{11} = (\mu_0/2\pi) I_1 \ln (2H/D_s)$, where $D_s =$ self-distance or GMR. For a bundle-conductor, we will observe that an equivalent radius of the bundle, equation (3.12), has to be used.

Now consider the current in conductor 2 only and the flux linkage of conductor 1 due to this and the image of conductor 2 located below ground. For the present neglect the presence of all other currents. Then, the flux lines will be concentric about conductor 2 and only those lines beyond the aerial distance A_{12} from conductor 1 to conductor 2 will link conductor 1. Similarly, considering only the current $-I_2$ in the image of conductor 2, only those flux lines flowing beyond the distance I_{12} will link the aerial conductor 1. Consequently, the total flux linkage of phase conductor 1 due to current in phase 2 will be

$$\psi_{12} = \frac{\mu_0 \mu_r}{2\pi} \left[I_2 \int_{A_{12}}^{\infty} dx/x - I_2 \int_{I_{12}}^{\infty} dx/x \right] = \frac{\mu_0 \mu_r}{2\pi} I_2 \ln (I_{12}/A_{12}) \qquad (3.26)$$

The mutual Maxwell's coefficient between conductors 1 and 2 will be

$$P_{12} = \ln (I_{12}/A_{12})$$

In general, it is evident that the mutual Maxwell's coefficient for the flux linkage of conductor i with conductor j (and vice-versa) will be, with $i, j = 1, 2, \dots n$,

$$P_{ij} = \ln (I_{ij}/A_{ij}) \qquad (3.27)$$

Thus, for a system of n conductors (phases or poles) shown in Fig. 3.8, the flux-linkage matrix is

$$[\psi]_n = \frac{\mu_0 \mu_r}{2\pi} [P]_{nn} [I]_n = [L]_{nn} [I]_n \qquad (3.28)$$

where
$$[\psi]_n = [\psi_1, \psi_2, \dots, \psi_n]_t$$
$$[I_n] = [I_1, I_2, \dots, I_n]_t,$$

and the elements of Maxwell's coefficient matrix are

$$P_{ii} = \ln(2H/r_{eq}) \quad \text{and} \quad P_{ij} = P_{ji} = \ln (I_{ij}/A_{ij}), \; i \neq j \qquad (3.29)$$

The diagonal elements of the inductance matrix $[L]_{nn}$ represent the self-inductances, and the off-diagonal elements the mutual-inductances.

3.4.3 Bundled Conductor Lines: Use of Equivalent Radius, r_{eq}
In this section we will show that for a bundle conductor consisting of N sub-conductors, the denominator in the self Maxwell's coefficient is to

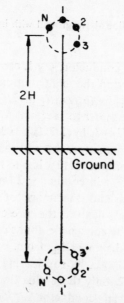

Fig. 3.9. Calculation of equivalent radius of bundle.

be taken as r_{eq} of equation (3.12). This is done under the following basic assumptions:

(1) The bundle spacing B between adjacent sub-conductors or the bundle radius R is very small compared to the height H of the phase conductors above ground. This allows the use of $2H$ as the distance between any sub-conductor of the bundle and the image of all the other $(N-1)$ subconductors below ground, as shown in Fig. 3.9. This means thut

$$I_{11} = I_{12} = I_{13} = \ldots = I_{1N} = 2H$$

(2) The total current carried by the bundle is I and that of each sub-conductor is $i = I/N$.

(3) Internal flux linkages are omitted, but can be included if the problem warrants it.

Consider the flux linkage of conductor 1, which is

$$\psi_1 = \frac{\mu_0 \mu_r}{2\pi} \frac{I}{N} \left[\ln \frac{2H}{r} + \ln \frac{2H}{2R \sin \frac{\pi}{N}} + \ln \frac{2H}{2R \sin \frac{4\pi}{N}} + \ldots \right.$$
$$\left. + \ln \frac{2H}{2R \sin \frac{N-1}{N} \pi} \right]$$

$$= \frac{\mu_0 \mu_r}{2\pi} \frac{I}{N} \ln \frac{(2H)^N}{r(2R)^{N-1} \cdot \sin \frac{\pi}{N} \cdot \sin \frac{2\pi}{N} \ldots \sin \frac{N-1}{N} \pi}$$

$$= \frac{\mu_0 \mu_r}{2\pi} I \ln \frac{2H}{r_{eq}} \tag{3.30}$$

where r_{eq} is precisely what is given in equation (3.12). The self-inductance of the entire bundle is

$$L = \psi_1/I = \frac{\mu_0 \mu_r}{2\pi} \ln (2H/r_{eq}) \tag{3.31}$$

while the inductance of each subconductor will be

$$L_c = \psi_1/i = \frac{\mu_0 \mu_r}{2\pi} N \ln (2H/r_{eq}) \tag{3.32}$$

which is also N times the bundle inductance since all the subconductors are in parallel. The Maxwell's coefficient for the bundle is $P_b = \ln (2H/r_{eq})$, as for a single conductor with equivalent radius r_{eq}.

Example 3.7: The dimensions of a 3-phase 400-kV horizontal line, Fig. 3.10, are: $H=15$ m, $S=11$ m phase separation, conductor 2×3.18 cm dia, and $B = 45.72$ cm. Calculate

(a) the matrix of inductances per km, for untransposed configuration, and

(b) the same when there is complete transposition.

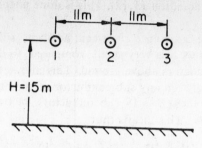

Fig. 3.10. 400-kV line for Example 3.7.

Solution; $r_{eq} = \sqrt{r.B} = \sqrt{0.0159 \times 0.4572} = 0.0853$ m

$P_{11} = P_{22} = P_{33} = \ln(2 \times 15/0.0853) = 5.863$

$P_{12} = P_{21} = P_{23} = P_{32} = \ln(\sqrt{4H^2 + S^2}/S) = 1.0664$

$P_{13} = P_{31} = \ln(\sqrt{4H^2 + 4S^2}/2S) = 0.525$

(a) $[L]_{ut} = 0.2 \begin{bmatrix} 5.863, & 1.0664, & 0.525 \\ 1.0664, & 5.863, & 1.0664 \\ 0.525, & 1.0664, & 5.863 \end{bmatrix} \begin{array}{l} \mu\text{H/m} \\ \text{mH/km} \end{array}$

$= \begin{bmatrix} 1.173, & 0.213, & 0.105 \\ & 1.173, & 0.213 \\ & & 1.173 \end{bmatrix} \text{mH/km}$

(b) For the completely transposed line, since each phase occupies each of the 3 positions for 1/3 the distance, the average mutual inductance will be $0.2(1.0664 + 1.0664 + 0.525)/3 = 0.177$ mH/km.

$[L]_t = \begin{bmatrix} L_s, & L_m, & L_m \\ L_m, & L_s, & L_m \\ L_m, & L_m, & L_s \end{bmatrix} = \begin{bmatrix} 1.173, & 0.177, & 0.177 \\ 0.177, & 1.173, & 0.177 \\ 0.177, & 0.177, & 1.173 \end{bmatrix} \text{mH/km}$

Note that the self inductance of usual e.h.v. lines is in the neighbourhood of 1 mH/km. We observed that as the number of sub-conductors is increased, the geometric mean radius or equivalent radius of bundle increases. Since r_{eq} divides $2H$ in the logarithm, bundling will reduce the series inductance of a line, which will increase the power-handling capacity. Table I in Chapter 2 shows this property where, as the system kV increases, the need for using more subconductors is urgent from considerations of current-carrying capacity as well as reducing the voltage gradient on conductor surface. This also brings the benefit of decrease in series reactance and improvement in power-handling capacity of a single circuit. Figure 3.5 must again be referred for details.

3.5 Line Capacitance Calculation

Consider two conductors of equal radii r located with their centres $2H$ apart, as shown in Fig. 3.11. The charges on each is Q coulombs/metre and of opposite polarity. On a unit positive test charge located at point F at a distance x from the centre of the conductor on the left with positive charge Q, the total force exerted will be

$$E_F = \frac{Q}{2\pi e_0}\left(\frac{1}{x} + \frac{1}{2H - x}\right) \text{ Newtons} \qquad (3.33)$$

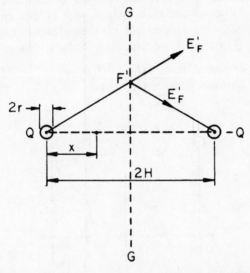

Fig. 3.11. Single-phase line for capacitance calculation.

(This results from Gauss's Law and the reader is referred to Chapter 4 if it is not already known from previous courses devoted to Electrostatics.) Consequently, the potential difference between the two conductors is

$$V = \frac{Q}{2\pi e_0}\int_r^{2H-r}[1/x + 1/(2H - x)]\,dx = \frac{Q}{\pi e_0}\ln\frac{2H - r}{r} \qquad (3.34)$$

If $2H \gg r$ as is usual in e.h.v. lines, $V = \frac{Q}{\pi e_0}\ln(2H/r)$. By symmetry, the mid-plane $G - G$ will be at 0.5 V and the p.d. between the positive conductor and $G - G$ is $V_g = V/2$.

Therefore

$$V_g = \frac{Q}{2\pi e_0}\ln\frac{2H}{r} = \frac{Q}{2\pi e_0}\ln\frac{\text{Dist. of cond. from negative charge}}{\text{Dist. of cond. from positive charge}} \qquad (3.35)$$

Since the factor $\ln(2H/r)$ multiplies the charge coefficient $(Q/2\pi e_0)$, which is in volts, it is called Maxwell's Potential coefficient. We encountered the same factor in inductance calculation also. The mid-plane $G - G$ is an equipotential surface since the electric force is everywhere perpendicular to it as can be observed from the vector field intensity E'_F at a point F', whose components along $G-G$ are equal and opposite.

We again observe that when the ground plane $G-G$ is considered an equipotential surface for capacitance calculations, its effect can be considered by using an image conductor with a charge equal to the charge on the aerial conductor but of opposite polarity. From equation (3.35), we write the self potential coefficient as

$$P_{11} = \ln (2H/r) \qquad (3.36)$$

The mutual potential coefficients between phases are determined by placing the conductors and their images with proper charges as shown in Fig. 3.12. Following equation (3.35) the potential of conductor 1 due to the charges Q_2 and $-Q_2$ of conductor 2 and its image will be

$$V_{12} = \frac{Q_2}{2\pi e_0} \ln \frac{\text{Distance from } -Q_2}{\text{Distance from } +Q_2} = \frac{Q_2}{2\pi e_0} \ln(I_{12}/A_{12}) = \frac{Q_2}{2\pi e_0} \cdot P_{12} \qquad (3.37)$$

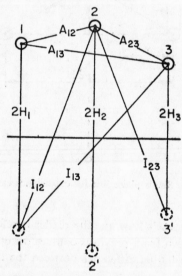

Fig. 3.12. Multi-conductor line for calculation of Maxwell's potential coefficients.

For a system of n conductors (phases or poles) above ground, the potentials of conductors will be

$$V_1 = \frac{Q_1}{2\pi e_0} \ln \frac{2H_1}{r} + \frac{Q_2}{2\pi e_0} \ln \frac{I_{12}}{A_{12}} + \dots + \frac{Q_n}{2\pi e_0} \ln \frac{I_{1n}}{A_{1n}}$$

$$\vdots$$

$$V_n = \frac{Q_1}{2\pi e_0} \ln \frac{I_{1n}}{A_{1n}} + \frac{Q_2}{2\pi e_0} \ln \frac{I_{2n}}{A_{2n}} + \dots + \frac{Q_n}{2\pi e_0} \ln \frac{2H_n}{r}$$

In matrix form,

$$[V]_n = [P]_{nn} [Q/2\pi e_0]_n \qquad (3.38)$$

where $[V]_n = [V_1, V_2, ..., V_n]_t =$ the potentials with respect to ground, and $[Q]_n = [Q_1, Q_2, ..., Q_n]_t =$ the conductor charges and the elements of the potential coefficient matrix are

$$P_{ii} = \ln(2H_i/r), \ P_{ij} = \ln(I_{ij}/A_{ij}), \ i \neq j, \ i, j = 1, 2, ..., n \quad (3.39)$$

The capacitance matrix of the n-conductor system is

$$[C]_{nn} = 2\pi e_0 \, [P]_{nn}^{-1} = 2\pi e_0 \, [M] \quad (3.40)$$

We derived the inductance matrix for the n-conductor system to be, equation (3.28),

$$[L]_{nn} = \frac{\mu_0}{2\pi} [P]_{nn} \quad (3.41)$$

If internal flux linkages be neglected there results the important relation

$$[L] \, [C] = \mu_0 e_0 \, [U] = \frac{1}{g^2} [U] \quad (3.42)$$

where $[U] =$ unit matrix, and $g =$ velocity of light $= 3 \times 10^5$ km/sec.

Potential Coefficients for Bundled-Conductor Lines

When each phase or pole comprises of N sub-conductors we once again arrive at the concept of geometric mean radius or equivalent radius of bundle r_{eq} exactly in the same manner as for inductance calculations. In the self potential coefficient, equation (3.39), r_{eq} will be used instead of r. This is derived as follows, on the assumptions that

(1) the bundle dimensions B and R are small compared to line height H, and

(2) B and R are small compared to the spacing S from the centre of one phase to another.

Figure 3.9 is again referred from which, with each conductor having a charge $q = Q/N$ per unit length, the potential of conductor 1 is

$$V_1 = \frac{Q}{2\pi e_0} \frac{1}{N} \left[\ln \frac{2H}{r} + \ln \frac{2H}{2R \sin \frac{\pi}{N}} + \ln \frac{2H}{2R \sin \frac{2\pi}{N}} + \cdots \right.$$
$$\left. + \ln \frac{2H}{2R \sin \frac{N-1}{N}\pi} \right]$$

$$= \frac{Q}{2\pi e_0} \ln (2H/r_{eq})$$

Thus, $$P_{ii} = \ln (2H/r_{eq}). \quad (3.43)$$

For calculating the mutual potential coefficients we can assume that the total bundle charge is concentrated at the centre of the bundle conductor, or else use the two assumptions above that the bundle dimensions are small compared to H and S.

Example 3.8: Calculate the capacitance matrix of the 3-phase 400 kV line of Example 3.7.

Solution: (a) For the untransposed configuration, the matrix of Maxwell's potential coefficients was found to be

$$[P]_{ut} = \begin{bmatrix} 5.863, & 1.0664, & 0.525 \\ 1.0664, & 5.863, & 1.0664 \\ 0.525, & 1.0664, & 5.863 \end{bmatrix}$$

Its inverse is

$$[M]_{ut} = [P]_{ut}^{-1} = \begin{bmatrix} 0.176, & -0.0298, & -0.0104 \\ -0.0298, & 0.1805, & -0.0298 \\ -0.0104, & -0.0298, & 0.176 \end{bmatrix}$$

The resulting capacitance matrix will be

$$[C] = 2\pi e_0 [M] = \frac{10^{-9}}{18} [M], \text{ Farad/metre}$$

$$[C]_{ut} = \begin{bmatrix} 9.77, & -1.65, & -0.58 \\ -1.65, & 10.02, & -1.65 \\ -0.58, & -1.65, & 9.77 \end{bmatrix} \text{ nF/km}$$

(b) For the completely transposed line,

$$[C] = \begin{bmatrix} c_s, & c_m, & c_m \\ c_m, & c_s, & c_m \\ c_m, & c_m, & c_s \end{bmatrix}, \text{ with } c_s = 9.85 \text{ nF/km and} \\ c_m = -1.29 \text{ nF/km}$$

We observe that the self capacitance of e.h.v. lines is in the neighbourhood of 10 nF/km. All the self-capacitance coefficients are positive while the mutual capacitance coefficients are negative. For the horizontal configuration of phases, the centre-phase has a slightly higher self-capacitance (capacitance to ground) due to the increase in number of dielectric lines of force terminating on it. The negative sign in mutual-capacitance follows from the physical consideration that a charge of one polarity placed on a conductor induces a charge of opposite polarity on another, i.e. the two conductors form the positive and negative electrodes of a capacitor.

Bundling increases the capacitance from that of a single conductor having the same cross-sectional area because of the equivalent bundle radius which depends on the bundle radius which is larger than the conductor radius. We will consider these aspects under voltage-gradient calculations in a later section.

3.6 Sequence Inductances and Capacitances

The use of Symmetrical Components for analyzing 3-phase problems has made it possible to solve very extensive network problems. It depends upon obtaining mutually-independent quantities from the original phase quantities that have mutual interaction. Following this concept, we will now resolve the inductances, capacitances, charges, potentials etc. into independent quantities by a general method. This procedure will be used for many types of excitations other than power-frequency later on. The basis for such transformations is to impress suitable driving functions and obtain the resulting responses.

Inductance Transformation to Sequence Quantities

We observed that for a fully-transposed 3-phase ac line, the flux-linkage equation is

$$[\psi]_3 = [L]_{33} [I]_3 \tag{3.44}$$

The inductance matrix is symmetric for a transposed line for which the symmetrical-components theory will be used. For zero-sequence, the currents in the 3 phases are equal and in phase so that $I_1 = I_2 = I_3 = I_0$. The resulting flux linkage is

$$[\psi_0] = \begin{bmatrix} L_s, & L_m, & L_m \\ L_m, & L_s, & L_m \\ L_m, & L_m, & L_s \end{bmatrix} \begin{bmatrix} 1 \\ 1 \\ 1 \end{bmatrix} I_0 = \begin{bmatrix} 1 \\ 1 \\ 1 \end{bmatrix} (L_s + 2L_m) I_0$$

Consequently, the inductance offered to zero-sequence currents is

$$L_0 = L_s + 2L_m \tag{3.45}$$

When positive-sequence currents are impressed,

$$I_1 = I_m \sin Wt, \ I_2 = I_m \sin (Wt - 120°) \text{ and } I_3 = I_m \sin (Wt + 120°)$$

$$[\psi_1] = \begin{bmatrix} L_s, L_m, L_m \\ L_m, L_s, L_m \\ L_m, L_m, L_s \end{bmatrix} \begin{bmatrix} \sin Wt \\ \sin (Wt - 120°) \\ \sin (Wt + 120°) \end{bmatrix} I_m$$

$$= \begin{bmatrix} \sin Wt \\ \sin (Wt - 120°) \\ \sin (Wt + 120°) \end{bmatrix} (L_s - L_m) I_m$$

Hence, the inductance offered to positive-sequence current is

$$L_1 = L_s - L_m \tag{3.46}$$

Similarly, for negative-sequence currents, the inductance is also

$$L_2 = L_s - L_m \tag{3.47}$$

Sequence Capacitances

In a similar manner, we can evaluate the zero-, positive-, and negative sequence capacitances by re-writing equation (3.38) as

$$[Q] = 2\pi e_0 [P]^{-1} [V] = [C][V] \tag{3.48}$$

For zero-sequence voltages, $[V_0] = [1, 1, 1] V$, and we obtain

$$C_0 = C_s + 2C_m \tag{3.49}$$

For positive-sequence voltages, $[V_+] = [\sin Wt, \sin (Wt - 120°),$ $\sin (Wt + 120°)] V$, and $C_1 = C_s - C_m$. Similarly, $C_2 = C_s - C_m$ for negative-sequence voltages.

From the above resolution of phase inductances into sequence inductances, and the capacitances, we can observe the following properties which are very important for modelling of transmission lines.

(1) The zero-sequence inductance L_0 is higher than the self inductance L_s, while the positive and negative sequence inductances are lower than L_s.

(2) The converse holds for capacitances since C_m is a negative quantity.

$$C_0 < C_s \text{ and, } C_1 \text{ and } C_2 > C_s.$$

(3) From equation (3.45) and (3.46), the self and mutual inductances can be found from the sequence impedances.

$$L_s = \tfrac{1}{3} (L_0 + 2L_1) \text{ and } L_m = \tfrac{1}{3} (L_0 - L_1) \tag{3.50}$$

(4) Similarly, $C_s = \tfrac{1}{3} (C_0 + 2C_1)$ and $C_m = \tfrac{1}{3} (C_0 - C_1)$ (3.51)

Figure 3.13 shows representative models of transmission lines used for system study on Network Analyzers or Digital Computers. It is easy to impress zero-sequence and positive-sequence voltages and currents and measure the responses in a practical test setup.

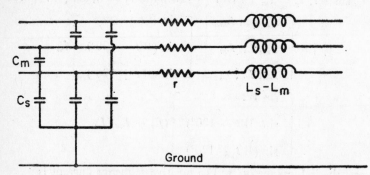

Fig. 3.13. Model of transmission lines for TNA study and Digital-Computer Calculation

The sequence inductances and capacitances can be obtained from the phase quantities in a very general manner by matrix diagonalization procedure which will be described in the next section. Referring to the symmetrical component theory, the phase quantities are transformed to symmetrical components by the matrix multiplication

$$\begin{bmatrix} V_0 \\ V_1 \\ V_2 \end{bmatrix} = \tfrac{1}{3} \begin{bmatrix} 1 & 1 & 1 \\ 1 & a & a^2 \\ 1 & a^2 & a \end{bmatrix} \begin{bmatrix} V_a \\ V_b \\ V_c \end{bmatrix} \tag{3.52}$$

where $a = 1 \angle 120°$.

The converse of combining the symmetrical components is carried out through the matrices

$$\begin{bmatrix} V_a \\ V_b \\ V_c \end{bmatrix} = \begin{bmatrix} 1 & 1 & 1 \\ 1 & a^2 & a \\ 1 & a & a^2 \end{bmatrix} \begin{bmatrix} V_0 \\ V_1 \\ V_2 \end{bmatrix} \tag{3.53}$$

These follow the original relations formulated by Dr. C.L. Fortescue as far back as 1915. We now denote

$$[T] = \begin{bmatrix} 1 & 1 & 1 \\ 1 & a^2 & a \\ 1 & a & a^2 \end{bmatrix} \text{ and } [T]^{-1} = \tfrac{1}{3} \begin{bmatrix} 1 & 1 & 1 \\ 1 & a & a^2 \\ 1 & a^2 & a \end{bmatrix} \tag{3.54}$$

If now we carry out the multiplications $[T]^{-1} [L] [T]$ and $[T]^{-1} [C] [T]$ for a completely transposed line, the result is

$$[T]^{-1} [L] [T] = \tfrac{1}{3} \begin{bmatrix} 1 & 1 & 1 \\ 1 & a & a^2 \\ 1 & a^2 & a \end{bmatrix} \begin{bmatrix} L_s, & L_m, & L_m \\ L_m, & L_s, & L_m \\ L_m, & L_m, & L_s \end{bmatrix}$$

$$\begin{bmatrix} 1 & 1 & 1 \\ 1 & a^2 & a \\ 1 & a & a^2 \end{bmatrix} = \begin{bmatrix} L_s + 2L_m, & & \\ & L_s - L_m, & \\ & & L_s - L_m \end{bmatrix} \tag{3.55}$$

The diagonal elements are the inductances for zero-sequence, positive-sequence and negative-sequence networks.

Similarly,

$$[T]^{-1} [C] [T] = \begin{bmatrix} C_s + 2C_m, & & \\ & C_s - C_m, & \\ & & C_s - C_m \end{bmatrix} \tag{3.56}$$

Such a general procedure is very convenient for de-coupling mutually-interacting quantities and then combining them suitably. For a general problem encountered with e.h.v. transmission lines, they are given the generic name 'Modes of Propagation'.

3.7 Line Parameters for Modes of Propagation

The sequence parameters given in the previous section apply to steady-state conditions and use phasor algebra. The quantity $a = 1 \angle 120° = -.5 + j\,0.866$ is a complex number. This is not always convenient when solving equations encountered with wave propagation on the phase or pole conductors which are characterized by (a) velocity of propagation, (b) attenuation, and (c) surge impedance. Following the ideas propounded by Dr. Fortescue, the waves on multi-conductor lines can also be resolved into 'modes of propagation'. The transformation matrix $[T]$ and its inverse $[T]^{-1}$ have to be evaluated for a given problem through standard set of rules which eventually diagonalize the given matrix of inductances, capacitances, resistances, surge impedances, and other parameters which govern the propagation characteristics. For a fully-transposed line, analytical expressions in closed form can be obtained for the transformation matrix and its inverse using real numbers. But for untransposed lines the evaluation of $[T]$ and $[T]^{-1}$ can also be carried case by case when numerical values for the inductances, etc. are given. These will be discussed in detail below.

3.7.1 Diagonalization Procedure

The resolution of mutually-interacting components of voltage, current, charge, or energy in waves propagating on the multi-conductors depends upon diagonalization of the $n \times n$ impedance matrix. A general procedure is given here while their application to Radio Noise, Switching Surges, etc. will be discussed in later chapters when we consider these problems individually.

First consider the diagonalization of the inductance matrix of a transposed line

$$[L] = \begin{bmatrix} L_s, & L_m, & L_m \\ L_m, & L_s, & L_m \\ L_m, & L_m, & L_s \end{bmatrix}$$

The following steps have to be followed for diagonalization.

Step 1. We evaluate the 'characteristic roots' or 'eigenvalues' (λ) of the given matrix according to the determinantal equation

$$|\,\lambda[U] - [L]\,| = 0, \text{ which gives } \begin{vmatrix} \lambda - L_s, & -L_m, & -L_m \\ -L_m, & \lambda - L_s, & -L_m \\ -L_m, & -L_m, & \lambda - L_s \end{vmatrix} = 0 \quad (3.57)$$

This gives $\lambda^3 - 3L_s\lambda^2 + 3\,(L_s^2 - L_m^2)\,\lambda - (L_s^3 - 3L_sL_m^2 + 2L_m^3) = 0$

or, $\qquad (\lambda - L_s - 2L_m)\,(\lambda - L_s + L_m)^2 = 0 \qquad (3.58)$

The three eigenvalues are

$$\lambda_1 = L_s + 2L_m, \ \lambda_2 = L_s - L_m, \ \lambda_3 = L_s - L_m \tag{3.59}$$

Step 2.   For each of these eigenvalues in turn, we evaluate the 'eigenvector' $[X]$, which is a column matrix, according to the equation

$$\{[U]\lambda_n - [L]\}[X] = [0] \tag{3.60}$$

Considering $\lambda_1 = L_s + 2L_m$, there results the explicit equation

$$\{\lambda_1[U] - [L]\}\begin{bmatrix} X_1 \\ X_2 \\ X_3 \end{bmatrix} = L_m \begin{bmatrix} 2, & -1, & -1 \\ -1, & 2, & -1 \\ -1, & -1, & 2 \end{bmatrix}\begin{bmatrix} X_1 \\ X_2 \\ X_3 \end{bmatrix} = \begin{bmatrix} 0 \\ 0 \\ 0 \end{bmatrix},$$

which in turn yields the three equations for X_1, X_2, X_3 to be

$$\left. \begin{aligned} 2X_1 - X_2 - X_3 &= 0 \\ -X_1 + 2X_2 - X_3 &= 0 \\ -X_1 - X_2 + 2X_3 &= 0 \end{aligned} \right\} \tag{3.61}$$

By choosing $X_1 = 1$, there results $X_2 = X_3 = X_1 = 1$. The corresponding eigenvector is $[1, 1, 1]_t$ with the normalized form $[1, 1, 1]_t (1/\sqrt{3})$.

By following a similar procedure for $\lambda_2 = L_s - L_m$, there results

$$\{\lambda_2[U] - [L]\}\begin{bmatrix} y_1 \\ y_2 \\ y_3 \end{bmatrix} = -L_m \begin{bmatrix} 1, & 1, & 1 \\ 1, & 1, & 1 \\ 1, & 1, & 1 \end{bmatrix}\begin{bmatrix} y_1 \\ y_2 \\ y_3 \end{bmatrix} = \begin{bmatrix} 0 \\ 0 \\ 0 \end{bmatrix}.$$

The three equations for y_1, y_2, y_3 are all equal to $y_1 + y_2 + y_3 = 0$. Once again letting $y_1 = 1$, we have $y_2 + y_3 = -1$. Now we have an infinite number of choices for the values of y_2 and y_3, and we make a judicial choice for them based on practical engineering considerations and utility.

As a first choice, let $y_2 = 0$. Then $y_3 = -1$. The resulting eigenvector and its normalized form are

$$\begin{bmatrix} 1 \\ 0 \\ -1 \end{bmatrix} \quad \text{and} \quad \frac{1}{\sqrt{2}}\begin{bmatrix} 1 \\ 0 \\ -1 \end{bmatrix}$$

Since the third eigenvalue λ_3 is also the same as λ_2, we obtain the same equations for the components of the eigenvector which we can designate as $[z_1, z_2, z_3]_t$. By choosing $z_1 = 1$ and $z_3 = 1$, we obtain $z_2 = -2$. The third eigenvector and its normalized form are

$$[1, -2, 1]_t \quad \text{and} \quad 1/\sqrt{6}[1, -2, 1]_t.$$

Step 3.   Formulate the complete 3×3 eigenvector matrix (or in general, $n \times n$) for the n eigenvalues and call it the inverse of the transformation matrix $[T]^{-1}$. For the problem under consideration,

$$[T]^{-1} = \begin{bmatrix} \sqrt{2}, & \sqrt{2}, & \sqrt{2} \\ \sqrt{3}, & 0, & -\sqrt{3} \\ 1, & -2, & 1 \end{bmatrix} 1/\sqrt{6} \qquad (3.62)$$

Step 4.   The transformation matrix $[T]$ will be

$$[T] = \begin{bmatrix} \sqrt{2}, & \sqrt{3}, & 1 \\ \sqrt{2}, & 0, & -2 \\ \sqrt{2}, & -\sqrt{3}, & 1 \end{bmatrix} 1/\sqrt{6} \qquad (3.63)$$

which turns out to be the transpose of $[T]^{-1}$. Their determinant is -6.

Step 5.   The given inductance matrix is diagonalized by the relation

$$[T]^{-1}[L][T] = \begin{bmatrix} L_s + 2L_m, & & \\ & L_s - L_m, & \\ & & L_s - L_m \end{bmatrix} = [\lambda] \qquad (3.64)$$

This is a diagonal matrix whose elements are equal to the three eigenvalues, which might be seen to be the same as the sequence inductances presented to the voltage and current. But now they will be called the inductances for the three modes of propagation of electromagnetic energy of the waves generating them.

3.7.2   Interpretation of the Eigenvectors

We observe that the eigenvector corresponding to the first eigenvalue $\lambda_1 = L_s + 2L_m$ consists of $[1, 1, 1]$, which can be interpreted as follows:

In the first mode of behaviour (or travel of all quantities on the three conductors), the voltages, currents, charges, and accompanying energies are all equal on the three conductors. Also, they are all of the same sign or polarity. In this mode of propagation, the return current flows in the ground and the resulting attenuation of energy etc. are high because of ground resistance. It is usually called the 'Line-to-Ground' mode of propagation or the 'homopolar' mode.

The eigenvector $[1, 0, -1]$ corresponds to the second eigenvalue $\lambda_2 = L_s - L_m$. In this case the propagation can be seen to take place between the outer phase-conductors only, with the centre phase being idle. Also, because this is a closed system involving the two outer phases, ground is not involved in propagation so that attenuation is lower than in the line-to-ground mode. The second mode is called 'Line-to-Line Mode of the 1st kind', or simply 'phase-phase' mode.

The last eigenvector is [1, −2, 1] corresponding to the repeated eigen-value $\lambda_3 = \lambda_2 = L_s - L_m$. We observe that these eigenvectors have depended upon the choice made by us for the relative values of the components of the eigenvector in step 2 of the diagonalization procedure. There are infinite set of such values for the eigenvector components, and the choice recommended here is felt to be the most convenient from the point of view of interpretation of the physical mechanism of behaviour of the several quantities of interest in the wave-propagation phenomenon involving the three phase-conductors and ground. The eigenvector [1, −2, 1] can be interpreted by noting that the outer phases form the 'go' for the current and the centre-phase the 'return'. Once again the system is closed if we assume charges of +1 and +1 on the outers and −2 on the centre phase. Therefore, ground is not involved in the propagation in this mode. It is called the 'Line-to-Line Mode of the 2nd kind', or the 'inter-phase mode'.

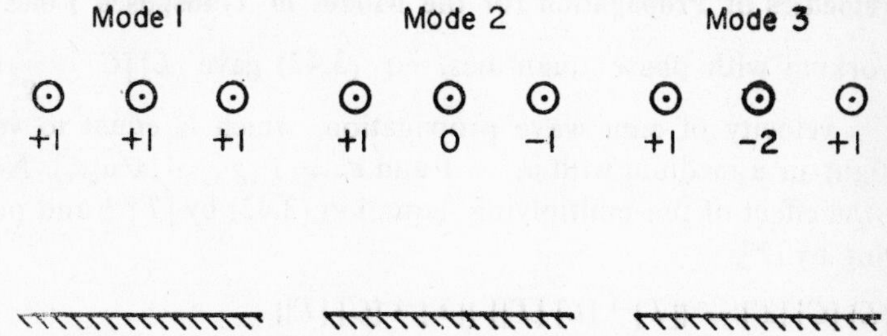

Mode 1	Mode 2	Mode 3
⊙ ⊙ ⊙	⊙ ⊙ ⊙	⊙ ⊙ ⊙
+1 +1 +1	+1 0 −1	+1 −2 +1

Fig. 3.14   Pictorial representation of conditions of Modes of Propagation on 3-phase line.

These three modes can be represented pictorially as shown in Fig. 3.14. Once the mutually-interacting quantities are resolved into the three independent modes of propagation, in a manner identical to symmetrical-component analysis of which we are familiar, the behaviour of all quantities in each mode can be analyzed and the phase quantities obtained finally by the inverse procedure. In general, resolution into modes requires pre-multiplication by $[T]^{-1}$ of the voltages, currents, and charges on the phase conductors, while combining the modal quantities to obtain phase quantities requires pre-multiplication of the modal voltages, currents, charges and other responses to these by $[T]$, the transformation matrix.

The concept of modes of propagation is very useful for:

(a) design of carrier equipment for speech and protection where the attenuation of signals and their distortion is of primary concern in determining the transmitter and receiver powers;

(b) propagation of switching and lightning surges on the lines which cause over-voltages and control the design of insulation clearances; and

(c) radio-interference levels generated by corona pulses on the phase-conductors which propagate on the conductors over a ground plane.

In the above illustration, the diagonalization procedure was applied to the matrix of inductances of a transposed line. A similar diagonalization of the capacitance matrix for the fully-transposed line will yield the same transformation matrices. In this case, the resulting eigenvalues for both inductance and capacitance are equal to the zero-, positive-, and negative-sequence quantities obtained from Fortescue's transformation using phasors in the transformation matrix. Here, we have used only real numbers to effect the diagonalization procedure. The resulting $[T]$ and $[T]^{-1}$ are called 'Modified Clarke Transformation' matrices after the eminent lady engineer Dr. Edith Clarke, who is also known for her $\alpha - \beta - \gamma$ or $\alpha - \beta - 0$ components of Machines Theory.

3.7.3 Velocities of Propagation for the Modes in Transposed Lines

When working with phase quantities, eq. (3.42) gave $[L][C] = \dfrac{1}{g^2}[U]$ where g = velocity of e.m. wave propagation, which is equal to velocity of light in a medium with $\mu_r = 1$ and $e_r = 1$ $[g = 1/\sqrt{\mu_0 e_0}]$. Now, consider the effect of pre-multiplying equation (3.42) by $[T]^{-1}$ and post-multiplying by $[T]$:

$$[T]^{-1}[L][C][T] = \{[T]^{-1}[L][T]\}\,\{[T]^{-1}[C][T]\}$$

$$= [T]^{-1}[U][T]\frac{1}{g^2} = \frac{1}{g^2}[U] \tag{3.65}$$

i.e.,
$$\begin{bmatrix} L_s + 2L_m, & & \\ & L_s - L_m, & \\ & & L_s - L_m \end{bmatrix}\begin{bmatrix} C_s + 2C_m, & & \\ & C_s - C_m, & \\ & & C_s - C_m \end{bmatrix}$$

$$= \frac{1}{g^2}\begin{bmatrix} 1 & & \\ & 1 & \\ & & 1 \end{bmatrix}$$

This gives $(L_s + 2L_m)(C_s + 2C_m) = L_0 C_0 = 1/g^2$
and $\qquad\quad (L_s - L_m)(C_s - C_m) = L_1 C_1 = 1/g^2$ $\quad\Big\}$ (3.66)

This shows that the velocities of propagation of waves in all three modes are equal to the velocity of light. This is the case when ground-return inductance is not taken into account. Usually, in the first or line-to-ground mode (homopolar mode) inductance of ground return reduces the velocity of propagation of that modal component to about $2 \sim 2.5 \times 10^5$ km/s (nearly 70–85% of g) as will be discussed after we evaluate the ground-return inductance in Section 3.8.

3.7.4   Untransposed Line: Modes of Propagation

The eigenvalues and eigenvectors resulting in transformation matrix and its inverse for an untransposed line must be worked out on a case-by-case basis using the numerical values of the inductance and capacitance. No general expressions can be given although the general programme for the diagonalization is the same as outlined before. Thus, a digital computer can handle any type of matrix diagonalization. We will illustrate the procedure through an example.

Example 3.9:   The inductance and capacitance matrices for the 400-kV horizontal line were worked out in Examples 3.7 and 3.8. Diagonalize the capacitance matrix of the untransposed line.

Solution: $$[C]_{ut} = \begin{bmatrix} 9.77, & -1.65, & -0.58 \\ -1.65, & 10.02, & -1.65 \\ -0.58, & -1.65, & 9.77 \end{bmatrix} \text{nF/km}$$

Step 1. $$|\lambda[U] - [C]| = \begin{vmatrix} \lambda-9.77, & 1.65, & 0.58 \\ 1.65, & \lambda-10.02, & 1.65 \\ 0.58, & 1.65, & \lambda-9.77 \end{vmatrix}$$

$$= \lambda^3 - 29.65\,\lambda^2 + 285.46\,\lambda - 896.67 = 0$$

The eigenvalues are $\lambda_1 = 11.9755$, $\lambda_2 = 10.35$, $\lambda_3 = 7.2345$. Note that the characteristic roots are now distinct.

Step 2. $$\{\lambda_1[U]-[C]\}[X] = \begin{bmatrix} 2.2055, & 1.65, & 0.58 \\ 1.65, & 1.9555, & 1.65 \\ 0.58, & 1.65, & 2.2055 \end{bmatrix} \begin{bmatrix} X_1 \\ X_2 \\ X_3 \end{bmatrix} = \begin{bmatrix} 0 \\ 0 \\ 0 \end{bmatrix}$$

Therefore, aking $X_1 = 1$, there result $X_2 + 0.3515\,X_3 = -1.3367$

and $$X_2 + 0.8438\,X_3 = -0.8438$$

Thus we obtain the components of the first eigenvector to be $X_1 = 1$, $X_2 = -1.6887$, and $X_3 = 1.00127 \approx 1$. Length of vector = 2.2032. The normalized form is [0.454, −0.7665, 0.454]. Similarly, for $\lambda_2 = 10.35$, there results the normalized eigenvector [0.7071, 0, −0.7071], and for $\lambda_3 = 7.2345$, [0.542, 0.6422, 0.542].

Step 3. $$[T]^{-1} = \begin{bmatrix} 0.454, & -0.7655, & 0.454 \\ 0.7071, & 0, & -0.7071 \\ 0.542, & 0.6422, & 0.542 \end{bmatrix}$$

$$\text{and} \quad [T] = \begin{bmatrix} 0.454, & 0.7071, & 0.542 \\ -0.7655, & 0, & 0.6422 \\ 0.454, & -0.7071, & 0.542 \end{bmatrix} = [T]_t^{-1}$$

Step 4. $[T]^{-1} [C] [T] = \begin{bmatrix} 11.954 & & \\ & 10.36 & \\ & & 7.234 \end{bmatrix} \begin{matrix} \text{nF/km,} \\ \text{which is nearly the matrix} \\ \text{of eigenvalues.} \end{matrix}$

Figure 3.15 shows the distribution of currents in the three modes. We note that in only one mode ground is not involved in the propagation.

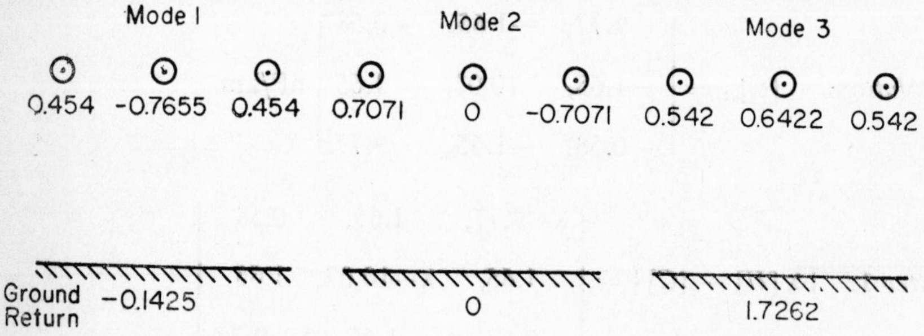

Fig. 3.15.　Distribution of currents in the three modes of propagation of Example 3.9.

Diagonalization of $[L]_{ut}$

We will show that the same transformation matrices obtained for diagonalizing the capacitance matrix of the untransposed configuration will also diagonalize the inductance matrix when $[L] [C] = \dfrac{1}{g^2} [U]$, i.e. when $[L]$ and $[C]$ are calculated on the basis of light-velocity theory.

We observe that

$$\{[T]^{-1} [L] [T]\} \{[T]^{-1} [C] [T]\} = \frac{1}{g^2} [U] \qquad (3.67)$$

so that $\quad [T]^{-1} [L] [T] = \dfrac{1}{g^2} \{[T]^{-1} [C] [T]\}^{-1} = \dfrac{1}{g^2} [\lambda]^{-1} \qquad (3.68)$

The three eigenvalues of $[L]$ are

$$\mu_1 = 1/g^2 \lambda_1, \ \mu_2 = 1/g^2 \lambda_2 \ \text{ and } \ \mu_3 = 1/g^2 \lambda_3 \qquad (3.69)$$

Example 3.10: The eigenvalues of the capacitance matrices are 11.954×10^{-9} F/km, 10.36×10^{-9} F/km, and 7.234×10^{-9} F/km.

Therefore, the eigenvalues of the inductance matrix will turn out to be

$$\mu_1 = \frac{1}{(3 \times 10^5)^2 \times 11.954 \times 10^{-9}} = \frac{1}{1076} \text{ Henry/km} = 0.93 \text{ mH/km.}$$

$$\mu_2 = \frac{1}{(3 \times 10^5)^2 \times 10.36 \times 10^{-9}} = \frac{1}{932.4} = 1.0725 \text{ mH/km}$$

and $\mu_3 = \dfrac{1}{(3 \times 10^5)^2 \times 7.234 \times 10^{-9}} = \dfrac{1}{651} = 1.536 \text{ mH/km}$

Example 3.11: Taking the inductance matrix $[L]_{ut}$ of example 3.7 and the transformation matrix $[T]$ and its inverse $[T]^{-1}$ from example 3.9, show that $[T]^{-1} [L]_{ut} [T]$ is diagonalized and the eigenvalues are nearly μ_1, μ_2, μ_3 obtained in example 3.10.

Solution:

$$\begin{bmatrix} .454, & -.7655, & .454 \\ .7071, & 0, & -.7071 \\ .542, & .6422, & .542 \end{bmatrix} \begin{bmatrix} 1.173, & .213, & .105 \\ .213, & 1.173, & .213 \\ .105, & .213, & 1.173 \end{bmatrix}$$

$$\begin{bmatrix} .454, & .7071, & .542 \\ -.7655, & 0, & .6422 \\ .454, & -.7071, & .542 \end{bmatrix} = \begin{bmatrix} 0.92, & 0, & 0 \\ 0, & 1.068, & 0 \\ 0, & 0, & 1.535 \end{bmatrix} \text{ mH/km,}$$

which check with μ_1, μ_2, μ_3.

3.8 Resistance and Inductance of Ground Return

Under balanced operating conditions of a transmission line, ground-return currents do not flow. However, many situations occur in practice when ground currents have important effect on system performance. Some of these are:

(a) Flow of current during short circuits involving ground. These are confined to single line to ground and double line to ground faults. During three phase to ground faults the system is still balanced;

(b) Switching operations and lightning phenomena;

(c) Propagation of waves on conductors;

(d) Radio Noise studies.

The ground-return resistance increases with frequency of the current while the inductance decreases with frequency paralleling that of the resistance and inductance of a conductor. In all cases involving ground, the soil is inhomogeneous and stratified in several layers with different values of electrical conductivity. In this section, the famous formulas of J.R. Carson (B.S.T.J. 1926) will be given for calculation of ground resistance and inductance at any frequency in a homogeneous single-layer soil. The problem was first applied to telephone transmission but we will restrict its use to apply to e.h.v. transmission lines.

The conductivity of soils has the following order of magnitudes:

10^0 mho/metre for moist soil, 10^{-1} for loose soil, 10^{-2} for clay, and 10^{-3} for bed rock.

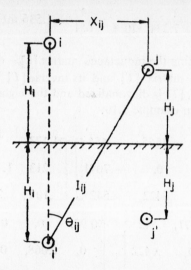

Fig. 3.16. Geometrical parameters for calculation of
ground-return resistance and inductance.

Figure 3.16 describes the important parameters involved in the calculation showing two conductors i and j above ground together with their images. We take

c_s = soil conductivity in mho/m,

f = frequency of current, Hz,

G = 1.7811, Euler's number, $\mu_0 = 4\pi \times 10^{-7}$ H/m,

I_{ij} = distance of conductor j from image of conductor i, metre.

θ_{ij} = arc tan $[X_{ij}/(H_i - H_j)]$, radians

The most important parameter involved in the calculation is

$$F_{ij} = I_{ij} \sqrt{2\pi\mu_0 f c_s} \qquad (3.70)$$

For usual e.h.v. configurations, $F_{ij} < 1$.

[When c.g.s. units are used, I_{ij} is in cm, and c_s is also in c.g.s. e.m. units. To convert ohm/m into c.g.s. units, the multiplying factor is 10^{-11}. Then, $F_{ij} = I_{ij}\sqrt{8\pi^2 f c_s}$.]

Having calculated F_{ij}, the ground resistance and inductance are

$$R_g = 8\pi f\, J_r \cdot 10^{-4}, \text{ ohm/km} \qquad (3.71)$$

and

$$L_g = 4 \cdot J_i \cdot 10^{-4}, \text{ Henry/km} \qquad (3.72)$$

where J_r and J_i are calculated as follows:

$$J_r = (1-S_4)\,\frac{\pi}{8} + 0.5 \cdot S_2 \cdot \ln\,(2/G\,F_{ij}) + 0.5\ \theta_{ij} T_2 - \frac{1}{2}\,W_1 + \frac{1}{2\sqrt{2}}\,W_2 \qquad (3.73)$$

$$J_i = 0.25 + 0.5 \, (1 - S_4) . \ln (2/G \, F_{ij}) - 0.5\theta_{ij}T_4 - \frac{\pi}{8} \, S_2$$

$$+ \frac{1}{\sqrt{2}} \, (W_1 + W_3) - 0.5 \, W_4 \qquad (3.74)$$

There are several quantities above, (S, T, W), which are given by Carson by the following infinite series when $F_{ij} < 1$. For most calculations, only two or three leading terms will be sufficient as will be shown by an example of a horizontal 400-kV line.

$$S_2 = \sum_{k=0}^{\infty} (-1)^k . (F_{ij}/2)^{2(2k+1)} \frac{1}{(2k+1)!(2k+2)!} \cos (2k+1) \, 2\theta_{ij}$$
$$(3.75)$$

$$T_2 = \text{same as } S_2 \text{ with cosine changed to sine} \qquad (3.76)$$

$$S_4 = \sum_{k=1}^{\infty} (-1)^{k-1} \left(\frac{F_{ij}}{2}\right)^{4k} \frac{1}{(k+1)! \, (k+2)!} \cos (4k)\theta_{ij} \qquad (3.77)$$

$$T_4 = \text{same as } S_4 \text{ with cosine changed to sine} \qquad (3.78)$$

$$W_1 = \sum_{k=1}^{\infty} (-1)^{k-1} F_{ij}^{(4k-1)} \frac{1}{1^2 . 3^2 . 5^2 \ldots (4k-1)} \cos (4k-3) \, \theta_{ij} \quad (3.79)$$

$$W_2 = 1.25 \, S_2, \quad W_4 = \frac{5S_4}{3} \qquad (3.80)$$

$$W_3 = \sum_{k=1}^{\infty} (-1)^{k-1} F_{ij}^{(4k-1)} \frac{1}{3^2 5^2 \ldots (4k+1)} \cos (4k-1) \, \theta_{ij} \quad (3.81)$$

The important and interesting properties of R_g and L_g for a 3-phase line are illustrated by taking an example of the horizontal 400-kV line. These properties come out to be

$$[R_g] \approx R_g \begin{bmatrix} 1, & 1, & 1 \\ 1, & 1, & 1 \\ 1, & 1, & 1 \end{bmatrix} \text{ and } [L_g] \approx L_g \begin{bmatrix} 1, & 1, & 1 \\ 1, & 1, & 1 \\ 1, & 1, & 1 \end{bmatrix} \qquad (3.82)$$

Example 3.12: Figure 3.17 shows all major dimensions of a 400-kV line. Calculate the matrices of ground-return resistance and inductance per km at $f = 1$ KHz for $c_s =$ earth conductivity $= 10^{-2}$ mho/m.

Solution: The frequency of 1 KHz is useful for switching-surge propagation studies. The required parameters are:

$$\theta_{12} = \theta_{23} = \text{arc tan } (11/26) = 0.4014 \text{ radian.}$$

$$\theta_{13} = \text{arc tan } (22/26) = 0.7025 \text{ radian.}$$

$$\theta_{11} = \theta_{22} = \theta_{33} = 0. \; I_{12} = 30 \text{ m}, I_{13} = 34 \text{ m.}$$

$$F_{ij} : F_{11} = 0.234, \; F_{12} = 0.27, \; F_{13} = 0.306$$

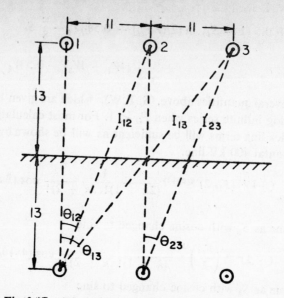

Fig. 3.17. Dimensions of 400 kV line for example 3.12

Self resistances: $\qquad\qquad R_{11}, R_{22}, R_{33}, \theta_{ij} = 0.$

$$S_2 = \left(\frac{0.234}{2}\right)^2 \frac{1}{2!} - \left(\frac{0.234}{2}\right)^6 \frac{1}{3!\,4!} + \ldots \approx 0.00685$$

$$T_2 = 0, \; W_2 = 1.25, \; S_2 = 0.00856$$

$$S_4 = \frac{0.117^4}{2 \times 6}, \text{ negligible.} \quad T_4 = 0. \quad W_4 \text{ is negligible.}$$

$$W_1 = \frac{0.234^3}{1^2.3} = 0.0043, \; W_3 = \frac{0.234^3}{3^2.5} = 0.00028$$

Therefore

$$J_r = \frac{\pi}{8} + 0.5 \times 0.00685 \times \ln\left(\frac{2}{1.7811 \times 0.234}\right) - 0.00215 + \frac{0.0086}{2\sqrt{2}}$$

$$= 0.4$$

$$J_i = 0.25 + 0.5 \times 1.61 - \frac{\pi}{8} \times 0.00685 + \frac{0.0046}{\sqrt{2}}$$

$$= 1.055$$

Therefore

$$R_{ll} = 8\pi \times 10^3 \times 10^{-4} \times 0.4 = 1 \text{ ohm/km}$$

$$L_{ll} = 4 \times 1.055 \times 10^{-4} = 0.422 \text{ mH/km}$$

Mutual between 1 and 2 (outer and inner)

Calculations yield $J_r = 0.3983 \approx 0.4$ and $J_i = 0.952$

Therefore $R_{12} = R_{23} = 1$ ohm/km, $L_{12} = L_{23} = 0.38$ mH/km

Mutual between 1 and 3 (outer and outer)

$J_r = 0.3925$, $J_i = 0.89$, giving $R_{13} \approx 1$ ohm/km and

$$L_{13} = 0.356 \text{ mH/km}.$$

$$\therefore \quad [R_g] = \begin{bmatrix} 1, & 1, & 1 \\ 1, & 1, & 1 \\ 1, & 1, & 1 \end{bmatrix} \text{ ohm/km}$$

and

$$[L_g] = \begin{bmatrix} 0.422, & 0.38, & 0.356 \\ 0.38, & 0.422, & 0.38 \\ 0.356, & 0.38, & 0.422 \end{bmatrix} \text{ mH/km}$$

The mutual inductances are quite close to an average value of

$$(0.422 \times 3 + 0.38 \times 4 + 0.356 \times 2)/9 = 0.39 \text{ mH/km}$$

We see that the resistance and inductance of ground return are given approximately as

$$[R_g] = \begin{bmatrix} 1, & 1, & 1 \\ 1, & 1, & 1 \\ 1, & 1, & 1 \end{bmatrix} R_g, \text{ where } R_g = 1 \text{ ohm/km}$$

and $\quad [L_g] = \begin{bmatrix} 1, & 1, & 1 \\ 1, & 1, & 1 \\ 1, & 1, & 1 \end{bmatrix} L_g, \text{ where } L_g = 0.39 \text{ mH/km.}$

We will denote

$$[D] = \begin{bmatrix} 1, & 1, & 1 \\ 1, & 1, & 1 \\ 1, & 1, & 1 \end{bmatrix} \tag{3.83}$$

Then, $[R_g] = R_g [D]$, and $[L_g] = L_g [D]$ $\hspace{2cm}$ (3.84)

This property will be used throughout the book whenever ground-return effects are to be considered.

Diagonalization of [D]

It is interesting to observe the properties of $[D]$ by diagonalizing it. The 3 eigenvalues come out to be $\lambda_1 = 3$, $\lambda_2 = \lambda_3 = 0$. One set of $[T]$ and $[T]^{-1}$ which will diagonalize $[D]$ turn out to be exactly the same as Clarke's Modified Transformation, equations (3.62) and (3.63), as can be worked out by the reader. Also,

$$[T]^{-1}[D][T] = \frac{1}{\sqrt{6}} \begin{bmatrix} \sqrt{2}, & \sqrt{2}, & \sqrt{2} \\ \sqrt{3}, & 0, & -\sqrt{3} \\ 1, & -2, & 1 \end{bmatrix} \begin{bmatrix} 1, & 1, & 1 \\ 1, & 1, & 1 \\ 1, & 1, & 1 \end{bmatrix}$$

$$\begin{bmatrix} \sqrt{2}, & \sqrt{3}, & 1 \\ \sqrt{2}, & 0, & -2 \\ \sqrt{2}, & -\sqrt{3}, & 1 \end{bmatrix} \frac{1}{\sqrt{6}} = \begin{bmatrix} 3 & 0 & 0 \\ 0 & 0 & 0 \\ 0 & 0 & 0 \end{bmatrix} = \begin{bmatrix} \lambda_1 & & \\ & \lambda_2 & \\ & & \lambda_3 \end{bmatrix} \quad (3.85)$$

We could have applied this property for diagonalizing the matrices of $[L]_t$ and $[C]_t$ of the transposed line instead of the lengthy procedure in step 1 of Section 3.7 for finding the three eigenvalues or characteristic roots. Observe that

$$[L]_t = \begin{bmatrix} L_s, & L_m, & L_m \\ L_m, & L_s, & L_m \\ L_m, & L_m, & L_s \end{bmatrix} = (L_s - L_m)[U] + L_m[D] \quad (3.86)$$

Any transformation matrix which diagonalizes $[D]$ will also diagonalize $(L_s - L_m)[U]$ which is already in diagonalized form.

Complete Line Parameters with Ground Return

We can now combine the resistances, inductances and capacitances of the phase conductors with those of ground return to formulate the complete line-parameters for a transposed line.

(a) *Resistance*

The conductor resistance matrix is $[R_c] = R_c[U]$

Ground-return resistance matrix $[R_g] = R_g[D]$

Therefore

$$[R] = [R_c] + [R_g] = R_c[U] + R_g[D] = \begin{bmatrix} R_c + R_g, & R_g, & R_g \\ R_g, & R_c + R_g, & R_g \\ R_g, & R_g, & R_g + R_g \end{bmatrix} \quad (3.87)$$

(b) *Inductance*

$$[L_c] = \begin{bmatrix} L_s, & L_m, & L_m \\ L_m, & L_s, & L_m \\ L_m, & L_m, & L_s \end{bmatrix} = (L_s - L_m)[U] + L_m[D] \quad (3.88)$$

$$[L_g] = L_g[D]$$

$$\therefore \quad [L] = [L_c] + [L_g] = (L_s - L_m)[U] + (L_g + L_m)[D] \quad (3.89)$$

(c) *Capacitance*

$$[C] = (C_s - C_m)[U] + C_m[D] \quad (3.90)$$

Diagonalization of any of these matrices is easily carried out through $[T]$ and $[T]^{-1}$ of Clarke's modified transformation matrices.

$$[T]^{-1}[R][T] = R_c[U] + R_g \begin{bmatrix} 3 & & \\ & 0 & \\ & & 0 \end{bmatrix} = \begin{bmatrix} R_c+3R_g & & \\ & R_c & \\ & & R_c \end{bmatrix} \quad (3.91)$$

Note that the ground contributes $(3R_g)$ to the first mode or line-to-ground mode of propagation. The term $R_0 = R_c + 3R_g$ may still be called the zero-sequence resistance.

$$[T]^{-1}[L][T] = (L_s-L_m)[U] + (L_g + L_m) \begin{bmatrix} 3 & & \\ & 0 & \\ & & 0 \end{bmatrix}$$

$$= \begin{bmatrix} L_s + 2L_m + 3L_g, & & \\ & L_s-L_m, & \\ & & L_s-L_m \end{bmatrix} \quad (3.92)$$

Comparing with equation (3.64) we observe that ground has contributed an inductance of $3L_g$ to the first or line-to-ground mode of propagation since in this mode ground-return current is equal to the sum of the currents flowing in the 3-phase conductors above ground.

On the other hand, the ground has not contributed to the line-to-line modes of propagation as there is no return current in the ground in these two modes. We also observe that $L_g + L_m = (L_0-L_1)/3$.

The capacitance transformation is

$$[T]^{-1}[C][T]=(C_s-C_m)[U]+C_m \begin{bmatrix} 3 & & \\ & 0 & \\ & & 0 \end{bmatrix} = \begin{bmatrix} C_s+2C_m & & \\ & C_s-C_m & \\ & & C_s-C_m \end{bmatrix}$$

$$(3.93)$$

This is the same as equation (3.64) with inductance replaced by capacitance. Therefore, ground has not added any capacitance to any of the three modes of propagation.

We can observe the further property that in the presence of ground activity,

$$[T]^{-1}[L][C][T]= \begin{bmatrix} (L_s+2L_m+3L_g)(C_s+2C_m), & & \\ & (L_s-L_m)(C_s-C_m) & \\ & & (L_s-L_m)(C_s-C_m) \end{bmatrix}$$

$$(3.94)$$

This clearly shows that the velocity of wave propagation in the second and third modes, the two line-to-line modes, is still the velocity of light as discussed in Section 3.7.3, equation (3.66). However, the velocity of propagation in the line-to-ground mode is

$$v_1 = 1/\sqrt{(L_s + 2L_m + 3L_g)(C_s + 2C_m)} \qquad (3.95)$$

Since L_g is a positive quantity, $v_1 < g$, the velocity of light.

Example 3.13: A 400-kV line (Example 3.7) gave the following inductance matrix

$$[L]_t = \begin{bmatrix} 1.173, & 0.177, & 0.177 \\ 0.177, & 1.173, & 0.177 \\ 0.177, & 0.177, & 1.173 \end{bmatrix} \text{mH/km}$$

Take $(C_s + 2C_m) = (1/g^2)(L_s + 2L_m)^{-1}$ and $(C_s - C_m)(L_s - L_m) = 1/g^2$. The ground return contributes 0.39 mH/km as calculated in Example 3.12. Calculate the velocities in the three modes.

Solution: $C_s + 2C_m = (1/g^2)[(1.173 + 2 \times 0.177) \times 10^{-3}]^{-1} = \dfrac{1}{g^2} \dfrac{10^3}{1.527}$

$L_s + 2L_m + 3L_g = 2.697 \times 10^{-3}$

$\therefore \qquad (L_s + 2L_m + 3L_g)(C_s + 2C_m) = 1.7662/g^2 = (1.329/g)^2$

and $\qquad v_1 = g/1.329 = 0.7524g = 2.2572 \times 10^5$ km/sec.

The velocity of propagation in the line-to-ground mode is now 75.24% of light velocity.

Equivalent Circuit of Line Model for Network Studies

From equations (3.91) and (3.92), we obtain the quantities

$$R_0 = R_c + 3R_g, \; R_1 = R_c, \; L_0 = L_s + 2L_m + 3L_g, \text{ and } L_1 = L_s - L_m \qquad (3.96)$$

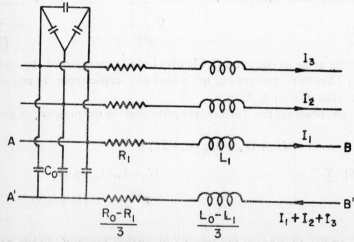

Fig. 3.18. Line model with ground-return parameters included.

The ground-return resistance is then

$$R_g = (R_0 - R_1)/3 \tag{3.97}$$

and there also results

$$L_g + L_m = (L_0 - L_1)/3 \tag{3.98}$$

The quantities R_0 and L_0 can be considered as the zero-sequence quantities while R_1 and L_1 are positive-sequence quantities obtained under symmetrical component concept of Fortescue. Using these, a line model can be constructed similar to Figure 3.13. This is shown in Figure 3.18. The three lines are represented by their positive-sequence quantities R_1 and L_1. The ground-return part constitutes R_g and $(L_g + L_m)$ which are denoted as $(R_0 - R_1)/3$, and $(L_0 - L_1)/3$ respectively. The capacitance network consists of a Y-connected set equal to $C_s = (C_0 + 2C_1)/3$ and a Δ-connected set of $C_m = (C_0 - C_1)/3$. The Δ-connected set may be converted to an equivalent Y if necessary each leg having a capacitance of $3C_m = (C_0 - C_1)$. Such a representation of a line section has been used for setting up miniature models on a Transient Network Analyzer for switching-surge studies and in mathematical models for computation of transient performance of transmission lines using the Digital Computer.

The reader can prove that the voltage drop from A to A' is

$$\{(R_c + R_g) + s(L_s + L_g)\} I_1(s) + \{R_g + (L_m + L_g)s\}$$
$$I_2(s) + \{R_g + (L_m + L_g)s\} I_3(s) \tag{3.99}$$

This is left as an exercise at the end of the chapter.

Review Questions and Problems

1. A Moose conductor has the following details—Outer dia $= 31.8$ mm, Area of Al $= 515.7$ mm². Calculate the resistance of 1 km of a double-Moose bundled conductor at 50°C given that $\rho_a = 2.7 \times 10^{-8}$ ohm-m at 20°C and temperature coefficient of Al $= 4.46 \times 10^{-3}/°C$. (Increase length by 5% for stranding.)

2. The closest conductor to Moose of North-American manufacture is Bluejay with area $= 1.113 \times 10^6$ cir-mil, outer dia $= 1.259''$. Its resistance is listed in tables of conductors as 0.0155 ohm/1000 feet at 20°C for dc and 0.0941 ohm/mile at 50°C and 50/60 Hz.
   (a) Verify these values.
   (b) Find % increase due to skin effect.

3. A 750 kV line has the details given below. Calculate the temperature rise of the conductor under given conditions. Conductor—4×0.03 m ACSR (area $= 954,000$ cir-mils). Power carried 2000MW. $\rho_a = 2.7 \times 10^{-8}$ ohm-m at 20°C, $\alpha = 0.0045$ ohm/°C, ambient $t_a = 45°C$, $e = 0.5$, $p = 1$, $v = 1.2$ m/s, solar irradiation 1 Kw/m², $s_a = 0.8$.

4. A 3-phase 750 kV horizontal line has minimum height of 12 m, sag at midspan = 12 m. Phase spacing $S = 15$ m. Conductors are 4×0.035 m with bundle spacing of $B = 0.4572$ m. Calculate per kilometre:

   (a) The matrix of Maxwell's Potential coefficients for an untransposed configuration.
   (b) The inductance and capacitance matrices for untransposed and transposed configurations.
   (c) The zero-, positive-, and negative-sequence inductances and capacitances for transposed line.
   (d) The ground-return resistance and inductance matrices at 750 Hz taking $\rho_s = 100$ ohm-metre.

5. Repeat problem 3.4 for a 1150-kV delta configuration of the 3-phases with average height of 18 m for the lower conductors, 36 m for the top conductor, and spacing of 24 m between bottom conductors. Bundle radius = 0.6 m and conductor size = 6×0.046 m diameter. $f = 1000$ Hz and $\rho_s = 50$ ohm-metre.

6. Diagonalize the matrix

$$[D] = \begin{bmatrix} 1 & 1 & 1 \\ 1 & 1 & 1 \\ 1 & 1 & 1 \end{bmatrix}.$$

   Give eigenvalues and eigen-vector matrices.

7. Discuss the convenience offered by using modes of propagation and possible uses of this technique.

8. The capacitance matrix of a 750-kV horizontal configuration line is

$$[C] = \begin{bmatrix} 10.20 & -1.45 & -0.35 \\ -1.45 & 10.40 & -1.45 \\ -0.35 & -1.45 & 10.20 \end{bmatrix} \text{nF/km}$$

   (a) Find the 3 eigenvalues of the matrix, $(\lambda_1, \lambda_2, \lambda_3)$.
   (b) Diagonalize the matrix by evaluating suitable transformation matrix $[T]$ and its inverse $[T]^{-1}$.
   (c) Then prove that

$$[T]^{-1}[C][T] = \begin{bmatrix} \lambda_1 & & \\ & \lambda_2 & \\ & & \lambda_3 \end{bmatrix}$$

9. In problems 3.4 and 3.5 calculate the charging current supplied. Assume full transposition and place all the capacitance at the line entrance across the source. $L = 400$ km.

10. In Figure 3.18 show that the voltage drop from A to B and B' to A' add to

$$\{(R_c + R_g) + s(L_s + L_g)\}\, I_1 + (R_g + sL_m + sL_g)\, I_2$$
$$+ (R_g + sL_m + sL_g)\, I_3,$$

where s = the Laplace-Transform operator.

11. (a) Using the transformation matrices for diagonalizing the matrix $[D]$, prove without multiplying, that the same transformation matrices will diagonalize the inductance or capacitance matrices of a fully-transposed line of the type

$$[L] = \begin{bmatrix} L_s & L_m & L_m \\ L_m & L_s & L_m \\ L_m & L_m & L_s \end{bmatrix}$$

(b) If λ_1, λ_2, λ_3 are the eigenvalues of matrix $[L]$ and given that $[L]\,[C] = [U]\tfrac{1}{2}/g^2$, prove that the eigenvalues of $[C]$ will be $\mu_1 = 1/g^2\lambda_1$, $\mu_2 = 1/g^2\lambda_2$ and $\mu_3 = 1/g^2\lambda_3$. In general, prove that if λ_1, λ_2, λ_3 are eigenvalues of a matrix $[M]$, then the eigenvalues of its inverse are the reciprocals of λ_1, λ_2, λ_3.

CHAPTER 4

Voltage Gradients of Conductors

4.1 Electrostatics

Conductors used for e.h.v. transmission lines are thin long cylinders which are known as 'line charges'. Their charge is described in coulombs/ unit length which was used for evaluating the capacitance matrix of a multi-conductor line in chapter 3. The problems created by charges on the conductors manifest themselves as high electrostatic field in the line vicinity from power frequency to TV frequencies through audio frequency, carrier frequency and radio frequency (PF, AF, CF, RF, TV, F). The attenuation of travelling waves is also governed in some measure by the increase in capacitance due to corona discharges which surround the space near the conductor with charges. When the macroscopic proper- ties of the electric field are studied, the conductor charge is assumed to be concentrated at its centre, even though the charge is distributed on the surface. In certain problems where proximity of several conductors affects the field distribution, or where conducting surfaces have to be forced to become equipotential surfaces (in two dimensions) in the field of several charges it is important to replace the given set of charges on the conductors with an infinite set of charges. This method is known as the Method of Successive Images. In addition to the electric-field properties of long cylinders, there are other types of important electrode configurations useful for extra high voltage practice in the field and in laboratories. Examples of this type are sphere-plane gaps, sphere-to- sphere gaps, point-to-plane gaps, rod-to-plane gaps, rod-rod gaps, con- ductor-to-tower gaps, conductor-to-conductor gap above a ground plane, etc. Some of these types of gaps will also be dealt with in this chapter which may be used for e.h.v. measurement, protection, and other functions. The coaxial-cylindrical electrode will also be discussed in great detail because of its importance in corona studies where the bundle of N sub-conductors is strung inside a 'cage' to simulate the surface voltage gradient on the conductors in a setup which is smaller in dimensions than an actual outdoor transmission line.

4.1.1 Field of a Point Charge and Its Properties

The properties of electric field of almost all electrode geometries will ultimately depend on that of a point charge. The laws governing the behaviour of this field will form the basis for extending them to other geometries. Consider Fig. 4.1 which shows the source point S_1 where a point charge $+Q$ coulombs is located. A second point charge q coulomb is located at S_2 at a distance r metre from S_1. From Coulomb's Law, the force acting on either charge is

$$F = Q.q/4\pi\ e_0 e_r\ r^2, \text{ Newton} \tag{4.1}$$

$$[e_0 = (\mu_0 g^2)^{-1} = 1000/36\pi\ \mu\mu\text{F/m} = 8.84194\ \mu\mu\text{F/m}$$

$$e_r = \text{relative permittivity of the medium} = 1 \text{ in air}]$$

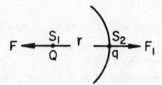

Fig. 4.1. Point charge Q and force on test charge q.

When q is very small ($q \to 0$), we define the electric field produced by Q at the location of q as

$$E = \operatorname*{Lim}_{q\to 0} (F/q) = Q/4\pi\ e_0 e_r\ r^2, \text{ Newton/Coulomb} \tag{4.2}$$

The condition $q \to 0$ is necessary in order that q might not disturb the electric field of Q. Equation (4.2) may be re-written as

$$(4\pi r^2)\ (e_0 e_r E) = 4\pi r^2\ D = Q \tag{4.3}$$

Here we note that $4\pi r^2 = $ surface area of a sphere of radius r drawn with centre at S_1. The quantity $D = e_0 e_r E$ is the dielectric flux density. Thus, we obtain Gauss's Law which states that 'the surface integral of the normal component of dielectric flux density over a closed surface equals the total charge in the volume enclosed by the surface.' This is a general relation and is valid for all types of electrode geometries.

Some important properties of the field of a point charge can be noted:

(a) The electric field intensity decreases rapidly with distance from the point charge inversely as the square of the distance, ($Q \propto 1/r^2$).

(b) E is inversely proportional to e_r, ($E \propto 1/e_r$).

(c) The potential of any point in the field of the point charge Q, defined as the work done against the force field of Q in bringing q from ∞ to S_2, is

$$\psi = \int_\infty^r - E\ dr = \frac{Q}{4\pi e_0\ e_r} \int - \frac{dr}{r^2} = \frac{Q}{4\pi e_0 e_r}\ \frac{1}{r}, \text{ volt} \tag{4.4}$$

(d) The potential difference between two points at distances r_1 and r_2 from S_1 will be

$$\psi_{12} = \frac{Q}{4\pi e_0 e_r}\left(\frac{1}{r_1} - \frac{1}{r_2}\right), \text{ volt} \tag{4.5}$$

For a positive point charge, points closer to the charge will be at a higher positive potential than a further point.

(e) The capacitance of an isolated sphere is

$$C = Q/\psi = 4\pi e_0 e_r \, r, \text{ Farad} \tag{4.6}$$

This is based on the assumption that the negative charge of $-Q$ is at infinity.

These properties and concepts can be extended in a straightforward manner to apply to the field of a line charge and other electrode configurations.

Example 4.1: A point charge $Q = 10^{-6}$ coulomb ($1 \, \mu C$) is kept on the surface of a conducting sphere of radius $r = 1$ cm, which can be considered as a point charge located at the centre of the sphere. Calculate the field strength and potential at a distance of 0.5 cm from the surface of the sphere. Also find the capacitance of the sphere, $\epsilon_r = 1$.

Solution: The distance of S_2 from the centre of the sphere S_1 is 1.5 cm.

$$E = 10^{-6} \Big/ \left[4\pi \times \frac{1000}{36\pi} \times 10^{-12} \times 1.5^2 \times 10^{-4} \right]$$

$$= 10^{-6}/(0.25 \times 10^{-13})$$

$$= 40 \times 10^6 \text{ V/m} = 40 \text{ MV/m} = 400 \text{ KV/cm}.$$

$$\psi = E.r = 600 \text{ KV [using equations 4.2 and 4.4]}$$

$$C = 4\pi e_0 e_r r = 1.111 \, \mu\mu\text{F with } r = 0.01.$$

Example 4.2: The field strength on the surface of a sphere of 1 cm radius is equal to the corona-inception gradient in air of 30 KV/cm. Find the charge on the sphere.

Solution:　30 KV/cm $= 3000$ KV/m $= 3 \times 10^6$ V/m.

$$3 \times 10^6 = Q \Big/ \left[4\pi \times \frac{1000}{36\pi} \times 10^{-12} \times 10^{-4} \right]$$

$$= 9 \times 10^{13} Q_0$$

giving　　　　　$Q_0 = 3.33 \times 10^{-8}$ coulomb $= 0.033 \, \mu C.$

The potential of the sphere is $V = 3 \times 10^6 \times 10^{-2} = 30$ KV.

4.2 Field of Sphere Gap

A sphere-sphere gap is used in h.v. laboratories for measurement of extra high voltages and for calibrating other measuring apparatus. If the gap spacing is less than the sphere radius, the field is quite well determined and the sphere gap breaks down consistently at the same voltage with a dispersion not exceeding $\pm 3\%$. This is the accuracy of such a measuring gap, if other precautions are taken suitably such as no collection of dust or proximity of other grounded objects close by. The sphere-gap problem also illustrates the method of successive images used in electrostatics.

Figure 4.2 shows two spheres of radii R separated by a centre-centre distance of S, with one sphere at zero potential (usually grounded) and the other held at a potential V. Since both spheres are metallic, their surfaces are equipotentials. In order to achieve this, it requires a set of infinite number of charges, positive inside the left sphere at potential V and negative inside the right which is held at zero potential. The magnitude and position of these charges will be determined from which the voltage gradient resulting on the surfaces of the sphere on a line join-ing the centres can be determined. If this exceeds the critical disruptive voltage, a spark break-down will occur. The voltage required is the breakdown voltage.

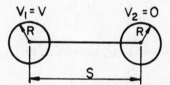

Fig. 4.2. The sphere gap.

Consider two point charges Q_1 and Q_2 located with a separation D, Fig. 4.3. At a point $P(x, y)$ with coordinates measured from Q_1, the potentials are as follows:

Potential at P due to $Q_1 = \dfrac{Q_1}{4\pi e_0} \dfrac{1}{r_1}$ with $r_1 = \sqrt{x^2 + y^2}$

Potential at P due to $Q_2 = \dfrac{Q_2}{4\pi e_0} \dfrac{1}{r_2}$ with $r_2 = \sqrt{(D-x)^2 + y^2}$

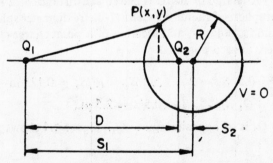

Fig. 4.3. Point charge Q_1 and sphere of radius R.

The total potential at P is $V_P = \dfrac{1}{4\pi e_0} (Q_1/r_1 + Q_2/r_2)$.

If this is to be zero, then $Q_2/Q_1 = -r_2/r_1$ (4.7)

This clearly shows that Q_1 and Q_2 must be of opposite polarity.

From (4.7), $r_2^2/r_1^2 = \dfrac{(D-x)^2 + y^2}{x^2 + y^2} = Q_2^2/Q_1^2$, giving

$$\left\{x - \frac{D}{1-(Q_2/Q_1)^2}\right\}^2 + y^2 = D^2(Q_2/Q_1)^2/\{1-(Q_2/Q_1)^2\}^2 \tag{4.8}$$

This is an equation to a circle in the two-dimensional plane and is a sphere in three-dimensional space.

The radius of the sphere is

$$R = D(Q_2/Q_1)/\{1 - (Q_2/Q_1)^2\} \tag{4.9}$$

This requires Q_2 to be less than Q_1 if the denominator is to be positive. The centre of the zero-potential surface is located at $(S_1, 0)$, where

$$S_1 = D/\{1 - (Q_2/Q_1)^2\} = \frac{Q_1}{Q_2} R \tag{4.10}$$

This makes $S_2 = S_1 - D = \dfrac{D(Q_2/Q_1)^2}{1-(Q_2/Q_1)^2} = \dfrac{Q_2}{Q_1} R$ (4.11)

Therefore, the magnitude of Q_2 in relation to Q_1 is

$$Q_2 = Q_1 \frac{S_2}{R} = Q_1 \frac{R}{S_1} \tag{4.12}$$

Also, $S_1 S_2 = R^2$ (4.13)

These relations give the following important property:

'Given a positive charge Q_1 and a sphere of radius R, with Q_1 located external to the sphere, whose centre is at a distance S_1 from Q_1, the sphere can be made to have a zero potential on its surface if a charge of opposite polarity and magnitude $Q_2 = (Q_1 R/S_1)$ is placed at a distance $S_2 = R^2/S_1$ from the centre of the given sphere towards Q_1.'

Example 4.3: A charge of 10 μC is placed at a distance of 2 metres from the centre of a sphere of radius 0.5 metre (1-metre diameter sphere). Calculate the magnitude, polarity, and location of a point charge Q_2 which will make the sphere at zero potential.

Solution: $R = 0.5,\ S_1 = 2$ $\therefore$ $S_2 = R^2/S_1 = 0.125$ m

         $Q_2 = Q_1 R/S_1 = 10 \times 0.5/2 = 2.5\ \mu C$

The charge Q_2 is of opposite sign to Q_1. Figure 4.4 shows the sphere, Q_1 and Q_2.

Example 4.4: An isolated sphere in air has a potential V and radius R. Calculate the charge to be placed at its centre to make the surface of the sphere an equipotential.

Solution: From equation (4.4), $Q = 4\pi e_0 VR$.

We now are in a position to analyze the system of charges required to

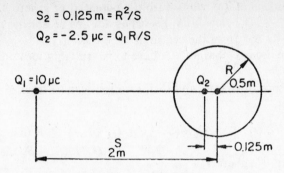

Fig. 4.4. Location of image charge Q_2 inside sphere to make sphere potential zero.

make one sphere at potential V and a second sphere at zero potential as is the case in a sphere-sphere gap with one sphere grounded. Figure 4.5 shows the two spheres separated by the centre-centre distance S. The sphere 1 at left has potential V and that at right zero potential. Both spheres have equal radii R.

In order to hold sphere 1 at potential V, a charge $Q_1 = 4\pi e_0\, VR$ must be placed at its centre. In the field of this charge, sphere 2 at right can be made a zero potential surface if an image charge Q_2 is placed inside this sphere. From the discussion presented earlier, $Q_2 = -\,Q_1 R/S$ and $S_2 = R^2/S$, as shown in Fig. 4.5. However, locating Q_2 will disturb the potential of sphere 1. In order to keep the potential of sphere 1 undisturbed, we must locate an image charge $Q_1' = -\,Q_2 R/(S - S_2)$ inside sphere 1 so that in the field of Q_2 and Q_1' the potential of sphere 1 is zero leaving its potential equal to V due to Q_1. The charge Q_1' is located at $S_1' = R^2/(S - S_2)$ from Q_1 (centre of sphere 1). It is now easy to see

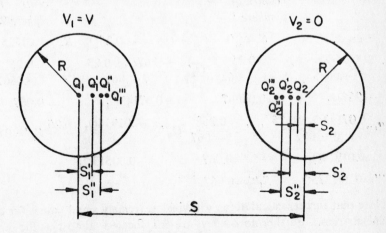

Fig. 4.5. Location of successive image charges to maintain sphere at potentials V and zero.

that the presence of Q_1' will disturb the potential of sphere 2. An image charge $Q_2' = - Q_1' R/(S - S_1')$ is called for inside sphere 2 located at $S_2' = R^2/(S - S_1')$ from the centre of sphere 2.

Successive image charges will have to be suitably located inside both spheres.

The sequence of charges and their locations can be tracked in a tabular form.

$$\left.\begin{array}{l}
Q_1 = 4\pi e_0 VR, \; Q_1' = - Q_2 R/(S - S_2), \; Q_1'' = - Q_2' R/(S - S_2'), \\
\qquad\qquad\qquad Q_1''' = - Q_2'' R/(S - S_2'), \text{ etc.} \\
\text{(From centre of sphere 1)} \\
S_1' = R^2/(S - S_2), \; S_1'' = R^2/(S - S_2'), \; S_1''' = R^2/(S - S_2'), \text{ etc.} \\
Q_2 = - Q_1 R/S, \; Q_2' = - Q_1' R/(S - S_1'), \; Q_2'' = - Q_1'' R/(S - S_1''), \\
\qquad\qquad\qquad Q_2''' = - Q_1''' R/(S - S_1'''), \text{ etc.} \\
\text{(From centre of sphere 2)} \\
S_2 = R^2/S, \; S_2' = R^2/(S - S_1'), \; S_2'' = R^2/(S - S)_1'', \\
\qquad\qquad\qquad S_2''' = R^2/(S - S_1'''), \text{ etc.}
\end{array}\right\} (4.14)$$

Example 4.5: A sphere gap consists of two spheres with $R = 0.25$ m each. The gap between their surfaces is 0.5 m. Calculate the charges and their locations to make the potentials 1 and 0.

Solution: The distance S between centres of the spheres = 1 m.

$$R^2/S = 0.25^2/1 = 0.0625 \text{ m}$$

Charges inside sphere 1		*Charges inside sphere 2*	
V = 1		*V = 0*	
Magnitude	*Distance from centre of sphere 1*	*Magnitude*	*Distance from centre of sphere 2*
$Q_1 = \pi e_0$	$S_1 = 0$	$Q_2 = - 0.25 Q_1$	$S_2 = 0.0625$ m
$Q_1' = \dfrac{0.25 Q_1}{1 - .0625}$	$S_1' = \dfrac{0.25^2}{1 - .0625}$	$Q_2' = \dfrac{-.267 Q_1 \times 0.25}{1 - .067}$	$S_2' = \dfrac{0.25^2}{1 - .0667}$
$= .267 Q_1$	$= 0.06667$	$= -0.07143 Q_1$	$= 0.067$
$Q_1'' = \dfrac{0.07143 Q_1 \times 0.25}{1 - .067}$	$S_1'' = \dfrac{0.25^2}{1 - .067}$	$Q_2'' = \dfrac{-.01914 Q_1 \times 0.25}{1 - .067}$	$S_2'' = 0.067$
$= 0.01914 Q_1$	$= 0.067$	$= -0.00513 Q_1$	
$Q_1''' = 0.001375 Q_1$	$S_1''' = 0.067.$		

Note that further calculations will yield extremely small values for the image charges. Furthermore they are all located almost at the same points. The charges reduce successively in the ratio $0.25/0.933 = 0.268$; i.e. $Q_2^n = - 0.268 \, Q_1^n$ and $Q_1^{n+1} = - 0.268 \, Q_2^n$. The electric field at any point X along the line joining the centres of the two spheres is now found from the expression

$$E = \frac{Q_1}{4\pi e_0}\frac{1}{X^2} + \frac{Q_1'}{4\pi e_0}\frac{1}{(X-S_1')^2} + \frac{Q_1''}{4\pi e_0}\frac{1}{(X-S_1'')^2} + \cdots$$

$$- \frac{Q_2}{4\pi e_0}\frac{1}{(S-X-S_2)^2} - \frac{Q_2'}{4\pi e_0}\frac{1}{(S-X-S_2')^2} - \frac{Q_2''}{4\pi e_0}\frac{1}{(S-X-S_2'')^2}\cdots \quad (4.15)$$

Since Q_2, Q_2', Q_2'' etc. are opposite in polarity to Q_1, Q_1', Q_1'' etc., the force on a unit test charge placed at X will be in the same direction due to all charges. The most important value of X is $X = R$ on the surface of sphere 1. If the value of E at $X = R$ exceeds the critical gradient for breakdown of air (usually 30 kV/cm peak at an air density factor $\delta = 1$), the gap breakdown commences.

Example 4.6: Calculate the voltage gradient at $X = 0.25$ m for the sphere gap in Example 4.5.

Solution: $S = 1$ m, $X = R = 0.25$ m, $S - X = 0.75$ m

$$E = \frac{Q_1}{4\pi e_0}\left[\frac{1}{0.25^2} + \frac{0.267}{(0.25-0.067)^2} + \frac{0.01914}{(0.25-0.067)^2} + \cdots\right]$$

$$+ \frac{Q_1}{4\pi e_0}\left[\frac{0.25}{(0.75-0.065)^2} + \frac{0.07143}{(0.75-0.067)^2} + \frac{0.00513}{(0.75-0.067)^2} + \cdots\right]$$

$$= \frac{Q_1}{4\pi e_0}[24.585 + 0.693] = 25.278\frac{Q_1}{4\pi e_0}$$

$$= 6.3195 \text{ V/m per volt}$$

The contribution of charges inside the grounded sphere amount to

$$\frac{0.693}{25.278} \times 100 = 2.74\%.$$

Example 4.7: In the previous example calculate the potential difference between the spheres for $E = 30$ kV/cm $= 3000$ kV/m, peak.

Solution: $V = 3000/6.3193$ KV $= 474.72$ KV $\cong 475$ KV.

Thus, the 0.5 metre diameter spheres with a gap spacing of 0.5 metre experience disruption at 475 KV, peak. The breakdown voltage is higher than this value.

Sphere-Plane Gap

When a sphere of radius R and potential V is placed above a ground plane at zero potential, the problem is similar to the sphere-to-sphere gap problem. It is clear that in order to place zero potential on the plane, an image sphere with potential $-V$ is necessary. The problem is first solved with the given sphere at $+V$ and the image sphere at 0 potential, then keeping the image sphere at $-V$ and the given sphere at 0 potential. The system of charges required will now be the same as with the sphere-to-sphere gap but the total charge inside the given sphere and its image are equal and amount to the sum of the charges Q_1, Q_1', Q_1'', Q_2, Q_2', Q_2'' ... with all charges having the same sign. Their locations are also the same as before. Charges inside the given sphere have positive polarity and those inside the image sphere are negative.

4.3   Field of Line Charges and their Properties

Figure 4.6 shows a line charge of q coulomb/metre and we will calculate the electric field strength, potential, etc. in the vicinity of the conductor. First, enclose the line charge by a Gaussian cylinder, a cylinder of radius r and length 1 metre. On the flat surfaces the field will not have an outward normal component since for an element of charge dq located at S, there can be found a corresponding charge located at S' whose fields (force exerted on a positive test charge) on the flat surface F will yield only a radial component. The components parallel to the line charge will cancel each other out. Then, by Gauss's Law, if E_p = field strength normal to the curved surface at distance r from the conductor,

$$(2\pi r)\ (e_0 e_r E_p) = q \tag{4.16}$$

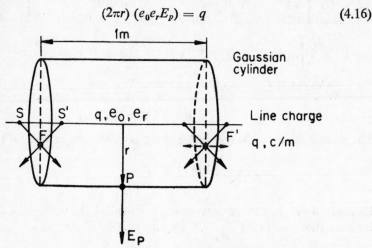

Fig. 4.6.   Line charge with Gaussian cylinder.

The field strength at a distance r from the conductor is

$$E_p = (q/2\pi e_0 e_r)\ (1/r),\ \text{Volts/metre} \tag{4.17}$$

This is called the $(1/r)$-field as compared to the $(1/r^2)$-field of a point charge.

Let a reference distance r_0 be chosen in the field. Then the potential difference between any point at distance r and the reference is the work done on a unit test charge from r_0 to r.

Thus, $V_r = \dfrac{q}{2\pi e_0 e_r} \displaystyle\int_{r_0}^{r} -\dfrac{1}{\rho}\ d\rho = \dfrac{q}{2\pi e_0 e_r}\ (\ln r_0 - \ln r)$ \hfill (4.18)

In the case of a line charge, the potential of a point in the field with respect to infinity cannot be defined as was done for a point charge because of logarithmic term. However, we can find the p.d. between two points at distances r_1 and r_2, since

(p.d. between r_1 and r_2) = (p.d. between r_1 and r_0)

$$- \text{(p.d. between } r_2 \text{ and } r_0)$$

i.e. $\quad V_{12} = \dfrac{q}{2\pi e_0 e_r}\ (\ln r_2 - \ln r_1) = \dfrac{q}{2\pi e_0 e_r}\ \ln\dfrac{r_2}{r_1}$ \hfill (4.19)

In the field of a positive line charge, points nearer the charge will be at a higher positive potential than points farther away ($r_2 > r_1$).

The potential (p.d. between two points, one of them being taken as reference r_0) in the field of a line charge is logarithmic. Equipotential lines are circles. In a practical situation, the charge distribution of a transmission line is closed, there being as much positive charge as negative.

4.3.1 2-Conductor Line: Charges $+q$ and $-q$

Consider a single-phase line, Fig. 4.7, showing two parallel conductors each of radius ρ separated by centre-to-centre distance of $2d$ with each conductor carrying a charge of q coulombs/metre but of opposite polarities. Place a unit test charge at point P at a distance X from the centre of one of the conductors. Then the force acting on it is the field strength at X, which is

$$E_p = \frac{q}{2\pi e_0 e_r}\left(\frac{1}{X} + \frac{1}{2d - X}\right) \text{ Newton/coulomb or V/m} \quad (4.20)$$

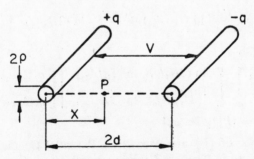

Fig. 4.7. Single-phase line.

The potential difference between the conductors is

$$V = \frac{q}{2\pi e_0 e_r}\int_{\rho}^{2d-\rho}\left(\frac{1}{X} + \frac{1}{2d - X}\right)dX$$

$$= \frac{q}{\pi e_0 e_r}\ \ln\frac{2d - \rho}{\rho} \approx \frac{q}{\pi e_0 e_r}\ \ln\frac{2d}{\rho}, \text{ if } 2d \gg \rho \quad (4.21)$$

Hence, the capacitance of 1 metre length of both conductors is

$$C = q/V = \pi e_0 e_r / \ln (2d/\rho) \qquad (4.22)$$

On the surface of any one of the conductors, the voltage gradient is, from equations (4.20) and (4.21)

$$E = \frac{q}{2\pi e_0 e_r} \left(\frac{1}{\rho} + \frac{1}{2d - \rho} \right) \approx \frac{V}{2\rho \ln (2d/\rho)} \qquad (4.23)$$

This is true on the basis of neglecting the effect of the charge of the other conductor if it is far away to make the separation between conductors much greater than their radii.

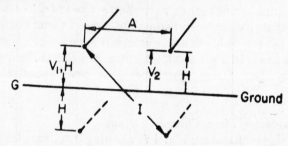

Fig. 4.8.   Two-conductor line above ground plane and image conductors.

A transmission line in practice is strung above a ground plane and we observed in Chapter 3 that its effect can be taken into account by placing image charges, as shown in Fig. 4.8. The charges on the aerial conductors are q_1 and q_2 coulombs/metre and their potentials with respect to ground are V_1 and V_2. Then,

$$\left. \begin{array}{l} V_1 = \dfrac{q_1}{2\pi e_0 e_r} \ln (2H_1/\rho_1) + \dfrac{q_2}{2\pi e_0 e_r} \ln (I_{12}/A_{12}) \\[2mm] V_2 = \dfrac{q_1}{2\pi e_0 e_r} \ln (I_{12}/A_{12}) + \dfrac{q_2}{2\pi e_0 e_r} \ln (2H_2/\rho_2) \end{array} \right\} \qquad (4.24\ a)$$

In matrix form,

$$\begin{bmatrix} V_1 \\ V_2 \end{bmatrix} = \begin{bmatrix} \ln (2H_1/\rho_1), & \ln (I_{12}/A_{12}) \\ \ln (I_{12}/A_{12}), & \ln (2H_2/\rho_2) \end{bmatrix} \begin{bmatrix} q_1/2\pi e_0 e_r \\ q_2/2\pi e_0 e_r \end{bmatrix} \qquad (4.24b)$$

or $\qquad [V] = [P] \, [q/2\pi e_0 e_r] \qquad (4.24c)$

The elements of $[P]$ are Maxwell's Potential Coefficients which we have encountered in Chapter 3. For a single-phase line above ground, $V_1 = -V_2 = V$. Also, let $H_1 = H_2 = H$ and $\rho_1 = \rho_2 = \rho$. Then obviously, $q_1 = -q_2 = q$. Let $A_{12} = A$.

$$V = \frac{q}{2\pi e_0 e_r} \, [\ln (2H/\rho) - \ln (\sqrt{4H^2 + A^2}/A)]$$

$$= \frac{q}{2\pi e_0 e_r} \ln [2HA/\rho \, \sqrt{4H^2 + A^2}], \text{ Volts} \qquad (4.25)$$

This gives the capacitance per unit length of *each* conductor to ground to be

$$C_g = q/V = 2\pi e_0 e_r / [\ln (A/\rho) - \ln (\sqrt{4H^2 + A^2}/H)], \text{ Farads} \qquad (4.26)$$

The capacitance between the two conductors is one-half of this since the two capacitances to ground are in series. Comparing equations (4.26) and (4.28) with $2d = A$, we observe that in the presence of ground, the capacitance of the system has been increased slightly because the denominator of (4.26) is smaller than that in (4.22).

The charging current of each conductor is

$$I_c = 2\pi f \, C_g \, V, \text{ Amperes} \qquad (4.27)$$

and the charging reactive power is

$$Q_c = 2\pi f \, C_g \, V^2, \text{ vars.} \qquad (4.28)$$

Example 4.8: A single-conductor e.h.v. transmission line strung above ground is used for experimental purposes to investigate high-voltage effects. The conductor is expanded ACSR with diameter of 2.5 inches (0.0635 m) and the line height is 21 metres above ground.

(a) Calculate the voltage to ground which will make its surface voltage gradient equal to corona-inception gradient given by Peek's Formula:

$$E_{or} = \frac{30}{\sqrt{2}} \frac{1}{m} \left(1 + \frac{0.301}{\sqrt{\rho}} \right), \text{ KV/cm, r.m.s., where } m = 1.3 \text{ required}$$

for stranding effect, and the conductor radius is in cm.
(b) Find the charging current and MVAR of the single-phase transformer for exciting 1 km length of the experimental line.

Solution: Refer to Fig. 4.9. $\rho = 0.03176 \text{ m} = 3.176 \text{ cm}$.

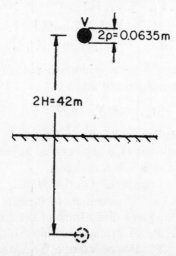

Fig. 4.9.   Single-phase experimental line above ground for Example 4.8.

From equations (4.24) and in the absence of a second conductor,

$$V = \frac{q}{2\pi e_0} \ln(2H/\rho) \text{ and } E = \frac{q}{2\pi e_0} \frac{1}{\rho}$$

Now, $E_{or} = \dfrac{30 \times 10^2}{1.3\sqrt{2}}\left(1 + \dfrac{0.301}{\sqrt{3.176}}\right) = 1907.4 \text{ KV/m}$

(a) $V = 1907.4 \times 0.03176 \times \ln(42/0.03176) = 435.4 \text{ KV, r.m.s.}$

(b) Capacitance $C_g = qL/V = 2\pi e_0 \, L/\ln(2H/\rho) = 7.747 \text{ nF/km}$

Charging current at 50 Hz is

$$I_c = 2\pi \times 50 \times 7.747 \times 10^{-9} \times 435.4 \times 10^3 \, A = 1.06 \text{ Ampere}$$

Charging KVAR $Q_c = 1.06 \times 435.4 = 461.5.$

The design of such an experimental 1 km line with 2.5 inch diameter conductor strung at an average height of 21m above ground will need a 500 KV single-phase transformer rated for 1.1 A and 550 KVA. In all such experimental projects, a research factor of 1.3 may be required so that the actual rating may be 565 KV at 1.38 A giving nearly 780 KVA.

4.4 Charge-Potential Relations for Multi-Conductor Lines

Section 3.5 in the last Chapter 3, equations (3.38) to (3.40) describe the charge-potential relations of a transmission line with n conductors on a tower. The effect of a ground plane considered as an equipotential surface gave rise to Maxwell's Potential coefficients and the general equations are

$$[V] = [P][Q/2\pi e_0] \tag{4.29}$$

where the elements of the three matrices are, for $i = 1, 2, ..., n$

$$\left.\begin{array}{l} [V] = [V_1, V_2, V_3, ..., V_n] \\ [Q/2\pi e_0] = (1/2\pi e_0)[Q_1, Q_2, ..., Q_n] \\ P_{ii} = \ln(2H_i/r_{eq}(i)), \; P_{ij} = \ln(I_{ij}/A_{ij}), \; i \neq j \end{array}\right\} \tag{4.30}$$

The equivalent radius or geometric mean radius of a bundled conductor has already been discussed and is

$$r_{eq} = R(N \cdot r/R)^{1/N} \tag{4.31}$$

where R = bundle radius = $B/2 \sin(\pi/N)$,

B = bundle spacing (spacing between adjacent conductors)

r = radius of each sub-conductor,

and N = number of conductors in the bundle.

The elements of Maxwell's potential coefficients are all known since they depend only on the given dimensions of the line-conductor configuration on the tower. In all problems of interest in e.h.v. transmission, it is required to find the charge matrix from the voltage since this is also known. The charge-coefficient matrix is evaluated as

$$[Q/2\pi e_0] = [P]^{-1} [V] = [M] [V] \tag{4.32}$$

or, if the charges themselves are necessary,

$$[Q] = 2\pi e_0 [M] [V] \tag{4.33}$$

In normal transmission work, the quantity $Q/2\pi e_0$ occurs most of the time and hence equation (4.32) is more useful than (4.33). The quantity $Q/2\pi e_0$ has units of volts and the elements of both $[P]$ and $[M] = [P]^{-1}$ are dimensionless numbers.

On a transmission tower, there are p phase conductors or poles and one or two ground wires which are usually at or near ground potential. Therefore, inversion of $[P]$ becomes easier and more meaningful if the suitable rows and columns belonging to the ground wires are eliminated. Some examples will serve to illustrate the procedure and observe the effect of ground wires on the line-conductor charges, voltage gradients, etc. On a 3-phase ac line, the phase voltages are varying in time so that the charges are also varying at 50 Hz or power frequency. This will be necessary in order to evaluate the electrostatic field in the line vicinity. But for Radio Noise and Audible Noise calculations, high-frequency effects must be considered under suitable types of excitation of the multi-conductors. Similarly, lightning and switching-surge studies also require unbalanced excitation of the phase and ground conductors.

4.4.1 Maximum Charge Condition on a 3-Phase Line

E.H.V. transmission lines are mostly single-circuit lines on a tower with one or two ground wires. For preliminary consolidation of ideas, we will restrict our attention here to 3 conductors excited by a balanced set of positive-sequence voltages under steady state. This can be extended to other line configurations and other types of excitation later on.

The equation for the charges is

$$\frac{1}{2\pi e_0} \begin{bmatrix} Q_1 \\ Q_2 \\ Q_3 \end{bmatrix} = \begin{bmatrix} M_{11}, & M_{12}, & M_{13} \\ M_{21}, & M_{22}, & M_{23} \\ M_{31}, & M_{32}, & M_{33} \end{bmatrix} \begin{bmatrix} V_1 \\ V_2 \\ V_3 \end{bmatrix} \tag{4.34}$$

For a 3-phase ac line, we have

$$V_1 = \sqrt{2} V \sin (wt + \phi),$$
$$V_2 = \sqrt{2} V \sin (wt + \phi - 120°)$$

and $V_3 = \sqrt{2} V \sin (wt + \phi + 120°)$ with

$V = $ r.m.s. value of line-to-ground voltage and $w = 2\pi f$,

$f = $ power frequency in Hz. The angle ϕ denotes the instant on V_1 where $t = 0$. If $wt + \phi$ is denoted as θ, there results

$$Q_1/2\pi e_0 = \sqrt{2} V [M_{11} \sin \theta + M_{12} \sin (\theta - 120°) + M_{13} \sin (\theta + 120°)]$$
$$= \sqrt{2} V [\{M_{11} - 0.5 (M_{12} + M_{13})\} \sin \theta$$
$$+ 0.866 (M_{13} - M_{12}) \cos \theta] \tag{4.35}$$

Differentiating with respect to θ and equating to zero gives

$$\frac{d}{d\theta}\left(Q_1/2\pi e_0\right) = 0$$

$$= \sqrt{2}V\left[(M_{11} - 0.5M_{12} - 0.5M_{13})\cos\theta - \frac{\sqrt{3}}{2}(M_{13} - M_{12})\sin\theta\right]$$

This gives the value of $\theta = \theta_m$ at which Q_1 reaches its maximum or peak value. Thus,

$$\theta_m = \text{arc tan}\,[\sqrt{3}\,(M_{13} - M_{12})/(2M_{11} - M_{12} - M_{13})] \qquad (4.36)$$

Substituting this value of θ in equation (4.35) yields the maximum value of Q_1.

An alternative procedure using phasor algebra can be devised. Expand equation (4.35) as

$$Q_1/2\pi e_0 = \sqrt{2}V\sqrt{(M_{11} - .5M_{12} - .5M_{13})^2 + 0.75\,(M_{13} - M_{12})^2}$$
$$\sin(\theta + \psi) \qquad (4.37)$$

The amplitude or peak value of $Q_1/2\pi e_0$ is

$$(Q_1/2\pi e_0)_{\text{max}} = \sqrt{2}V\,[M_{,1}^{2} + M_{12}^2 + M_{13}^2 - (M_{11}M_{12} + M_{12}M_{13}$$
$$+ M_{13}M_{11})]^{1/2} \qquad (4.38)$$

It is left as an exercise to the reader to prove that substituting θ_m from equation (4.36) in equation (4.35) gives the amplitude of $(Q_1/2\pi e_0)_{\text{max}}$ in equation (4.38).

Similarly for Q_2 we take the elements of 2nd row of $[M]$.

$$(Q_2/2\pi e_0)_{\text{max}} = \sqrt{2}V\,[M_{22}^2 + M_{21}^2 + M_{23}^2 - (M_{21}M_{22} + M_{22}M_{23}$$
$$+ M_{23}M_{21})]^{1/2} \qquad (4.39)$$

For Q_3 we take the elements of the 3rd row of $[M]$.

$$(Q_3/2\pi e_0)_{\text{max}} = \sqrt{2}V\,[M_{33}^2 + M_{31}^2 + M_{32}^2 - (M_{31}M_{32} + M_{32}M_{33}$$
$$+ M_{33}M_{31})]^{1/2} \qquad (4.40)$$

The general expression for any conductor is, for $i = 1, 2, 3$,

$$(Q_i/2\pi e_0)_{\text{max}} = \sqrt{2}V\,[M_{i1}^2 + M_{i2}^2 + M_{i3}^2 - (M_{i1}M_{i2} + M_{i2}M_{i3}$$
$$+ M_{i3}M_{i1})]^{1/2} \qquad (4.41)$$

4.4.2 Numerical Values of Potential Coefficients and Charge of Lines

In this section, we discuss results of numerical computation of potential coefficients and charges present on conductors of typical dimensions from 400 KV to 1200 KV whose dimensions are given in chapter 3, Fig. 3.5. For one line, the effect of considering or neglecting the presence of ground wires on the charge coefficient will be discussed, but in a digital-computer programme the ground wires can be easily accommodated without difficulty. In making all calculations we must remember that the height H_i of conductor i is to be taken as the average height. It will be quite adequate to use the relation

$$H_{av} = H_{min} + Sag/3 \qquad (4.42)$$

This relation is proved below.

Average Line Height for Inductance Calculation*

The shape assumed by a freely hanging cable of length L_c over a horizontal span S between supports is a catenary. We will approximate the shape to a parabola for deriving the average height which holds for small sags. Figure 4.10 shows the dimensions required. In this figure,

H = minimum height of conductor at midspan

S = horizontal span,

and d = sag at midspan.

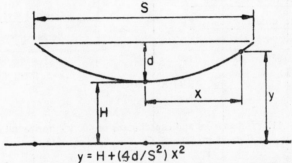

$$y = H + (4d/S^2) X^2$$

Fig. 4.10. Calculation of average height over a span S with sag d.

The equation to the parabolic shape assumed is

$$y = H + (4d/S^2) X^2 \qquad (4.43)$$

The inductance per unit length at distance X from the point of minimum height is

$$L = 0.2 \, Ln(2y/r) = 0.2[Ln \, 2y - Ln \, r] \qquad (4.44)$$

Since the height of conductor is varying, the inductance also varies with it. The average inductance over the span is

$$L_{av} = \frac{1}{S} \int_{-S/2}^{S/2} 0.2 \, (\ln 2y - \ln r) \, dx$$

$$= \frac{0.4}{S} \int_{0}^{S/2} (\ln 2y - \ln r) \, dX \qquad (4.45)$$

Now, $\ln 2y = \ln (2H + 8dX^2/S^2) = \ln(8d/S^2) + \ln(X^2 + S^2H/4d)$ (4.46)

Let $a^2 = S^2H/4d$. Then,

$$\frac{2}{S} \int_{0}^{S/2} \ln (X^2 + a^2) \, dX = \ln [(1 + H/d) \, S^2/4]$$

---

*The author is indebted to Ms. S. Ganga for help in making this analysis.

$$-2 + 2\sqrt{H/d}\ \tan^{-1}\sqrt{d/H} + \ln(8d/S^2)$$
$$= \ln 2d + \ln(1 + H/d) - 2 + 2\sqrt{H/d}\tan^{-1}\sqrt{d/H} \qquad (4.47)$$

If the right-hand side can be expressed as $\ln(2H_{av})$, then this gives the average height for inductance calculation. We now use some numerical values to show that H_{av} is approximately equal to

$$(H + \tfrac{1}{3}d) = H_{min} + \tfrac{1}{3}\text{ Sag.}$$

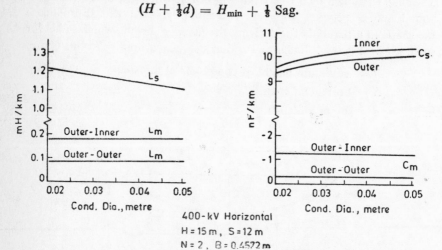

Fig. 4.11. Inductances and capacitances of 400 kV horizontal line. $H = 15$ m, $S = 12$ m, $N = 2$, $B = .4572$ m.

(a) Consider $H = 10$, $d = 10$. Using these in equation (4.47) gives
$$\ln 20 + \ln(1 + 1) - 2 + 2\tan^{-1} 1 = 3 + 0.6931 - 2 + \pi/2$$
$$= 3.259 = \ln 26 = \ln(2H_{av})$$
$$H_{av} = 26/2 = 13 = 10 + 3 = H + 0.3d$$

(b) $H = 10$, $d = 8$. $\ln 16 + \ln 2.25 - 2 + 2\sqrt{1.25}\ \tan^{-1}\sqrt{0.8}$
$$= \ln 24.9$$
$$H_{av} = 12.45 = H + 2.45 = H + 0.306d$$

(c) $H = 14$, $d = 10$. $H_{av} = 17.1 = 14 + 3.1 = H + 0.3d$

These examples appear to show that a reasonable value for average height is $H_{av} = H_{min} + \text{sag}/3$. A rigorous formulation of the problem is not attempted here.

Figures 4.11, 4.12, and 4.13 show inductances and capacitances of conductors for typical 400 kV, 750 kV and 1200 kV lines.

4.5 Surface Voltage Gradient on Conductors

The surface voltage gradient on conductors in a bundle governs generation of corona on the line which have serious consequences causing audible noise and radio interference. They also affect carrier communication and signalling on the line and cause interference to television

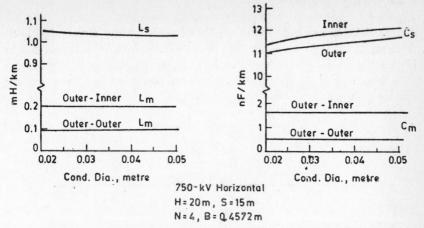

750-kV Horizontal
H= 20m, S = 15m
N = 4, B = 0.4572m

Fig. 4.12. *L* and *C* of 750 kV horizontal line. $H = 20$ m, $S = 15$ m, $N = 4$, $B = 0.4572$ m.

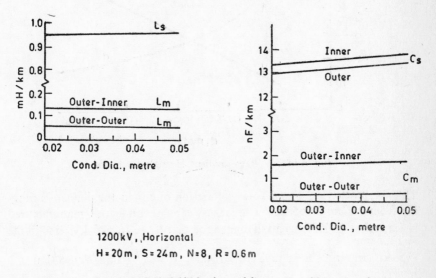

1200kV, Horizontal
H = 20m, S = 24m, N = 8, R = 0.6m

Fig. 4.13. *L* and *C* of 1200 kV horizontal line. $H = 20$ m, $S = 24$ m, $N = 8$, $R = 0.6$ m.

reception. The designer of a line must eliminate these nuisances or reduce them to tolerable limits specified by standards, if any exist. These limits will be discussed at appropriate places where *AN*, *RI* and other interfering fields are discussed in the next two chapters. Since corona generation depends on the voltage gradient on conductor surfaces, this will be taken up now for e.h.v. conductors with number of sub-conductors in a bundle ranging from 1 to *N*. The maximum value of *N* is 8 at present but a general derivation is not difficult.

4.5.1 Single Conductor

Figure 4.9 can be used for a single conductor whose charge is q coulomb/metre. We have already found the line charges or the terms $(Q_i/2\pi e_0)$ in terms of the voltages V_i and the Maxwell's Potential Coefficient matrix $[P]$ and its inverse $[M]$, where $i = 1, 2, ..., n$, the number of conductors on a tower. For the single conductor per phase or pole, the surface voltage gradient is

$$E_c = \frac{q}{2\pi e_0} \frac{1}{r} \text{ volts/metre} \tag{4.48}$$

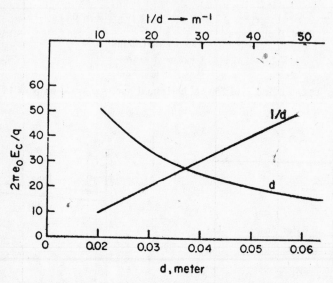

Fig. 4.14. Voltage gradient of single conductor.

This is plotted in Fig. 4.14 as a function of conductor diameters ranging from 0.02 to 0.065 m. The largest single conductor manufactured is 2.5 inches (0.0635 m) in diameter for the B. P. A. 525 kV line in the

U.S.A. In terms of voltage to ground, $V = \frac{q}{2\pi e_0} \ln(2H/r)$ so that

$$E_c = \frac{V}{r \cdot \ln(2H/r)} \text{ volts/metre} \tag{4.49}$$

The factor $E_c/(q/2\pi e_0)$ is also plotted against the reciprocal of diameter and yields a straight line.

4.5.2 2-Conductor Bundle (Fig. 4.15)

In this case, the charge Q obtained from equation (4.33) is that of the total bundle so that the charge of each sub-conductor per unit length is $q = Q/2$. This will form one phase of an ac line or a pole of a dc line. In calculating the voltage gradient on the surface of a sub-conductor, we will make the following assumptions:

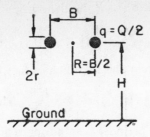

Fig. 4.15.   2-conductor bundle above ground for voltage
gradient calculation.

(1)   The conductors of the other phases or poles are very far from
the bundled conductor under examination, i.e. $S \gg B$ or $2R$.

(2)   The image conductors are also very far, i.e. $2H \gg B$ or $2R$.

This allows us to ignore all other charges except that of the conductors
in the bundle. Now, by definition, the electric field intensity is the force
exerted on a unit positive test charge placed at the point where the field
intensity is to be evaluated, which in this case is a point on the sub-conductor surface. Consider point P_i on the inside of the bundle. The
force on a test charge is

E_i = Force due to conductor charge − Force due to the charge on
second conductor of bundle

At the point P_0 on the outside of the bundle, the two forces are directed
in the same sense. It is clear that it is here that the maximum surface
voltage gradient occurs.

Now, the force due to conductor charge $= \dfrac{q}{2\pi e_0} \dfrac{1}{r}$. In computing the

force due to the charge of the other sub-conductor there is the important point that the conductors are metallic. When a conducting cylinder
is located in the field of a charge, it distorts the field and the field intensity is higher than when it is absent. If the conducting cylinder is placed
in a uniform field of a charge q, electrostatic theory shows that stress-
doubling occurs on the surface of the metallic cylinder. In the present
case, the left cylinder is placed at a distance B from its companion at
right. Unless $B \gg r$, the field is non-uniform. However, for the sake
of calculation of surface voltage gradients on sub-conductors in a
multi-conductor bundle, we will assume that the field is uniform and
stress-doubling takes place. Once again, this is a problem in successive
images but will not be pursued here.

Therefore, $E_i = \dfrac{q}{2\pi e_0}\left(\dfrac{1}{r} - \dfrac{2}{B}\right) = \dfrac{q}{2\pi e_0} \cdot \dfrac{1}{r}\,(1 - r/R)$

$$= \dfrac{Q}{2\pi e_0}\,\dfrac{1}{2}\cdot\dfrac{1}{r}\,(1 - r/R) \qquad (4.50)$$

where Q = total bundle charge.

On the other hand, at point P_0 on the outside of bundle

$$E_0 = \frac{Q}{2\pi e_0} \frac{1}{2} \frac{1}{r} (1 + r/R) \qquad (4.51)$$

These are the minimum and maximum values, and they occur at $\theta = \pi$

and $\theta = 0$. The average is $E_{av} = \dfrac{Q}{2\pi e_0} \dfrac{1}{2} \dfrac{1}{r}$. The variation of surface

voltage gradient on the periphery can be approximated to a cosine curve

$$E(\theta) = \frac{Q}{2\pi e_0} \frac{1}{2} \frac{1}{r} \left(1 + \frac{r}{R} \cos \theta\right) = E_{av} \left(1 + \frac{r}{R} \cos \theta\right) \qquad (4.52)$$

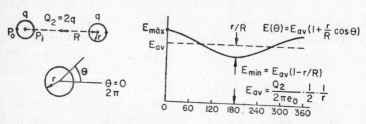

Fig. 4.16.   Distribution of voltage gradient on 2-conductor bundle illustrating the cosine law.

This is shown in Fig. 4.16.   This is called the "cosine law" of variation of E with θ.   We are now encountering three terms: Maximum gradient, Minimum gradient and Average gradient.   Since we have neglected the charges on the other phases and the image conductors, the surface voltage gradient distribution on both sub-conductors of the bundle is identical.   The concepts given above can be easily extended to bundles with more sub-conductors and we will consider N from 3 to 8.

Example 4.9: The dimensions of a ± 400 kV dc line are shown in Fig. 4.17.   Calculate

(a) the charge coefficient $Q/2\pi e_0$ for each bundle,
(b) the maximum and minimum surface gradient on the conductors by
   (i) omitting the charges of the second pole and image conductors,
   (ii) considering the charge of the second pole but omitting the charge of the image conductors,
(c) the average maximum surface voltage gradient of the bundle under case b (ii).

Solution: The potential coefficients are first calculated.

$$r_{eq} = \sqrt{0.0175 \times 0.45} = 0.08874 \text{ m}$$

$$P_{11} = P_{22} = \ln (24/0.08874) = 5.6$$

$$P_{12} = P_{21} = \ln (\sqrt{24^2 + 9^2}/9) = 1.047.$$

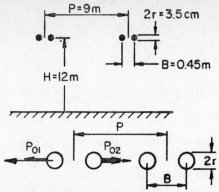

Fig. 4.17.   Maximum and minimum values of voltage gradients
on 2-conductor bundles.

$$[P] = \begin{bmatrix} 5.6, & 1.047 \\ 1.047, & 5.6 \end{bmatrix} \text{ giving } [M] = [P]^{-1} = \begin{bmatrix} 0.185, & -0.0346 \\ -0.0346, & 0.185 \end{bmatrix}$$

(a) $\dfrac{1}{2\pi e_0}\begin{bmatrix} Q_1 \\ Q_2 \end{bmatrix} = \begin{bmatrix} 0.185, & -0.0346 \\ -0.0346, & 0.185 \end{bmatrix} \begin{bmatrix} 400 \\ -400 \end{bmatrix} \times 10^3$

$$= \begin{bmatrix} 1 \\ -1 \end{bmatrix} 87.84 \times 10^3$$

The charge of each bundle is $2\pi e_0 \times 87.84 \times 10^3 = \dfrac{87.84}{18}$ μc/m
$= 4.88$ μc/m.

The charge of each sub-conductor is $q = 2.44$ μc/m length.

(b) (i) Maximum and minimum gradient $= \dfrac{q}{2\pi e_0} \dfrac{1}{r} (1 \pm r/R)$

Maximum $E_0 = 87.84 \times 10^3 \dfrac{1}{2} \dfrac{1}{0.0175} (1 + 0.0175/0.225)$

$$= 2705 \text{ kV/m} = 27.05 \text{ kV/cm}$$

Minimum $E_i = 23.15$ kV/cm $= 2315$ kV/m

Average gradient $= \dfrac{q}{2\pi e_0} \dfrac{1}{r} = \dfrac{87.84}{2} \dfrac{1}{.0175} = 2510$ kV/m

$$= 25.1 \text{ kV/cm}$$

(ii) Consider the 2 sub-conductors on the left.
At P_{01}, the force on a positive test charge are as shown in Fig. 4.17.

$$E_{01} = \dfrac{q}{2\pi e_0} \dfrac{1}{r} (1 + r/R) - \dfrac{q}{2\pi e_0} \left(\dfrac{2}{9.0175} + \dfrac{2}{9.4675} \right)$$

$$= 2705 - 19 = 2686 \text{ kV/m} = 26.86 \text{ kV/cm}$$

$$E_{02} = \dfrac{q}{2\pi e_0} \dfrac{1}{r} (1 + r/R) + \dfrac{q}{2\pi e_0} \left(\dfrac{2}{8.4825} + \dfrac{2}{8.9325} \right)$$

$$= 2705 + 20.2 = 2725.2 \text{ kV/m} = 27.25 \text{ kV/cm}$$

$$E_{l_1} = \frac{q}{2\pi e_0}\frac{1}{r}(1 - r/R) + \frac{q}{2\pi e_0}\left(\frac{2}{8.9825} + \frac{2}{9.4325}\right)$$

$$= 2315 + 19.1 = 2334 \text{ kV/m} = 23.34 \text{ kV/cm}$$

$$E_{l_2} = \frac{q}{2\pi e_0}\frac{1}{r}(1 - r/R) - \frac{q}{2\pi e_0}\left(\frac{2}{8.5725} + \frac{2}{9.0225}\right)$$

$$= 2315 - 20 = 2295 \text{ kV/m} = 22.95 \text{ kV/cm}.$$

The maximum gradients on the two sub-conductors have now become 26.86 kV/cm and 27.25 kV/cm instead of 27.05 kV/cm calculated on the basis of omitting the charges of other pole.

(c) The average maximum gradient is defined as the arithmetic average of the two maximum gradients.

$$E_{\text{avm}} = \tfrac{1}{2}(26.86 + 27.25) = 27.055 \text{ kV/cm}$$

This is almost equal to the maximum gradient obtained by omitting the charges of the other pole.

For a 3-phase ac line, the effect of charges on other phases can usually be ignored because when the charges on the conductor of one phase is at peak value, the charges on the other phases are passing through 50% of their peak values but of opposite polarity. This has an even less effect than what has been shown for the bipolar dc line where the charge of the second pole is equal and opposite to the charge of the conductor under consideration. However, in a digital computer programme, all these could be incorporated. It is well to remember that all calculations are based on the basic assumption that ground is an equipotential surface and that the sub-conductor charges are concentrated at their centres. Both these assumptions are approximate. But a more rigorous analysis is not attempted here.

4.5.3 Maximum Surface Voltage Gradients for $N \geqslant 3$

The method described before for calculating voltage gradients for a twin-bundle conductor, $N = 2$, can now be extended for bundles with more than 2 sub-conductors. A general formula will be obtained under the assumption that the surface voltage gradients are only due to the charges of the N sub-conductors of the bundle, ignoring the charges of other phases or poles and those on the image conductors. Also, the sub-conductors are taken to be spaced far enough from each other so as to yield a uniform field at the location of the sub-conductor and hence the concept of stress-doubling will be used.

Figure 4.18 shows bundles with $N = 3, 4, 6, 8$ sub-conductors and the point P where the maximum surface voltage gradient occurs. The forces exerted on a unit positive test charge at P due to all N conductor-charges q are also shown as vectors. The components of these forces along the vector force due to conductor charge will yield the maximum surface

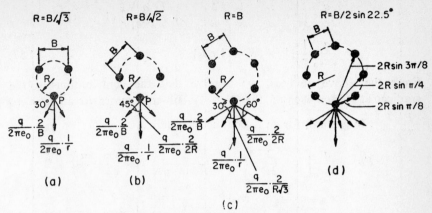

Fig. 4.18. Distribution of surface voltage gradients on 3-, 4-, 6-, and 8-conductor bundles.

voltage gradient. Due to symmetry, the components at right angles to this force will cancel each other, as shown on the figures.

$N = 3$. $B = \sqrt{3}R$ for equilateral spacing.

$$E_P = \frac{q}{2\pi e_0} \frac{1}{r} \left(1 + 2 \times r \cdot \frac{2}{B} \cos 30° \right) = \frac{q}{2\pi e_0} \frac{1}{r} (1 + 2r/R) \quad (4.53)$$

The sub-conductor charge is $q = Q/3$ where Q is obtained from equation (4.32), $[Q/2\pi e_0] = [M][V]$ as discussed earlier.

$N = 4$. $B = \sqrt{2}R$ for quadrilateral spacing.

$$E_P = \frac{q}{2\pi e_0} \left(\frac{1}{r} + \frac{2}{2R} + 2 \times \frac{2}{R\sqrt{2}} \cos 45° \right)$$

$$= \frac{q}{2\pi e_0} \frac{1}{r} (1 + 3r/R) \quad (4.54)$$

Note the emergence of a general formula

$$E_P = \frac{q}{2\pi e_0} \frac{1}{r} [1 + (N-1)r/R] \quad (4.55)$$

$N = 6$. $B = R$ for hexagonal spacing.

$$E_P = \frac{q}{2\pi e_0} \left[\frac{1}{r} + \frac{2}{2R} + 2 \times \frac{2}{R} \cos 60° + 2 \times \frac{2}{R\sqrt{3}} \cos 30° \right]$$

$$= \frac{q}{2\pi e_0} \frac{1}{r} (1 + 5r/R) = \frac{q}{2\pi e_0} \frac{1}{r} [1 + (N-1)r/R] \quad (4.56)$$

$N = 8$. $B = 2R \sin 22°.5$ for octagonal spacing

$$E_P = \frac{q}{2\pi e_0} \left[\frac{1}{r} + \frac{2}{2R} + 2 \times \frac{2}{2R \sin \pi/8} \cos \frac{3\pi}{8} + 2 \times \frac{2}{2R \sin \pi/4} \right.$$

$$\left. \times \cos \frac{\pi}{4} + 2 \frac{2}{2R \sin 3\pi/8} \cos \frac{\pi}{8} \right]$$

$$= \frac{q}{2\pi e_0} \left(\frac{1}{r} + \frac{1}{R} + \frac{2}{R} + \frac{2}{R} + \frac{2}{R} \right) = \frac{q}{2\pi e_0} \frac{1}{r} (1 + 7r/R)$$

$$= \frac{q}{2\pi e_0} \frac{1}{r} [1 + (N - 1)r/R] \qquad (4.57)$$

From the above analysis, we observe that the contributions to the gradient at P from each of the $(N - 1)$ sub-conductors are all equal to $\frac{q}{2\pi e_0} \frac{1}{R}$. In general, for an N-conductor bundle,

$$E_P = \frac{q}{2\pi e_0} \left[\frac{1}{r} + \frac{2}{2R} + 2 \times \frac{2}{2R \sin \pi/N} \cos \left(\frac{\pi}{2} - \frac{\pi}{N} \right) + 2 \right.$$

$$\times \frac{2}{2R \sin 2\pi/N} \cos \left(\frac{\pi}{2} - \frac{2\pi}{N} \right) + \ldots + 2$$

$$\left. \times \frac{2}{2R \sin (N - 1)\pi/N} \cos \left(\frac{\pi}{2} - \frac{N-1}{N} \pi \right) \right]$$

$$= \frac{q}{2\pi e_0} \left[\frac{1}{r} + \frac{1}{R} + \frac{(N - 2)}{R} \right] = \frac{q}{2\pi e_0} \frac{1}{r} [1 + (N-1)r/R] \quad (4.58)$$

Figures 4.19 4.20 and 4.21 show typical results of maximum surface voltage gradients for 400 kV, 750 kV, and 1200 kV lines whose dimensions are shown in Fig. 3.5.

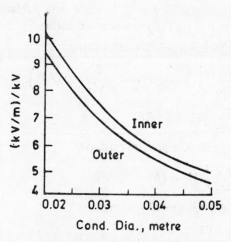

400kV, Horizontal

H = 15m, S = 12m, N = 2,

B = 0.4572 m

Fig. 4.19.   Surface voltage gradient on conductors of 400 kV line.
See Fig. 4.11 for dimensions.

4.5.4   Mangoldt (Markt-Mengele) Formula

In the case of a 3-phase ac line with horizontal configuration of phases, a convenient formula due to Mangoldt can be derived. This is

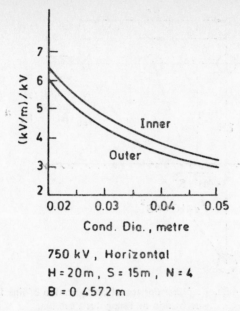

750 kV , Horizontal
H = 20m , S = 15m , N = 4
B = 0 4572 m

Fig. 4.20.   Surface voltage gradient on conductors of 750 kV line.
See Fig. 4.12 for dimensions.

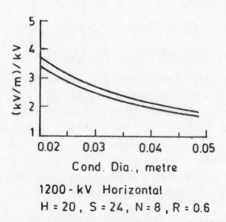

1200 - kV Horizontal
H = 20 , S = 24 , N = 8 , R = 0.6

Fig. 4.21.   Surface voltage gradients on 1200 kV lines.
See Fig. 4.13 for dimensions.

also known as the Markt-Mengele Formula by some others.   Referring
to Fig. 4.22, let Q_1, Q_2, Q_3 be the instantaneous charges on the bundles.
The Maxwell's Potential coefficients are

$$P_{11} = P_{22} = P_{33} = \ln (2H/r_{eq}), \text{ where } r_{eq} = R(Nr/R)^{1/N}$$

$$P_{12} = P_{21} = P_{23} = P_{32} = \ln (\sqrt{4H^2 + S^2}/S) = \ln\sqrt{1 + (2H/S)^2}$$

$$P_{13} = P_{31} = \ln (\sqrt{4H^2 + 4S^2}/2S) = \ln \sqrt{1 + (H/S)^2} \qquad (4.59)$$

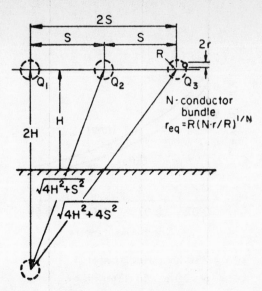

Fig. 4.22.   3-phase horizontal configuration of line for derivation of Mangoldt Formula.

The voltage-charge relations are

$$
\left.
\begin{aligned}
V_1 &= P_{11} \cdot Q_1/2\pi e_0 + P_{12} \cdot Q_2/2\pi e_0 + P_{13} \cdot Q_2/2\pi e_0 \\
V_2 &= P_{21} \cdot Q_1/2\pi e_0 + P_{22} \cdot Q_2/2\pi e_0 + P_{23} \cdot Q_3/2\pi e_0 \\
V_3 &= P_{31} \cdot Q_1/2\pi e_0 + P_{32} \cdot Q_2/2\pi e_0 + P_{33} \cdot Q_3/2\pi e_0
\end{aligned}
\right\}
\qquad (4.60)
$$

Both the voltages and the charges are sinusoidally varying at power frequency and at every instant of time, $V_1 + V_2 + V_3 = 0$ and $Q_1 + Q_2 + Q_3 = 0$. When the charge of any phase is passing through its peak value, the charges of the remaining two phases are negative but of magnitude 0.5 peak. From symmetry, the peak values of charges on the two outer phases will be equal. If we assume the peak values of Q_1, Q_2, Q_3 to be approximately equal, then, combining equations (4.59) and (4.60) we obtain the following equations:

For the Outer Phases

$$
V_1 = \frac{Q_1}{2\pi e_0} \, (P_{11} - 0.5 \, P_{12} - 0.5 \, P_{13})
$$

$$
= \frac{Q_1}{2\pi e_0} \ln \frac{2H}{r_{eq}} \frac{1}{\left[\left\{1 + \left(\frac{2H}{S}\right)^2\right\}\left\{1 + \left(\frac{H}{S}\right)^2\right\}\right]^{1/4}} \qquad (4.61)
$$

Therefore, $\dfrac{Q_1}{2\pi e_0}$ is found from the given voltage. If the r.m.s. value of phase voltage to ground is used, $\dfrac{Q_1}{2\pi e_0}$ is also the r.m.s. value of the charge coefficient and the resulting surface voltage gradient will also be in kV(r.m.s.)/metre, if V is in kV. The maximum surface voltage gradient will then be according to equation (4.58),

$$E_{0m} = \frac{Q_1}{2\pi e_0} \frac{1}{N} \frac{1}{r} \left[1 + (N-1)\frac{r}{R} \right]$$

$$= \frac{1 + (N-1)\,r/R}{N.r.\ln \dfrac{2H}{r_{eq}} \dfrac{1}{[\{1 + (2H/S)^2\}\{1 + (H/S)^2\}]^{1/4}}} \, V, \qquad (4.62)$$

Similarly, for the centre phase,

$$\frac{Q_2}{2\pi e_0} = [P_{22} - 0.5\,(P_{21} + P_{23})]^{-1}\, V$$

and $\qquad E_{cm} = \dfrac{1 + (N-1)\,r/R}{.r.\ln \dfrac{2H}{r_{eq}} \dfrac{1}{[1 + (2H/S)^2]^{1/2}}} \, V \qquad (4.63)$

Equations (4.62) and (4.63) are known as Mangoldt or Markt-Mengele Formulas. They were first derived only for the centre phase which gives a higher maximum voltage gradient than the outer phases. With corona assuming lot more importance since this formula was derived, we have extended their thinking to the outer phases also.

Example 4.10: For a 400-kV line, calculate the maximum surface voltage gradients on the centre and outer phases in horizontal configuration at the maximum operating voltage of 420 kV, r.m.s., line-to-line. The other dimensions are

$$H = 13 \text{ m}, S = 11 \text{ m}, N = 2, r = 0.0159 \text{ m}, B = 0.45 \text{ m}.$$

Solution: $\qquad 2H/S = 26/11 = 2.364$ and $H/S = 1.182$

$[\{1 + (2H/S)^2\}\{1 + (H/S)^2\}]^{1/4} = 1.982, \sqrt{1 + (H/S)^2} = 2.567$

$r_{eq} = R\,(N.\,r/R)^{1/N} = 0.225\,(2 \times 0.0159/0.225)^{1/2} = 0.0846$ m

Also, $r_{eq} = \sqrt{0.0159 \times 0.45} = 0.0846$ m, $2H/r_{eq} = 307.3$

(a) Outer Phases.

$$E_{0m} = \frac{(1 + 0.0159/0.225)\,420/\sqrt{3}}{2 \times 0.0159\ Ln\,(307.3/1.982)} = 1619 \text{ kV/m} = 16.19 \text{ kV/cm}$$

(b) Centre Phase.

$$E_{cm} = \frac{(1 + 0.0159/0.225)\,420/\sqrt{3}}{2 \times 0.0159\ Ln\,(307.3/2.567)} = 1707 \text{ kV/m} = 17.07 \text{ kV/cm}$$

The centre phase gradient is higher than that on the outer phases by

$$\frac{17.07 - 16.19}{16.19} \times 100 = 5.44\%$$

For a bipolar dc line, it is easy to show that the maximum surface voltage gradient on the sub-conductor of a bundle is

$$E_m = \frac{1 + (N - 1) \, r/R}{N.r. \ln \left[\dfrac{2H}{r_{eq}} \dfrac{1}{\sqrt{1 + (2H/P)^2}} \right]} \quad V \qquad (4.64)$$

where H = height of each pole above ground

P = pole spacing

and V = voltage to ground

Example 4.11: Using the data of Example 4.9, and using equation (4.64), calculate the maximum surface voltage gradient on the 2-conductor bundle for $\pm$ 400 kV dc line.

Solution: $H = 12, P = 9, r = 0.0175, N = 2, R = 0.225$

$r_{eq} = 0.08874,$

$$E_m = \frac{(1 + 0.0175/0.225) \, 400}{2 \times 0.0175 \ln \left[\dfrac{24}{.08874} \dfrac{1}{\sqrt{1 + (24/9)^2}} \right]}$$

$=2705 \, \text{kV/m} = 27.05 \, \text{kV/cm}$

4.6 Examples of Conductors and Maximum Gradients on Actual Lines

Several examples of conductor configurations used on transmission lines following world-wide practice are given in the Table 4.1. The maximum surface voltage gradients are also indicated. These are only examples and the reader should consult the vast literature (CIGRE Proceedings, etc.) for more details. Most conductor manufacturers use British units for conductor sizes and the SI units are given only for calculation purposes. These details are gathered from a large number of sources listed in the bibliography at the end of the book.

The conductor sizes given in the table are not the only ones used. For example, the following range of conductor sizes are found on the North American continent.

345 kV.   Single conductor—1.424, 1.602, 1.737, 1.75, 1.762 inches dia.

2-conductor bundle—1.108, 1.165, 1.196, 1.246 inches dia.

500 kV.   Single conductor—2.5 inches dia.

2-bundle—1.602, 1.7, 1.75, 1.762, 1.82 inches dia (ACAR).

3-bundle—1.165 inches dia.

4-bundle—0.85, 0.9, 0.93 inches dia.

735-765 kV.   4-bundle—1.165, 1.2, 1.382 inches dia.

4.7 Gradient Factors and Their Use

From the Mangoldt (Markt-Mengele) formula given in Section 4.5, it is observed that the maximum surface voltage gradient in the centre phase of a horizontal 3-phase ac line is a function of the geometrical dimensions and the maximum operating voltage V. As shown in the previous table, the maximum operating voltages show a wide variation. It is therefore advantageous to have a table or graph of the normalized value called the 'gradient factor' in kV/cm per kV or V/m per volt or

Table 4.1. Conductor Details and Maximum Surface Voltage Gradients Used in EHV Lines

Country	Maximum Operating Voltage, kV, RMS	Conductor Details No. × dia in inches (cm)	Maximum Gradient kV/cm, RMS
(1) India	420	2×1.258 (3.18)	17.0
(2) Canada	315	2×1.382 (3.51)	17.4
	380	2×1.108 (2.814)	17.5—18.0
	525	4×0.93 (2.362)	18.8
	735	4×1.2, 1.382 (3.05, 3.51)	17.7—20.4
	1200	6×1.84, 2.0 (4.674, 5.08)	—
		8×1.65, 1.84 (4.19, 4.674)	—
(3) U.S.A.	355—362	1×1.602 (4.07)	16.6
		2×1.175, 1.196 (2.985, 3.04)	15—16
	500	2×1.65 (4.19)	16.9
		3×1.19 (3.02)	16.4
	550	2×1.6 (4.07)	16.9
		1×2.5 (6.35)	16.7
	765	4×1.165 (2.96)	20.4
	1200	8×1.602 (4.07)	13.5
(4) U.S.S.R.	400	3×1.19 (3.02)	13.6
	525	3×1.19 (3.02)	18.0
	1200	8×0.96 (2.438)	21.4
(5) U.K.	420	2×1.09 (2.77)	19.6
		4×1.09 (2.77)	13.5
(6) France	420	2×1.04 (2.64)	19.0
(7) Germany	380	4×0.827 (2.1)	15.7
	420	4×0.854 (2.17)	16.7
(8) Italy	380	2×1.168 (2.97)	15.0
	1050	4×1.76, 1.87 (4.47, 4.75)	17.1—19.8
		6×1.5 (3.81)	—
(9) Sweden	380	3×1.25 (3.18)	12.5
	400	2×1.25 (3.18)	16.5
	800	4×1.6 (4.06)	17.6

other units which will be independent of the voltage. The gradient factor is denoted by $g = E_{cm}/V$ and its value is

$$g = E_{cm}/V = \frac{1 + (N - 1)\, r/R}{N.r.\ \ln\left[\dfrac{2H}{r_{eq}}\dfrac{1}{\sqrt{1 + (2H/S)^2}}\right]} \qquad (4.65)$$

By varying the parameters (r, N, R, H, and S) over a large range curves can be plotted for g against the desired variable. From product of g and the maximum operating line-to-ground voltage of the line, the maximum surface voltage gradient is readily obtained. Such curves are shown in Fig. 4.23 for $N = 1$, 2, 3, 4, and for conductor diameters ranging from 0.7″ to 2.5″ (1.78 cm to 6.35 cm). The height H in all cases has been fixed at $H = 50'$ (15 m) and the phase spacing S ranging from 20′ to 50′ (6 to 15 m). Calculations have shown that a variation of height H from 10 to 30 metres does not change g by more than 1%. The abscissa has been chosen as the reciprocal of the diameter and the resulting variation of gradient factor with (1/d) is nearly a straight line, which is very convenient. In equation (4.65) it is observed that the conductor radius occurs in the denominator and this property has been used in plotting Fig. 4.23.

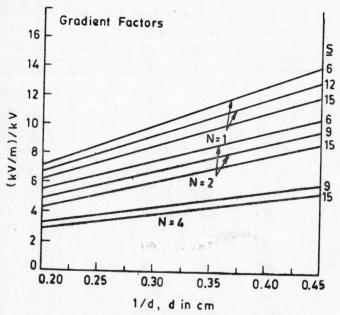

Fig. 4.23.   Gradient factors of conductors (g in V/m/Volt).

Such graphs can also be prepared for the maximum surface voltage gradient factors for the outer phases or bipolar dc lines, etc., in a design office.

4.8 Distribution of Voltage Gradient on Sub Conductors of Bundle

While discussing the variation of surface voltage gradient on a 2-conductor bundle in section 4.5.2, it was pointed out that the gradient distribution follows nearly a cosine law, equation (4.52). We will derive rigorous expressions for the gradient distribution and discuss the approximations to be made which yields the cosine law. The cosine law has been verified to hold for bundled conductors with up to 8 sub-conductors. Only the guiding principles will be indicated here through an example of a 2-conductor bundle and a general outline for $N \geqslant 3$ will be given which can be incorporated in a digital-computer programme.

Figure 4.24 shows a detailed view of 2-conductor bundle where the charges q on the two sub-conductors are assumed to be concentrated at the conductor centres. At a point P on the surface of a conductor at angle θ from the reference direction, the field intensities due to the two conductor charges are, using stress doubling effect,

$$E_1 = \frac{q}{2\pi e_0} \frac{1}{r} \text{ and } E_2 = \frac{q}{2\pi e_0} \frac{2}{B'} \tag{4.66}$$

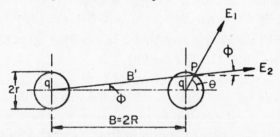

Fig. 4.24. Gradient distribution calculation on 2-conductor bundle.

The total horizontal and vertical components of E_1 and E_2 at P will be

$$E_h = E_1 \cos \theta + E_2 \cos \phi, \text{ and } E_v = E_1 \sin \theta + E_2 \sin \phi \tag{4.67}$$

The total field-intensity is $E_p = \sqrt{E_h^2 + E_v^2}$ $\qquad$ (4.68)

$$\text{Now, } B' = \sqrt{(B + r \cos \theta)^2 + (r \sin \theta)^2}, \text{ with } B = 2R, \left.\begin{array}{l} \\ \sin \phi = r \sin \theta/B' \text{ and } \cos \phi = (B + r \cos \theta)/B' \end{array}\right\} \tag{4.69}$$

$$\therefore \quad E_p^2 = \left(\frac{q}{2\pi e_0}\right)^2 \frac{1}{r^2} (1 + 8r^2/(B')^2 + 4rB \cos \theta/(B')^2) \tag{4.70}$$

Now $(B')^2 = B^2 [1 + (r^2 + 2Br \cos \theta)/B^2]^2$

and if $r \ll B$, $B' \approx B + r \cos \theta$ and $(B')^2 \approx B^2$

Then, $E_p^2 = \left(\frac{q}{2\pi e_0}\right)^2 \frac{1}{r^2} (1 + 8r^2/B^2 + 4r \cos \theta/B)$

$$\approx \left(\frac{q}{2\pi e_0}\right)^2 \frac{1}{r^2} (1 + 4r \cos \theta/B)$$

$$\therefore \quad E_p \approx \frac{q}{2\pi e_0 r} \left(1 + \frac{2r}{B} \cos \theta\right) = \frac{q}{2\pi e_0 r} \left(1 + \frac{r}{R} \cos \theta\right) \tag{4.71}$$

where, for $X \ll 1$, $(1 + X)^{1/2} \approx 1 + X/2$

For $N \geqslant 3$, the general expressions become very lengthy and it is best to write a programme for a digital computer. The procedure is illustrated here for a 6-conductor bundle whose dimensions are shown in Fig. 4.25 with all relevant parameters for the calculation.

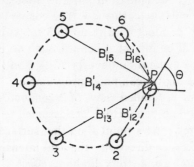

Fig. 4.25. Six-conductor bundle and gradient on sub-conductor.

We first evaluate the distances B'_{12} to B'_{16}.

$$
\left.
\begin{aligned}
B'_{12} &= \sqrt{[(R \sin 60° + r \sin \theta)^2 + (R \cos 60° + r \cos \theta)^2]} \\
B'_{13} &= \sqrt{[(\sqrt{3}R \sin 30° + r \sin \theta)^2 + (\sqrt{3}R \cos 30° + r \cos \theta)^2]} \\
B'_{14} &= \sqrt{[(r \sin \theta)^2 + (2R + r \cos \theta)^2]} \\
B'_{15} &= \sqrt{[(\sqrt{3}r \sin 30° - r \sin \theta)^2 + (\sqrt{3}R \cos 30° + r \cos \theta)^2]} \\
B'_{16} &= \sqrt{[(R \sin 60° - r \sin \theta)^2 + (R \cos 60° + r \cos \theta)^2]}
\end{aligned}
\right\} \quad (4.72)
$$

Next the horizontal and vertical components of the field intensity at P are evaluated. The factor $(q/2\pi e_0)$ is omitted in writing for the present but will be included at the end.

Conductor i ($i=1,2,3$ $4,5,6$)	Horizontal component $E_h(i)$	Vertical component $E_v(i)$
1.	$\dfrac{1}{r} \cos \theta$	$\dfrac{1}{r} \sin \theta$
2.	$2(R \cos 60° + r \cos \theta)/(B'_{12})^2$	$2(R \sin 60° + r \sin \theta)/(B'_{12})^2$
3.	$2(\sqrt{3}R \cos 30° + r \cos \theta)/(B'_{13})^2$	$2(\sqrt{3}R \sin 30° + r \sin \theta)/(B'_{13})^2$
4.	$2(2R + r \cos \theta)/(B'_{14})^2$	$2r \sin \theta/(B'_{14})^2$
5.	$2(\sqrt{3}R \cos 30° + r \cos \theta)/(B'_{15})^2$	$-2(\sqrt{3}R \sin 30° - r \sin \theta)/(B'_{15})^2$
6.	$2(R \cos 60° + r \cos \theta)/(B'_{16})^2$	$-2(R \sin 60° - r \sin \theta)/(B'_{16})^2$

Total field intensity

$$
E_P(\theta) = \frac{q}{2\pi e_0} [\{ \sum_{i=1}^{6} E_h(i)\}^2 + \{ \sum_{i=1}^{6} E_v(i)\}^2]^{1/2}
$$

Figure 4.26 shows examples of surface voltage gradient distributions on bundled conductors with $N = 2, 4, 6$ subconductors.

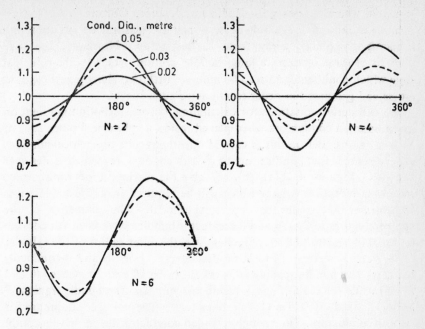

Fig. 4.26. Distribution of voltage gradient on bundle conductors for $N=2$, 4, 6. Variation with conductor diameter.

4.9 Design of Cylindrical Cages for Corona Experiments

The effects of high voltage-gradients on bundled conductors are evaluated all over the world by "cages". In the simplest of these arrangements a large metallic cylinder at ground or near ground potential forms the outer cage with the bundled conductor strung inside. The centres of the cylinder and the bundle are coincident while the subconductors themselves are displaced off-centre, except for $N = 1$, a single conductor. Several examples are shown in Fig. 4.27, in which a square cage is also included. When the dimensions of the outer cage become very large or where the length of conductor and weight is large with a resulting sag, a square cage made with mesh can be contoured to follow the sag. The cage arrangement requires lower voltage for creating the required surface voltage gradient on the conductors than in an overhead line above ground. Also artificial rain equipment can be used if necessary to obtain quick results. Measuring instruments are connected to ground both from the conductor at high voltage and the cage at near ground

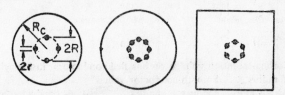

Fig. 4.27. Configuration of 'cages' used for corona studies.

potential where necessary for Radio Interference, Corona Loss, charge, etc., measurements. Audible noise is usually measured as radiation into a microphone placed away from the test setup. Normally, the cage consists of three sections, a long middle section which is the principal cage which could extend up to 60 metres, with two short guard cages at either end grounded in order to minimize edge effects.

We will derive equations for calculating corona-inception gradient on given bundled conductors when placed inside a cylindrical outer cage of a given radius. Before this is carried out, the results of such calculations are presented for bundles up to 8 sub-conductors which a designer can use. Figures 4.28(a), (b), (c) give the variation of r.m.s. value of corona-inception voltage which will be required from the single-phase transformer that excites the entire experiment. The diameters of the sub-conductors in 2, 4, 6 and 8 conductor bundles have been varied from 0.02 to 0.055 metre (0.8 to 2.2 inches), and the outer radius of the cylinder from 2 to 5 metres, (4 to 10 m diameters). For 2- and 4-conductor bundles, the bundle spacing B is taken to be 16 and 20 inches (40 and 50 cm) while for the 6-, and 8-conductor bundles the bundle radius R has been fixed at 0.6 m (1.2 m diameter). The voltages shown are for smooth conductors. On a rough stranded conductor the corona-inception occurs at a lower voltage and a suggested roughness factor is 1.4. Thus, the ordinates must be divided by this factor to yield the required corona-inception voltage in practical cases.

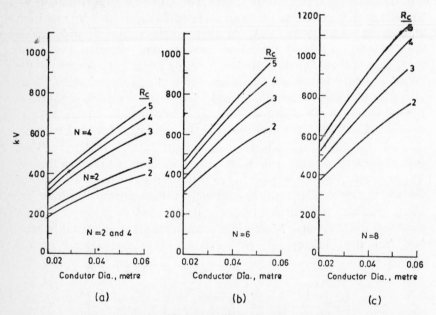

Fig. 4.28. Corona-inception voltage on bundled conductors inside-cylindrical cage of radius R_c. Smooth-conductor values.
(a) 2- and 4- conductor bundles.   (b) 6- conductor bundle.
(c) 8- conductor bundle.

4.9.1 Single Conductor Concentric with Cylinder

Figure 4.29 shows a coaxial cylindrical cage with outer cylinder radius R_c and inner conductor radius r. The problem is the same as a coaxial cable with air dielectric. At any radius X, the field strength is

$$E_X = \frac{q}{2\pi e_0} \frac{1}{X} \text{ and the voltage is } V = \frac{q}{2\pi e_0} \int_r^{R_c} dX/X = \frac{q}{2\pi e_0} \ln \frac{R_c}{r}$$

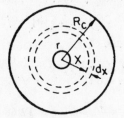

Fig. 4.29. Coaxial cylindrical arrangement of conductor and cage.

This gives the surface voltage gradient to be

$E_r = V/r \ln (R_c/r)$, which should equal the corona-inception gradient $E_0 = 21.92 \times 10^2 (1 + 0.0308/\sqrt{r})$ kV/m which is valid for a coaxial cylindrical electrode geometry.

∴ The required voltage at which corona-inception takes place on the smooth conductor is

$$V_0 = 2192 (1 + 0.0308/\sqrt{r}).r. \ln (R_c/r), \text{ kV} \qquad (4.74)$$

The gradient factor of the electrode arrangement is

$$g = E_r/V = 1/r \ln (R_c/r) \qquad (4.75)$$

The capacitance of the arrangement is

$$C = q/V = 2\pi e_0/\ln (R_c/r) \qquad (4.76)$$

and the resulting characteristic impedance is

$$Z_0 = \frac{1}{vC} = \frac{1}{2\pi e_0 v} \ln (R_c/r) = \frac{1}{2\pi} \sqrt{\frac{\mu_0}{e_0}} \ln \left(\frac{R_c}{r}\right)$$

$$= 60 \ln (R_c/r), \text{ ohms} \qquad (4.77)$$

where v = velocity of light = $1/\sqrt{\mu_0 e_0}$

$\mu_0 = 4\pi \times 10^{-7}$ Henry/m

and $\sqrt{\mu_0/e_0}$ = characteristic impedance of free space = 120π. The capacitance and surge or characteristic impedance of a cage system are important from the point of view of determining the charge-voltage relationships under corona discharge and for terminating the conductor suitably to prevent standing waves from disturbing the actual phenomena.

Example 4.12: A conductor 5 cm diameter is strung inside an outer cylinder of 2 metre radius. Find

(a) The corona-inception gradient on the conductor, kV/cm,

  (b)   The corona-inception voltage in kV, rms,

  (c)   The gradient factor for the electrode arrangement,

  (d)   The capacitance of the coaxial arrangement per metre, and

  (e)   the surge impedance.

Solution:

  (a) $E_0 = 21.92(1 + 0.0308/\sqrt{0.025}) = 26.19$ kV/cm, r.m.s.

  (b) $V_0 = E_0 r \ln(R_c/r) = 26.19 \times 2.5 \ln(200/2.5) = 26.19 \times 10.955$ kV

       $= 286.9$ kV, r.m.s.

  (c)   The gradient factor is $26.19/286.9 = 0.0913$ kV/cm per KV

     Also, $1/r \ln(R_c/r) = 0.0913$.   or V/cm per V

  (d) $C = \dfrac{2\pi e_0}{\ln(R_c/r)} = \dfrac{10^{-9}}{18 \times \ln(80)} = 12.68 \ \mu\mu F/m.$

  (e) $Z_0 = 60 \ln(R_c/r) = 263$ ohms

     Also, $Z_0 = 1/vc = 1/(3 \times 10^8 \times 12.68 \times 10^{-12}) = 263$ ohms

Such an arrangement with a single conductor is not of much use in extra high voltage investigations since the conductor is invariably a bundled conductor without exception. However, when the cage consists of shapes other than cylinders, or in a large hall of a high voltage laboratory, it is very important to have an idea of the gradient factor and an equivalent radius of the cage or hall. This is determined experimentally by stringing a smooth aluminium or copper tube of known diameter for which corona-inception gradient is known. A radio noise meter is used to determine the corona-inception and the voltage initiating the reading on the meter is accurately determined. This yields the gradient factor from which the equivalent radius R_c of the outer electrode can be easily calculated. The smooth tube must be preferably strung at the centre of the hall or where normal RIV (Radio Influence Voltage) measurements are carried out. In a square cage, the smooth tube is accurately centred. The equivalent radius can be used for measurements and interpretation of results obtained with bundled conductors. This will form the contents of further discussion.

4.9.2   Bundled Conductors Inside a Cylinder

When a bundle with N sub-conductors is centred inside a cylinder, the conductors themselves are eccentric with respect to the cylinder. We now examine the properties of the electric field when a conducting cylinder is placed off-centre inside a larger conducting cylinder. This problem parallels the sphere-gap problem discussed earlier.

4.9.2.1   *Single conductor with eccentricity*

Figure 4.30 shows a cylinder of radius R_c in which is located a conductor of radius r_c with its centre displaced by a distance R from the centre of

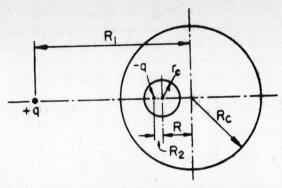

Fig. 4.30. Outer cylinder with eccentric conductors.

the outer cylinder. Both the cylindrical electrodes are equipotentials with a voltage V applied between them. In nearly all cases, the outer cylinder is grounded and its potential is zero. The two circles (in two dimensions) of radii r_c and R_c can be made equipotential lines under the influence of two equal but opposite charges $+q$ and $-q$ per unit length which are located as shown in Fig. 4.30. An important property of the field, which will be proved later in order not to interrupt continuity at this stage and which will be used for cage designs, is

$$\left\{\begin{array}{l}\text{Product of distances to} \\ \text{the two charges from the} \\ \text{centre of a given cylinder}\end{array}\right\} = \text{square of the cylinder radius} \quad (4.78)$$

Thus, $\qquad R_1 (R_2 + R) = R_c^2$
$$\left.\begin{array}{l}\end{array}\right\} \qquad\qquad (4.79)$$
and $\qquad R_2 (R_1 - R) = r_c^2$

Since we are given r_c, R_c, and R, the radii and the eccentricity, it is easy to determine R_1 and R_2 as follows:

Let $D = (R_c^2 + R^2 - r_c^2)/R$. Then,

$$R_1 = 0.5 (D + \sqrt{D^2 - R_c^2}) \approx R_c^2/R \text{ when } R_c^2 + R^2 \gg r_c^2,$$

and $\quad R_2 = r_c^2/(R_1 - R)$.

The voltage required in terms of the charge q is

$$V = \frac{q}{2\pi e_0} \ln \frac{(R_1 - R) (R_2 + R)}{R_c r_c} \qquad (4.80)$$

This will also be derived later. The maximum surface voltage gradient on the inner conductor is $(q/2\pi e_0 r_c)$, if we assume the other charge to be located very far. For corona-inception, this gradient should equal the corona-inception gradient

$$E_0 = 2192 (1 + 0.0308/\sqrt{r_c}), \text{ kV/m}$$

$$\therefore \quad V_0 = 2192(1 + 0.0308/\sqrt{r_c}).r_c.\ln \frac{(R_1 - R)(R_2 + R)}{R_c r_c} \qquad (4.81)$$

Note that when the inner conductor is concentric with the cylinder, $R = 0$ and $R_1 R_2 = R_c^2$. This reduces to equation (4.74) derived before. All quantities in equation (4.81) are known so that the corona-inception voltage for a smooth conductor can be determined. The capacitance of the eccentric conductor and cylinder is obviously

$$C = q/V = 2\pi e_0 / \ln \left[\frac{(R_1 - R)(R_2 + R)}{R_c\, r_c} \right], \text{ Farad/metre} \qquad (4.82)$$

The surge impedance is

$$Z_0 = 1/3 \times 10^8 C = 60 \ln \frac{(R_1 - R)(R_2 + R)}{R_c\, r_c} \qquad (4.83)$$

The above analysis becomes very cumbersome when there are more than one eccentric conductor. We will develop a very general method in the next section based on the ideas developed here for one eccentric conductor.

Example 4.13: A stranded conductor 3.5 cm diameter $(r_c = 0.0175$ m) is displaced by $R = 22.5$ cm from the centre of a cylinder 3.2 metres in diameter $(R_c = 1.6$ m). Taking a roughness factor $m = 1.4$, calculate

(a) the charge coefficient $q/2\pi e_0$ at corona inception,
(b) the location of charges q and $-q$,
(c) the corona-inception voltage V_0, and
(d) the capacitance per metre and surge impedance based on light velocity.

Solution: Because of surface roughness, corona-inception takes place at a gradient equal to $(1/1.4)$ times that on a smooth conductor.

$$E_{or} = \frac{21.92}{1.4} \ (1 + 0.0308/\sqrt{0.0175}) = 19.3 \text{ kV/cm} = 1930 \text{ kV/m}$$

(a) $q/2\pi e_0 = r_c \cdot E_{or} = 33.81$ kV

(b) $D = (R_c^2 + R^2 - r_c^2)/R = (160^2 + 22.5^2 - 1.75^2)/22.5 = 1161$ cm

$\quad R_1 = 0.5\,(D + \sqrt{D^2 - R_c^2}) = 1155$ cm $= 11.55$ m

$\quad R_2 = r_c^2/(R_1 - R) = 2.7 \times 10^{-3}$ cm $= 2.7 \times 10^{-5}$ m

Note that the negative charge is displaced from the centre of the inner conductor by only 2.7×10^{-3} cm while the positive charge is located very far (11.55 m) from the centre of the outer cylinder.

(c) Corona-inception voltage

$$V_0 = 19.32 \times 1.75 \ln [(1155 - 22.5)\ 22.5/(160 \times 1.75)]$$

$$= 33.81 \ln 91 = 33.81 \times 4.51 = 152.5 \text{ kV}.$$

(d) Capacitance $C = \dfrac{10^{-9}}{18}/\ln 91 = 12.316\ \mu\mu\text{F/m}$

Surge impedance $Z_0 = 1/Cv = 10^{12}/12.316 \times 3 \times 10^8 = 271$ ohms.

Example 4.14: Repeat the previous problem if the inner conductor is concentric with the outer cylinder.

Solution: (a) $q/2\pi e_0 = 33.81$ kV; (b) $R_1 = \infty$, $R_2 = 0$

(c) $V_0 = r_c.E_{or} \ln (R_c/r_c) = 152.6$ kV

(d) $C = 2\pi e_0/\ln (R_c/r_c) = 12.3 \ \mu\mu F/m$

$Z_0 = 10^{12}/12.3 \times 3 \times 10^8 = 271$ ohms.

The displacement of the conductor by 22.5 cm in a cylinder of radius 160 cm i.e. 14% eccentricity has not made a noticeable difference in the corona-inception voltage, etc., for the dimensions given (This is one conductor of a 2-conductor bundle with 45 cm bundle spacing placed inside a cylinder 10 ft in diameter).

We now give a proof for equations (4.78) and (4.79), which are also given in any standard text book on electrostatics (see Bewley or Zahn, 4 and 27 under Books in Bibliography). Figure 4.31 shows two line charges $+q$ and $-q$ coulomb/metre separated by a distance $2S$. At a point midway between them the potential is zero and we can choose this as the origin of a coordinate system. The potential of a point $P(x, y)$ with respect to the reference point 0 will be, according to equation (3.35)

$$V = \frac{q}{2\pi e_0} \ln \frac{\text{Distance from negative charge}}{\text{Distance from positive charge}} = \frac{q}{2\pi e_0} \ln \frac{(x-S)^2+y^2}{(x+S)^2+y^2} \quad (4.84)$$

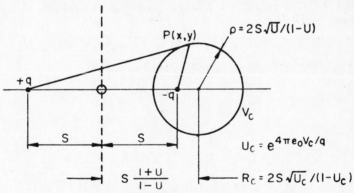

Fig. 4.31. Location of charges and an equipotential surface of radius R_c with potential V_c.

If we denote $U = \exp (4\pi e_0 \ V/q)$, a constant-potential line in the field of the two charges will satisfy the equation

$$(x - S)^2 + y^2 = U[(x + S)^2 + y^2]$$

or

$$\left(x - S \frac{1 + U}{1 - U}\right)^2 + y^2 = 4S^2U/(1-U)^2 \quad (4.85)$$

This is a circle with radius $\rho = |2S\sqrt{U}/(1-U)|$ and centre at $\left(S\frac{1 + U}{1 - U}, 0\right)$

on the line joining the two charges.

A circle of radius R_c has the potential given by

$$R_c^2 = 4S^2\, U_c/(1-U_c)^2 = \frac{2SU_c}{1-U_c}\frac{2S}{1-U_c} \tag{4.86}$$

Now, referring to Fig. 4.31 and equation (4.86), we observe that the distances of the two charges from the centre of the circle (a cylinder in 3 dimensions) are

$$\rho_1 = S\frac{1+U_c}{1-U_c} - S = \frac{2SU_c}{1-U_c}, \text{ and } \rho_2 = S\frac{1+U_c}{1-U_c} + S = \frac{2S}{1-U_c}$$

Their product is $\rho_1\rho_2 = \dfrac{4S^2 U_c}{(1-U_c)^2}$ which is R_c^2 according to equation (4.86). Therefore equation (4.78) results.

In order to derive equation (4.80) we utilize the following relations. Let the potential of outer cylinder R_c be V_c, and that of the inner conductor r_c be v_c. Also let, $V_c - v_c = V$, the p.d. Let

$$U_c = \exp(4\pi e_0\, V_c/q) \text{ and } u_c = \exp(4\pi e_0\, v_c/q)$$

$$\therefore\ \frac{U_c}{u_c} = \exp(4\pi e_0\, V/q) \text{ so that } \frac{q}{2\pi e_0} = \frac{V}{\ln\sqrt{U_c/u_c}} \tag{4.87}$$

We note that since the outer cylinder is closer to the positive charge, its potential V is more positive than v_c. Now, the following relations are valid from Fig. 4.31:

$$\left.\begin{array}{l} 2S = R_1 - R_2 - R,\ R_1(R_2+R) = R_c^2,\ R_2(R_1-R) = r_c^2, \\[4pt] R_c^2 = 4S^2 U_c/(1-U_c)^2, \text{ and } r_c^2 = 4S^2 u_c/(1-u_c)^2 \end{array}\right\} \tag{4.88}$$

Let F

$$= U_c/(1-U_c)^2. \text{ Then } F = R_c^2/4S^2 = R_1(R_2+R)/(R_1-R_2-R)^2$$

$$\therefore\ U_c^2 - (2+1/F)U_c + 1 = 0 \text{ giving}$$

$$U_c = 0.5[(2+1/F) - \sqrt{(2+1/F)^2 - 4}]$$

Similarly, if

$$F_0 = u_c/(1-u_c)^2 = r_c^2/4S^2 = R_2(R_1-R)/(R_1-R_2-R)^2.$$

$$\therefore\ u_c = 0.5[(2+1/F_0) - \sqrt{(2+1/F_0)^2 - 4}]$$

Now, $\ 2 + 1/F = 2 + \dfrac{(R_1-R_2-R)^2}{R_1(R_2+R)} = \dfrac{R_1}{R_2+R} + \dfrac{R_2+R}{R_1}$

and $\ \sqrt{(2+1/F)^2 - 4} = \dfrac{R_1}{R_2+R} - \dfrac{R_2+R}{R_1}$

$$\therefore\ U_c = 0.5[(2+1/F) - \sqrt{(2+1/F)^2 - 4}] = (R_2+R)/R_1$$

Similarly, $\quad 2+1/F_0 = \dfrac{R_1-R}{R_2} + \dfrac{R_2}{R_1-R}$

and $\quad \sqrt{(2 + 1/F_0)^2 - 4} = \dfrac{R_1 - R}{R_2} - \dfrac{R_2}{R_1 - R}$

$$u_c = 0.5[(2 + 1/F_0) - \sqrt{(2 + 1/F_0)^2 - 4}] = R_2/(R_1 - R)$$

Consequently, $\qquad U_c/u_c = \dfrac{(R_2 + R)(R_1 - R)}{R_1 R_2}$

But, $R_1 R_2 (R_2 + R)(R_1 - R) = R_c^2\, r_c^2$ from (4.88)

Or, $\qquad\qquad R_1 R_2 = R_c^2\, r_c^2/(R_2 + R)(R_1 - R).$

$\therefore \quad \sqrt{U_c/u_c} = (R_2 + R)(R_1 - R)/R_c r_c \qquad (4.89)$

Consequently, from equations (4.89) and (4.87), the potential difference between the two cylinders is

$$V = \frac{q}{2\pi e_0} \ln \sqrt{U_c/u_c} = \frac{q}{2\pi e_0} \ln \frac{(R_1 - R)(R_2 + R)}{R_c r_c} \qquad (4.80)$$

4.9.2.2 Multi-conductors with eccentricity inside cylinder

The corona-inception voltage of a single conductor of radius r_c placed off-centre inside a cylinder of radius R_c was derived before, Fig. 4.30. In order to handle more conductors, we give a different method. This uses a circle called the "Circle of Inversion" whose radius is $2R_c$ and touches the cylinder at one point.

Consider Fig. 4.32 in which is shown a circle of radius R_c with centre at O, and another with radius $2R_c$ and centre O' with both circles touching at O''. A straight line FF is drawn through O'' tangent to both circles. Taking a point P on circle R_c, extend O' to P to cut FF in P'. Then,

$$O'P \times O'P' = (2R_c)^2 = 4R_c^2 \qquad (4.90)$$

which is proved as follows: Let $\angle O''O'P = \phi$. Then

$$\angle O'PO'' = O'O''P' = 90°,\ O'P = 2R_c \cos \phi,\ O'P' = 2R_c/\cos \phi$$

giving $O'P \times O'P' = 4R_c^2$.

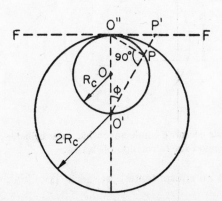

Fig. 4.32. Circle of inversion and outer cylinder of cage arrangement.

This is true of all points such as P on the circle of radius R_c. Thus, the straight line FF is the locus of all points on the circle R_c about the circle of inversion with radius $2R_c$ and centre at O'. If the circle is the outer cylinder of a "cage" arrangement, then it has been transformed into a straight line and every point on it is the inverse point on the circle. Any other small cylinder of radius r_c placed with an eccentricity R can also be transformed into a cylinder about the circle of inversion. Then Maxwell's Potential coefficients can be utilized to formulate the charge equation in terms of the potential of the conductor. This can be done for all the conductors of the bundle through matrix $[P]$. This is carried out as follows: For the point P_0, the centre of conductor, an inverse point P_0' can be found about the circle of inversion from the relation $O'P_0' = 4R_c^2/O'P_0 = 4R_c^2/(R_c + R)$, Fig. 4.33. If the charge of the inner conductor is located at its centre P_0, then the imaged charge about the circle of inversion will be located at P_0'. Its distance from the flat surface FF will be

$$H = O'P_0' - 2R_c = 2R_c(R_c - R)/(R_c + R) \qquad (4.91)$$

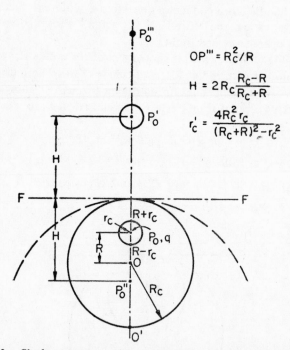

$$OP''' = R_C^2/R$$

$$H = 2R_C \frac{R_C - R}{R_C + R}$$

$$r_C' = \frac{4R_C^2 r_C}{(R_C + R)^2 - r_C^2}$$

Fig. 4.33. Single eccentric conductor inside outer cylinder and circle of inversion for calculation of images.

Since the surface is an equipotential at zero potential, the potential of the conductor is

$$V = (q/2\pi e_0) \ln (2H/r_c') \qquad (4.92)$$

where $r_c' = $ radius of the image of r_c with centre at P_0'. This is determined as follows:

Choosing two points on a diameter of r_c at distances $(R - r_c)$ and $(R + R_c)$, their inverse points are located with respect to O' at distances $4R_c^2/(R_c + R - r_c)$ and $4R_c^2/(R_c + R + r_c)$. Therefore,

$$r_c' = \tfrac{1}{2} \cdot 4R_c^2[1/(R_c + R - r_c) - 1/(R_c + R + r_c)]$$

$$= r_c \cdot 4R_c^2 /[(R_c + R)^2 - r_c^2] \tag{4.93}$$

$$\therefore \quad V = \frac{q}{2\pi e_0} \ln \frac{(R_c + R)^2 - r_c^2}{R_c r_c} \frac{R_c - R}{R_c + R} \tag{4.94}$$

In practice, $R_c + R \gg r_c$ so that

$$V = \frac{q}{2\pi e_0} \ln \frac{R_c^2 - R^2}{R_c r_c} \tag{4.95}$$

Having located the charge q at P_0', the image charge with respect to FF is situated at P_0'' at a depth H whose distance from O' is

$$O'P_0'' = 2R_c - H = 2R_c - \frac{2R_c(R_c - R)}{R_c + R} = \frac{4R_c R}{R_c + R} \tag{4.96}$$

$\therefore$ The inverse point of P_0'' is located at

$$O'P_0''' = 4R_c^2/O'P_0'' = R_c(R_c + R)/R \tag{4.97}$$

Furthermore, $OP_0''' = O'P_0''' - R_c = R_c^2/R \tag{4.98}$

There are now two charges $+q$ and $-q$ located at P_0 and P_0''' at the points (O, R) and $(O, R_c^2/R)$ in whose field the outer cylinder of radius R_c and inner conductor of radius r_c are equipotential surfaces. The maximum surface voltage gradient on the surface of the inner conductor is

$$E = (q/2\pi e_0) [1/r_c + 2R/(R_c^2 - R^2)] \tag{4.99}$$

since the distance of P_0''' from the conductor centre is

$$OP''' - R = (R_c^2 - R^2)/R.$$

But in terms of potential, equation (4.95), $\dfrac{q}{2\pi e_0} = V/\ln \dfrac{R_c^2 - R^2}{R_c r_c}$

The relation between voltage and surface voltage gradient becomes

$$V = E. \left(\ln \frac{R_c^2 - R^2}{R_c r_c} \right) \Big/ [1/r_c + 2R/(R_c^2 - R^2)] \tag{4.100}$$

For corona-inception to occur, $E = E_{or} = \dfrac{21.92}{m} (1 + 0.0308/\sqrt{r_c})$ where $m = $ roughness factor.

Example 4.15: Repeat Example 4.13 using the modified method of images, equation (4.100).

Solution: $r_c = 0.0175$ m, $R_c = 1.6$ m, $R = 0.225$ m.

$$E_{or} = 19.3 \text{ kV/cm} = 1930 \text{ kV/m}.$$

$$(R_c^2 - R^2)/R_c r_c = 90, \, 2Rr_c/(R_c^2 - R^2) = 1/318.7$$

which may be neglected in comparison to 1.

$$\therefore \quad V_0 = 1930 \times 0.0175 \times \ln 90 = 152 \text{ kV}$$

In Example 4.13, the value obtained was $V_0 = 152.5$ kV.

From this calculation we observe that the effect of image charge located at P_0''' is practically nil on the surface voltage gradient of the conductor ($1/318.7 = 0.3\%$). Hence its presence will be neglected in all further calculations.

N-conductor Bundle

For several sub-conductors located off-centre inside the outer cylinder with eccentricity R from O, we must obtain the location of images of all sub-conductors about the circle of inversion so that Maxwell's Potential Coefficient Matrix will yield the relation between voltage and charge. The computation is lengthy and must be performed step by step. We shall first develop a programme suitable for a digital computer commencing with a 2-conductor bundle and extend it to a general case. Figures 4.28 (a), (b), (c) were obtained according to this method and a designer can use them.

In Fig. 4.34, the presence of an image charge is indicated for the 2-conductor bundle, but as mentioned before this can be omitted in all cases. The quantities to be calculated are listed in a tabular form and will be the sequence of statements in a digital computer programme.

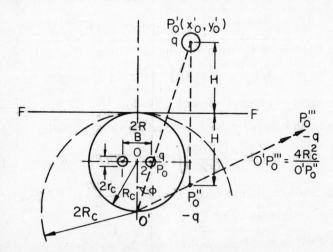

Fig. 4.34.   2-conductor bundle inside outer cylinder and its image conductors about the circle of inversion

Quantity Indicated	Conductor No. 1	Conductor No. 2

(1) *Coordinates of Conductor*

X_0	$A_1 = -R$	$B_1 = R$
Y_0	$A_2 = 0$	$B_2 = 0$
$O'P_0$	$A_3 = \sqrt{R_c^2 + R^2}$	$B_3 = A_3$

(2) *Inverse*

$O'P_0'$	$R_1 = 4R_c^2/A_3$	$R_2 = R_1$
$\sin \phi$	$S_1 = A_1/A_3$	$S_2 = B_1/B_3$
$\cos \phi$	$S_3 = R_c/A_3$	$S_4 = R_c/B_3$
x_0'	$A_4 = R_1 . S_1$	$B_4 = R_2 . S_2$
y_0'	$A_5 = R_1 S_3 - R_c$	$B_5 = R_2 S_4 - R_c$

(3) *Image about FF*

x_0''	A_4	B_4
y_0''	$A_6 = 2R_c - A_5$	$B_6 = 2R_c - B_5$
$O'P_0''$	$R_3 = \sqrt{A_4^2 + (A_6 + R_c)^2}$	$R_4 = \sqrt{B_4^2 + (B_6 + R_c)^2}$

(4) *Inverse of P_0''*

$O'P_0'''$	$A_7 = 4R_c^2/R_3$	$B_7 = 4R_c^2/R_4$
$\sin \theta'''$	$T_1 = (A_6 + R_c)/R_3$	$T_2 = (B_6 + R_c)/R_4$
$\cos \theta'''$	$T_3 = A_4/R_3$	$T_4 = B_4/R_4$
x_0'''	$A_8 = A_7 T_3$	$B_8 = B_7 T_4$
y_0'''	$A_9 = A_7 T_1 - R_c$	$B_9 = B_7 T_2 - R_c$
$H = y_0' - R_c$	$H_1 = A_5 - R_c$	$H_2 = H_1$
r_c'	$G_1 = 4R_c^2 r_c/(A_3^2 - r_c^2)$	$G_2 = G_1$

(5) *Potential coefficients and charges*

$$P(1, 1) = P(2, 2) = \ln (2H_1/G_1),$$

$$P(1, 2) = P(2, 1) = 0.5 \ln \frac{(B_2 + A_5 - 2R_c)^2 + (B_4 - A_4)^2}{(B_5 - A_5)^2 + (B_4 - A_4)^2}.$$

$$V = \frac{q}{2\pi e_0} [P(1, 1) + P(1, 2)] = \frac{q}{2\pi e_0} L, \; L = P(1, 1) + P(1, 2).$$

For unit voltage, $\dfrac{q}{2\pi e_0} = 1/L$.

(6) *Potential Gradient*

Due to positive charge on conductors, the unit gradient is

$$K_1 = (q/2\pi e_0)\,(1/r_c + 1/R).$$

Due to the image charge at P_0''' the gradient is

$$K_2 = -(q/2\pi e_0)\left[\frac{A_9 - A_2 + r_c}{(A_9 - A_2 + r_c)^2 + (A_8 - A_1)^2} + \frac{B_9 - A_2 + r_c}{(B_9 - A_2 + r_c)^2 + (B_8 - A_1)^2}\right]$$

The total gradient for unit applied voltage, the gradient factor is

$g = K_1 + 2K_2$ by taking $q/2\pi e_0 = 1/L = 1/[P(1,1) + P(1,2)]$.

The corona-inception gradient is

$$E_0 = \frac{2919}{m}\,(1 + 0.0308/\sqrt{r_c}),\ \text{kV/m.}$$

Then, the corona-inception voltage becomes finally,

$$V_0 = E_0/g.$$

Figure 4.28(a) shows results of corona-inception voltage on a 2-conductor bundle as the inner conductor radius is varied, and for several values of outer-cylinder radius R_c. These are for smooth conductors for which $m = 1$.

For 4-, 6-, and 8-conductor bundles, the following steps are used.

Computer Programme (for any number of sub-conductors N)

(1) *Inputs.* R_c = radius of outer cylinder

$\qquad\qquad r_c$ = radius of sub-conductor of bundle

$\qquad\qquad R$ = bundle radius = eccentricity

$\qquad\qquad N$ = number of sub-conductors of bundle.

(2) *Coordinates of conductors*

x — coordinates: $A_1(I)$ $\left.\right\}$ $I = 1, 2, \ldots, N$
y — coordinates: $A_2(I)$

$O'P_0 \qquad\qquad : A_3(I) = \sqrt{A_1^2(I) + (A_2(I) - R_c)^2}$

(3) *Inverses about the circle of inversion*

$O'P_0' : A_4(I) = 4R_c^2/A_3(I)$

$\sin\phi : S_1(I) = A_1(I)/A_3(I)$

$\cos\phi : S_2(I) = (A_2(I) + R_c)/A_3(I)$

$x_0' \quad : A_5(I) = S_1(I).A_4(I)$

$y_0' \quad : A_6(I) = S_2(I).A_4(I) - R_c$

$H \quad : H(I) = A_6(I) - R_c$

$r_c' \quad : R_1(I) = 4R_c^2 r_c/(A_3^2(I) - r_c^2)$

(4) *Image of P_0' about FF*

$$x_0'' = x_0': A_5(I)$$

$$y_0'' = \quad : A_7(I) = 2R_c - A_6(I)$$

$$O'P_0'' \quad : A_8(I) = \sqrt{A_5^2(I) + (A_7(I) + R_c)^2}$$

(5) *Potential Coefficient Matrix*

Self coefficient: $P(I, I) = \ln(2H(I)/R_1(I))$

$I \neq J \begin{cases} \text{Image distance:} \ B(I, J) = \sqrt{(H(I) + H(J))^2 + (A_1(I) - A_1(J))^2} \\ \text{Aerial distance:} \ F(I, J) = \sqrt{(H(I) - H(J))^2 + (A_1(I) - A_1(J))^2} \end{cases}$

Mutual coefficient: $P(I, J) = \text{Ln}(B(I, J)/F(I, J))$

(6) *Charge Coefficient for Unit Voltage*

Capacitance: $[C] = $ **Inverse of** $[P] \times 2\pi e_0$

Charge   : $[Q] = [C][V]$

where $[V] = $ column matrix with all elements equal to 1.

(7) *Surface Voltage Gradient*

(a) **Due to conductor charges:**

$$g(I) = Q(I)/r_c + \sum_{J \neq I}^{N} Q(J)/R$$

Average maximum gradient for unit voltage

$$g = \frac{1}{N} \sum_{I=1}^{N} g(I)$$

(b) *Due to image charges.* Surface voltage gradient is neglected.

(8) *Corona-inception Voltage.* $\delta = $ air-density factor.

$$E_0 = \frac{2193}{m}(1 + 0.0308/\sqrt{r_c}), \ \text{kV/m, r.m.s., for } \delta = 1.$$

$\therefore$ corona-inception voltage $V_0 = E_0/g$, kV, r.m.s.

Results of calculation are shown in Figs. 4.28(a), (b), (c) for smooth conductors. The designer can use a roughness factor for stranded conductors. A suggested value is $m = 1.4$ so that the voltages in Figs. 4.28 must be divided by 1.4.

Review Questions and Problems

1. A sphere gap with the spheres having radii $R = 0.5$ m has a gap of 0.5 m between their surfaces.

   (a) Calculate the required charges and their locations to make the potentials 100 and 0.

(b) Then calculate the voltage gradient on the surface of the high-voltage sphere.

(c) If the partial breakdown of air occurs at 30 kV/cm peak, calculate the disruptive voltage between the spheres.

2. A 735-kV line has $N = 4$, $r = 0.0176$ m, $B = 0.4572$ m for the bundled conductor of each phase. The line height and phase spacing in horizontal configuration are $H = 15$ m, $S = 15$ m. Calculate the maximum surface voltage gradients on the centre phase and outer phases using Mangoldt formula. Compare them with values given in table in Section 4.6.

3. The corona-inception gradient for a smooth conductor by Peek's

Formula is $E_{0s} = \dfrac{30\delta}{\sqrt{2}}\,(1 + .0301/\sqrt{r\delta})$, kV/cm, r.m.s. For a stranded conductor $E_{0r} = E_{0s}/1.4$. Take the air-density factor $\delta = 1$. Then find the % difference between E_{0r} and the gradients calculated in problem 2. Give your conclusions.

4. A transmission line traverses hills and plains in Northern India. Calculate E_{0s} for a 400-kV line with conductor radius $r = 0.0159$ m for the following sets of elevation in metres and temperature in °C. (i) 0, 25°, (iii) 1000, 15°, (iv) 2000, 5°, (v) 3000, 0°C. Use $\delta = b_m(273 + t_0)/1013\,(273 + t)$, where $b_m =$ barometric pressure in millibars. Assume $b_m = 1013$ at sea level and to drop by 10 millibars for every 100-metre increase in elevation. $t_0 = 20$°C.

Discuss which gives the lowest value for E_{0s} and how a line should be designed for such variations.

5. If corona-inception gradient is measured in a h.v. testing laboratory at an elevation of 1000 metres and 25°C, give correction factors to be used when the equipment is used at (a) sea level at 35°C, and (b) 2000 m elevation at 15°C. Use conductor radius $= r$ metre.

6. The outer cylinder of a cage is 2 metres in diameter. A single conductor 4 cm in diameter is strung concentric with it and the arrangement is 30 m long. Calculate (a) the surface voltage gradient (gradient factor) on the conductor, (b) the capacitance of the arrangement, and (c) the surge impedance.

7. If in problem 6 the inner conductor is a 2-conductor bundle with each sub-conductor 4 cm in diameter at a bundle spacing of 40 cm, repeat the three parts. Note—Calculate Maxwell's Potential coefficients about the flat surface obtained from the circle of inversion.

CHAPTER 5

Corona Effects-I : Power Loss and Audible Noise

5.1 I²R Loss and Corona Loss

In chapter 2, the average power-handling capacity of a 3-phase e.h.v. line and percentage loss due to I^2R heating were discussed. Representative values are given below for comparison purposes.

System kV	400		750		1000		1150	
Line Length, Km	400	800	400	800	400	800	400	800
3-Phase MW/circuit $(P = 0.5V^2/xL)$	640	320	2860	1430	6000	3000	8640	4320
% Power Loss = 50 r/x	4.98%		2.4%		0.8%		0.6%	
KW/Km Loss, 3-phase	80	20	170	42.5	120	30	130	32.5

When compared to the I^2R heating loss, the average corona losses on several lines from 345 kV to 750 kV gave 1 to 20 Kw/Km in fair weather, the higher values referring to higher voltages. In foul-weather, the losses can go up to 300 Kw/Km. Since, however, rain does not fall all through the year (an average is 3 months of precipitation in any given locality) and precipitation does not cover the entire line length, the corona loss in Kw/Km cannot be compared to I^2R loss directly. A reasonable estimate is the yearly average loss which amounts to roughly 2 Kw/Km to 10 Kw/Km for 400 Km lines, and 20-40 Kw/Km for 800 Km range since usually higher voltages are necessary for the longer lines. Therefore, cumulative annual average corona loss amounts only to 10% of I^2R loss, on the assumption of continuous full load carried. With load factors of 60 to 70%, the corona loss will be a slightly higher percentage. Nonetheless, during rainy months, the generating station has to supply the heavy corona loss and in some cases it has been the experience that generating stations have been unable to supply full rated load to the transmission line. Thus, corona loss is a very serious aspect to be considered in line design.

When a line is energized and no corona is present, the current is a pure sine wave and capacitive. It leads the voltage by 90°, as shown in Fig. 5.1(a). However, when corona is present, it calls for a loss component and a typical waveform of the total current is as shown in Fig. 5.1(b).

When the two components are separated, the resulting inphase component has a waveform which is not purely sinusoidal, Fig. 5.1(c). It is still a current at power frequency, but only the fundamental component of this distorted current can result in power loss.

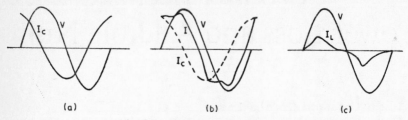

(a)　　　　　　　　　　(b)　　　　　　　　　　(c)

Fig. 5.1. Corona current waveform.

The mechanism of corona generation and its properties have been very extensively investigated and the reader is referred to the bibliography at the end of the book. Of vital importance is the generation of pulses which causes interference to radio, carrier communication, and gives rise to TV interference. These aspects will be discussed in the next chapter. In this chapter, engineering aspects of corona loss and audible noise will be described and data useful for design of lines based on these two phenomena will be discussed.

5.2 Corona-Loss Formulas

5.2.1 List of Formulas

Corona-loss formulas were initiated by F.W. Peek Jr., in 1911 derived empirically from most difficult and painstaking experimental work. Since then a horde of formulas have been derived by others, both from experiments and theoretical analysis. They all yield the power loss as a function of (a) the corona-inception voltage, V_0; (b) the actual voltage of conductor, V; (c) the excess voltage $(V - V_0)$ above V_0; (d) conductor surface voltage gradient, E; (e) corona-inception gradient, E_0; (f) frequency, f; (g) conductor size, d, and number of conductors in bundle, N, as well as line configuration; (h) atmospheric conditions, chiefly rate of rainfall, ρ, and (i) conductor surface condition.

The available formulas can be classified as follows: (see Bewley and *EHV* Reference Books in Bibliography)

A. **Those based on voltages**

(i) *Linear relationship*: Skilling's formula 1931.

$$P_c \propto V - V_0 \tag{5.1}$$

(ii) *Quadratic relationship*
  (a) Peek's formula (1911):

$$P_c \propto (V - V_0)^2 \tag{5.2}$$

(b)   Ryan and Henline formula (1924):

$$P_c \propto V(V - V_0) \tag{5.3}$$

(c)   Peterson's formula (1933):

$$P_c \propto V^2 . F(V/V_0) \tag{5.4}$$

where F is an experimental factor.

(iii) *Cubic Relationship*

(a)   Foust and Menger formula (1928):

$$P_c \propto V^3, (V > V_0) \tag{5.5}$$

(b)   Prinz's formula (1940):

$$P_c \propto V^2 (V - V_0) \tag{5.6}$$

B.   Those based on voltage gradients

(a)   Nigol and Cassan formula (1961):

$$P_c \propto E^2 \ln (E/E_0) \tag{5.7}$$

(b)   Project *EHV* formula (1966):

$$P_c \propto V.E^m, m = 5 \tag{5.8}$$

In order to obtain corona-loss figures from e.h.v. conductor configurations, outdoor experimental projects are established in countries where such lines will be strung. The resulting measured values pertain to individual cases which depend on local climatic conditions existing at the projects. It is therefore difficult to make a general statement concerning which formula or loss figures fit corona losses universally. In addition to equations (5.1) to (5.8), the reader is referred to the work carried out in Germany at the Rheinau 400 kV Research Project, in France at the Les Renardières Laboratory, in Russia published in the CIGRE Proceedings from 1956-1966, in Japan at the CRIEPI, in Sweden at Uppsala, and in Canada by the IREQ and Ontario Hydro.

We will here quote some formulas useful for evaluating 3-phase corona loss in Kw/Km, which are particularly adopted for e.h.v. lines, and some which are classic but cannot be used for e.h.v. lines since they apply only to single conductors and not to bundles. There is no convincing evidence that the total corona loss of a bundled conductor with N conductors is N times that of a single conductor.

(1)   *Nigol and Cassan Formula* (Ontario Hydro, Canada).

$$P_c = K.f.r^3.\theta.E^2.\ln (E/E_0) \tag{5.9}$$

where f = frequency in Hz, r = conductor radius in cm.,

θ = angular portion in radians of conductor surface where the voltage gradient exceeds the critical corona-inception gradient.

E = effective surface gradient at operating voltage V, kV/cm, r.m.s.

E_0 = corona-inception gradient for given weather and conductor surface condition, kV/cm, r.m.s.

and K = a constant which depends upon weather and conductor surface condition.

Many factors are not taken into account in this formula such as the number of sub-conductors in bundle, etc.

(2) *Anderson, Baretsky, McCarthy Formula* (Project EHV, USA)
An equation for corona loss in rain giving the excess loss above the fair-weather loss in Kw/3-phase Km is:

$$P_c = P_{FW} + 0.3606 \ K.V.r^2 . \ln (1 + 10\rho) . \sum_1^{3N} E^5 \qquad (5.10)$$

Here, P_{FW} = total fair-weather loss in Kw/Km,

= 1 to 5 Kw/Km for 500 kV, and 3 to 20 Kw/Km for 700 kV,

K = 5.35×10^{-10} for 500 to 700 kV lines,

= 7.04×10^{-10} for 400 kV lines (based on Rheinau results),

V = conductor voltage in kV, line-line, r.m.s.,

E = surface voltage gradient on the underside of the conductor, kV/cm, peak,

ρ = rain rate in mm/hour,

r = conductor radius in cm,

and N = number of conductors in bundle of each phase [The factor $0.3606 = 1/1.609 \ \sqrt{3}$].

Formulas are not available for (a) snow, and (b) hoar frost which are typical of Canadian and Russian latitudes. The EHV Project suggests $K = 1.27 \times 10^{-9}$ for snow, but this is a highly variable weather condition ranging from heavy to light snow. Also, the conductor temperature governs in a large measure the condition immediately local to it and it will be vastly different from ground-level observation of snow.

The value of ρ to convert snowfall into equivalent rainfall rate is given as follows:

Heavy snow : $\rho = 10\%$ of snowfall rate;

Medium snow : $\rho = 2.5\%$ of snowfall rate;

Light snow : $\rho = 0.5\%$ of snowfall rate.

The chief disadvantage in using formulas based upon voltage gradients is the lack of difinition by the authors of the formulas regarding the type of gradient to be used. As pointed out in Chapter 4, there are several types of voltage gradients on conductor surfaces in a bundle, such as nominal smooth-conductor gradient present on a conductor of the same outer radius as the line conductor but with a smooth surface, or the gradient with surface roughness taken into account, or the average gradient, or the average maximum gradient, and so on. It is therefore evident that for Indian conditions, an outdoor e.h.v. project is the only way of obtaining meaningful formulas or corona-loss figures applicable to local conditions.

Later on in Section 5.3, we will derive a formula based upon charge-voltage relations during the presence of a corona discharge.

5.2.2 The Corona Current

The corona loss P_c is expressed as

$P_c = 3 \times$ line-to-ground voltage $\times$ in-phase component of current.

From the previously mentioned expressions for P_c, we observe that different investigators have different formulas for the corona current. But in reality the current is generated by the movement of charge carriers inside the envelope of partial discharge around the conductor. It should therefore be very surprising to a discerning reader that the basic mechanism, being the same all over the world, has not been unified into one formula for this phenomenon. We can observe the expressions for current according to different investigators below.

1. *Peek's Law.* F.W. Peek, Jr., was the forerunner in setting an example for others to follow by giving an empirical formula relating the loss in watts per unit length of conductor with nearly all variables affecting the loss. For a conductor of radius r at a height H above ground,

$$P_c = 5.16 \times 10^{-3} f \sqrt{r/2H} \, V^2 (1 - V_0/V)^2, \text{ Kw/Km} \qquad (5.11)$$

where V, V_0 are in kV, r.m.s., and r and H are in metres. The voltage gradients are at an air density of δ,

$$E = V/r \ln (2H/r), \text{ and } E_0 = 21.4 \, \delta(1 + 0.0301/\sqrt{r\delta}) \qquad (5.12)$$

Example 5.1: For $r = 1$ cm, $H = 5$ m, $f = 50$ Hz, calculate corona loss P_c according to Peek's formula when $E = 1.1 E_0$, and $\delta = 1$.

Solution: $E_0 = 21.4(1 + 0.0301/\sqrt{0.01}) = 27.84$ kV/cm, r.m.s.

$\therefore \quad E = 1.1 E_0 = 30.624$ kV/cm. $\sqrt{r/2H} = 0.0316$

$V = 30.624 \ln (10/0.01) = 211.4$ kV (line-to-line voltage

$= 211.4 \sqrt{3} = 366$ kV)

$P_c = 5.16 \times 10^{-3} \times 50 \times 0.0316 \times 211.4^2 (1 - 1/1.1)^2 = 2.954$ Kw/Km

$\cong 3$ Kw/Km.

The expression for the corona-loss current is

$$i_c = P_c/V = 5.16 \times 10^{-3} f \sqrt{r/2H} \, V(1 - V_0/V)^2, \text{ Amp/Km} \qquad (5.13)$$

For this example, $i_c = 3000$ watts/211.4 kV $= 14$ mA/Km.

2. Ryan-Henline Formula

$$P_c = 4fCV(V - V_0) \cdot 10^6, \text{ Kw/Km} \qquad (5.14)$$

Here C = capacitance of conductor to ground, Farad/m

$$= 2\pi e_0 / \ln(2H/r)$$

and V, V_0 are in kV, r.m.s.

We can observe that the quantity (CV) is the charge of conductor per unit length. The corona-loss current is

$$i_c = 4fC(V - V_0) \cdot 10^6, \text{ Amp/Km} \qquad (5.15)$$

Example 5.2: For the previous example 5.1, compute the corona loss P_c and current i_c using Ryan-Henline formulas, equations (5.14) and (5.15).

Solution: $P_c = 4 \times 50 \times 211.4^2 (1 - 1/1 \cdot 1) \times 10^{-3}/18 \ln(1000)$

$$= 6.47 \text{ Kw/Km}$$

$$i_c = 6.47/211.4 = 3.06 \times 10^{-2} \text{ Amp/Km} = 30.6 \text{ mA/Km}$$

3. Project EHV Formula. Equation (5.10).

Example 5.3: The following data for a 750 kV line are given. Calculate the corona loss per kilometre and the corona loss current.

Rate of rainfall $\rho = 5$ mm/hr. $K = 5.35 \times 10^{-10}$, $P_{FW} = 5$ Kw/Km
$V = 750$ kV, line-to-line. $H = 18$ m, $S = 15$ m phase spacing
$N = 4$ sub-conductors each of $r = 0.0175$ m with bundle spacing

$$B = 0.4572 \text{ m}.$$

(Bundle radius $R = B/\sqrt{2} = 0.3182$ m). Use surface voltage gradient on centre phase for calculation.

Solution: From Mangoldt Formula, the gradient on the centre phase conductor will be

$$E = V \cdot [1 + (N - 1)r/R]/[N.r. \text{ Ln } \{2H/r_{eq} \sqrt{(2H/S)^2 + 1}\}]$$

where $r_{eq} = R(N.r/R)^{1/N}$. Using the values given,

$$E = 18.1 \text{ kV/cm, r.m.s.,} = 18.1\sqrt{2} \text{ peak} = 25.456$$

$$P_c = 5 + 0.3606 \times 5.35 \times 10^{-10} \times 750 \times 0.175^2 \times \ln(1 + 50)$$
$$\times 12 \times (25.456)^5$$

$$= 5 + 229 = 234 \text{ Kw/Km, 3-phase}$$

The current is

$$i_c = P_c/\sqrt{3}V = 234/750\sqrt{3} = 0.54 \text{ A/Km}.$$

Note that the increase in loss under rain is nearly 46 times that under fair weather. Comparing with Table I, the corona loss is much higher than the I^2R heating loss.

5.3 Charge-Voltage ($q - V$) Diagram and Corona Loss

5.3.1 Increase in Effective Radius of Conductor and Coupling Factors

The partial discharge of air around a line conductor is the process of creation and movement of charged particles and ions in the vicinity of a conductor under the applied voltage and field. We shall consider a simplified picture for conditions occurring when first the voltage is passing through the negative half-cycle and next the positive half-cycle, as shown in Fig. 5.2.

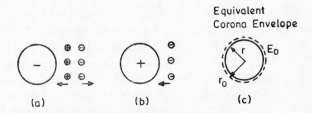

Fig. 5.2. Space-charge distribution in corona and increase in effective radius of conductor.

In Fig. 5.2(a), free electrons near the negative conductor when repelled can acquire sufficient energy to form an electron avalanche. The positive ions (a neutral molecule which has lost an electron) are attracted towards the negative conductor while the electrons drift into lower fields to attach themselves to neutral atoms or molecules of Nitrogen and Oxygen to form negative ions. Some recombination could also take place. The energy imparted for causing initial ionization by collision is supplied by the electric field. During the positive half cycle, the negative ions are attracted towards the conductor, but because of local conditions not all ions drift back to the conductor. A space charge is left behind and the hysteresis effect gives rise to the energy loss. Furthermore, because of the presence of charged particles, the effective charge of the conductor-ground electrode system is increased giving rise to an increase in effective capacitance. This can be interpreted in an alternative manner by assuming that the conductor diameter is effectively increased by the conducting channel up to a certain extent where the electric field intensity decreases to a value equal to that required for further ionization, namely, the corona-inception gradient, Fig. 5.2(c).

Example 5.4: A single conductor 1 cm in radius is strung 5 metres above ground; using Peek's formula for corona-inception gradient, find
(a) the corona-inception voltage,
(b) the equivalent radius of conductor to the outside of the corona envelope at 20% overvoltage. Take $\delta = 1$.

Solution:

(a) $E_0 = 21.4 (1 + 0.0301/\sqrt{r}) = 27.84$ kV/cm $= 2784$ kV/m.

$\therefore \quad V_0 = E_0 \cdot r \cdot \ln (2H/r) = 2784 \times 0.01 \times \ln (10/0.01)$

$\quad\quad = 192.3$ kV, r.m.s.

(b) Let $r_0 = $ effective radius of the corona envelope

Then, $\quad E_{0c} = 2140 (1 + 0.0301/\sqrt{r_0})$, and

$1.2 \times 192.3 = E_{0c} \cdot r_0 \cdot \ln (2H/r_0)$ giving

$\quad\quad 230.8 = 2140 (1 + 0.0301/\sqrt{r_0}) \cdot r_0 \cdot \ln (10/r_0)$.

A trial solution yields $r_0 = 0.0126$ m $= 1.26$ cm.

So far we have considered power-frequency excitation and worked with effective or r.m.s. values of voltage and voltage gradient. There are two other very important types of voltage, namely the lightning impulse and switching surge, which give rise to intense corona on the conductors. The resulting energy loss helps to attenuate the voltage magnitudes during travel from a source point to other points far away along the overhead line. The resulting attenuation or decrease in amplitude, and the distortion or waveshape will be discussed in detail in the next section. We mention here that at power frequency, the corona-inception gradient and voltage are usually higher by 10 to 30% of the operating voltage in fair weather so that a line is not normally designed to generate corona. However, local conditions such as dirt particles etc. do give rise to some corona. On the other hand, under a lightning stroke or a switching operation, the voltage exceeds twice the peak value of corona-inception voltage. Corona plumes have been photographed from actual lines which extend up to 1.2 metres from the surface of the conductor. Therefore, there is evidence of a very large increase in effective diameter of a conductor under these conditions.

Corona-inception gradients on conductors under impulse conditions on cylindrical conductors above a ground plane are equal to those under power frequency but crest values have to be used in Peek's formula. The increase in effective radius will in turn change the capacitance of the conductor which has an influence on the voltage coupled to the other phase-conductors located on the same tower. The increased coupling factor on mutually-coupled travelling waves was recognized in the 1930's and 40's under lightning conditions. At present, switching surges are of great concern in determining insulation clearance between conductor and ground, and conductor to conductor. We will consider the increase in diameter and the resulting coupling factors under both types of impulses.

Example 5.5: A single conductor 2.5 inch in diameter of a 525-kV line (line-to-line voltage) is strung 13 m above ground. Calculate (a) the corona-inception voltage and (b) the effective radius of conductor at an overvoltage of 2.5 p.u. Consider a stranding factor $m = 0.8$ for roughness. (c) Calculate the capacitance of conductor to ground with and without corona. (d) If a second conductor is strung 10 m away at the same height, calculate the coupling factors in the two cases. Take $\delta = 1$.

Solution: $r = 0.03176$ m, $H = 13$.

(a) $E_{or} = 2140 \times 0.8(1 + 0.0301/\sqrt{0.03176})$

$= 2001$ kV/m $= 20$ kV/cm.

$V_0 = E_0 . r . \ln (2H/r) = 532.73$ kV r.m.s.

$= 753.4$ kV peak.

At 525 kV, r.m.s., line-to-line, there is no corona present.

(b) 2.5 p.u. voltage $= 2.5 \times 525\sqrt{2/3} = 1071.65$ kV, crest.

Therefore, corona is present since the corona-onset voltage is 753.4 kV, crest.

When considering the effective radius, we assume a smooth surface for the envelope so that

$$1071.65 = 3000(1 + 0.0301/\sqrt{r_0}) . r_0 . \ln (26/r_0).$$

A trial and error solution yields $r_0 = 0.05$ metre. This is an increase in radius of $0.05 - 0.03176 = 0.01824$ metre or 57.43%.

(c) $C = 2\pi e_0 / \ln (2H/r)$.

Without corona, $C = 8.282$ nF/Km;

With corona, $C = 8.88$ nF/Km.

(d) The potential coefficient matrix is

$$[P] = \begin{bmatrix} P_{11}, & P_{12} \\ P_{21}, & P_{22} \end{bmatrix} \quad \text{and} \quad [M] = [P]^{-1} = \begin{bmatrix} P_{22}, & -P_{12} \\ -P_{21}, & P_{11} \end{bmatrix} \frac{1}{\Delta}$$

where $\Delta = |[P]|$, the determinant.

The self-capacitance is $2\pi e_0 P_{11}/\Delta$ while the mutual capacitance is $-2\pi e_0 P_{12}/\Delta$. The coupling factor is $K_{12} = -P_{12}/P_{11}$.

Without corona, $K_{12} = -\ln (\sqrt{26^2 + 10^2}/10)/\ln (26/0.03176)$

$= -1.0245/6.7076 = -0.15274$

With corona, $K_{12} = -1.0245/\ln (26/.05) = -0.1638$.

This is an increase of 7.135%.

For bundled conductors, coupling factors between 15% to 25% are found in practice. Note that with a switching surge of 1000 kV crest, the second conductor experiences a voltage of nearly 152 kV crest to ground so that the voltage between the conductors could reach 850 kV, crest.

5.3.2 Charge-Voltage Diagram with Corona

When corona is absent the capacitance of a conductor is based on the physical radius of the metallic conductor. The charge-voltage relation is a straight line OA as shown in Fig. 5.3 and $C = q_0/V_0$, where

V_0 = the corona-inception voltage and q_0 the corresponding charge. However, beyond this voltage there is an increase in charge which is more rapid than given by the slope C of the straight-line $q_0 - V_0$ relation. This is shown as the portion AB which is nearly straight. When the voltage is decreased after reaching a maximum V_m there is a hysteresis effect and the $q - V$ relation follows the path BD. The slope of BD almost equals C showing that the space-charge cloud near the conductor has been absorbed into the conductor and charges far enough

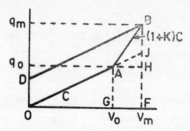

Fig. 5.3. Charge-voltage diagram of corona.

away from the conductor are not entirely pulled back. The essential properties of this $q - V$ diagram for one half-cycle of an ac voltage or the unipolar lightning and switching impulses can be obtained from the trapezoidal area $OABD$ which represents the energy loss.

Let the slope of AB equal $(1 + K)C$ where K is an experimental factor which lies between 0.6 and 0.8 having an average value of 0.7. The maximum charge corresponding to V_m is denoted as q_m. The area of $OABD$ equals $\frac{1}{2} KC(V_m^2 - V_0^2)$ as shown below.

Area $OABD$ = (Area $DOFB$)−(Area OAG)−(Area $GAHF$)−Area (AHB)

$$= \text{(Area } DOFB) - \tfrac{1}{2}q_0V_0 - q_0(V_m-V_0)-\tfrac{1}{2}(q_m-q_0)(V_m-V_0)$$

$$= \text{Area } DOFB - \tfrac{1}{2}\{q_0V_m + q_m(V_m - V_0)\} \tag{5.16}$$

Now, $BH = q_m - q_0 = (1 + K)q_0(V_m - V_0)/V_0,$

$\qquad JH = q_0(V_m - V_0)/V_0,$

and $BF = q_m = q_0 + (1 + K)q_0(V_m - V_0)/V_0,$

$\therefore \quad DO = BJ = BH - JH = Kq_0(V_m - V_0)/V_0. \tag{5.17}$

Area $DOFB = \tfrac{1}{2}(DO+BF)V_m = Kq_0(V_m-V_0)V_m/V_0+\tfrac{1}{2}q_0V_m^2/V_0 \tag{5.18}$

Area $OAG = \tfrac{1}{2}q_0V_0$; Area $AGFH = q_0(V_m - V_0)$.

Area $AHB = \tfrac{1}{2}(q_m - q_0)(V_m - V_0) = \tfrac{1}{2}q_0(V_m^2 - V_0^2)/V_0$
$$+ \tfrac{1}{2}Kq_0(V_m - V_0)^2/V_0 \tag{5.19}$$

$\therefore$ Areas $(OAG+AGFH+AHB) = \tfrac{1}{2}(q_0/V_0)[V_m^2+K(V_m - V_0)^2]. \tag{5.20}$

Finally, Area $OABD = \tfrac{1}{2}K.q_0(V_m - V_0)(V_m + V_0)/V_0$

$$= \tfrac{1}{2}KC(V_m^2 - V_0^2) \tag{5.21}$$

For a unipolar waveform of voltage the energy loss is equal to equation

(5.21).   For an ac voltage for one cycle, the energy loss is twice this value.

$$\therefore \quad W_{ac} = KC(V_m^2 - V_0^2) \tag{5.22}$$

The corresponding power loss will be

$$P_c = fW_{ac} = fKC(V_m^2 - V_0^2) \tag{5.23}$$

If the maximum voltage is very close to the corona-inception voltage V_0, we can write

$V_m^2 - V_0^2 = (V_m + V_0)(V_m - V_0) = 2V_m(V_m - V_0)$, so that

$$P_c = 2f\,KC\,V_m(V_m - V_0) \tag{5.24}$$

where all voltages are crest values.   When effective values for V_m and V_0 are used,

$$P_c = 4f\,KC\,V(V - V_0) \tag{5.25}$$

This is very close to the Ryan-Henline formula, equation (5.14) with $K = 1$.   For lightning and switching impulses the energy loss is equation (5.21) which is $W = \frac{1}{2}KC(V_m^2 - V_0^2)$ where all voltages are crest values.

Example 5.6:   An overhead conductor of 1.6 cm radius is 10 m above ground.   The normal voltage is 133 kV r.m.s. to ground (230 kV, line-to-line).   The switching surge experienced is 3.5 p.u.   Taking $K = 0.7$, calculate the energy loss per Km of line.

Solution. $C = 2\pi e_0/\ln(2H/r) = 7.79$ nF/Km

$\quad\quad E_0 = 30(1 + 0.0301/\sqrt{0.016}) = 37.14$ kV/cm, crest

$\therefore \quad V_0 = E_0.r.\ln(2H/r) = 423.8$ kV, crest

$\quad\quad V_m = 3.5 \times 133\ \sqrt{2} = 658.3$ kV, crest

$\therefore \quad W = 0.5 \times 0.7 \times 7.79 \times 10^{-9}\,(658.3^2 - 423.8^2) \times 10^6$ Joule/Km

$\quad\quad = 0.7$ kJ/Km.

This property will be used in deriving attenuation of travelling waves caused by lightning and switching in later sections.

5.4   Attenuation of Travelling Waves Due to Corona Loss

A voltage wave incident on a transmission line at an initial point $x = 0$ will travel with a velocity v such that at a later time t the voltage reaches a point $x = vt$ from the point of incidence, as shown in Fig. 5.4.   In so doing if the crest value of voltage is higher than the corona-disruptive voltage for the conductor, it loses energy while it travels and its amplitude decreases corresponding to the lower energy content.   In addition to the attenuation or decrease in amplitude, the waveshape also shows distortion.   In this section, we will discuss only attenuation since distortion must include complete equations of travelling waves

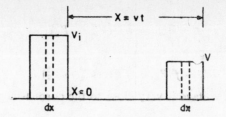

Fig. 5.4.   Attenuation of voltage on a transmission line.

caused by inductance and capacitance as well as conductor and ground-return resistance.   This is dealt with in Chapter 8.

The energy of the wave is stored in both electromagnetic form and electrostatic form.   The time rate of loss of stored energy is equal to the power loss due to corona, whose functional relationship with voltage has been given in Section 5.2.   The total energy in a differential length dx of the wave will be

$$dw = \tfrac{1}{2}(C.dx) V^2 + \tfrac{1}{2}(L.dx) I^2 \qquad (5.26)$$

where L and C are inductance and capacitance per unit length of line. For a travelling wave, the voltage V and current I are related by the surge impedance $Z = \sqrt{L/C}$, and the wave velocity in $v = 1/\sqrt{LC}$. Consequently, $I^2 = V^2/Z^2 = V^2C/L$.   Thus, equation (5.26) becomes

$$dw = C.dx.V^2. \qquad (5.27)$$

The rate of dissipation of energy, assuming the capacitance does not change with voltage for the present analysis,

$$dw/dt = d(CV^2.dx)/dt = 2CV.dx.dV/dt. \qquad (5.28)$$

Now, the power loss over the differential length dx is

$$P_c = f(V).dx, \text{ so that } 2CV.dV/dt = -P_c = -f(V). \qquad (5.29)$$

For different functional relations $P_c = f(V)$, equation (5.29) can be solved and the magnitude of voltage after a time of travel t (or distance $x = vt$) can be determined.   We will illustrate the procedure for a few typical values of $f(V)$, but will consider the problem later on by using equation (5.21) in Chapter 8.

(a) *Linear Relationship*

Let $f(V) = K_s(V - V_0)$.   Then, with $V_i =$ initial voltage, $2CV.dV/dt = -K_s(V - V_0)$.   By separating variables and using the initial condition $V = V_i$ at $t = 0$ yields

$$(V - V_0).e^{V/V_0} = (V_i - V_0).e^{(V_i - \alpha t)/V_0}, \qquad (5.30)$$

where $\alpha = K_s/2C$ and $V_0 =$ corona-inception voltage.

Also, the voltages in excess of the corona-inception voltage at any time t or distance $x = vt$ will be

$$(V - V_0)/(V_i - V_0) = e^{(V_i - V - \alpha t)/V_0}. \qquad (5.31)$$

This expression yields an indirect method of determining $\alpha = K_s/2C$ by experiment, if distortion is not too great. It requires measuring the incident wave magnitude V_i and the magnitude V after a time lapse of t or distance $x = vt$ at a different point on the line, whose corona-inception voltage V_0 is known. When C is also known, the constant K_s is calculated.

Example 5.7: A voltage with magnitude of 500 kV crest is incident on a conductor whose corona-inception voltage is 100 kV, crest and capacitance $C = 10$ nF/Km. After a lapse of 120 μs (36 Km of travel at light velocity) the measured amplitude is 110 kV. Calculate α and K_s.

Solution: $V_i/V_0 = 5, V - V_0 = 10, V_i - V_0$
$$= 400.\ V_0 = 100\ \text{kV} = 10^5\ \text{volts}$$

$\therefore\quad 10/400 = \exp(500 - 110)/100 . \exp(-120 \times 10^{-6} \times \alpha/10^5)$

This gives $\alpha = K_s/C = 6.325 \times 10^9$ volt-sec^{-1}.

$\therefore\quad K_s = 2 \times 6.325 \times 10^9 \times 10 \times 10^{-9} = 126.5$ watts/Km-volt or **Amp/Km.**

(b) *Quadratic Formula*
(i) Let the loss be assumed to vary as
$$f(V) = K_R V(V - V_0) \tag{5.32}$$
so that
$$2CV\,(dV/dt) = -K_R V(V - V_0).$$
With $V = V_i$ at $t = 0$, integration gives
$$\frac{V - V_0}{V_i - V_0} = \exp[-(K_R/2C)t] \tag{5.33}$$

The voltage in excess of corona-inception value behaves as if it is attenuated by a resistance R per unit length as given by the formula
$$(V - V_0) = (V_i - V_0).\exp[-(R/2L)t]$$
$$\therefore\qquad K_R = RC/L = R/Z^2.$$
For the previous example,
$$K_R = \frac{2 \times 10^{-9}}{120 \times 10^{-6}} . \ln\frac{400}{10} = 61.5 \times 10^{-6}.$$
The units are watts/Km-volt^2.
(ii) If the loss is assumed to vary as
$$f(V) = K_Q(V^2 - V_0^2) \tag{5.34}$$
then $2CV(dV/dt) = -K_Q(V^2 - V_0^2)$ with $V = V_i$ at $t = 0$.
$$\therefore\qquad \frac{V^2 - V_0^2}{V_i^2 - V_0^2} = \exp[-(K_Q/C)t].$$
For the previous example,
$$K_Q = \frac{10 \times 10^{-9}}{120 \times 10^{-6}} \ln\frac{500^2 - 100^2}{110^2 - 100^2} = 395 \times 10^{-6}.$$

(iii) Let $f(V) = K_P (V - V_0)^2$. Then,

$$2CV.(dV/dt) = - K_P(V - V_0)^2 \text{ with } V = V_i \text{ at } t = 0.$$

This gives

$$\ln \frac{V_i - V_0}{V - V_0} + \frac{V_0(V_i - V)}{(V_i - V_0)(V - V_0)} = (K_P/2C)t \qquad (5.35)$$

For $V_0 = 100$ kV, $V_i = 500$ kV, $V = 110$ kV, $C = 10 \times 10^{-9}$ F/Km, and $t = 120 \times 10^{-6}$ sec, there results $K_P = 224 \times 10^{-6}$.

(c) *Cubic Relation*

If $f(V) = K_c.V^3$, $2CV(dV/dt) = - K_c V^3$

giving $V_i/V = 1 + (K_c/2C) V_i t$. $\qquad (5.36)$

For the previous example, $K_c = 1.182 \times 10^{-9}$ watts/Km-volt³.

5.5 Audible Noise: Generation and Characteristics

When corona is present on the conductors, e.h.v. lines generate audible noise which is especially high during foul weather. The noise is broad-band, which extends from very low frequency to about 20 KHz. Corona discharges generate positive and negative ions which are alternately attracted and repelled by the periodic reversal of polarity of the ac excitation. Their movement gives rise to sound-pressure waves at frequencies of twice the power frequency and its multiples, in addition to the broadband spectrum which is the result of random motions of the ions, as shown in Fig. 5.5. The noise has a pure tone super-imposed on the broadband noise. Due to differences in ionic motion between ac and dc excitations, dc lines exhibit only a broadband noise, and furthermore, unlike for ac lines, the noise generated from a dc line is nearly equal in both fair and foul weather conditions. Since audible noise (AN) is man-made, it is measured in the same manner as other types of man-made noise such as aircraft noise, automobile ignition noise, transformer hum, etc. We will describe meters used and methods of AN measurements in a subsequent section 5.7.

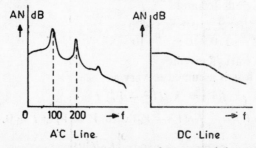

Fig. 5.5. Audible Noise frequency spectra from ac and dc transmission lines.

Audible noise can become a serious problem from 'psycho-acoustics' point of view, leading to insanity due to loss of sleep at night to inha-bitants residing close to an e.h.v. line. This problem came into focus in the 1960's with the energization of 500 kV lines in the USA. Regulatory bodies have not as yet fixed limits to AN from power transmission lines since such regulations do not exist for other man-made sources of noise. The problem is left as a social one which has to be settled by public opinion. The proposed limits for AN are discussed in the next section,

5.6 Limits for Audible Noise

Since no legislation exists setting limits for AN for man-made sources, power companies and environmentalists have fixed limits from public-relations point of view which power companies have accepted from a moral point of view. In doing so, like other kinds of interference, human beings must be subjected to listening tests. Such objective tests are performed by every civic-minded power utility organization. The first such series of tests performed from a 500-kV line of the Bonneville Power Administration in the U.S.A. is known as Perry Criterion. The AN limits are as follows:

No complaints : Less than 52.5 dB (A)
Few complaints : 52.5 dB (A) to 59 dB (A),
Many complaints: Greater than 59 dB (A).

The reference level for audible noise and the dB relation will be explained later. The notation (A) denotes that the noise is measured on a meter on a filter designated as A-weighting network. There are several such networks in a meter.

Design of line dimensions at e.h.v. levels is now governed more by the need to limit AN levels to the above values. The selection of width of line corridor or right-of-way (R–O–W), where the nearest house can be permitted to be located, if fixed from AN limit of 52.5 dB(A) will be found adequate from other points of view at 1000 to 1200 kV levels. The design aspect will be considered in Section 5.8. The audible noise generated by a line is a function of the following factors.

(a) the surface voltage gradient on conductors,
(b) the number of sub-conductors in the bundle,
(c) conductor diameter,
(d) atmospheric conditions, and
(e) the lateral distance (or aerial distance) from the line conductors to the point where noise is to be evaluated.

The entire phenomenon is statistical in nature, as in all problems related to e.h.v. line designs, because of atmospheric conditions.

While the Perry criterion is based upon actual listening experiences on test groups of human beings, and guidelines are given for limits for AN from an e.h.v. line at the location of inhabited places, other man-made sources of noise do not follow such limits. A second criterion for setting limits and which evaluates the nuisance value from man-made sources of AN is called the 'Day-Night Equivalent' level of noise. This is based not only upon the variation of AN with atmospheric conditions but also with the hours of the day and night during a 24-hour period. The reason is that a certain noise level which can be tolerated during the waking hours of the day, when ambient noise is high, cannot be tolerated during sleeping hours of the night when little or no ambient noises are present. This will be elaborated upon in Section 5.9. According to the Day-Night Criterion, a noise level of 55 dB(A) can be taken as the limit instead of 52.5 dB(A) according to the Perry Criterion. From a statistical point of view, these levels are considered to exist for 50% of the time during precipitation hours. These are designated as L_{50} levels.

5.7 AN Measurement and Meters

5.7.1 Decibel Values in AN and Addition of Sources

Audible noise is caused by changes in air pressure or other transmission medium so that it is described by Sound Pressure Level (SPL). Alexander Graham Bell established the basic unit for SPL as 20×10^{-6} Newton/m² or 20 micro Pascals [2×10^{-5} micro bar]. All decibel values are referred to this basic unit. In telephone work there is a flow of current in a set of headphones or receiver. Here the basic units are 1 milliwatt across 600 ohms yielding a voltage of 775 mV and a current of 1.29 mA. For any other SPL, the decibel value is

$$\text{SPL(dB)} = 10 \, \text{Log}_{10} \, (\text{SPL}/20 \times 10^{-6} \, \text{Pascals}) \tag{5.37}$$

This is also termed the 'Acoustic Power Level', denoted by PWL, or simply the audible noise level, AN.

Example 5.8: The AN level of one phase of a 3-phase transmission line at a point is 50 dB. Calculate (a) the SPL in Pascals; (b) if a second source of noise contributes $48 dB$ at the same location, calculate the combined AN level due to the two sources.

Solution.
(a) $10 \, \text{Log}_{10} \, (\text{SPL}_1/2 \times 10^{-5}) = 50$, which gives
$\text{SPL}_1 = 2 \times 10^{-5} \times 10^{50/10} = 2$ Pascals
(b) Similarly, $\text{SPL}_2 = 2 \times 10^{-5} \times 10^{48/10} = 1.262$ Pascals
$\therefore$ Total SPL $= \text{SPL}_1 + \text{SPL}_2 = 3.262$ Pascals.
The decibel value will be
$\text{AN} = 10 \, \text{Log}_{10} \, (\text{SPL}/2 \times 10^{-5}) = 52.125 \text{dB}.$

Consider N sources whose decibel values, at a given point where AN level is to be evaluated, are $\text{AN}_1, \text{AN}_2, ..., \text{AN}_N$. In order to add these sources and evaluate the resultant SPL and dB values, the procedure is as follows:

The individual sound pressure levels are

$$\text{SPL}_1 = 20 \times 10^{-6} \times 10^{\text{AN}_1/10}, \, \text{SPL}_2 = 20 \times 10^{-6} \times 10^{\text{AN}_2/10} \text{ etc.}$$

$$\therefore \quad \text{The total SPL} = \text{SPL}_1 + \text{SPL}_2 + \dots = 2 \times 10^{-5} \sum_{i=1}^{N} 10^{\text{AN}_i/10} \quad (5.38)$$

The decibel value of the combined sound pressure level is

$$\text{AN} = 10 \, \text{Log}_{10} \, (\text{SPL}/2 \times 10^{-5}) = 10 \, \text{Log}_{10} \sum_{i=1}^{N} 10^{0.1\text{AN}_i} \quad (5.39)$$

Example 5.9. A 3-phase line yields AN levels from individual phases to be 55 dB, 52 dB, and 48 dB. Find the resulting AN level of the line.

Solution.
$$10^{0.1\text{AN}_1} + 10^{0.1\text{AN}_2} + 10^{0.1\text{AN}_3} = 10^{5.5} + 10^{5.2} + 10^{4.8}$$
$$= 5.382 \times 10^5$$
$$\therefore \quad \text{AN} = 10 \, \text{Log}_{10} \, (5.382 \times 10^5) = 57.31 \text{ dB}.$$

5.7.2 Microphones

Instruments for measurement of audible noise are very simple in construction in so far as their principles are concerned. They would conform to standard specifications of each country, as for example, ANSI, ISI, or I.E.C., etc. The input end of the AN measuring system consists of a microphone as shown in the block diagram, Fig. 5.6. There are three types of microphones used in AN measurement from e.h.v. lines and equipment. They are (i) air-condenser type; (ii) ceramic; and (iii) electret microphones. Air condenser microphones are very stable and exhibit highest frequency response. Ceramic ones are the most rugged of the three types. Since AN is primarily a foul-weather phenomenon, adequate protection of microphones from weather is necessary. In addition, the electret microphone requires a polarization voltage so that a power supply (usually battery) will also be exposed to rain and must be protected suitably.

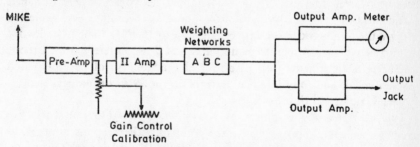

Fig. 5.6.   Block diagram of AN Measuring Circuit.

Some of the microphones used in AN measurement from e.h.v. lines are General Radio Type 1560-P, or 1971-9601, or Bruel and Kjor type 4145 or 4165, and so on. The GR type has a weather protection. Since AN level from a transmission line is much lower than, say, aircraft or ignition noise, 1" (2.54 cm) diameter microphones are used although some have used $\frac{1}{2}$" ones, since these have more sensitivity than 1" microphones. Therefore, size is not the determining factor.

The most important characteristic of a microphone is its frequency response. In making AN measurements, it is evident that the angle between the microphone and the source is not always 90° so that the grazing angle determines the frequency response. Some typical characteristics are as shown in Fig. 5.7.

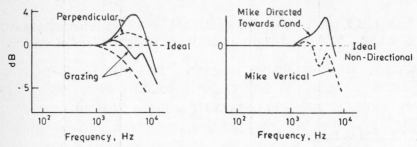

Fig. 5.7. Response of microphone for grazing and perpendicular incidence.

AN levels are statistical in nature and long-term measurements are carried out by protecting the microphone from rain, wind, animals, and birds. Some types of shelters use wind-screens with a coating of silicone grease. Foam rubber wind-covers have also been used which have negligible attenuation effect on the sound, particularly on the A-weighted network which will be described below. Every wind-cover must also be calibrated and manufacturers supply this data. Foam-rubber soaks up rain and must be squeezed out periodically and silicone grease applied.

5.7.3 Weighting Networks

There are 5 weighting networks designated as A to E in Sound Pressure Level Meters. The 'A' weighted network has been designed particularly to have nearly the same response as the human ear, while the 'C' weighted network has a flat response up to 16 KHz. The 'A' network is also least susceptible to wind gusts. It is also preferred by Labour Relations Departments for assessing the adversity of noise-created psychological and physiological effects in noisy environments such as factories, power stations, etc.

Typical frequency response of the A, B, C weighting networks are sketched in Fig. 5.8(a) while 5.8(b) compares the responses of A, D networks. The A-weighting network is widely used for relatively non-directional sources. From these curves it is seen that the C-network provides essentially flat response from 20 Hz to 10 KHz. The human ear exhibits such flat response for sound pressure levels up to 85 dB or more. At lower SPL, the human ear does not have a flat response with frequency and the A and B networks are preferred. The A network is used for SPL up to 40 dB and the B for SPL up to 70 dB. Sometimes, the A-weighted network is known as the 40-dB network. It is also used for transformer noise measurements.

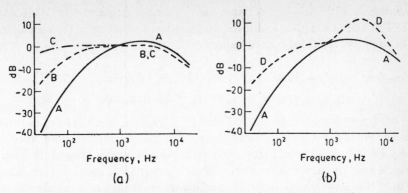

Fig. 5.8. Frequency responses of (a) A, B, C weighting networks, (b) A, D weighting networks.

5.7.4 Octave Band and Third Octave Band

It was mentioned earlier that in addition to the broadband noise generated by corona, pure tones at double the power frequency and its multiples exist. These discrete-frequency components or line spectra are measured on octave bands by selective filters. Figure 5.9 shows a schematic diagram of the switching arrangement for use with a 50-Hz line.

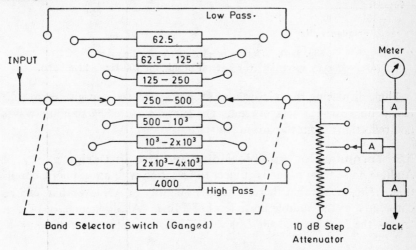

Fig. 5.9. Octave band AN meter circuit.

The octave band consists of a centre frequency f_0. Let f_1 and f_2 be the upper and lower frequencies of the bands. Then $f_0 = \sqrt{f_1 f_2}$. An octave band extends from the lower frequency $f_2 = f_0/\sqrt{2}$ to the upper frequency $f_1 = \sqrt{2} f_0 = 2 f_2$.

A third-octave band extends from the lower frequency $f_3 = f_0/(2)^{1/6}$ $= 0.891 f_0$ to an upper frequency $f_4 = (2)^{1/6}$. $f_0 = 1.1225 f_0 = (2)^{1/3} f_3$. The octave and third-octave band SPL is the integrated SPL of all the frequency components in the band.

Example 5.10: An octave band has a centre frequency of 1000 Hz. (a) Calculate the upper and lower frequencies of the band. (b) Calculate the same for the third-octave band.

Solution: (a) $f_0 = 1000.$ $\therefore f_1 = \sqrt{2} f_0 = 1414$ Hz, $f_2 = 707$ Hz.

(b) $f_4 = (2)^{1/6} f_0 = 1122$ Hz, $f_3 = 1000/(2)^{1/6} = 891$ Hz.

All frequency components radiated by a transmission line have to propagate from the conductor to the meter and therefore intervening media play an important role, particularly reflections from the ground surface. The lowest-frequency octave band is most sensitive to such disturbances while the A-weighted network and higher frequency octave bands give a flat overall response. Examples are shown in Fig. 5.10.

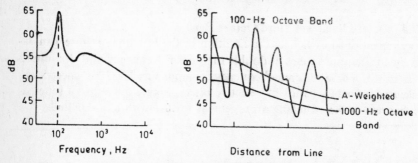

Fig. 5.10. Octave band response for line AN.
(a) Frequency spectrum, (b) Lateral profile for 100-Hz Octave band.

Thus, displacing the microphone even a few metres can give erroneous results on a 100-Hz octave band. The curve is caused by standing waves and reflections from the ground surface.

5.8 Formulae for Audible Noise and Use in Design
Audible noise from a line is subject to variation with atmospheric condition. This means that there is no one quantity or AN level that can be considered as the audible noise level of a line. All designers accept two levels—the L_{50} level and the L_5 level. These are defined as follows:

L_{50} *Level:* This is the AN level as measured on the A-weighted network which is exceeded 50% of the time during periods of rain, usually extending over an entire year.

L_5 *Level:* Similar to L_{50}, but exceeded only 5% of the total time.
The L_5 level is used for describing the noise levels in heavy rain which are generated in artificial rain tests. These are carried out in 'cage tests' where artificial rain apparatus is used, as well as from short outdoor experimental lines equipped with such apparatus. Conversion from L_5 to L_{50} levels are carried out suitably as will be described below.

Many empirical formulas exist for calculating the AN level of an e.h.v. line [see IEEE Task Force paper, October 1982]. However, we will discuss the use of the formula developed by the B.P.A. of the USA. It is applicable for the following conditions:

(a) All line geometries with bundles having up to 16 sub-conductors.
(b) Sub-conductor diameters in the range 2 cm to 6.5 cm.
(c) The AN calculated is the L_{50} level in rain.
(d) Transmission voltages are 230 kV to 1500 kV, 3-phase ac.

Referring to Fig. 5.11, the AN level of each phase at the measuring point M is, with $i = 1, 2, 3$,

$$\text{AN}(i) = 120 \log_{10} E_{am}(i) + 55 \log_{10} d - 11.4 \log_{10} D(i) - 115.4, \text{ dB}(A) \quad (5.40)$$

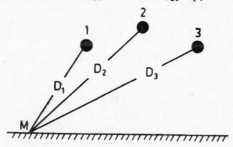

Fig. 5.11. Calculation of AN level of line by B.P.A. Formula.

It applies for $N < 3$ sub-conductors in the bundles. For $N \geqslant 3$, the formula becomes

$$\text{AN}(i) = 120 \log_{10} E_{am}(i) + 55 \log_{10} d - 11.4 \log_{10} D(i) +$$
$$+ 26.4 \log_{10} N - 128.4, \text{ dB}(A) \quad (5.41)$$

Here, $E_{am}(i)$ = average maximum surface voltage gradient on bundle belonging to phase i in kV/cm, r.m.s.

d = diameter of sub-conductor in cm.,

N = number of sub-conductors in bundle,

and $D(i)$ = aerial distance from phase i to the location of the microphone in metres.

When all dimensions are in metre units, the above give

$$N < 3: \text{AN}(i) = 120 \log E_m(i) + 55 \log d_m - 11.4 \log D(i)$$
$$+ 234.6, \text{ dB}(A) \quad (5.42)$$

$$N \geqslant 3: \text{AN}(i) = 120 \log E_m(i) + 55 \log d_m - 11.4 \log D(i)$$
$$+ 26.4 \log N + 221.6, \text{ dB}(A) \quad (5.43)$$

Having calculated the AN level of each phase, the rule for addition of the three levels follows equation (5.39),

$$\text{AN} = 10 \log_{10} \sum_{i=1}^{3} 10^{0.1 \text{ AN}(i)}, \text{ dB}(A) \quad (5.44)$$

For a double-circuit line, the value of i extends from 1 to 6.

We observe that the AN level depends on the following four quantities:

(i) the surface voltage gradient on conductor,

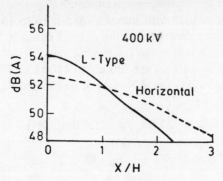

Fig. 5.12. AN level of 400 kV line (calculated).

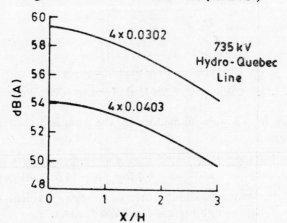

Fig. 5.13. AN level of 735 kV line (calculated).

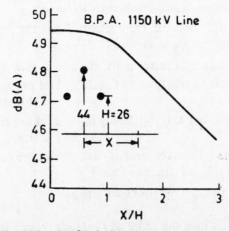

Fig. 5.14. AN level of 1150 kV line (calculated).

(ii) the conductor diameter,

(iii) the number of sub-conductors in bundle, and

(iv) the aerial distance to the point of measurement from the phase conductor under consideration. Atmospheric condition is included in having prescribed this as the L_{50}-level while the weighting network is described by the notation dB(A) on a Sound Level Meter.

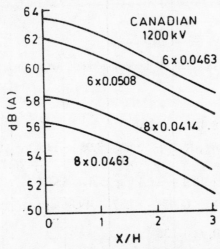

Fig. 5.15.   AN level of 1200 kV horizontal line (calculated).

A model for the generation of AN under rain has been developed very recently by Kirkham and Gajda to which the reader is referred for very thoughtful ideas on the basic mechanisms involved in AN generation.

L_{50} levels of AN from several representative lines from 400 kV to 1300 kV are plotted in Figs. 5.12 to 5.15, calculated according to the B.P.A. formula, equations (5.40) to (5.44). In all cases, the average maximum gradient does not differ from the maximum gradient in the bundle by more than 4%, so that only the maximum surface voltage gradient is used in the above figures.   This is

$$E_{max} = \frac{q}{2\pi e_0} \frac{1}{N} \frac{1}{r} [1 + (N-1) r/R] \qquad (5.45)$$

as described in chapter 4.

Example 5.11:   A 735 kV line has the following details: $N = 4$, $d = 3.05$ cm, B = bundle spacing = 45.72 cm, Height H = 20 m, phase separation S = 14 m in horizontal configuration.   By the Mangoldt formula, the maximum conductor surface voltage gradients are 20kV/cm and 18.4 kV/cm for the centre and outer phases, respectively.   Calculate the SPL or AN in dB(A) at a distance of 30 m along ground from the centre phase (line centre).   Assume that the microphone is kept at ground level.   See Fig. 5.16.

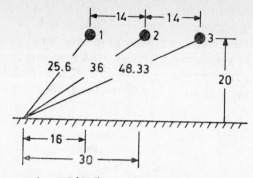

Fig. 5.16. 735 kV line configuration for Example 5.11.

Solution:

Phase 1. $E_m = 18.4$, $D_1 = 25.6$

$AN_1 = 120 \log_{10} 18.4 + 55 \log_{10} 3.05 - 11.4 \log_{10} 25.6$
$$+ 26.4 \log_{10} 4 - 128.4$$

$= 120 \log 18.4 - 11.4 \log 25.6 - 85.87$

$= 151.78 - 16.05 - 85.87 = 49.86 \text{ dB}(A)$

Phase 2. $E_m = 20$, $D_2 = 36$

$AN_2 = 120 \log 20 - 11.4 \log 36 - 85.87 = 52.5 \text{ dB}(A)$

Phase 3. $E_m = 18.4$, $D_3 = 48.33$

$AN_3 = 120 \log 18.4 - 11.4 \log 48.33 - 85.87 = 46.71 \text{ dB}(A)$

$\therefore \quad AN = 10 \log_{10} (10^{4.986} + 10^{5.25} + 10^{4.671})$
$$= 10 \log (32.15 \times 10^4) = 55 \text{ dB}(A)$$

This is within the range of low-complaint region according to the Perry Criterion which is 52.5 to 59 dB(A).

5.9 Relation Between Single-Phase and 3-Phase AN Levels

Obtaining data of AN and other quantities from e.h.v. lines involves great expense in setting up full-scale outdoor 3-phase experimental lines. Most of the design data can be obtained at less cost from a single-phase outdoor line or from cage experiments. The quantities of interest in so far as interference from e.h.v. lines are concerned are AN, Radio Interference and Electrostatic Field at 50 Hz. It is therefore worth the effort to consider what relation, if any, exists between experimental results obtained from 1-phase lines and an actual 3-phase line. If such a relation can be found, then 1-phase lines can be used for gathering data which can then be extrapolated to apply to 3-phase lines. We first consider a horizontal line.

The AN level from any phase at the measuring point M consists of a constant part and a variable part which can be seen from equation (5.40) to (5.43). They are written as, for $N \geqslant 3$,

$$AN_1 = (55 \log d + 26.4 \log N - 128.4) + 120 \log E_1 - 11.4 \log D_1$$
$$= K + 120 \log E_1 - 11.4 \log D_1$$

Similarly, $\qquad AN_2 = K + 120 \log E_2 - 11.4 \log D_2$

and $\qquad AN_3 = K + 120 \log E_1 - 11.4 \log D_3$

Let the centre-phase gradient be written as $E_2 = (1 + m) E_1$ and the ratios $k_2 = D_2/D_1$ and $k_3 = D_3/D_1$. Then, total AN level of the 3-phases obtained after combining the AN levels of the 3 individual phases is

$$AN_T = 10 \log_{10} \sum_{i=1}^{3} 10^{0.1 \, AN(i)}$$

$$= 10 \log_{10} [10^{0.1K} . 10^{12 \log E_1} \{10^{-1.14 \log D_1} + 10^{-1.14 \log D_2} + 10^{-1.14 \log D_3 + 12 \log (1+m)}\}]$$

$$= K + 120 \log E_1 - 11.4 \log D_1 + 10 \log [1 + k_3^{-1.14} + (1 + m)^{12} . k_2^{-1.14}] \qquad (5.46)$$

For a single-phase line with the same surface voltage gradient E_2 as the centre-phase conductor of the 3-phase configuration, and at a distance D_2, the noise level is

$$AN_s = K + 120 \log E_1 - 11.4 \log D_1 + 10 \log_{10} [k_2^{-1.14} (1 + m)^{12}] \qquad (5.47)$$

Therefore, the difference in AN levels of equation (5.46) and (5.47) is

$$AN_T - AN_s = 10 \log_{10} \frac{1 + k_3^{-1.14} + k_2^{-1.14} (1 + m)^{12}}{k_2^{-1.14} (1 + m)^{12}} \qquad (5.48)$$

This is the decibel adder which will convert the single-phase AN level to that of a 3-phase line.

Example 5.12: Using equation (5.46) compute AN_T for the 735 kV of example 5.11.

Solution: $K = - 85.87$, $1 + m = 20/18.4 = 1.087$.

$11.4 \log D_1 = 16.05$, $120 \log E_1 = 151.8$, $k_2 = 36/25.6 = 1.406$,

$\qquad k_3 = 48.33/25.6 = 1.888$

$\qquad \therefore \quad 1 + k_3^{-1.14} + k_2^{-1.14}(1 + m)^{12} = 3.33$

$AN_T = - 85.86 + 151.8 - 16.05 + 10 \log 3.33 = 55.1 \, dB(A)$.

Example 5.13: Using equation (5.48), compute the decibel adder to convert the single-phase AN level of the centre phase to the three-phase AN level of examples 5.11 and 5.12.

Solution:

$$\frac{1 + k_3^{-1.14} + k_2^{-1.14}(1 + m)^{12}}{k_2^{-1.14}(1 + m)^{12}} = \frac{3.33}{1.8444} = 1.8054$$

$$\therefore \quad \text{dB adder} = 10 \log 1.8054 = 2.566.$$

The AN level of the centre phase was 52.5 dB(A) at 30 m from the conductor along ground. This will be the level of a single-phase line with the same surface voltage gradient and distance to the location of microphone.

$$\therefore \quad AN_T = AN_s + dB \text{ adder} = 52.5 + 2.566 = 55.07 \text{ dB}(A)$$

5.10 Day-Night Equivalent Noise Level

In previous discussions the AN level of a transmission line has been chosen as the L_{50} value or the audible noise in decibels on the A-weighted network that is exceeded for 50% of the duration of precipitation. This has been assumed to give an indication of the nuisance value. However, another criterion which is actively followed and applied to man-made AN sources is called the Day-Night Equivalent Noise level. This has found acceptance to aircraft noise levels, heavy road traffic noise, ignition noise, etc., which has led to litigation among many aggrieved parties and the noise makers. According to this criterion, certain sound level might be acceptable during day-time hours when ambient noises will be high. But during the night-time hours the same noise level from a power line or other man-made sources could be found objectionable because of the absence of ambient noises. The equivalent annoyance during nights is estimated by adding 10 dB(A) to the day-time AN level, or, in other words, by imposing a 10 dB(A) penalty.

Consider an L_{50} level of a power line to be AN and the day-time to last for D hours. Then, the actual annoyance level for the entire 24 hours is computed as a day-night equivalent level as follows:

$$L_{dn} = 10 \log_{10}\left[\frac{1}{24}\left\{D \cdot 10^{0.1 \text{AN}} + (24 - D) \cdot 10^{0.1(\text{AN}+10)}\right\}\right], \text{dB}(A) \quad (5.49)$$

This is under the assumption the level AN is present throughout the 24 hours.

Example 5.14: The L_{50} level of a line is 55 dB(A). The day-light hours are 15 and night-time is 9 hours in duration. Calculate the day-night equivalent and the decibel adder to the day-time AN level.

Solution. $L_{dn} = 10 \log_{10}\left[\frac{1}{24}(15 \times 10^{5.5} + 9 \times 10^{6.5})\right] = 61.4 \text{ dB}(A)$

The decibel adder is $61.4 - 55 = 6.4 \text{ dB}(A)$.

An addition of 6.4 dB increases the SPL by 4.365 lines [10 Log 4.365 = 6.4]. The nuisance value of the line has been increased by 6.4 dB(A) by adding 10 dB(A) penalty for night hours.

If the day-night hours are different from 15 and 9, a different decibel adder will be necessary. The 10 dB(A) penalty added to night time contributes 6 times the AN value as the day-time level, since

$$9 \times 10^{6.5}/15 \times 10^{5.5} = 90/15 = 6.$$

In evaluating the nuisance value of AN from an e.h.v. line, we are only concerned with the duration of rainfall during a day and not the total day-night hours or 24 hours. If rain is not present over the entire 24 hours but only for a certain percentage of the day and night, then the day-night equivalent value of AN is calculated as shown below. Let it be assumed that

$p_d = \%$ of duration of rainfall during the day time,

and $\quad p_n = \%$ of duration of rainfall during the night.

Then,

$$L_{dn} = 10 \, \text{Log}_{10} \, [(1/24) \, \{(Dp_d/100) \cdot 10^{0.1\text{AN}} + (24 - D)(p_n/100)$$
$$10^{0.1(\text{AN}+10)}\}] \qquad (5.50)$$

Example 5.15: The following data are given for a line: $L_{50} = 55\text{dB}(A)$. $D = 15$, $p_d = 20$, $p_n = 50$. Calculate the day-night equivalent of AN and the dB-adder.

Solution: Duration of rain is 3 hours during the day and 4.5 hours during the night.

$$\therefore \quad L_{dn} = 10 \, \text{Log}_{10} \, [(1/24) \, (15 \times 0.2 \times 10^{5.5} + 9 \times 0.5 \times 10^{6.5})]$$
$$= 58\text{dB}(A).$$

The decibel adder is 3dB(A).

Now, the night-time contribution is 15 times that during the day time $(4.5 \times 10^{6.5}/3 \times 10^{5.5} = 45/3 = 15)$.

In the above equation (5.50), it was assumed that the L_{50} levels for both day and night were equal, or in other words, the precipitation characteristics were the same. If this is not the case, then the proper values must be used which are obtained by keeping very careful record of rainfall rates and AN levels. Such experiments are performed with short outdoor experimental lines strung over ground, or in 'cages'.

5.11 Some Examples of AN Levels From EHV Lines

It might prove informative to end this chapter with data on the performance of some e.h.v. line designs based upon AN limits from all over the world.

(1) The B.P.A. in the U.S.A. has fixed 50 dB(A) limit for the L_{50} noise level at 30 m from the line centre in rain for their 1150 kV line operating at 1200 kV.

(2) The A.E.P., U.H.V. Project of the E.P.R.I., and several other designs fall very close to the above values.

(3) Operating 750 kV lines of the A.E.P. in U.S.A. gave 55.4 dB(A) at 760 kV. The same company performed experiments from short outdoor line at Apple Grove and obtained 56.5 dB(A) at 775 kV, proving that short-line data can be relied upon to provide adequate design values.

(4) The Hydro-Quebec Company of Canada has given the following calculated AN levels at 30.5 m from the centres of their proposed line designs.

Voltage, kV	525		735	
Conductor size	$2 \times 1.602''$	$3 \times 1.302''$	$4 \times 1.382''$	$4 \times 1.2''$
Bundle spacing	18''	18''	18''	18''
Phase spacing	34'	34'	50'	45'
AN Level, dB(A)	57	52	55	58.5

Review Questions and Problems

1. Describe the behaviour of space-charge effects inside a corona envelope and discuss why load current cannot flow in a conductor inside this envelope even though it is a conducting zone.

2. Derive the expression $P_c = \frac{1}{2} KC (V_m^2 - V_0^2)$ for the energy loss from the charge-voltage diagram, Fig. 5.3.

3. It was theorized by E.W. Boehne, the famous electrical engineer and professor at M.I.T., that the increase in effective radius of conductor and consequent increase in capacitance was partly responsible for attenuation of travelling waves on conductors due to lightning. From the material given in this chapter, discuss why this theory is valid.

4. A 400 kV line supplies a load of 600 MW over a distance of 400 Km. Its conductors are 2×3.18 cm dia. with a resistance of 0.03 ohm/ Km per phase. It carries an average load of 400 MW over the year (66.7% load factor).
   (a) Calculate annual energy loss of the line.
   (b) If the average corona loss is 20 KW/Km for the 3-phases for 2 months of the year, calculate the annual energy loss due to corona.
   (c) Calculate the % corona energy loss as compared to the I^2R heating loss of the line.

5. Describe the difference between a line spectrum and band spectrum for noise. What is the difference between a pure tone and broadband spectrum?

6. For the 735-kV line of problem 2 in chapter 4, calculate the AN level at a distance of 15 metres along ground from the outer phase.

7. Using this as the AN level of the line, calculate the day-night equivalent if daytime is 15 hours and the penalty for night is 8 dB(A) instead of the suggested 10 dB(A) in the text.

CHAPTER 6

Corona Effects-II: Radio Interference

6.1 Corona Pulses: Their Generation and Properties

There are in general two types of corona discharge from transmission-line conductors: (i) Pulseless or Glow Corona; (ii) Pulse Type or Streamer Corona. Both these give rise to energy loss, but only the pulse-type of corona gives interference to radio broadcast in the range of 0.5 MHz to 1.6 MHz. In addition to corona generated on line conductors, there are spark discharges from chipped or broken insulators and loose guy wires which interfere with TV reception in the 80–200 MHz range. Audible noise has already been discussed in Chapter 5 which is caused by rain drops and high humidity conditions. Corona on conductors also causes interference to Carrier Communication and Signalling in the frequency range 30 KHz to 500 KHz.

In the case of Radio and TV interference the problem is one of locating the receivers far enough from the line in a lateral direction such that noise generated by the line is low enough at the receiver location in order to yield a satisfactory quality of reception. In the case of carrier interference, the problem is one of determining the transmitter and receiver powers to combat line-generated noise power.

In this section we discuss the mechanism of generation and salient characteristics of only pulse-type corona in so far as they affect radio reception. As in most gas discharge phenomena under high impressed electric fields, free electrons and charged particles (ions) are created in space which contain very few initial electrons. We can therefore expect a build up of resulting current in the conductor from a zero value to a maximum or peak caused by the avalanche mechanism and their motion towards the proper electrode. Once the peak value is reached there is a fall in current because of lowering of electric field due to the relatively heavy immobile space charge cloud which lowers the velocity of ions. We can therefore expect pulses to be generated with short crest times and relatively longer fall times. Measurements made of single pulses by the author in co-axial cylindrical arrangement are shown in Fig. 6.1 under dc excitation. Similar pulses occur during the positive and negative half-cycles under ac excitation. The best equations that fit the observed wave shapes are also given on the figures. It will be assumed that positive corona pulses have the equation

$$i_+ = k_+ i_p \left(e^{-\alpha t} - e^{-\beta t}\right) \tag{6.1}$$

while negative pulses can be best described by

$$i_- = k_- i_p \, t^{-3/2} . e^{-\gamma/t - \delta t} \tag{6.2}$$

These equations have formed the basis for calculating the response of bandwidth-limited radio receivers (noise meters), and for formulating mathematical models of the radio-noise problem. In addition to the waveshape of a single pulse, their repetition rate in a train of pulses is also important.

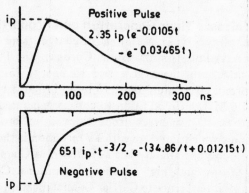

Fig. 6.1. Single positive and negative pulses

$$i_+ = k_+ i_p \left(e^{-\alpha t} - e^{-\beta t}\right); \quad i_- = k_- i_p . t^{-1.5}, \quad e^{(-\gamma/t - \delta t)}$$

Referring to Fig. 6.2, when a conductor is positive with respect to ground, an electron avalanche moves rapidly into the conductor leaving the heavy positive-ion charge cloud close to the conductor which drifts away. The rapid movement of electrons and motion of positive ions gives the steep front of the pulse, while the further drift of the positive-ion cloud will form the tail of the pulse. It is clear that the presence of positive charges near the positive conductor lowers the field to an extent that the induced current in the conductor nearly vanishes. As soon as the positive ions have drifted far enough due to wind or neutralized by other agencies such as free electrons by recombination, the electric field in the vicinity of the conductor regains sufficiently high value for pulse formation to repeat itself. Thus, a train of pulses results from a point in corona on the conductor. The repetition rate of pulses is governed by factors local to the conductor. It has been observed that only one pulse usually occurs during a positive half cycle in fair weather

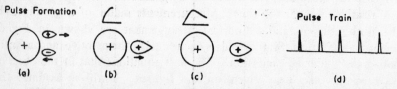

Fig. 6.2. Formation of pulse train from positive polarity conductor.

and could increase to about 10 in rain where the water spray resulting from breaking raindrops under the applied field control electrical conditions local to the conductor.

The situation when the conductor is negative with respect to ground is the reverse of that described above. The electron avalanche moves away from the conductor while the positive-ion cloud moves towards the negatively-charged conductor. However, since the heavy positive ions are moving into progressively higher electric fields, their motion is very rapid which gives rise to a much sharper pulse than a positive pulse. Similarly, the lighter electrons move rapidly away from the conductor and the electric field near the conductor regains its original value for the next pulse generation quicker than for the positive case. Therefore, negative pulses are smaller in amplitude, have much smaller rise and fall times but much higher repetition rates than positive pulses. It must at once be evident that all the properties of positive and negative pulses are random in nature and can only be described through random variables.

Typical average values or pulse properties are as follows:

Type	Time to crest	Time to 50% on Tail	Peak value of current	Repetition Rate Pulses per Second	
				A.C.	D.C.
Positive	50 ns	200 ns	100 mA	Power Freq.	1,000
Negative	20 ns	50 ns	10 mA	100 × P.F.	10,000

Pulses are larger as the diameter of conductor increases because the reduction in electric field strength a one moves away from the conductor is not as steep as for a smaller conductor so that conditions for longer pulse duration are more favourable. In very small wires, positive pulses can be absent and only a glow corona can result, although negative pulses are present when they are known as Trichel Pulses named after the first discoverer of the pulse-type discharge. Negative pulses are very rarely important from the point of view of radio interference as will be described under "Radio Noise Meter Response" to corona pulses in Section 6.2. Therefore, only positive polarity pulses are important because of their larger amplitudes even though their repetition rate is lower than negative pulses.

6.1.1 Frequency Spectrum

The frequency spectrum of radio noise measured from long lines usually corresponds to the Fourier Amplitude Spectrum (Bode Amplitude Plot) of single pulses. These are shown in Fig. 6.3. The Fourier integral for a single double-exponential pulse is

$$F(jw) = \int_{-\infty}^{\infty} f(t) \cdot e^{-jwt} \cdot dt = \int_{0}^{t_0} K i_p (e^{-\alpha t} - e^{-\beta t}) \cdot e^{-jwt} \cdot dt$$

$$= K\,i_p\,[1/(\alpha + jw) - 1/(\beta + jw)]$$
$$= K\,i_p\,(\beta - \alpha)/(\alpha + jw)\,(\beta + jw)$$

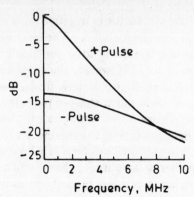

Frequency, MHz

Fig. 6.3.   Bode frequency plot of positive and negative corona pulses.

The amplitude is

$$A\,(w) = K\cdot i_p\cdot(\beta - \alpha)/\sqrt{(\alpha^2 + w^2)\,(\beta^2 + w^2)} \qquad (6.4)$$

At low frequencies when $w \ll \alpha$ and β, the amplitude varies as $K\,i_p\,(\beta - \alpha)/\alpha\beta$, showing that it is independent of frequency. At high frequencies when $w \gg \alpha$ and β, $A(w) = K\,i_p\,(\beta - \alpha)/w^2$. On a log-log plot this has a slope of -2. Assuming that both the positive and negative pulses are double-exponential in shape with the timings 50/150 and 20/50 ns, the values of α and β are:

Positive Pulse : $-\alpha = 10.5 \times 10^6$, $\beta = 34.65 \times 10^6$

Negative Pulse: $-\alpha = 38.3 \times 10^6$, $\beta = 83 \times 10^6$

Table 6.1 shows details of calculation of amplitude vs. frequency from 0 to 10 MHz by taking the reference (0 dB) at $f = 0$ for the positive pulse. At $f = 0$, the amplitude of negative pulse frequency spectrum is 13.5 dB below that for positive pulse.

Table 6.1   Amplitude-Frequency Spectrum of Pulses

f	0	0.5	1	1.5	2	4	6	8	10	MHz
Positive	0	−.4	−1.47	−2.9	−4.4	−10	−15	−19	−22	dB
Negative	−13.5	−13.54	−13.64	−13.8	−14	−15.4	17.3	−19.2	−21.1	dB

These are plotted in Fig. 6.3. Evaluation of K, α and β values with given crest time and time to 50% value on tail is carried out in Chapter 13. The equations are however given here.

Let $x = \beta/\alpha$ and $y = t_i/t_p$ which is given

Then, $x\,(y - 1)\,.\ln(x) = (x - 1)\,.\ln\{2\,(x^y - 1)/(x - 1)\}$ \qquad (6.5)

$$\alpha = \ln (x)/(x - 1) \, t_p \qquad (6.6)$$

and $K = 1/(e^{-\alpha t_p} - e^{-\beta t_p})$ $\qquad\qquad\qquad$ (6.7)

First, equation (6.5) is solved by trial and error for x and α found from (6.6). Then $\beta = x \, \alpha$. Finally, K is determined from calculated values of α, β and the known value of crest time t_p.

Example 6.1:   A double-exponential pulse has a crest time of $t_p = 50$ ns, and time to 50% value on tail equal to $t_t = 150$ ns.   Calculate α, β and K, and write the equation to the pulse in terms of the peak value i_p.

Solution: $y = t_t/t_p = 3$.   Hence the equation for $x = \beta/\alpha$ is

$$2x \ln (x) = (x - 1) \ln \{2 \, (x^3 - 1)/(x - 1)\}.$$

A trial and error solution yields $x = \beta/\alpha = 3.45$.

$$\alpha = \ln (x)/t_p (x - 1) = \ln (3.45)/2.45 \times 50 \times 10^{-9} = 10 \times 10^6.$$

$\therefore$ $\qquad\qquad\qquad\qquad$ $\beta = 34.5 \times 10^6$.

Also, $\qquad\qquad\qquad$ $\alpha t_p = 0.50546$ and $\beta t_p = 1.7438$.

Finally, $\qquad\qquad\qquad$ $K = 1/(e^{-\alpha t_p} - e^{-\beta t_p}) = 2.3346$.

The equation to the pulse is $i \, (t) = 2.3346 \, i_p \, (e^{-10^7 t} - e^{-3.45 \times 10^7 t})$. If time is measured in nanoseconds,

$$i(t) = 2.3346 i_p \, (e^{-0.01t} - e^{-0.0345t}).$$

6.2   Properties of Pulse Trains and Filter Response

Radio Interference (RI) level is governed not only by the amplitude and waveshape of a single pulse but also by the repetitive nature of pulses in a train.   On an ac transmission line, as mentioned earlier, positive pulses from one single point in corona occur once in a cycle or at the most 2 or 3 pulses are generated near the peak of the voltage.   In rain, the number of pulses in one positive half cycle shows an increase. Therefore, the number of pulses per second on a 50 Hz line from a single point in fair weather range from 50 to 150 and may reach 500 in rain.   Since there exist a very large number of points in corona, and the pulses occur randomly in time without correlation, the frequency spectrum is band-type and not a line spectrum.   It is also found that in fair weather there exists a certain shielding effect when one source in corona does not permit another within about 20 to 50 cm.   This is verified by photographs taken at night when plumes of bluish discharges occur at discrete points.   However, in rain, there is a continuous lumi- nous envelope around a conductor.   It is therefore a matter of some difficulty in actually ascertaining the repetition rate of pulses as seen by the input end of a noise meter, which is either a rod antenna or a loop antenna.   Fortunately, this is not as serious as it looks, since the integrated response of a standard noise-meter circuit is practically independent of the pulse repetition rate if the number of pulses per second (pps) is less than the bandwidth frequency of the filter in the meter weighting circuit. This is discussed below for a simple case of rectangular pulses with periodic repetition. The analysis can be extended to include finally the actual case of randomly-occurring pulse trains with double-exponential shape. But this is a highly advanced topic suitable for experts involved in design of noise meters.

Consider Fig. 6.4 showing rectangular pulses of amplitude A and width τ having a periodic time T and repetition frequency $f = 1/T$ pulses per second. This is an even function and the Fourier Series for this type of pulse train contains only cosine terms. The amplitude of any harmonic is

$$F(k) = \frac{4}{T} \int_0^\tau A \cos 2\pi kft \, dt = \frac{4A}{2\pi k} \sin \frac{2\pi k\tau}{T} \qquad (6.8)$$

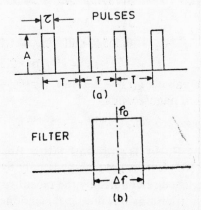

Fig. 6.4. (a) Pulse train: Amplitude A, width τ, period T.

(b) Ideal bandwidth-limited filter with centre frequency f_0 and bandwidth Δf.

The Fourier Series is

$$F(w) = \sum_{k=1}^\infty \left(\frac{4A}{2\pi k} \sin \frac{2\pi k\tau}{T} \right) \cos kwt \qquad (6.9)$$

Let such a signal be passed through an ideal filter with bandwidth Δf and steep cut-off, as shown in Fig. 6.4(b). Then, the number of harmonics passed will be

$$N = \Delta f / f = \Delta f . T. \qquad (6.10)$$

If the meter is tuned to a centre frequency $f_0 = nf$, the output will contain harmonic terms from $k = (n - \frac{1}{2}N)$ to $(n + \frac{1}{2}N)$. The output of the filter is then

$$\left[\frac{4A}{2\pi} \frac{1}{n - \frac{1}{2}N} \sin 2\pi \left(n - \frac{N}{2} \right) f\tau \right] \text{ to } \left[\frac{4A}{2\pi} \frac{1}{n + (N/2)} \sin 2\pi \left(n + \frac{N}{2} \right) f\tau \right]$$

Certain approximations can be made to obtain a workable expression when we consider what happens in an actual situation in practice. The tuned frequency is about $f_0 = 1$ MHz and the repetition frequency f can be considered as $f = 1000$ pps. The meter bandwidth is 5 KHz for ANSI meters and 9 KHz for CISPR meters of European design. Therefore,

$$n = f_0/f = 10^6/10^3 = 1000.$$

Since the bandwidth is 5 KHz and the harmonics of the pulse train are separated by 1000 Hz, only 5 or 6 harmonic components will pass in the 5 KHz bandwidth. For $\Delta f = 9$ KHz, about 9 or 10 harmonic components will pass. Thus, $(n - N/2)$ and $(n + N/2)$ range from 997 to 1003 and we can approximate both these with a value of 1000 which is f_0/f. The output of the filter will then be the sum of harmonic terms such as

$$\frac{4A}{2\pi} \frac{f}{f_0} \left[\sin 2\pi \left(f_0 - \frac{\Delta f}{2} \right) \tau \text{ to } \sin 2\pi \left(f_0 + \frac{\Delta f}{2} \right) \tau \right]$$

where the pulse width τ is of the order of 100 ns $= 10^{-7}$.

Now, $\quad \sin \left(2\pi f_0 \tau - 2\pi \frac{\Delta f}{2} \tau \right) = \sin 2\pi f_0 \tau \cdot \cos 2\pi \frac{\Delta f}{2} \tau -$

$$\cos 2\pi f_0 \tau \cdot \sin 2\pi \frac{\Delta f}{2} \tau, \qquad (6.11)$$

Since $\quad 2\pi \dfrac{\Delta f}{2} \tau \approx 2\pi \times 2500 \times 10^{-7} = 157 \times 10^{-5}$,

we can write $\cos 2\pi \dfrac{\Delta f}{2} \tau = 1$ and $\sin 2\pi \dfrac{\Delta f}{2} \tau = 0$. Then, the output of the filter will be nearly

$$\frac{4A}{2\pi} \cdot \frac{f}{f_0} \cdot N \cdot \sin 2\pi f_0 \tau \cdot \cos 2\pi f_0 t \qquad (6.12)$$

where we have introduced the time function of equation (6.9). The final output can also be written as

$$\frac{4A}{2} \cdot \frac{f}{f_0} \cdot \sin 2f_0 \tau \cdot \cos 2f_0 t \qquad (6.13)$$

since $Nf = \Delta f$, the bandwidth, according to (6.10).

The output of the filter is a cosine wave at frequency f_0, the tuned frequency (1 MHz, say) and is modulated by the amplitude

$$\frac{2A}{\pi} \cdot \frac{\Delta f}{f_0} \cdot \sin 2\pi f_0 \tau = \frac{2A}{\pi} \cdot \Delta f \cdot \frac{\sin 2\pi f_0 \tau}{2\pi f_0 \tau} (2\pi \tau)$$

$$= 4(A\tau) \cdot (\Delta f) \cdot \frac{\sin 2\pi f_0 \tau}{2\pi f_0 \tau} \qquad (6.14)$$

For low tuned frequencies, the sigma factor $\dfrac{\sin 2\pi f_0 \tau}{2\pi f_0 \tau}$ is nearly 1. Thus, at low frequencies f_0, the response of the filter is nearly flat and rolls off at higher frequency. The following salient properties can also be noted from equation (6.14).

(1) The response of the filter is directly proportional to $(A\tau)$ = area of the pulse.

(2) The response is proportional to (Δf) = bandwidth, provided the number of harmonics passed is very low.

(3) The response is proportional to the Si-factor.

That the response of the filter is flat up to a certain repetition frequency of pulses is not surprising since as the repetition frequency increases, the tuned frequency becomes a lower order harmonic of the fundamental frequency with resulting higher amplitude. But there is a corresponding reduction in the number of harmonics passed in the bandwidth of the filter giving an output which is nearly equal to that obtained at a lower tuned frequency. Therefore, changes in repetition frequency of the pulses in a train affect the noise level only to a small degree.

Since the amplitude-duration product of the pulse determines the output, it is evident that a positive corona pulse yields much higher noise level than a negative corona pulse. In practice, we omit negative-corona-generated radio interference.

6.3 Limits for Radio Interference Fields

Radio Interference (RI) resulting from a transmission line is a man-made phenomenon and as such its regulation should be similar to other man-made sources of noise as mentioned in Chapter 5, such as audible noise, automobile ignition noise, aircraft noise, interference from welding equipment, r − f heating equipment, etc. Some of these are governed by *IS* 6842. Legislation for fixing limits to all these noise sources is now gaining widespread publicity and awareness in public in order to protect the environment from all types of pollution, including noise. Interference to communication systems is described through Signal-to-Noise Ratio designated as S/N Ratio, with both quantities measured on the same weighting circuit of a suitable standard meter. However, it has been the practice to designate the signal from a broadcast station in terms of the average signal strength called the Field Intensity (FI) setting of the field-strength meter, while the interference signal to a radio receiver due to line noise is measured on the Quasi-Peak (QP) detector circuit. The difference in weighting circuits will be discussed later on. There are proposals to change this custom and have both signal and noise measured on the same weighting circuit. This point is mentioned here in order that the reader may interpret S/N ratios given by public utility organizations in technical literature or elsewhere since interference problems result in expensive litigations between contesting parties.

As mentioned earlier, it is the duty or responsibility of a designer to keep noise level from a line below a limiting value at the edge of the right-of-way (R-O-W) of the line corridor. The value to be used for this RI limit is causing considerable discussion and we shall describe two points of view currently used in the world. Some countries, particularly in Europe, have set definite limits for the RI field from power lines, while in North America only the minimum acceptable S/N ratio at the receiver location has been recommended. We will examine the rationale of the two points of view and the steps to be followed in line design.

When a country is small in size with numerous towns with each having its own broadcast station and a transmission line runs close by, it is easy on the design engineer of the line if a definite RI limit is set and station signal strengths are increased by increasing the transmitter power in order to yield satisfactory quality of radio reception to all receivers located along the line route. The following Table 6.2, gives limits set by certain European countries.

Table 6.2.   RI Limits in Various Countries of the World

Country	Distance from Line		RI Limit	Frequency	Remarks
1) Switzerland	20 m from outermost phase		200 μV/m (46 dB above 1 μV/m)	500 KHz	Dry weather 10°C
2) Poland	20 m from outerphase		750 μV/m (57.5 dB)	500 KHz ±10 KHz	Air humidity <80% Temp. 5°C
3) Czechoslovakia	Voltage kV	Distance from line centre			
	220	50 m			
	400	55	40 dB	500 KHz	Air humidity =70%
	750	70			Dry weather
4) U.S.S.R.	100 m from outerphase		40 dB	500 KHz	For 80% of the year limit should not be exceeded

One immediate observation to make is that there is no uniformity even in a small area such as Europe, excluding the U.S.S.R. In countries like France where a large rural population exists, no set RI limit is specified since broadcast stations are located far from farming communities who have to be assured satisfactory S/N ratio.

154 EXTRA HIGH VOLTAGE AC TRANSMISSION

In North America the following practice is adopted in the U.S.A. and Canada.

U.S.A. Recommended practice is to guarantee a minimum S/N ratio of 24 dB at the receiver for broadcast signals having a minimum strength of 54 dB at the receiver.

Canada. For satisfactory reception, a S/N ratio of 22 dB or better must be provided in fair weather in suburban regions for stations with a mean signal strength of 54 dB (500 μV/m). In urban regions, this limit can be increased by 10 dB, and in rural areas lowered by 3 dB.

Based on these two points of view, namely, (1) setting a definite RI limit, and (2) providing a minimum S/N ratio at the receiver, the procedure required for line design will be different, which are outlined here.

(1) When RI limit in dB or μV/m is specified at a particular frequency and weather condition, it is only necessary to calculate the lateral decrement or profile of RI. This is the attenuation of the noise signal as one proceeds away from the line for an assumed line configuration (The procedure for calculation of lateral profile will be outlined in Section 6.6 and following sections). By taking a large number of alternative line designs, a choice can be made of the most suitable conductor configuration.

(2) When line design is based upon S/N ratio, measurement of broadcast-station signals must be carried out all along a proposed line route of all stations received. This can also be calculated provided the station power, frequency, and distance to the receiver are known. The allowable noise from the line at these frequencies is then known from the S/N ratio value. The R-O-W can be specified at every receiver location for a chosen size of conductor and line configuration (line height and phase spacing). The procedural difficulties involved in this method are illustrated as follows. Consider that at a farming community where a future line may pass nearby, the station field strengths are recorded and a S/N ratio of 24 dB must be allowed. The table below shows an example of station strengths and allowable noise at the station frequencies.

Frequency of Station	0.5	0.8	1	1.1	1.3	1.52 MHz
Received Signal Strength	55	60	50	75	57	52 dB
Allowable Noise Level (signal strength-24 dB)	31	36	26	51	33	28 dB

If we assume that corona-generated noise has a frequency spectrum such as shown in Fig. 6.3 which varies nearly as $f^{-1.5}$, then the noise level in relation to that at 1 MHz taken as reference can also be tabulated.

Since the allowable noise level at 1 MHz is 26 dB, the permissible noise at other frequencies are determined. When this is done, it will become clear that certain stations will not be guaranteed a minimum S/N ratio of 24 dB, as shown below.

Frequency, MHz	0.5	0.8	1.0	1.1	1.3	1.52 MHz
Corona Noise Adder	9	4	0	−1	−2	−3.6 dB
Allowed Noise	35	30	26	23	23	22.4 dB
S/N Ratio, dB	20	30	24	50	34	29.6 dB

Therefore, for the station broadcasting at 0.5 MHz, the recommended minimum S/N ratio of 24 dB cannot be guaranteed. The situation is represented pictorially in Fig. 6.5. This leads to the conclusion that at any given receiver location, with a chosen line design, all stations received cannot be guaranteed satisfactory quality of reception with a given width of R-O-W. It is highly uneconomical to increase the width of R-O-W to accommodate all radio stations. Therefore, regulatory bodies must also specify the number of stations (or percentage) received at a village or town for which satisfactory reception can be guaranteed. This may usually be 50% so that listeners have the choice of tuning into at least 50% of the stations for which satisfactory reception is guaranteed.

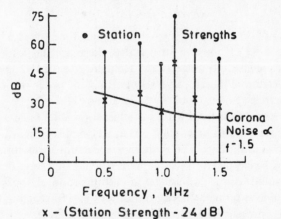

Fig. 6.5.   Station signal strength (·), −24dB(×), and corona-generated noise (—). Illustrating basis for design based upon minimum S/N ratio of 24 dB.

This process has to be repeated at all villages and towns or other locations such as military establishments, etc., along the proposed line route and a best compromise for line design arrived at. Once a definite RI limit is set at a fixed lateral distance from the line as controlled by the S/N ratio, which in the above example was 26 dB at 1 MHz at the edge of R–O–W, the line design follows similar lines as for case (1) where the limit is specified to start with under legislation of the country.

Referring to Fig. 6.5, any X-mark falling below the corona-generated noise curve represents a station for which the minimum S/N ratio of 24 dB cannot be obtained. Therefore, quality of reception for such a station at the receiver location will be unsatisfactory.

6.4 Frequency Spectrum of the RI Field of Line

The frequency spectrum of radio noise refers to the variation of noise level in μV or μV/m (or their dB values referred to 1 μV or 1 μV/m) with frequency of measurement. The frequency-spectrum of a single corona pulse of double-exponential shape was found in Section 6.1 to be

$$A(w) = K \cdot i_p \cdot (\beta - \alpha)/\sqrt{(\alpha^2 + w^2)(\beta^2 + w^2)} \tag{6.4}$$

On a long line, there exist a very large number of points in corona and a noise meter located in the vicinity of the line (usually at or near ground level) responds to a train of pulses originating from them. The width of a single pulse is about 200 ns (0.2 μs) while the separation of pulses as seen by the input end of the meter could be 1 μs or more. Therefore, it is unusual for positive pulses to overlap and the noise is considered as impulsive. When pulses overlap, the noise is random. Measurements indicate that from a long line, the RI frequency spectrum follows closely

$$RI(w) \propto f^{-1} \text{ to } f^{-1.5} \tag{6.15}$$

Thus, at 0.5 MHz the noise is 6–9 dB higher than at 1 MHz, while at 2 MHz it is 6–9 dB lower. In practice, these are the adders suggested to convert measured noise at any frequency to 1 MHz level. The frequency spectrum is therefore very important in order to convert noise levels measured at one frequency to another. This happens when powerful station signal interferes with noise measurements from a line so that measurements have to be carried out at a frequency at which no broadcast station is radiating. The frequency spectrum from corona-generated line noise is nearly fixed in its characteristic so that any deviation from it as measured on a noise meter is an indication of sources other than the line, which is termed "background noise". In case a strong source of noise is present nearby, which is usually a factory with motors that are sparking or a broken insulator on the tower, this can be easily recognized since these usually yield high noise levels up to 30 MHz and their frequency spectrum is relatively flat.

6.5 Lateral Profile of RI and Modes of Propagation

The most important aspect of line design from interference point of view is the choice of conductor size, number of sub-conductors in bundle, line height, and phase spacing. Next in importance is the fixing of the width of line corridor for purchase of land for the right-of-way. The lateral decrement of radio noise measured at ground level as one moves away from the line has the profile sketched in Fig. 6.6. It exhibits a characteristic double hump within the space between the conductors and then decreases monotonically as the meter is moved away from the outer phase. For satisfactory radio reception, a limit RI_l is determined as explained in the previous section. No receiver should be located within the distance d_0 from the outer phase or d_c from the line centre. Therefore, it becomes essential to measure or to be able to calculate at design stages the lateral profile very accurately from a proposed line in order to advise regulatory bodies on the location of receivers. In practice, many complaints are heard from the public who experience interference to radio broadcasts if the line is located too close to their homesteads when the power company routes an e.h.v. line wrongly. In such cases, it is the engineer's duty to recommend remedies and at times appear as witness in judicial courts to testify on the facts of a case.

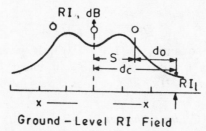

Fig. 6.6. Lateral profile of RI at ground level for fixing width of right-of-way of line.

We will discuss this lateral profile in great detail and dissect it into several components which belong to different modes of propagation, as discussed in Chapter 3, for the radio-frequency energy on the multi-conductor line. This is the basis for determining the expected noise profile from a chosen conductor size and line configuration in un-transposed and fully-transposed condition. We consider 6 preliminary cases of charge distribution on the line conductors after which we will combine these suitably for evaluating the total noise level of a line. In all these cases, the problem is to calculate the field strength at the location of a noise meter when the r-f charge distribution is known. Here, we consider the vertical component of ground-level field intensity which can be related to the horizontal component of magnetic field intensity by the characteristic impedance of free space. We restrict our attention to horizontal 3-phase line for the present. In every case, only the magnitude is of concern.

Case 1: Single Conductor above Ground

Consider the simplest of all cases of a single conductor carrying a charge of q coulombs/meter at radio frequency above a perfectly-conducting ground surface at height H, Fig. 6.7. As mentioned in Chapter 3, the effect of ground in all such problems is replaced by an image charge-q at depth H below the ground surface. It is desired to evaluate the vertical component of electric field strength at point M at a lateral distance d from the conductor on the ground surface.

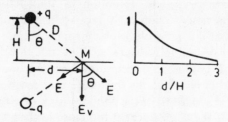

Fig. 6.7. Single conductor: (a) Vertical component of ground-level electric field and (b) lateral profile.

The vertical component due to $+q$ and $-q$ will be

$$E_v = 2 \times \frac{q}{2\pi e_0} \frac{H}{H^2 + d^2} = \frac{q}{\pi e_0 H} \frac{1}{1 + (d/H)^2}$$
(6.16)

The dimensionless quantity $[1/\{1 + (d/H)^2\}]$ is given the name "Field Factor", F. It has a value of 1 at $d = 0$ under the conductor and decreases to 0.1 at $d = 3H$ or $d/H = 3$, as shown in Fig. 6.7(b)

Case 2: *3-Phase AC Line—Charges* $(+ q, + q, + q)$
On a perfectly-transposed line, the line-to-ground mode carries equal charges q of the same polarity as described in Chapter 3, and shown in Fig. 6.8(a). These are obtained from the eigen-values and eigen-vector and their properties. Following the procedure for case 1 of a single conductor, the field factor for this case is

$$F_{la} = E_v/(q/\pi e_0) = \frac{1}{1 + (d+s)^2/H^2} + \frac{1}{1 + (d/H)^2} + \frac{1}{1 + (d-s)^2/H^2}$$
(6.17)

where s = phase spacing and d = the distance to the noise meter or radio-receiver from the line centre.

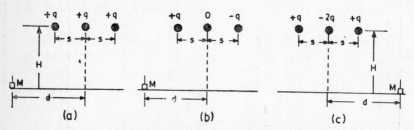

Fig. 6.8. Charge distributions at r-f on 3-phase lines:
(a) 1st or line-to-ground mode.
(b) 2nd or line-to-line mode of 1st kind.
(c) 3rd or line-to-line mode of 2nd kind.

Case 3:   *3-Phase AC Line—Charges* $(+ q, 0, -q)$
From Fig. 6.8(b), the field factor will be

$$F_{2a} = E_v/(q/\pi e_0) = \frac{1}{1 + (d - s)^2/H^2} - \frac{1}{1(d + s)^2/H^2} \qquad (6.18)$$

Case 4:   *3-Phase AC Line—Charges* $(+ q, - 2q, + q)$
The field factor for this case from Fig. 6.8(c) is

$$F_{3a} = E_v/(q/\pi e_0) = \frac{1}{1 + (d + s)^2/H^2} + \frac{1}{1 + (d - s)^2/H^2} - \frac{2}{1 + (d/H)^2} \qquad (6.19)$$

Case 5:   *Bipolar DC Line—Charges* $(+ q, + q)$
On a bipolar dc line, the two modes of propagation yield charge distributions $(+ q, + q)$ and $(+ q, - q)$ on the two conductors in each mode, as will be explained later.   Considering the first or line-to-ground mode with charges $(+ q, + q)$, as shown in Fig. 6.9(a), with pole spacing P, the field factor is

$$F_{ld} = E_v/(q/2\pi e_0) = \frac{1}{1 + (d + 0.5P)^2/H^2} + \frac{1}{1 + (d - 0.5P)^2/H^2} \qquad (6.20)$$

Case 6:   *Bipolar DC Line—Charges* $(+ q, - q)$
If the polarity of one of the charges is reversed, the resulting field factor is, see Fig. 6.9(b),

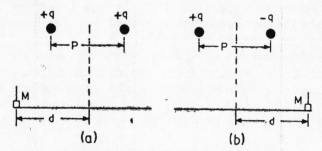

Fig. 6.9.   Charge distributions in the 2 modes of bipolar dc line:
(a) line-to-ground mode.   (b) line-to-line mode.

$$F_{2d} = \frac{1}{1 + (d - 0.5P)^2/H^2} - \frac{1}{1 + (d + 0.5P)^2/H^2} \qquad (6.21)$$

We will plot these in order to observe their interesting and salient properties. Fig. 6.10 shows such plots of only the magnitudes of the field factors since a rod antenna of a noise meter picks up these.   For purposes of illustration we take $S/H = P/H = 1$.

For cases 1, 2 and 5 where the charges on the conductors are of the same polarity, the vertical component of electric field decreases from a maximum under the line centre monotonically as the meter is moved

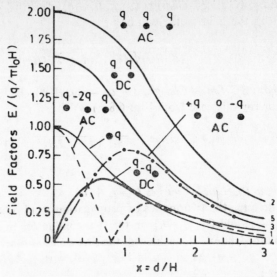

Fig. 6.10.   Plot of field factors for charge distributions of Figs. 6.7 to 6.9.

along the ground away from the line.   For cases 3 and 6 with charge distributions $(+ q, 0, - q)$ and $(+ q, - q)$, we observe that field is zero at the line centre, reaches a maximum value and then decreases monotonically.   A combination of field profiles of cases 2 and 3 (or 5 and 6) yield the characteristic double hump of Fig. 6.6.   For case 4 with the charge distribution $(+ q, - 2q, + q)$, the field commences at a high value under the line centre, reaches zero, and then after increasing to a maximum value decreases monotonically.

These different types of r-f charge distributions occur when corona-generated current, voltage, charge and power propagate on the line conductors which can be resolved into modes of propagation.   This has already been discussed in Chapter 3.   In the next sections we will give methods of calculating the total RI level of a line from the different modal voltages.

6.6 The CIGRE Formula

The problem of radio interference as a controlling factor in design of line conductors became very acute and came to focus in early 1950's with the planned 500 kV lines. The growth of transmission lines beyond 345 kV became very rapid and RI measurements also became very widespread from cage models, outdoor experimental lines and actual lines in operation.   In the 1960's the voltage level increased to 735 kV and 765 kV.   Based on all RI data gathered over a number of years and from lines of various configurations, the CIGRE and IEEE evolved an empirical formula relating most important line and atmospheric parameters with the radio noise level.   This has come to be known as the CIGRE Formula.   There are about eight empirical formulas available from experience gained by different countries, but we will deal only with the CIGRE Formula here. The important quantities involved in the empirical formula are:

(1) conductor radius, r, or diameter $d = 2r$;
(2) maximum surface voltage gradient on conductor, g_m;
(3) aerial distance from conductor to the point where RI is to be evaluated, D;
(4) other factors, such as frequency and climatic conditions.

The basic formula is, referring to Fig. 6.11,

$$RI_i(dB) = 3.5g_m + 6d - 33 \log_{10} (D_i/20) - 30 \qquad (6.22)$$

This RI level is from conductor i at an aerial distance D_i from conductor to the point M.

There are several restrictions on the use of this formula. It applies when

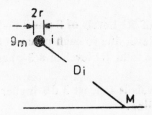

Fig. 6.11. CIGRE formula for evaluating RI.

(a) the values of g_m and d are in centimetre units;
(b) the aerial distance D_i is in metres and $D_i > 20$ m;
(c) the frequency is 0.5 MHz;
(d) the number of sub-conductors N in the bundle is less than or equal to 4. This is true of lines up to 765 kV;
(e) the ratio of bundle spacing B between sub-conductors to the conductor diameter lies between 12 and 20;
(f) the weather condition is average fair weather;
(g) the RI level has a dispersion of $\pm$ 6 dB.

Example 6.2: A 400-kV line has conductors in horizontal configuration at average height $H = 14$ m and phase spacing $S = 11$ m, as shown in Fig. 6.12. The conductors of each phase are 2×0.0318 m diameter at $B = 0.4572$ m spacing. Calculate the RI level of each phase at a distance of 30 m from the outer phases at ground level at 0.5 MHz at 420 kV using the CIGRE formula.

Solution. The first step is to calculate the maximum surface voltage gradient on the three phases at 420 kV using the Mangoldt formula.

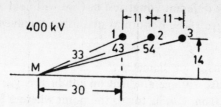

Fig. 6.12. Calculation of RI level of 400 kV line using CIGRE formula.

Calculations give $g_{mc} = 17.3$ kV/cm on the centre phase and $g_{mo} = 16.2$ kV/cm on the two outer phases.

Aerial distances: $D_1 = 33$ m, $D_2 = 43$ m, $D_3 = 54$ m.

RI (1) = $3.5 \times 16.2 + 6 \times 3.18 - 30 - 33 \log (33/20) = 38.6$ dB

RI (2) = $3.5 \times 17.3 + 6 \times 3.18 - 30 - 33 \log (43/20) = 38.7$ dB

RI (3) = $3.4 \times 16.2 + 6 \times 3.18 - 30 - 33 \log (54/20) = 31.6$ dB.

6.6.1 Rules for Addition of RI Levels of 3-phases—S/C Line
Having calculated the RI level due to each phase at the measuring point, the rules for evaluating the total RI level of a 3-phase single-circuit line are as follows:

(a) If one of the RI levels is at least 3 dB higher than the rest, then this is the RI level of the line.

(b) If only one of the three RI levels is at least 3 dB lower than the rest, then the RI level of line is

RI = (average of the two highest + 1.5) dB.

(c) At 1 MHz, the RI level is 6 dB lower.

(d) For evaluating the RI level in rain, add 17 dB.

Example 6.3: In the previous example, calculate the RI level of the line at the measuring point at 0.5 MHz and 1 MHz in fair weather.

Solution: RI (3) is lower than the others by more than 3 dB.

$\therefore$ RI$_{line} = \frac{1}{2}(38.6 + 38.7) + 1.5 = 40.15$ dB at 0.5 MHz and 34.15 dB at 1 MHz.

Example 6.4: In the above example, if the RI limit is given to be 40 dB at 1 MHz, calculate the width of right-of-way of the line corridor.

Solution: We observe that at 30 m from the outer phases, the RI level is 34.15 dB. Therefore, the value of 40 dB will be obtained at less than this distance. We will find the value of d where RI = 40 dB at 1 MHz by trial and error.

For the outer phases, $3.5 \times 16.2 + 6 \times 3.18 - 30 = 45.78$

For the centre phase, $3.5 \times 17.3 + 6 \times 3.18 - 30 = 49.63$.

Assume $d = 25, 20, 15$ metres and calculate the RI level of line at 1 MHz.

d	D_1	D_2	D_3	$33 \log$ $D_1/20$	$33 \log$ $D_2/20$	$33 \log$ $D_3/20$	$RI(1)$	$RI(2)$	$RI(3)$
25	28.7	38.6	49	5.15	9.43	12.83	40.6	40.2	33
20	24.4	34	44.3	2.85	2.85	11.4	42.9	42	34.4
15	20.5	29.5	39.6	0.367	5.58	9.97	45.4	44	36

The resulting RI levels are

$d = 25$, RI $= 41.9$ at 0.5 MHz, 35.9 dB at 1 MHz
$d = 20$, RI $= 44$, 38
$d = 15$, RI $= 46.2$, 40.2

$\therefore$ $d = 15$ metres at the edge of R-O-W, and the width of line corridor required is $2(d + s) = 52$ m. Since the CIGRE formula has a dispersion of ± 6 dB, further calculation may not be necessary. With this dispersion, a width of R-O-W giving 46 dB at 1 MHz at the edge of the line corridor may be acceptable.

6.6.2 Rules for Addition of RI Levels for a D/C Line

E.H.V. lines are usually single-circuit lines, but there are some double-circuit and four-circuit lines at 400 kV in the world. We will give rules that apply only for a double-circuit line here. The reader is recommended for advanced research papers and reports for 4-circuit line problems.

On a double-circuit line, there are two phase conductors belonging to each phase. Let RI_{A1} and RI_{A2} be the RI values at any point M on ground due to phase A which can be evaluated individually by the CIGRE formula. Then the resulting RI value due to the two will be given as

$$RI_A = \sqrt{RI_{A_1}^2 + RI_{A_2}} \tag{6.23}$$

Similarly $RI_B = \sqrt{RI_{B_1}^2 + RI_{B_2}^2}$ and $RI_C = \sqrt{RI_{C_1}^2 + RI_{C_2}^2}$ (6.24)

These three quantities are now treated as the contributions from the three phases and the rules for adding them are the same as for a single-circuit line given before.

The reason for quadratic addition is based on the property that the pulses causing the noise from any one phase are time correlated from its two conductors so that energies or powers are added arithmetically in the noise-meter circuitry. It is known from experiments that if there are N identical noise sources which are correlated in time, that is, they occur on the same conductor, then the resulting meter reading is

$$RI(N) = \sqrt{N} \times RI \text{ due to each source acting individually}$$

$\therefore$ $[RI(N)]^2 = N \times (RI)^2$ (6.25)

The same concept can be extended to any number of circuits on a tower.

6.7 The RI Excitation Function

With the advent of voltages higher than 750 kV, the number of sub-conductors used in a bundle has become more than 4 so that the CIGRE formula does not apply. Moreover, very little experience of RI levels of 750 kV lines were available when the CIGRE formula was evolved, as compared to the vast experience with lines for 230 kV, 345 kV, 400 kV and 500 kV. Several attempts were made since the 1950's to evolve a rational method for predicting the RI level of a line at the design stages before it is actually built when all the important line parameters are varied. These are the conductor diameter, number of sub-conductors, bundle spacing or bundle radius, phase spacing, line height, line configuration (horizontal or delta), and the weather variables. The most important concept resulting from such an attempt in recent years is the "Excitation Function" or the "Generating Function" of corona current injected at a given radio frequency in unit bandwidth into the conductor. This quantity is determined experimentally from measurements carried out with short lengths of conductor strung inside a cylindrical or rectangular cage, as described in Chapter 4, or from short outdoor overhead experimental lines. It can also be predicted from existing long-line measurements and extrapolated to other line configurations.

Consider Fig. 6.13 which shows a source of corona at S located at a distance X from one end of a line of length L. According to the method using the Excitation Function to predict the RI level with given dimensions and conductor geometry, the corona source at S on the conductor generates an excitation function I measured in $\mu A/\sqrt{m}$. The line has a surge impedance Z_0 so that the r-f power generated per unit length of line is

$$E = I^2 Z_0 \tag{6.26}$$

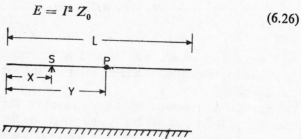

Fig. 6.13.   The excitation function and its propagation on line for RI calculation.

Under rain, a uniform energy or power per unit bandwidth is generated so that in a differential length dx, the power generated is $(E.dx)$. In this method, we calculate the RI level under rain first and deduct 17 dB to obtain fair-weather RI. This power will split equally in two directions and travel along the line to reach the point P at a distance $(y - x)$ from the source S. In doing so, it will attenuate to the value $e^{-2a(y-x)}$, where $a =$ attenuation factor for voltage in Nepers per unit length. Therefore, the total energy received at P due to all sources to the left of P will be

$$E_L = \int_0^y \frac{1}{2} (E \cdot dx) \cdot e^{-2a(y-x)} = \frac{E}{4a} (1 - e^{-2ay}) \qquad (6.27)$$

Similarly, the energy received at P due to all sources to its right will be

$$E_R = \frac{E}{4a} (1 - e^{-2a(L-y)}) \qquad (6.28)$$

For a line of finite length, repeated reflections occur from the ends, but for a very long line these are not of consequence. Also, unless the point P is located very close to the ends, the exponential terms can be neglected. Therefore, the total r-f energy received at P will be

$$E_P = E/2a \qquad (6.29)$$

which shows that all points on a long line receive the same r-f energy when the corona generation is uniform.

Example 6.5: For attenuation factors of 6, 1 and 0.17 dB/km, calculate the energy received at point P in terms of energy generated per unit length.

Solution: The conversion factor from dB to Nepers is 8.7. Thus, $a = 0.69$ N/km for 6 dB/km, 0.115 N/km for 1 dB/km, and 0.019 N/km for 0.17 dB/km.

$$\therefore \quad E_P = E/2a = 0.7246 \times \text{energy generated per km}$$

$$= 724.6 \times \text{energy generated per metre for } a = 6 \text{ dB/km}$$

$$E_P = 4.348 \times \text{energy generated per km for } a = 1 \text{ dB/km}$$

and $\quad E_P = 25.575 \times$ energy generated per km for $a = 0.17$ dB/km.

Associated with the current injected into the conductor per unit length, there is a voltage to ground which is known as the Radio Interference Voltage (RIV). Then,

$$E_P = E/2a = (\text{RIV})^2/Z_0 \qquad (6.30)$$

But $\quad E = I^2 Z_0$ so that $(\text{RIV}) = IZ_0/\sqrt{2a} \qquad (6.31)$

The ground level field depends upon the conductor charge per unit length and the field factor, as shown in Section 6.5. If the capacitance of line per unit length is C, the charge is

$$q = C(\text{RIV}) = ICZ_0/\sqrt{2a} \qquad (6.32)$$

However, for an overhead line, the velocity of propagation, capacitance, and surge impedance are related by

$$v = 1/C\,Z_0 \tag{6.33}$$

$$\therefore \quad q = I/v\,\sqrt{2a} \tag{6.34}$$

Since the charge is found in terms of the excitation function, velocity, and attenuation factor, the resulting RI level of line is

$$RI = \frac{q}{\pi e_0 H} \times \text{Field Factor} = IF/[\pi e_0 H\,v\sqrt{2a}] \tag{6.35}$$

Thus, the quantities involved in estimating the RI level of a line at a specific distance d from the line centre (or a corresponding distance from the outer phase) are the following:

(a) the field factor F which is a function of the line geometry (H, S, d);

(b) the line height H;

(c) the velocity of propagation v;

(d) the attenuation factor, a; and

(e) the injected current or the excitation function, I.

The velocity and attenuation factors are known either by performing suitable experiments on existing lines, or calculated if possible. The field factor is also calculated. However, the excitation function I can only be determined for the conductor under consideration from small-scale experiments using artificial rain apparatus and cages or outdoor overhead lines.

3-Phase Transmission Line

In order to apply the method described above to a 3-phase transmission line and calculate the RI level at a specified distance, the procedure involves resolving the r-f quantities into 3 modes of propagation. Let us consider the line to be perfectly transposed to illustrate the procedure. The transformation matrix $[T]$ and its inverse $[T]^{-1}$ which diagonalize the impedance matrix were found in Chapter 3 to be

$$[T] = \frac{1}{\sqrt{6}}\begin{bmatrix} \sqrt{2}, & \sqrt{3}, & 1 \\ \sqrt{2}, & 0, & -2 \\ \sqrt{2}, & -\sqrt{3}, & 1 \end{bmatrix} \text{ and } [T^{-1}] = \frac{1}{\sqrt{6}}\begin{bmatrix} \sqrt{2}, & \sqrt{2}, & \sqrt{2} \\ \sqrt{3}, & 0, & -\sqrt{3} \\ 1, & -2, & 1 \end{bmatrix} \tag{6.36}$$

For each mode of propagation, the charge on the conductors is

$$\begin{bmatrix} q(1) \\ q(2) \\ q(3) \end{bmatrix} = \begin{bmatrix} I(1)/v(1) & \sqrt{2a(1)} \\ I(2)/v(2) & \sqrt{2a(2)} \\ I(3)/v(3) & \sqrt{2a(3)} \end{bmatrix} \tag{6.37}$$

where the quantities q, I, v and a belong to the mode indicated in the brackets. The modal excitation function is obtained from those of the phases by the transformation

$$[I]_m = \begin{bmatrix} I(1) \\ I(2) \\ I(3) \end{bmatrix} = [T]^{-1}\,[I]_{ph} \tag{6.38}$$

Now, if all the three phase conductors are developing equal intensities of corona, the excitation functions will be equal. Let this be denoted by I. Then,

$$[I]_m = \begin{bmatrix} I(1) \\ I(2) \\ I(3) \end{bmatrix} = \frac{1}{\sqrt6}\begin{bmatrix} \sqrt2, & \sqrt2, & \sqrt2 \\ \sqrt3, & 0, & -\sqrt3 \\ 1, & -2, & 1 \end{bmatrix}\begin{bmatrix} I \\ I \\ I \end{bmatrix} = \begin{bmatrix} I/\sqrt3 \\ 0 \\ 0 \end{bmatrix} \tag{6.39}$$

This shows that only the first or line-to-ground mode has an injected current and the remaining two modes are not present.

$$\therefore \quad q(1) = I/\sqrt3 v(1)\cdot\sqrt{2a(1)},\; q(2) = 0,\; q(3) = 0.$$

Fig. 6.14 shows the modal charge distributions. Converting this back to phase quantities, the r-f charges on the three conductors are

$$\begin{bmatrix} q_1 \\ q_2 \\ q_3 \end{bmatrix} = [T]\begin{bmatrix} q(1) \\ q(2) \\ q(3) \end{bmatrix} = \frac{1}{\sqrt6}\begin{bmatrix} \sqrt2, & \sqrt3, & 1 \\ \sqrt2, & 0, & -2 \\ \sqrt2, & -\sqrt3, & 1 \end{bmatrix}\begin{bmatrix} q(1) \\ 0 \\ 0 \end{bmatrix} = \frac{1}{\sqrt3}\begin{bmatrix} 1 \\ 1 \\ 1 \end{bmatrix}q(1)$$

$$= \begin{bmatrix} 1 \\ 1 \\ 1 \end{bmatrix}[I/3v(1)\,\sqrt{2a(1)}] \tag{6.40}$$

This has the charge distribution (q, q, q) showing that the three conductor charges are equal and of the same polarity. The RI level at the meter

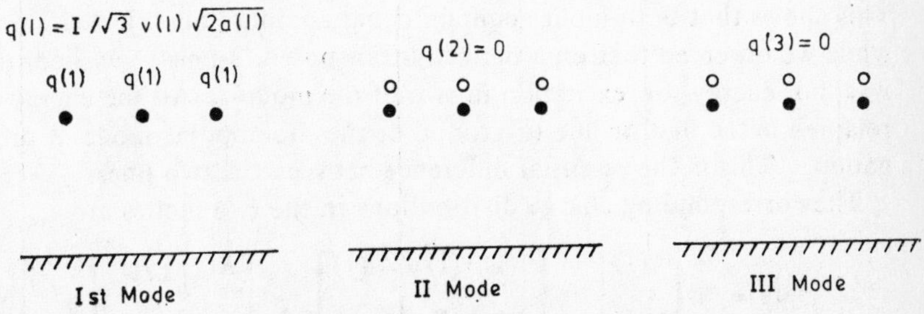

Fig. 6.14.   Modal charge distributions on a fully-transposed 3-phase ac line.

placed on ground will now be controlled by the field factors, as shown in Fig. 6.10.

Thus,
$$RI_1 = \frac{q}{\pi e_0 H} \frac{1}{1 + (d - s)^2/H^2}$$

$$RI_2 = \frac{q}{\pi e_0 H} \frac{1}{1 + (d/H)^2} \qquad (6.41)$$

and
$$RI_a = \frac{q}{\pi e_0 H} \frac{1}{1 + (d + s)^2/H^2}$$

The addition rule according to the CIGRE formula can now be applied to these three RI levels and the RI level of the line calculated. The quantities to be ascertained are I, v and a from measurements. The velocity and attenuation factor pertain to this mode of excitation while the excitation function applies to the conductor usually performed in a cage arrangement in single-phase configuration at the correct surface voltage gradient that exists on the overhead line.

Bipolar DC Line

It is interesting to observe the properties of modal excitation functions on a bipolar dc line and compare these with a perfectly transposed 3-phase ac line.

In Sections 6.1 and 6.2 it was mentioned that radio noise is almost entirely caused by positive corona pulses so that on a bipolar dc line, only the positive-polarity conductor develops measurable r-f energy. Thus, the excitation functions will be $[I]_p = [I_+, 0]_t$. The transformation matrix and its inverse used for diagonalizing the impedance of a bipolar dc line are

$$[T] = [T]^{-1} = \frac{1}{\sqrt{2}} \begin{bmatrix} 1, & 1 \\ 1, & -1 \end{bmatrix} \qquad (6.42)$$

The excitation function in the two modes will be

$$\begin{bmatrix} I(1) \\ I(2) \end{bmatrix} = [T]^{-1} [I]_p = \begin{bmatrix} 1 \\ 1 \end{bmatrix} (I_+/\sqrt{2}) \qquad (6.43)$$

This shows that both modes contain equal amounts of injected current, while we observed that on a perfectly-transposed 3-phase ac line there was no energy or excitation in two of the modes. All the energy was retained in the first or line-to-ground or the homopolar mode of propagation. This is the essential difference between the two lines.

The corresponding charge distributions in the two modes are

$$[q]_m = \begin{bmatrix} q(1) \\ q(2) \end{bmatrix} = \begin{bmatrix} I(1)/v(1)\sqrt{2a(1)} \\ I(2)/v(2)\sqrt{2a(2)} \end{bmatrix} = \begin{bmatrix} K(1) \cdot I(1) \\ K(2) \cdot I(2) \end{bmatrix} \qquad (6.44)$$

Figure 6.15 depicts the modal charge distributions. Converting this back q(1)   q(1)   q(2)   q(2)

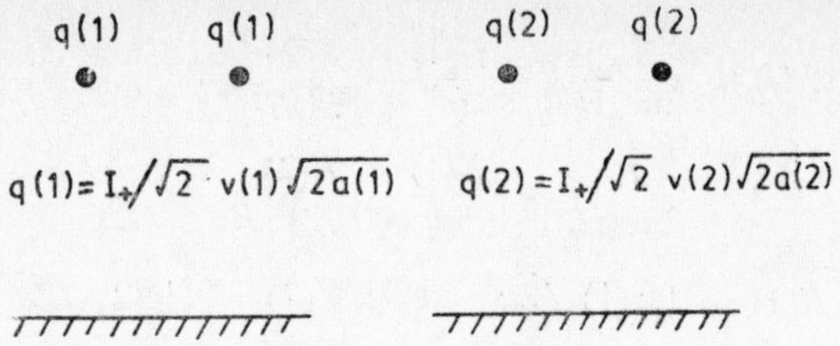

$q(1) = I_+/\sqrt{2} \; v(1)\sqrt{2a(1)}$ $q(2) = I_+/\sqrt{2} \; v(2)\sqrt{2a(2)}$

Fig. 6.15.   Modal charge distributions on a bipolar dc line.

to the charges residing on the poles by multiplying $[q]_m$ by $[T]$, there results

$$[q] = \begin{bmatrix} q_+ \\ q_- \end{bmatrix} = [T][q]_m$$

$$= \frac{I_+}{2} \begin{bmatrix} K(1) + K(2) \\ K(1) - K(2) \end{bmatrix} \tag{6.45}$$

We observe that even though the negative pole is not developing any r-f energy it has a charge due to mutual coupling from the positive pole. The RI level at ground level can be calculated at two representative points M_1, M_2 at the same lateral distance d from the line centre with M_1 closer to the positive pole and M_2 nearer the negative pole, Fig. 6.16.

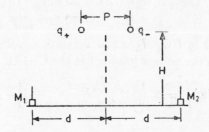

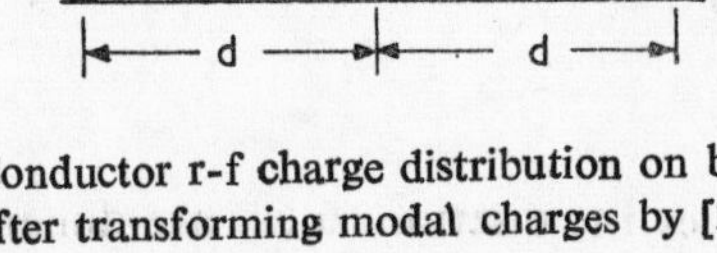

Fig. 6.16.   Conductor r-f charge distribution on bipolar dc line after transforming modal charges by $[T]$.

At M_1

Due to positive conductor,

$$RI_+(M_1) = \frac{q_+}{\pi e_0 H} \frac{1}{1 + (d - P/2)^2/H^2} = \frac{I_+}{2\pi e_0 H} \frac{K(1) + K(2)}{1 + (d - P/2)^2/H^2} \tag{6.46}$$

Due to the negative conductor

$$RI_-(M_1) = \frac{I_+}{2\pi e_0 H} \frac{K(1) - K(2)}{1 + (d + P/2)^2/H^2} \tag{6.47}$$

Similarly at M_2

$$RI_+(M_2) = \frac{I_+}{2\pi e_0 H} \frac{K(1) + K(2)}{1 + (d + P/2)^2/H^2} \tag{6.48}$$

$$RI_-(M_2) = \frac{I_+}{2\pi e_0 H} \frac{K(1) - K(2)}{1 + (d - P/2)^2/H^2} \tag{6.49}$$

where $\quad K(1) = 1/v(1)\sqrt{2a(1)}$ and $K(2) = 1/v(2)\sqrt{2a(2)}$ $\qquad$ (6.50)

In general, the velocities in the two modes will be different with $v(1)$ being about 80% light velocity since it involves ground return and $v(2)$ equal to light velocity. The attenuation factors $a(1)$ and $a(2)$ will also be different with $a(1)$ equal to about 6 dB/km (0.69 Neper/km) and $a(2)$ equal to about 1 dB/km (0.115 Neper/km).

Procedure for Obtaining Excitation Function from CIGRE Formula

Since the procedure for calculating RI level from the excitation function parallels the empirical formula given by the CIGRE, we will now examine the relation between the expressions given in equations (6.40) for the charge and the resulting RI with (6.22). For the charge given in (6.40), the RI level at ground will be

$$RI = \frac{1}{3v\sqrt{2a}} \frac{1}{\pi e_0 H} \frac{1}{1 + d^2/H^2} \tag{6.51}$$

The corresponding decibel value is

$$RI_{dB} = 20 \, Log_{10} \, (I/ve_0 H\sqrt{a}) - 20 \, Log_{10} \, (3\pi\sqrt{2})$$
$$- 20 \, Log_{10} \, [(d^2 + H^2)/H^2] \tag{6.52}$$

The first term $g = I/ve_0 \, H\sqrt{a}$ has dimension volt/metre and therefore corresponds to the voltage gradient in the CIGRE formula.

$$\left[(Amp/\sqrt{m}) \Big/ \left(\frac{m}{S} \times \frac{F}{m} \times \sqrt{\frac{Neper}{m}} \times m \right) \right] = \frac{Amp \cdot Sec}{Farad \cdot metre}$$
$$= \frac{Coulomb}{Farad \cdot metre} = \frac{volt}{metre}$$

The other quantities in equation (6.52) are dimensionless.

$$\therefore \quad RI_{dB} = 20 \, Log_{10} \, g - 22.5 - 40 \, Log_{10} \, (\sqrt{d^2 + H^2}/H) \tag{6.53}$$

The CIGRE formula is

$$RI_{dB} = 3.5 \, g_m - 12r - 30 - 33 \, Log \, (\sqrt{d^2 + H^2}/20) \tag{6.54}$$

By calculating the RI level for a given conductor using the CIGRE formula and equating it to equation (6.53), the value of g can be determined and thereby the excitation function I, as shown by an example.

Example 6.5: In example 6.2, the RI level at 30 m from outer phase due to the centre phase was calculated as 38.7 dB when $H = 14$ m,

$r = 1.59$ cm, $g_m = 17.3$ kV/cm and $d = 41$m. (a) Calculate the value of g in equation (6.53) by equating (6.54) with it. (b) Taking $v = 2.5 \times 10^8$ m/s and $a = 0.69 \times 10^{-3}$ Neper/m, calculate the excitation function and its dB value above 1 $\mu A/\sqrt{m}$.

Solution. $38.7 = 20 \, \text{Log} \, (g) - 22.5 - 40 \, \text{Log} \, (\sqrt{41^2 + 14^2}/14)$

$= 20 \, \text{Log} \, (g) - 42.1$

(a) $\therefore \quad g \quad = $ antilog $4.04 = 1.0965 \times 10^4$ volt/metre.

(b) $I = gve_0 H\sqrt{a} = 1.0965 \times 10^4 \times 2.5 \times 10^8 \times 8.842 \times 10^{-12}$

$\times 14\sqrt{0.69 \times 10^{-3}} = 8.984 \, \mu A/\sqrt{m}$.

$I_{dB} = 20 \, \text{Log}_{10} \, 8.984 = 19.07$ dB above 1 $\mu A/\sqrt{m}$.

Thus, the RI level is 19.63 dB higher than the dB value of the excitation function. The decibel adder to convert the excitation function to the RI level is therefore 19.63 dB for this example.

6.8 Measurement of RI, RIV, and Excitation Function

The interference to AM broadcast in the frequency range 0.5 MHz to 1.6 MHz is measured in terms of the three quantities: Radio Interference Field Intensity (RIFI or RI), the Radio Influence Voltage (RIV), and more recently through the Excitation Function. Their units are $\mu V/m$, μV, and $\mu A/\sqrt{m}$, or the decibel values above their reference values of 1 unit ($\mu V/m$, μV, $\mu A/m\sqrt{}$). The nuisance value for radio reception is governed by a quantity or level which is nearly equal to the peak value of the quantity and termed the Quasi Peak. A block diagram of a radio noise meter is shown in Fig. 6.17. The input to the meter is at radio frequency (r-f) which is amplified and fed to a mixer. The rest of the circuit works exactly the same as a highly sensitive super-heterodyne radio receiver. However, at the IF output stage, a filter with 5 KHz or 9 KHz bandwidth is present whose output is detected by the diode D. Its output charges a capacitance C through a low resistance R_c such that the charging time constant $T_c = R_c \, C = 1$ ms. A second resistance R_d is in parallel with C which is arranged to give a time constant $T_d = R_d C = 600$ ms in ANS I meters and 160 ms in CISPR or European standard meters. Field tests have shown that there is not considerable

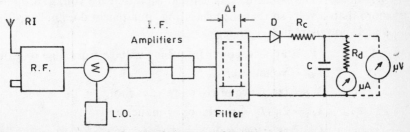

Fig. 6.17. Block diagram of Radio Noise meter.

difference in the output when comparing both time constants for line-generated corona noise. The voltage across the capacitor can either be read as a current through the discharge resistor R_d or a micro-voltmeter connected across it.

For radiated interference measurement RI, the front end of the meter is fitted with either a rod antenna of 0.5 to 2 metres in length or a loop antenna of this size of side. For conducted measurements, the interfering voltage RIV is fed through a jack. The input impedance of the meter is 50 ohms.

The following formulas due to Nigol apply to the various settings of the noise meter for repetitive pulses:

Peak Value: $V_p = \sqrt{2}.A.\tau.\Delta f$ (6.55)

Quasi Peak: $V_{qp} = KV_p$ (6.56)

Average: $V_{av} = \sqrt{2}.A.\tau.f_0$ (6.57)

R.M.S. Value: $V_{rms} = \sqrt{2}.A.\tau.\sqrt{f_0.\Delta f}$ (6.58)

where, A = amplitude of repetitive pulses,

       τ = pulse duration,

       Δf = bandwidth of meter,

       f_0 = repetition frequency of pulses, $< \Delta f$,

and K = a constant $\approx 0.9 - 0.95$.

Relations can be found among these four quantities if necessary.

Conducted RIV is measured by a circuit shown schematically in Fig. 6.18. The object under test, which could be an insulator string with guard rings, is energized by a high voltage source at power frequency or impulse. A filter is interposed such that any r-f energy produced by partial discharge in the test object is prevented from flowing into the source and all r-f energy goes to the measuring circuit. This consists of a discharge-free h.v. coupling capacitor of about 500 to 2000 pF in series at the ground level with a small inductance L. At 50 Hz, the coupling capacitor has a reactance of 6.36 Megohms to 1.59 Megohms. The value of L is chosen such that the voltage drop is not more than 5 volts so that the measuring equipment does not experience a high power-frequency voltage.

Let V = applied power frequency voltage from line to ground,

          V_L = voltage across L,

          X_c = reactance of coupling capacitor

and $X_L = 2\pi fL$ = reactance of inductor.

Then, $V_L = V.X_L/(X_c - X_L) \approx VX_L/X_c = 4\pi^2 f^2 LC.V$ (6.59)

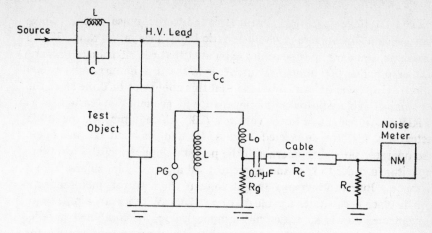

Fig. 6.18. Circuit for measuring Radio Influence Voltage (RIV).

Example 6.6: A test object for 400 kV is undergoing an RIV test. The coupling capacitor has 1000 pF and the voltage across the measuring system is to be 1 volt. Calculate the value of inductance required if

$$V = 420/\sqrt{3} \text{ kV} = 243.5 \text{ kV}.$$

Solution: $L = X_L/2\pi f = V_L/4\pi^2 f^2 \, CV$

$$L = 1/(4\pi^2 \times 50^2 \times 10^{-9} \times 242.5 \times 10^3) = 41.8 \text{ mH}.$$

[At 50 Hz, $X_L = 2\pi fL = 13.1$ ohm, $X_c = 3.185$ Megohm].

At radio frequencies, the inductance presents a very high impedance while the coupling capacitor has very low reactance. The capacitor is tuned at a fixed frequency, usually 1 MHz, with an r-f choke, L_0. There is a series R_g to ground. This r-f voltage is fed to the noise meter through a length of cable of 50 ohm characteristic impedance terminated in a 50 ohm resistance at the input end of the meter.

The value of L_0 is obtained from the equation

$$f = 1/2\pi\sqrt{L_0 \, C_0} \text{ or } L_0 = 1/4\pi^2 f^2 \, C_0 \qquad (6.60)$$

where f = measuring frequency.

Example 6.7: In the above example with $C_c = 1000$ pF, calculate
(a) the value of L_0 to tune the circuit to 1 MHz,
(b) the reactance of $L = 41.8$ mH at 1 MHz and
(c) that of the coupling capacitor C_c. Check with reactance of L_0.

Solution:
(a) $L_0 = 1/4\pi^2 \times 10^{12} \times 10^{-9} = 25.33 \ \mu\text{H}.$
(b) $X_L = 2\pi \times 10^6 \times 41.8 \times 10^{-3} = 262.64$ Kilohms.
(c) $X_c = 1/2\pi \times 10^6 \times 10^{-9} = 159$ ohms.
  $X_{L0} = 2\pi \times 10^6 \times 25.33 \times 10^{-6} = 159$ ohms.

The r-f voltage developed across R_g is fed to the noise meter. Since transmission lines have a characteristic impedance in the range 300 to 600 ohms, standard specifications stipulated that the r-f voltage must be measured across 600 ohms. Thus, R_g is in the neighbourhood of 600 ohms. However, it was obvious that this could not be done since the presence of cable will lower the impedance to ground. In earlier days of RIV measurement at lower voltages (230 kV equipment) the noise meter was directly connected across R_g and its input end was open. The operator sat right underneath the pedestal supporting the coupling capacitor in order to read the meter or used a pair of binoculars from a distance. But with increase in test voltage, the need for maintaining a safe distance necessitates a cable of 10m to 20m. With its surge impedance R_c connected across R_g, which has a higher value, the combined parallel impedance is lower than R_c. No coaxial cables are manufactured for high surge impedance, so that standard specifications allow RIV to be measured across 150 ohm resistance made up of R_g and R_c in parallel. It is clear that the measuring cable must have R_c greater than 150 ohms. The highest impedance cable has $R_c = 175$ ohms on the market. The value of R_g can be selected such that

$$R_c R_g/(R_c + R_g) = 150 \text{ ohms}$$

giving $\qquad R_g = 150 \, R_c/(R_c - 150)$ $\hfill (6.61)$

For $\qquad R_c = 175 \text{ ohms}, R_g = 6R_c = 1050 \text{ ohms}.$

The meter reading is then multiplied by a factor of 4 in order to give the RIV measured across 600 ohms, or by a factor of 2 for 300 ohms surge impedance.

6.9.1 Measurement of Excitation Function

The corona generating function or the excitation function caused by injected current at radio frequencies from a corona discharge is measured on short lengths of conductor strung inside "cages" as discussed earlier. The design of cages has been covered in great detail in Chapter 4. Some examples of measuring radio noise and injected current are shown in Fig. 6.19. In every case the measured quantity is RIV at a fixed frequency and the excitation function calculated as described later. The filter provides an attenuation of at least 25 dB so that the RI current is solely due to corona on conductor. The conductor is terminated in a capacitance C_c at one end in series with resistances R_1 and R_c, while the other end is left open. The conductor is strung with strain insulators at both ends which can be considered to offer a very high impedance at 1 MHz so that there is an open-termination. But this must be checked experimentally in situ. The coupling capacitor has negligible reactance at r-f so that the termination at the measuring end is nearly equal to $(R_1 + R_c)$, where $R_c =$ surge impedance of the cable to the noise meter. The resistance R_c is also equal to the input impedance of the noise meter.

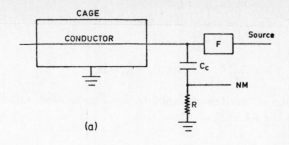

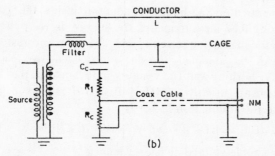

Fig. 6.19.   Cage setups for measuring excitation function with measuring circuit.

The excitation function is calculated as follows:

Let $J = $ RI current injected in $\mu A/\sqrt{m}$,

   $C = $ capacitance of conductor in cage, Farad/meter,

   $R = $ outer radius of cage,

   $r_{eq} = $ equivalent radius of bundle.

Then, the excitation function is

$$I = 2\pi e_0.J/C,\ \mu A/\sqrt{m}. \tag{6.62}$$

The injected current in terms of measured RIV is

$$J = 2(R_c + R_m)\ (RIV)/R_c\ R_m G \tag{6.63}$$

where $R_c R_m/(R_c + R_m) = $ resistance of R_c and meter in parallel,

   RIV $= $ measured noise reading on meter in μV,

and $G = $ an amplification factor caused by addition of the uniformly distributed r-f currents generated on the conductor.

The test is normally carried out under rain conditions from artificial rain apparatus so that corona is generated uniformly along the conductor. For a line terminated at one end in its surge impedance with the other end open, the amplification factor is, for a conductor length L,

$$G = \sqrt{\left[\int_0^L \cos^2 \frac{2\pi f}{v}\ x.dx\right]} = \sqrt{\left[\frac{L}{2} + \frac{v}{8\pi f}\sin\frac{4\pi fL}{v}\right]} \tag{6.64}$$

where v = velocity of propagation

and f = frequency of measurement.

Since $R_c = R_m$, there results

$$I = \frac{2\pi e_0}{C} \frac{4}{R_c\,G} \cdot (\text{RIV}), \ \mu A/\sqrt{m}. \tag{6.65}$$

This value of excitation function has been used in Section 6.7 for evaluating the RI level of a long line.

6.9.2 Design of Filter

When corona is generated on the bundle-conductor inside the cage or on a test object during RIV measurements, the energy in the pulses will be divided between the measuring circuit and the source transformer. The transformer can be assumed to offer a pure capacitive reactance consisting of the h.v. bushing and the winding inductance at the frequency of r-f measurement, usually 1 MHz. Fig. 6.20 shows a simple R-L filter, for which the attenuation factor is

$$A = |\,V_i/V_0\,| = [(1 - W^2\,L_f C_t)^2 + W^2\,R_f^2\,C_t^2]^{1/2} \tag{6.66}$$

This is obtained by simple voltage division which is

$$V_i/V_0 = [(R_f + jWL_f) + jWL_t/(1 - W^2 L_t C_t)]/[jWL_t/(1 - W^2 L_t C_t)] \tag{6.67}$$

where $W = 2\pi f$, C_t = transformer capacitance, L_t = transformer inductance, and R_f, L_f = filter resistance and inductance.

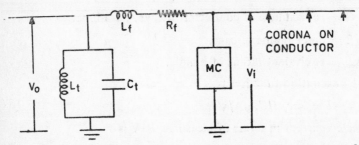

Fig. 6.20. Line filter for blocking corona energy from entering source transformer.

Let the upper and lower frequencies of measurement be f_2 and f_1, and the desired attenuation A_2 and A_1. The decibel values are

$$D_2 = 20\,\log_{10}\,A_2 \text{ and } D_1 = 20\,\log_{10}\,A_1.$$

Then the following equations hold from which we can calculate the products $X = L_f C_t$ and $Y = R_f C_t$. When the transformer capacitance is known, the values of filter elements are fixed.

$$(1 - W_2^2\,X)^2 + W_2^2\,Y^2 = A_2^2 \tag{6.68}$$

$$(1 - W_1^2\,X)^2 + W_1^2\,Y^2 = A_1^2 \tag{6.69}$$

$$\text{Then, } (W_2^2 - W_1^2)\,X^2 = (A_2^2 - 1)/W_2^2 - (A_1^2 - 1)/W_1^2 \tag{6.70}$$

and $\qquad Y^2 = [A_2^2 - (1 - W_2^2 X)^2]/W_2^2$ $\qquad$ (6.71a)

or, $\qquad Y^2 = [A_1^2 - (1 - W_1^2 X)^2]/W_1^2$ $\qquad$ (6.71b)

The Q of the filter at any frequency is,

$$Q = 2\pi f L_f/R_f = 2\pi f X/Y. \qquad (6.72)$$

Example 6.8: Design a filter to give 40 dB attenuation at 1 MHz and 25 dB at 0.4 MHz. Calculate Q of filter at 1 MHz.

Solution: Let $W_2 = 2\pi \times 10^6$, $W_1 = 2\pi \times 0.4 \times 10^6$, $A_2 = 100$,

$$A_1 = 10^{1.25} = 17.78.$$

$\therefore \quad 4\pi^2 \times 10^{12} (1^2 - 0.4^2) X^2 = (100^2 - 1)/4\pi^2 \times 10^{12}$
$$- (17.78^2 - 1)/4\pi^2 \times 0.16 \times 10^{12}$$

This gives $\quad X = L_f C_t = 5.76 \times 10^{-12}$

Then, $\quad Y^2 = [100^2 - (1 - 4\pi^2 \times 10^{12} \times 5.76 \times 10^{-12})^2]/4\pi^2 \times 10^{12}$

giving $\quad Y = 16.1 \times 10^{-6} = R_f C_t$.

$$Q = 2\pi f L_f/R_f = 2\pi f X/Y = 2.248 \text{ at 1 MHz.}$$

Example 6.9: The limiting frequencies for AM broadcast are 0.5 MHz and 1.6 MHz. Design filter elements L_f and R_f and find Q at 1 MHz for giving 40 dB at 1.6 MHz and 25 dB at 0.5 MHz.

Solution: Using equations (6.68) to (6.71), there result

$X = L_f C_t = 0.852 \times 10^{-12}$ and $Y = R_f C_t = 5.222 \times 10^{-6}$

$Q = 1.025$ at 1 MHz.

If the transformer capacitance is 1000 pF $= 10^{-9}$ Farad, the values are $L_f = 0.852$ mH and 5.222 kilohms.

The most important point to observe is that even though the filter elements are in series with the source transformer and conductor in the cage at the high voltage, they must be designed for the full working voltage. This is a very expensive item in the experimental set-up.

Review Questions and Problems

1. Describe the mechanism of formation of a positive corona pulse train.

2. The positive and negative corona pulses can be assumed to be square pulses of amplitudes 100 mA and 10 mA respectively. Their widths are 200 ns and 100 ns respectively. Their repetition rates are 1000 pps and 10,000 pps. The bandwidth of a filter is 5 KHz. Using equation (6.15) calculate the ratio of output of the filter for the two pulse trains at a tuned frequency $f_0 = 1$ MHz.

3. At a town near which a proposed e.h.v. line will run, the measured radio-station field strengths are as follows:

Frequency of
Station   0.5 0.75 0.9 1.0 1.1 1.2 1.3 1.4 1.5 MHz
Received Signal
Strength   50   60   75   70   65   52   80   75   60 dB

Take the corona noise from a line to vary as $f^{-1.5}$, and the minimum S/N ratio to be 22 dB. If the station at 1 MHz is to be received with this quality of reception, determine which of the stations will have a S/N ratio lower than the minimum allowable S/N ratio of 22 dB.

4. Calculate and plot the field factors for the 3 modes of propagation for a line with $H = 15$ m, $S = 12$ m as the distance from the line centre is varied from 0 to 3 H.

5. The height of conductors of a bipolar dc line are $H = 18$ m and the pole spacing $P = 12$ m. Calculate and plot the field factors for this line for the two modes of propagation as the distance d from line centre is varied from 0 to 3 H.

6. A 750-kV line in horizontal configuration has $H = 18$ m and phase spacing $S = 15$ m. The conductors are 4×0.03 metre diameter with bundle spacing of 0.4572 metre. Using Mangoldt's Formula and the CIGRE formula, compute the RI level at 15 metres at ground level from the outer phase at 1 MHz in average fair weather. Is the width of corridor of 60 metres sufficient from the RI point of view?

7. Design a filter with series R-L elements for a cage measurement to give an attenuation of 40 dB at 1 MHz and 30 dB at 0.5 MHz. (a) If the transformer bushing has a capacitance $C_t = 500$ pF, calculate the values of R_f and L_f required. (b) What are attenuations offered by this filter at 1.5 MHz and 0.8 MHz?

8. Why does line-generated corona noise not interfere with TV reception or FM radio reception? What causes interference at these frequencies?

CHAPTER 7

Electrostatic Field of EHV Lines

7.1 Electric Shock and Threshold Currents

Electrostatic effects from overhead e.h.v. lines are caused by the extremely high voltage while electromagnetic effects are due to line loading current and short-circuit currents. Hazards exist due to both causes of various degree. These are, for example, potential drop in the earth's surface due to high fault currents, direct flashover from line conductors to human beings or animals. Electrostatic fields cause damage to human life, plants, animals, and metallic objects such as fences and buried pipe lines. Under certain adverse circumstances these give rise to shock currents of various intensities.

Shock currents can be classified as follows:

(a) *Primary Shock Currents.* These cause direct physiological harm when the current exceeds about 6–10 mA. The normal resistance of the human body is about 2–3 Kilohms so that about 25 volts may be necessary to produce primary shock currents. The danger here arises due to ventricular fibrillation which affects the main pumping chambers of the heart. This results in immediate arrest of blood circulation. Loss of life may be due to (a) arrest of blood circulation when current flows through the heart, (b) permanent respiratory arrest when current flows in the brain, and (c) asphyxia due to flow of current across the chest preventing muscle contraction.

The 'electrocution equation' is $i^2t = K^2$, where $K = 165$ for a body weight of 50 kg, i is in mA and t is in seconds. On a probability basis death due to fibrillation condition occurs in 0.5% of cases. The primary shock current required varies directly as the body weight. For $i = 10$ mA, the current must flow for a time interval of 272 seconds before death occurs in a 50 kg human being.

(b) *Secondary Shock Currents.* These cannot cause direct physiological harm but may produce adverse reactions. They can be steady state 50 Hz or its harmonics or transient in nature. The latter occur when a human being comes into contact with a capacitively charged body such as a parked vehicle under a line. Steady state currents up to 1 mA cause a slight tingle on the fingers. Currents from 1 to 6 mA are classed as 'let go' currents. At this level, a human being has control of muscles to let the conductor go as soon as a tingling sensation occurs.

For a 50% probability that the let-go current may increase to primary shock current, the limit for men is 16 mA and for women 10 mA. At 0.5% probability, the currents are 9 mA for men, 6 mA for women, and 4.5 mA for children.

A human body has an average capacitance of 250 pF when standing on an insulated platform of 0.3 m above ground (1 ft.). In order to reach the let-go current value, this will require 1000 to 2000 volts. Human beings touching parked vehicles under the line may experience these transient currents, the larger the vehicle the more charge it will acquire and greater is the danger.

Construction crews are subject to hazards of electrostatic induction when erecting new lines adjacent to energized lines. An ungrounded conductor of about 100 metres in length can produce shock currents when a man touches it. But grounding both ends of the conductor brings the hazard of large current flow. A movable ground mat is generally necessary to protect men and machines. When stringing one circuit on a double-circuit tower which already has an energized circuit is another hazard and the men must use a proper ground. Accidents occur when placing or removing grounds and gloves must be worn. Hot-line techniques are not discussed here.

7.2   Capacitance of Long Object

Electrostatic induction to adjacent lines such as telephone lines can be determined by Maxwell's Potential Coefficients and their inverses. If ground resistance and inductance is to be considered, Carson's formulas given in chapter 3 are used. However, for a long object such as a lorry or vehicle parked parallel to a line under it, an empirical formula for its capacitance due to Comsa and René is given here. The object is replaced by an equivalent cylinder of diameter D and height h above ground

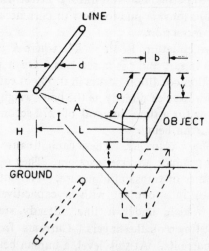

Fig. 7.1.   Calculation of capacitance of long object located near an e.h.v. line.

as shown below. Figure 7.1 gives the actual dimensions of the object where a = length of object, b = width, v = height, t = height of tyres. Then, $h = t + v - 0.5\,b$, and $D = b$. Other dimensions of line are shown on the figure. The capacitance of the vehicle, including end effects, is

$$C = a.C_1 + C_2 \tag{7.1}$$

where

$$C_1 = 2\pi e_0 \ln (I/A) \Big/ \Big[\Big(\ln \frac{4H}{d} \Big) \cdot \Big(\ln \frac{4h}{D} \Big)$$
$$+ \ln (I/A) \Big] \tag{7.2}$$

and

$$C_2 = 31.b, \text{ pF due to end effects.} \tag{7.3}$$

Example 7.1: The following details of truck parked parallel to a line are given. Find its capacitance. Length a = 8m, height of body v = 3m, width b = 3m, t = 1.5m, Height of line conductor H = 13m, dia. of conductor = 0.0406m, distance of parking L = 6m.

Solution. $h = t + v - 0.5b = 3$, $D = b = 3$, $I = 17.1$, $A = 11.66$,

$$\therefore \quad C_1 = \frac{10^{-9}}{18} \times \ln (17.1/11.66) \Big/ \Big[\ln \frac{52}{.0406} \cdot \ln \frac{12}{3} + \ln \frac{17.1}{11.66} \Big]$$

$$= 2.065 \text{ pF/metre length of truck.}$$

$$C_2 = 31\, b = 93 \text{ pF.}$$

$$\therefore \quad C = 8 \times 2.065 + 93 = 109.5 \text{ pF.}$$

Note that the edge effect is considerable.

7.3   Calculation of Electrostatic Field of A.C. Lines

7.3.1   Power-Frequency Charge of Conductors
In Chapter 4, we described the method of calculating the electrostatic

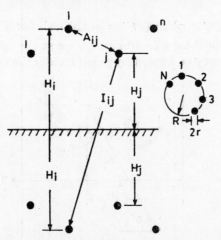

Fig. 7.2.   *n*-phase line configuration for charge calculation.

charges on the phase conductors from line dimensions and voltage. For n phases, this is, see Fig. 7.2, with q = total bundle charge and V = line to ground voltage,

$$\frac{1}{2\pi e_0}[q] = [P]^{-1}[V] = [M][V] \tag{7.4}$$

where $\qquad [q] = [q_1, q_2, q_3, \ldots, q_n]_t \tag{7.5}$

$\qquad\qquad [V] = [V_1, V_2, V_3, \ldots, V_n]_t$

$\qquad\qquad [P] = n \times n$ matrix of Maxwell's Potential coefficient

matrix with $P_{ii} = \ln(2H_i/r_{eq})$ and $P_{ij} = \ln(I_{ij}/A_{ij}) \tag{7.6}$

Here, $\qquad H_i$ = height of conductor i above ground = $H_{min} + \frac{1}{3}$ Sag,

$\qquad\qquad I_{ij}$ = distance between conductor i above ground and the image of conductor j below ground, $i \neq j$,

$\qquad\qquad A_{ij}$ = aerial distance between conductors i and j,

$\qquad\qquad r_{eq} = R(N.r/R)^{1/N}$ = equivalent bundle radius,

$\qquad\qquad R$ = bundle radius = $B/2 \sin(\pi/N)$,

$\qquad\qquad N$ = number of sub-conductors in bundle,

$\qquad\qquad r$ = radius of each sub-conductor,

and $\qquad i, j = 1, 2, 3, \ldots, n,$

Since the line voltages are sinusoidally varying with time at power frequency, the bundle charges q_1 to q_n will also vary sinusoidally. Consequently, the induced electrostatic field in the vicinity of the line also varies at power frequency and phasor algebra can be used to combine several components in order to yield the amplitude of the required field, namely, the horizontal, vertical or total vectors.

7.3.2   Electrostatic Field of Single-Circuit 3-Phase Line

Let us consider first a 3-phase line with 3 bundles on a tower and excited by the voltages

$$[V] = V_m[\sin(wt + \varphi), \sin(wt + \varphi - 120°), \sin(wt + \varphi + 120°)] \tag{7.7}$$

Select an origin O for a coordinate system at any convenient location. In general, this may be located on ground under the middle phase in a

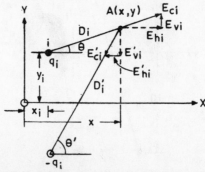

Fig. 7.3.   Calculation of e.s. field components near the line.

symmetrical arrangement. The coordinates of the line conductors are (x_i, y_i). A point $A(x, y)$ is shown where the horizontal, vertical, and total e.s. field components are required to be evaluated, as shown in Fig. 7.3. The field vector at A due to the charge of the aerial conductor is with

$$D_i^2 = (x - x_i)^2 + (y - y_i)^2,$$
$$E_c = (q_i/2\pi e_0) (1/D_i) \qquad (7.8)$$

Its horizontal and vertical components are

$$E_h = E_c \cos \theta = (q_i/2\pi e_0) (x - x_i)/D_i^2 \qquad (7.9)$$

and $\qquad E_v = E_c \sin \theta = (q_i/2\pi e_0) (y - y_i)/D_i^2 \qquad (7.10)$

Similarly, due to image charge of conductor i_s

$$E_c' = (q_i/2\pi e_0) (1/D_i')$$

where $\qquad (D_i')^2 = (x - x_i)^2 + (y + y_i)^2,$

$$\left. \begin{array}{l} E_h' = (q_i/2\pi e_0) (x - x_i)/(D_i')^2 \\ E_v' = (q_i/2\pi e_0) (y + y_i)/(D_i')^2 \end{array} \right\} \qquad (7.11)$$

We observe that the field components of E_c and E_c' are in opposite directions. Therefore, the total horizontal and vertical components at A due to both charges are

$$E_{hi} = (q_i/2\pi e_0 (x - x_i) [1/D_i^2 - 1/(D_i')^2] \qquad (7.12)$$
$$E_{vi} = (q_i/2\pi e_0) [(y - y_i)/D_i^2 - (y + y_i)/(D_i')^2] \qquad (7.13)$$

Consequently, due to all n phases, the sum of horizontal and vertical components of e.s. field at the point $A(x, y)$ will be

$$E_{hn} = \sum_{i=1}^{n} E_{ni}, \text{ and } E_{vn} = \sum_{i=1}^{n} E_{vi} \qquad (7.14)$$

The total electric field at A is

$$E_{tn} = (E_{hn}^2 + E_{vn}^2)^{1/2} \qquad (7.15)$$

We can write these out explicitly for a 3-phase line.

Let $\qquad J_i = (x - x_i) [1/D_i^2 - 1/(D_i')^2] \qquad (7.16)$

and $\qquad K_i = (y - y_i)/D_i^2 - (y + y_i)/(D_i')^2 \qquad (7.17)$

The bundle charges are calculated from equations (7.4), (7.5), and (7.6), so that from equations (7.12), (7.13), (7.16) and (7.17), there results

$$E_{h1} = (q_1/2\pi e_0) J_1 = V_m \cdot J_1 [M_{11} \sin (wt + \varphi) + M_{12} \sin (wt + \varphi - 120°) \\ + M_{13} \sin (wt + \varphi + 120°)]$$

$$E_{h2} = (q_2/2\pi e_0) J_2 = V_m \cdot J_2 [M_{21} \sin (wt + \varphi) + M_{22} \sin (wt + \varphi - 120°) \\ + M_{23} \sin (wt + \varphi + 120°)]$$

$$E_{h3} = (q_3/2\pi e_0) J_3 = V_m \cdot J_3 [M_{31} \sin (wt + \varphi) + M_{32} \sin (wt + \varphi - 120°) \\ + M_{33} \sin (wt + \varphi + 120°)]$$

∴ The total horizontal component is, adding vertically,

$$E_{hn} = V_m [(J_1 \cdot M_{11} + J_2 \cdot M_{21} + J_3 \cdot M_{31}) \sin (wt + \varphi)$$

$$+ (J_1 \cdot M_{21} + J_2 \cdot M_{22} + J_3 \cdot M_{23}) \sin (wt + \varphi - 120°)$$
$$+ (J_1 \cdot M_{31} + J_2 \cdot M_{32} + J_3 \cdot M_{33}) \sin (wt + \varphi + 120°)]$$
$$= V_m [J_{h1} \cdot \sin (wt + \varphi) + J_{h2} \cdot \sin (wt + \varphi - 120°)$$
$$+ J_{h3} \sin (wt + \varphi + 120°)]$$

and in phasor form,

$$E_{hn} = V_m [J_{h1} \angle \varphi + J_{h2} \angle \varphi - 120° + J_{h3} \angle \varphi + 120°] \qquad (7.18)$$

This is a simple addition of three phasors of amplitudes J_{h1}, J_{h2}, J_{h3} inclined at 120° to each other. Resolving them into horizontal and vertical components (real and j parts with $\varphi = 0$), we obtain

$$\text{real part} = J_{h1} - 0.5 J_{h2} - 0.5 J_{h3} \qquad (7.19)$$
$$\text{and imaginary part} = 0 - 0.866 J_{h2} + 0.866 J_{h3}$$

Consequently, the amplitude of electric field is

$$E_{hn} = [(J_{h1} - 0.5 J_{h2} - 0.5 J_{h3})^2 + 0.75 (J_{h3} - J_{h2})^2]^{1/2} V_m$$
$$= (J_{h1}^2 + J_{h2}^2 + J_{h3}^2 - J_{h1} J_{h1} - J_{h2} J_{h3} - J_{h3} J_{h1})^{1/2} \cdot V_m$$
$$= J_h \cdot V_m .$$

The r.m.s. value of the total horizontal component at A (x, y) due to all 3 phases will be

$$E_{hn} = \hat{E}_{hn}/\sqrt{2} = J_h \cdot V \qquad (7.20)$$

where $V =$ r.m.s. value of line to ground voltage.

In a similar manner, the r.m.s. value of total vertical component of field at A due to all 3 phases is

$$E_{vn} = K_v \cdot V = V (K_{v1}^2 + K_{v2}^2 + K_{v3}^2 - K_{v1} K_{v2} - K_{v2} K_{v3}$$
$$- K_{v3} K_{v1})^{1/2} \qquad (7.21)$$

where $K_{v1} = K_1 \cdot M_{11} + K_2 \cdot M_{21} + K_3 \cdot M_{31}$
$$K_{v2} = K_1 \cdot M_{12} + K_2 \cdot M_{22} + K_3 \cdot M_{32} \qquad (7.22)$$

and $K_{v3} = K_1 \cdot M_{13} + K_2 \cdot M_{23} + K_3 \cdot M_{33}$

where the values of K_1, K_2, K_3 are obtained from equation (7.17) for K_i with $i = 1, 2, 3$.

Example 7.2: Compute the r.m.s. values of ground-level electrostatic field of a 400-kV line at its maximum operating voltage of 420 kV (line-to-line) given the following details. Single circuit horizontal configuration. $H = 13$ m, $S = 12$ m, conductor 2×3.18 cm diameter, $B = 45.72$ cm. Vary the horizontal distance along ground from the line centre from 0 to $3 H$. See Fig. 7.4.

Solution. At the ground level, the horizontal component of e.s. field is zero everywhere since the ground surface is assumed to be an equipotential. Also, for every point on ground, the distances from aerial conductor and its image are such that D_i and D_i' are equal.

Step 1. $P_{ii} = \ln(2H/r_{eq})$, $P_{ij} = \ln(I_{ij}/A_{ij})$, $r_{eq} = R(N.r/R)^{1/N}$

$N = 2$, $R = B/2$. Then the [P] and [M] matrices are

$$[P] = \begin{bmatrix} 5.64, & 0.87, & 0.39 \\ 0.87, & 5.64, & 0.87 \\ 0.39, & 0.87, & 5.64 \end{bmatrix} \text{ and}$$

$$[M] = [P]^{-1} = \begin{bmatrix} 172.8, & -25.6, & -8 \\ -25.6, & 172.2, & -25.6 \\ -8, & -25.6, & 172.8 \end{bmatrix} 10^{-3}$$

Step 2. Coordinates of conductors with origin placed on ground under the centre-phase, see Fig 7.4, $x_1 = -12$, $x_2 = 0$, $x_3 = +12$, $y_1 = y_2 = y_3 = 13$. $y = 0$ on ground.

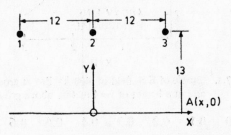

Fig. 7.4. Details of 400 kV line for of evaluation of ground-level e.s. field at point A.

Step 3. At a point $A(x, 0)$ along ground, from equation (7.17),

$$K_1 = -13/[(x + 12)^2 + (-13)^2] - 13/[(x + 12)^2 + (+13)^2]$$
$$= -26/[(x + 12)^2 + 169].$$

Similarly, $K_2 = -26/(x^2 + 169)$ and $K_3 = -26/[(x - 12)^2 + 169]$.

Step 4. $K_{v1} = K_1 M_{11} + K_2 M_{21} + K_3 M_{31}$

$$= \frac{-4.493}{(x + 12)^2 + 169} + \frac{0.666}{x^2 + 169} + \frac{0.208}{(x - 12)^2 + 169}$$

$K_{v2} = K_1 M_{12} + K_2 M_{22} + K_3 M_{32}$

$$= \frac{0.666}{(x + 12)^2 + 169} - \frac{4.607}{x^2 + 169} + \frac{0.666}{(x - 12)^2 + 169}$$

$K_{v3} = K_1 M_{13} + K_2 M_{23} + K_3 M_{33}$

$$= \frac{0.208}{(x + 12)^2 + 169} + \frac{0.666}{x^2 + 169} - \frac{4.493}{(x - 12)^2 + 169}$$

Step 5. $K_v = (K_{v1}^2 + K_{v2}^2 + K_{v3}^2 - K_{v1}K_{v2} - K_{v2}K_{v3} - K_{v3}K_{v1})^{1/2}$

Step 6. $E_v = K_v \cdot 420/\sqrt{3}$, kV/metre.

A computer programme written for x varying from 0 to $3H = 39$ m from the line centre has given the following results, which are also plotted in Fig. 7.5.

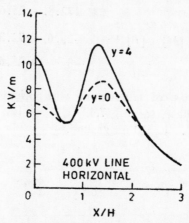

Fig. 7.5. Profile of E.S. field of 400 kV line at ground level and at a height of $y=4$ metres above ground.

X/H	0	0.1	0.2	0.3	0.4	0.5	0.6	0.7	0.8	0.9
E_v, kV/m	7	6.8	6.6	6.2	5.7	5.4	5.4	5.5	5.9	6.5
X/H	1.0	1.1	1.2	1.3	1.4	1.5	1.6	1.7	1.8	1.9
E_v	7.2	7.8	8.3	8.5	8.6	8.4	8	7.5	7	6.4
X/H	2.0	2.2	2.4	2.6	2.8	3.0				
E_v	5.8	4.7	3.8	3	2.4	2				

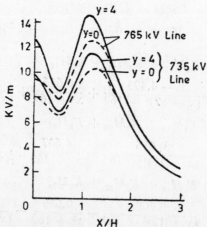

Fig. 7.6. E.S. field profile of 750 kV line (calculated).

We observe that the maximum value of ground-level field does **not** occur at the line centre but at $X/H = 1.4$ ($X = 18.2$ m). There is a double hump in the graph. Figure 7.5 also shows the e.s. field at 4 metres above ground which a truck might experience. Further examples for 750 kV, 1000 kV and 1200 kV lines are shown in Figs. 7.6 to 7.8 using typical dimensions. In all cases, the presence of overhead ground wires has been neglected. In a digital computer programme they can be included but their effect is negligible.

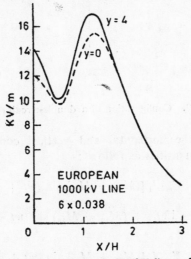

Fig. 7.7. E.S. field profile of 1000 kV line (calculated).

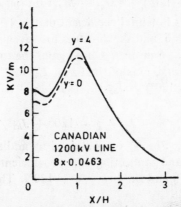

Fig. 7.8. E.S. field profile of 1200 kV line (calculated).

7.3.3 Electrostatic Field of Double-Circuit 3-phase A.C. Line

On a D/C line there are 6 conductors on a tower, neglecting ground wires above the line conductors [Some proposals for using shielding wires under the line conductors are made, but such problems will not be discussed in this book]. The e.s. field will depend on the phase configuration of the two circuits and for illustrating the procedure, the arrangement shown in Fig. 7.9 will be used. The positions occupied by the phases are numbered 1 to 6 and it is evident that (a) conductors 1 and 4 have the voltage $V_m \sin(wt + \varphi)$, (b) conductors 2 and 5 have voltages $V_m \sin(wt + \varphi - 120°)$ and (c) conductors 3 and 6 have voltages $V_m \sin(wt + \varphi + 120°)$.

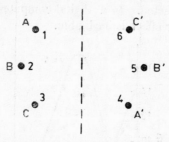

Fig. 7.9. Configuration of a double-circuit (D/C) line.

Consequently, the horizontal and vertical components of e.s. field will consist of 6 quantities as follows:

$$E_{h1} = \frac{q_1}{2\pi e_0} J_1 = V_m \cdot J_1 [(M_{11} + M_{14}) \sin(wt + \varphi) + (M_{12} + M_{15})$$

$$\sin(wt + \varphi - 120°) + (M_{13} + M_{16}) \sin(wt + \varphi + 120°)] \quad (7.23)$$

$$\vdots$$

$$E_{h6} = \frac{q_6}{2\pi e_0} J_6 = V_m \cdot J_6 [(M_{61} + M_{64}) \sin(wt + \varphi) + (M_{62} + M_{65})$$

$$\sin(wt + \varphi - 120°) + (M_{63} + M_{66}) \sin(wt + \varphi + 120°)] \quad (7.24)$$

where the Maxwell's Potential coefficient matrix $[P]$ and its inverse $[M]$ are now of order 6×6, and the J_i values are given from equation (7.16) by using $i = 1, 2, ..., 6$ in turn. Once again, the total horizontal component of e.s. field at $A(x, y)$ is of the form $E_{ht} = V_m [J_{h1} \sin(wt + \varphi) + J_{h2} \sin(wt + \varphi + 120°) + J_{h3} \cdot \sin(wt + \varphi + 120°)]$ and in phasor form

$$E_{ht} = V_m(J_{h1} \angle \varphi + J_{h2} \angle \varphi - 120° + J_{h3} \angle \varphi + 120°) \quad (7.25)$$

The quantities J_{h1}, J_{h2}, J_{h3}, are obtained by adding equations (7.23) to (7.24) vertically and collecting the coefficients of $\sin(wt + \varphi)$, $\sin(wt + \varphi - 120°)$ and $\sin(wt + \varphi + 120°)$. This is of the same form as equation (7.19).

Similarly, the total vertical component of e.s. field can be obtained by using equation (7.17) for calculating K_1 to K_6. Then,

$$E_{vt} = K_h \cdot V \quad (7.26)$$

where $K_h = (K_{h1}^2 + K_{h2}^2 + K_{h3}^2 - K_{h1} K_{h2} - K_{h2} K_{h3} - K_{h3} K_{h1})^{1/2}$

with $K_{h1} = K_1(M_{11} + M_{14}) + K_2(M_{21} + M_{24}) + K_3(M_{31} + M_{34})$

$$+ K_4(M_{41} + M_{44}) + K_5(M_{51} + M_{54}) + K_6(M_{61} + M_{66}) \quad (7.27)$$

$$\vdots$$

$$\ddot{K}_{h3} = K_1(M_{13} + M_{16}) + K_2(M_{23} + M_{26}) + K_3(M_{33} + K_{36})$$
$$+ K_4(M_{43} + M_{46}) + K_5(M_{53} + M_{56}) + K_6(M_{63} + M_{66}) \quad (7.28)$$

The total e.s. field at every point $A(x, y)$ will be

$$E_t = [E_{ht}^2 + E_{vt}^2]^{1/2} \quad (7.29)$$

7.3.4 Six-Phase A.C. Line

A recent advancement in medium-voltage lines is the use of high phase order lines with more than 3 phases on the same tower. We will complete the problem of e.s. field calculation with a 6-phase line where the voltages are now

$$V_1 = V_m \angle 0°, \ V_2 = V_m \angle -60°, \ V_3 = V_m \angle -120°, \ V_4 = V_m \angle -180°,$$
$$V_5 = V_m \angle -240° \text{ and } V_6 = V_m \angle -300°. \text{ Thus, } V_1 \text{ and } V_4 \text{ are in}$$
phase opposition, which is also true of V_2 and V_5, and V_3 and V_6. The arrangement of phase positions are shown in Fig. 7.10. It is clear that the Maxwell's Potential coefficient matrix $[P]$ and its inverse $[M]$ are the same as for a double-circuit 3-phase line of the previous section. The values of J_i and K_i are also the same as were obtained from equations (7.16) and (7.17), for $i = 1, 2, ..., 6$.

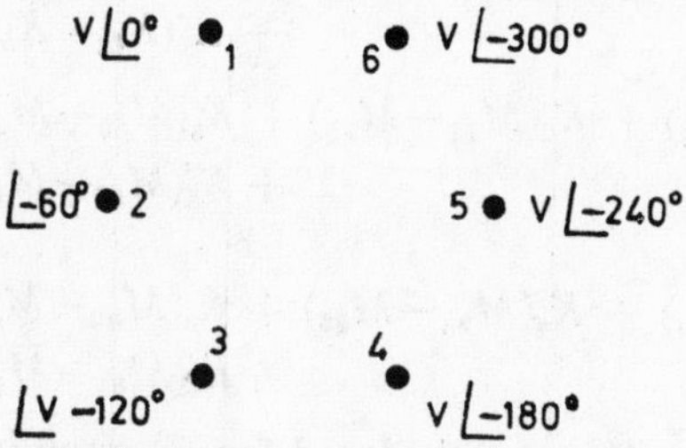

$$V \angle 0° \ \bullet_1 \qquad _6 \bullet \ V \angle -300°$$

$$V \angle -60° \bullet 2 \qquad 5 \bullet \ V \angle -240°$$

$$\begin{array}{cc} \bullet^3 & \bullet^4 \\ \angle V -120° & V \angle -180° \end{array}$$

Fig. 7.10. Details of 6-phase line.

Equations are now written down for the horizontal and vertical components of the e.s. field at a point $A(x, y)$ due to the six voltages as follows:

$$E_{h1} = J_1 \cdot \frac{q}{2\pi e_0} = J_1 \cdot V_m[M_{11} \sin(wt + \varphi) + M_{12} \sin(wt + \varphi - 60°) + M_{13}$$
$$\sin(wt + \varphi - 120°) + M_{14} \sin(wt + \varphi - 180°) + M_{15} \sin(wt + \varphi - 240°)$$
$$+ M_{16} \sin(wt + \varphi - 300°)]$$
$$= V_m \cdot J_1[(M_{11} - M_{14}) \sin(wt + \varphi) + (M_{12} - M_{15}) \sin(wt + \varphi - 60°)$$
$$+ (M_{13} - M_{16}) \sin(wt + \varphi - 120°)] \quad (7.30)$$

Similarly for the remaining 5 conductors. For conductor 6,

$$E_{h6} = V_m \cdot J_6[(M_{61} - M_{64}) \sin(wt + \varphi) + (M_{62} - M_{65}) \sin(wt + \varphi - 60°)$$
$$+ (M_{63} - M_{66}) \sin(wt + \phi - 120°)] \quad (7.31)$$

The total horizontal component of e.s. field at $A(x, y)$ will be of the form

$$E_{ht} = E_{h1} + E_{h2} + E_{h3} + E_{h4} + E_{h5} + E_{h6}$$

$$= V_m[J_{h1} \sin(wt + \varphi) + J_{h2} \sin(wt + \varphi - 60°) + J_{h3} \sin(wt + \varphi - 120°)]$$

$$= V_m(J_{h1} \angle 0° + J_{h2} \angle -60° + J_{h3} \angle -120°) \tag{7.32}$$

Its amplitude is obtained by separating the real part and j-part and taking the resulting vectorial sum.

The real part is $[J_{h1} + 0.5 (J_{h2} - J_{h3})]$ and the j-part is $0.866 (J_{h2} + J_{h3})$. The amplitude is

$$J_h = [\{J_{h1} + 0.5 (J_{h2} - J_{h3})\}^2 + 0.75 (J_{h2} + J_{h3})^2]^{1/2}$$

$$= (J_{h1}^2 + J_{h2}^2 + J_{h3}^2 + J_{h1} J_{h2} + J_{h2} J_{h3} - J_{h3} J_{h1})^{1/2} \tag{7.33}$$

The total horizontal component of e.s. field at the point $A(x, y)$ has the corresponding r. m. s. value

$$E_h = J_h . V, \text{ where } V = V_m / \sqrt{2} \tag{7.34}$$

Similarly, the total vertical component at $A(x, y)$ is

$$E_v = K_v . V \tag{7.35}$$

where $\quad K_v = (K_{v1}^2 + K_{v2}^2 + K_{v3}^2 + K_{v1} K_{v2} + K_{v2} K_{v3} - K_{v3} K_{v1})^{1/2} \tag{7.36}$

$$K_{v1} = K_1(M_{11} - M_{14}) + K_2(M_{21} - M_{24}) + K_3(M_{31} - M_{34}) + K_4(M_{41} - M_{44})$$
$$+ K_5(K_{51} - K_{54}) + K_6(M_{61} - M_{64})$$

$$K_{v2} = K_1(M_{12} - M_{15}) + K_2(M_{22} - M_{25}) + K_3(M_{32} - M_{35}) + K_4(M_{42} - M_{45})$$
$$+ K_5(M_{52} - M_{55}) + K_6(M_{62} - M_{65})$$

and

$$K_{v3} = K_1(M_{13} - M_{16}) + K_2(M_{23} - M_{26}) + K_3(M_{32} - M_{35}) + K_4(M_{43} - M_{46})$$
$$+ K_5(M_{55} - M_{56}) + K_6(M_{63} - M_{65})$$

The quantities K_1 to K_6 are calculated from equation (7.17).

7.4   Effect of High E.S. Field on Humans, Animals, and Plants

In section 7.1, a discussion of electric shock was sketched. The use of e.h.v. lines is increasing danger of the high e.s. field to (a) human beings, (b) animals, (c) plant life, (d) vehicles, (e) fences, and (f) buried pipe lines under and near these lines. It is clear from section 7.2 that when an object is located under or near a line, the field is disturbed, the degree of distortion depending upon the size of the object. It is a matter of some difficulty to calculate the characteristics of the distorted field, but measurements and experience indicate that the effect of the distorted field can be related to the magnitude of the undistorted field. A case-by-case study must be made if great accuracy is needed to observe the effect of the distorted field. The limits for the undistorted field will be discussed here in relation to the danger it poses.

(a) *Human Beings*

The effect of high e.s. field on human beings has been studied to a much greater extent than on any other animals or objects because of its grave and shocking effects which has resulted in loss of life. A farmer ploughing his field by a tractor and having an umbrella over his head for shade will be charged by corona resulting from pointed spikes. The vehicle is also charged when it is stopped under a transmission line traversing his field. When he gets off the vehicle and touches a grounded object, he will discharge himself through his body which is a pure resistance of about 2000 ohms. The discharge current when more than the let-go current can cause a shock and damage to brain.

It has been ascertained experimentally that the limit for the undisturbed field is 15 kV/m, r.m.s., for human beings to experience possible shock. An e.h.v. or u.h.v. line must be designed such that this limit is not exceeded. The minimum clearance of a line is the most important governing factor. As an example, the B.P.A. of the U.S.A. have selected the ʼmaximum e.s. field gradient to be 9 kV/m at 1200 kV for their 1150 kV line and in order to do so used a minimum clearance at midspan of 23.2 m whereas they could have selected 17.2 m based on clearanec required for switching-surge insulation clearance recommended by the National Electrical Safety Council.

(b) *Animals*

Experiments carried out in cages under e.h.v. lines have shown that pigeons and hens are affected by high e.s. field at about 30 kV/m. They are unable to pick up grain because of chattering of their beaks which will affect their growth. Other animals get a charge on their bodies and when they proceed to a water trough to drink water, a spark usually jumps from their nose to the grounded pipe or trough.

(c) *Plant Life*

Plants such as wheat, rice, sugarcane, etc., suffer the following types of damage. At a field strength of 20 kV/m (r.m.s.), the sharp edges of the stalk give corona discharges so that damage occurs to the upper portion of the grain-bearing parts. However, the entire plant does not suffer damage. At 30 kV/m, the by-products of corona, namely ozone and N_2O become intense. The resistance heating due to increased current prevents full growth of the plant and grain. Thus, 20 kV/m can be considered as the limit and again the safe value for a human being governs line design.

(d) *Vehicles*

Vehicles parked under a line or driving through acquire electrostatic charge if their tyres are made of insulating material. If parking lots are located under a line, the minimum recommended safe clearance is 17 m for 345 kV and 20 m for 400 kV lines. Trucks and lorries will require an extra 3 m clearance. The danger lies in a human being attempting to open the door and getting a shock thereby.

(e) *Others*

Fences, buried cables, and pipe lines are important pieces of equipment to require careful layout. Metallic fences parallel to a line must be grounded preferably every 75 m. Pipelines longer than 3 Km and larger than 15 cm in diameter are recommended to be buried at least 30 m laterally from the line centre to avoid dangerous eddy currents that could cause corrosion. Sail boats, rain gutters and insulated walls of nearby houses are also subjects of potential danger. The danger of ozone emanation and harm done to sensitive tissues of a human being at high electric fields can also be included in the category of damage to human beings living near e.h.v. lines.

7.5 Meters and Measurement of Electrostatic Fields

The principle on which a meter for measuring the e.s. field of an e.h.v. line is based is very simple. It consists of two conducting plates insulated from each other which will experience a potential difference when placed in the field. This can be measured on a voltmeter or an ammeter through a current flow. There exist 3 configurations for the electrodes in meters used in practice: (1) Dipole, (2) Spherical Dipole, and (3) Parallel Plates. The dimensions and other details are given in the following tabular form. Also see Fig. 7.11.

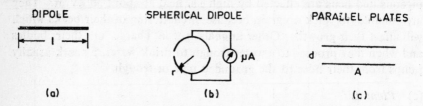

Fig. 7.11. Measurement of E.S. field. (a) Dipole. (b) Spherical dipole. (c) Parallel plates.

Table 7.1. Meters for E.S. Field Measurement

(1)	*Electrode Shape*	*Dipole*	*Spherical Dipole*	*Parallel Plates*
(2)	*Major Dimensions*	Effective length, *l*	Radius, *r*	Distance, *d*, Surface area, *A*
(3)	*Open-circuit Voltage* (as voltage source)	E.l	—	E.d
(4)	*Short-circuit Current* (as current source)	—	$3\pi r^2 W e_0 E$ $(C = 3\pi e_0 r^2)$	$W e_0 A E$ $(C = e_0 A/D)$

The formula for the spherical dipole, which consists of two hemi-spheres insulated from each other, is very accurate and has been recommended as a standard. The two insulated hemispheres are connected by a micro-ammeter whose scale is calibrated in terms of kV/m of the electric field. On the other hand, the parallel plate meter is very easy to fabricate. A small digital voltmeter can be used by attaching a copper-clad printed-circuit board with its insulated side placed on top of the casing. The input to the meter is taken from the top copper-clad side and the casing.

The procedure for measurement consists of attaching the meter to a long (2 m) insulated rod and placing it in the field at the desired height. The insertion of the meter and rod as well as the human-body should not distort the field.

The meters must be first calibrated in a high voltage laboratory. One procedure is to suspend two horizontal plane parallel electrodes with constant separation. A suggested type is 6 m × 6 m wire mesh electrodes separated by 1 m. These will be accurate to within 5%. The meter is placed in the centre of this parallel-plate arrangement and a known voltage is gradually applied until about 100 kV/m is reached. The following application rules may be followed:

1. The instrument should be capable of measuring at least the vertical component of electric field. Up to a height of 3 or 4 m above ground, the horizontal component is very small and the total 50 Hz field is nearly the same as its vertical component.

2. If the instrument is not held in hand, it should be mounted at about 2 m above ground.

3. It should be battery powered in case a power supply for the electronic circuitry is necessary.

4. The r.m.s. value is preferably measured by a full-wave rectifier circuit and suitably calibrated.

5. Analogue indication may be preferable to digital representation as it disturbs the field.

6. The instrument may have full-scale readings of 3, 10, 30, and 100 kV/m or other suitable ranges for legible reading.

7. It should be designed for outdoor use and portable. A pole at least 2 m long must go with the instrument if hand-held.

7.6 Electrostatic Induction in Unenergized Circuit of a D/C Line

We shall end this chapter with some discussion of electrostatic and electromagnetic induction from energized lines into other circuits. This is a very specialized topic useful for line crew, telephone line interference etc. and cannot be discussed at very great length. EHV lines must be provided with wide enough right-of-way so that other low-voltage lines are located far enough, or when they cross the crossing must be at right angles.

Consider Fig. 7.12 in which a double-circuit line configuration is shown with 3 conductors energized by a three-phase system of voltages $V_1 = V_m \sin wt$, $V_2 = V_m \sin (wt - 120^\circ)$ and $V_3 = V_m \sin (wt + 120^\circ)$. The other circuit consisting of conductors 4, 5 and 6 is not energized. We will calculate the voltages on these conductors due to electrostatic induction which a line-man may experience. Now,

$$V_4 = \frac{q_1}{2\pi e_0} \ln (I_{14}/A_{14}) + \frac{q_2}{2\pi e_0} \ln (I_{24}/A_{24}) + \frac{q_3}{2\pi e_0} \ln (I_{34}/A_{34}) \qquad (7.37)$$

$$= \frac{q_1}{2\pi e_0} P_{14} + \frac{q_2}{2\pi e_0} P_{24} + \frac{q_3}{2\pi e_0} P_{34}$$

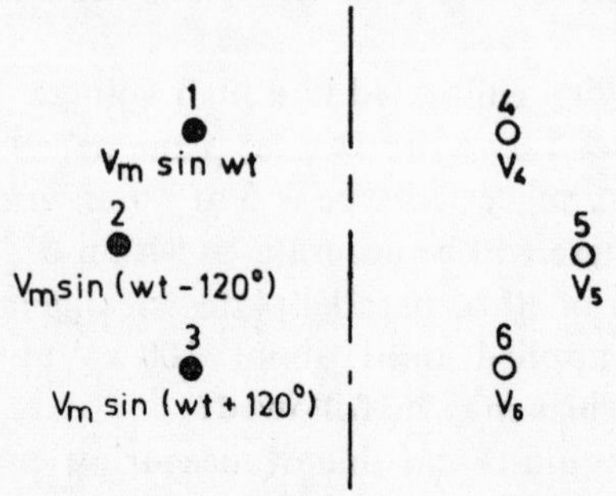

Fig. 7.12. D/C line: One line energized and the other unenergized to illustrate induction.

The charge coefficients q_1, q_2, q_3 are obtained from the applied voltages

$$\begin{bmatrix} V_1 \\ V_2 \\ V_3 \end{bmatrix} = \begin{bmatrix} P_{11}, & P_{12}, & P_{13} \\ P_{21}, & P_{22}, & P_{23} \\ P_{31}, & P_{32}, & P_{33} \end{bmatrix} \begin{bmatrix} q_1 \\ q_2 \\ q_3 \end{bmatrix} (1/2\pi e_0) = [P][q](1/2\pi e_0) \quad (7.38)$$

so that
$$[q/2\pi e_0] = [M][V] \qquad (7.39)$$

$$V_4 = V_m \cdot P_{14}[M_{11} \sin wt + M_{12} \sin (wt - 120^\circ) + M_{13} \sin (wt + 120^\circ)]$$

$$+ V_m P_{24}[M_{21} \sin wt + M_{22} \sin (wt - 120^\circ) + M_{23} \sin (wt + 120^\circ)]$$

$$+ V_m P_{34}[M_{31} \sin wt + M_{32} \sin (wt - 120^\circ) + M_{33} \sin (wt + 120^\circ)]$$

$$= V_m[(P_{14} M_{11} + P_{24}M_{21} + P_{34}M_{31}) \sin wt$$

$$+ (P_{14}M_{12} + P_{24}M_{22} + P_{34}M_{32}) \sin (wt - 120^\circ)$$

$$+ (P_{14}M_{13} + P_{24}M_{23} + P_{34}M_{33}) \sin (wt + 120^\circ)]$$

$$= V_m(\lambda_1 \sin wt + \lambda_2 \sin (wt - 120^\circ) + \lambda_3 \sin (wt + 120^\circ)] \qquad (7.40)$$

In phasor form,

$$V_4 = V [\lambda_1 \angle 0^\circ + \lambda_2 \angle -120^\circ + \lambda_3 \angle 120^\circ], \text{ r.m.s.} \qquad (7.41)$$

$$V_4 = V (\lambda_1^2 + \lambda_2^2 + \lambda_3^2 - \lambda_1\lambda_2 - \lambda_2\lambda_3 - \lambda_3\lambda_1)^{1/2} \qquad (7.42)$$

Similarly, $\quad V_5 = V(\lambda_4^2 + \lambda_5^2 + \lambda_6^2 - \lambda_4\lambda_5 - \lambda_5\lambda_6 - \lambda_6\lambda_4)^{1/2}$ (7.43)

and $\qquad V_6 = V(\lambda_7^2 + \lambda_8^2 + \lambda_9^2 - \lambda_7\lambda_8 - \lambda_8\lambda_9 - \lambda_9\lambda_7)^{1/2}$ (7.44)

where

$$\left.\begin{aligned}
\lambda_1 &= P_{14}M_{11} + P_{24}M_{21} + P_{34}M_{31} \\
\lambda_2 &= P_{14}M_{12} + P_{24}M_{22} + P_{34}M_{32} \\
\lambda_3 &= P_{14}M_{13} + P_{24}M_{23} + P_{34}M_{33} \\
\lambda_4 &= P_{15}M_{11} + P_{25}M_{21} + P_{35}M_{31} \\
\lambda_5 &= P_{15}M_{12} + P_{25}M_{22} + P_{35}M_{32} \\
\lambda_6 &= P_{15}M_{13} + P_{25}M_{23} + P_{35}M_{33} \\
\lambda_7 &= P_{16}M_{11} + P_{26}M_{21} + P_{36}M_{31} \\
\lambda_8 &= P_{16}M_{12} + P_{26}M_{22} + P_{36}M_{32} \\
\lambda_9 &= P_{16}M_{13} + P_{26}M_{23} + P_{36}M_{33}
\end{aligned}\right\}$$ (7.45)

and

Example 7.3: A 230-kV D/C line has the dimensions shown in Fig. 7.13. The phase conductor is a single Drake 1.108 inch (0.028 m) diameter. Calculate the voltages induced in conductors of circuit 2 when circuit 1 is energized assuming (a) no transposition, and (b) full transposition.

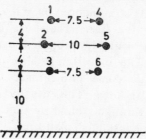

Fig. 7.13. D/C 230-kV line dimensions.

Solution. The Maxwell's Potential coefficients for the energized circuit are as follows:

$$P_{11} = \ln(36/0.014) = 7.852, \quad P_{22} = \ln(28/0.014) = 7.6,$$

$$P_{33} = \ln(20/0.014) = 7.264$$

$$P_{12} = \ln(32.02/4.19) = 2.0335, \quad P_{13} = \ln(28/8) = 1.253,$$

$$P_{23} = \ln(24.03/4.19) = 1.747.$$

$$\therefore \quad [P]_{ut} = \begin{bmatrix} 7.852, & 2.0335, & 1.253 \\ 2.0335, & 7.6, & 1.747 \\ 1.253, & 1.747, & 7.264 \end{bmatrix};$$

$$[M]_{ut} = [P]_{ut}^{-1} = \begin{bmatrix} 0.1385, & -.0334, & -.0159 \\ -.0334, & .1474, & -.0297 \\ -.0159, & -.0297, & .14755 \end{bmatrix}$$

$$P_{14} = \ln (36.77/7.5) = 1.59, \; P_{24} = \ln (33.175/9.62) = 1.238,$$
$$P_{34} = \ln (29/10.97) = 0.972$$
$$P_{15} = P_{24} = 1.238, \; P_{25} = \ln (29.73/10) = 1.09,$$
$$P_{35} = \ln (25.55/9.62) = 0.9765$$
$$P_{16} = P_{34} = 0.972, \; P_{26} = P_{35} = 0.9765,$$
$$P_{36} = \ln (21.36/7.5) = 1.0466.$$

These values give, from equation (7.42) to (7.45), with $V = 230/\sqrt{3}$,

$$\lambda_1 = 0.1634, \quad \lambda_2 = 0.1005, \quad \lambda_3 = 0.0813, \qquad \therefore \quad V_4 = 9.76 \, \text{kV.}$$
$$\lambda_4 = 0.1196, \quad \lambda_5 = 0.09035, \quad \lambda_6 = 0.092, \qquad \therefore \quad V_5 = 3.9 \, \text{kV.}$$
$$\lambda_7 = 0.0854, \quad \lambda_8 = 0.0804, \quad \lambda_9 = 0.11, \qquad \therefore \quad V_6 = 3.65 \, \text{kV.}$$

For the completely transposed line, the induced voltage is the average of these three voltages which amounts to $\frac{1}{3}(9.76 + 3.9 + 3.65) = 5.77$ kV.

7.7 Electromagnetic Interference

Electromagnetic currents are induced by e.h.v. transmission lines in grounded objects near the line. These are harmful to low-voltage circuits which must be suitably protected against damage by fuses, circuit breakers, etc. The induction is also important to line-repair crew who may accidentally come into contact with objects when body-currents causing harm can flow. This usually occurs when there is a return path for current through ground so that suitable insulation must be used such as gloves or insulated platforms. Therefore, the method of grounding is all important.

When low-voltage lines run parallel to existing e.h.v. lines, and a loop exists in the l.v. line, voltages up to 2000 volts/kilometre can exist under normal operating load currents and twice this value under short-circuit currents. The computation of current induced in the l.v. circuit involves mutual inductance between the e.h.v. and l.v. lines, and Carsons' equations for ground-return resistance and inductance have to be used. These were given in Chapter 3, Section 3.9. The usual hazards to workmen when they work on or near lines are of the following types and they may experience shocks by e.m. induction.

(a) High potential developed between an open-ended l.v. line and ground;

(b) High potential between a grounded line and a remote ground on which they stand;

(c) Step potential due to current flowing in ground.

We will here consider a very simple case in which a current I flowing in a high-voltage line, Fig. 7.14, exists and a low-voltage line is to be strung nearby or even a conductor on the same tower. One end of the line is grounded as shown. The mutual impedance between the two lines

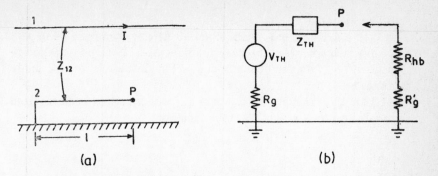

Fig. 7.14.   Electromagnetic induction (a) Overhead line with low-voltage line.
(b) Thevenin equivalent.

is Z_{12} per unit length $= 2\pi f L_{12} = 2\pi f \times 0.2 \ln (I_{12}/A_{12})$. When ground is involved, Carsons' equation for ground inductance must be used at 50 Hz. The open-circuit voltage for a length L of the line is $V_{OC} = I Z_{12} L$. The Thevenin impedance of the l.v. line is due to its self impedance, Z_{TH}. The resulting equivalent circuit is shown in Fig. 7.14 (b). The grounding resistance of the l.v. line is shown as R_g. For a hemisphere of radius R this is $R_g = 2\pi \rho_S/R$ where $\rho_S =$ soil resistivity in ohm-metre.

If now a human being contacts the open end of the l.v. line, his body resistance R_{hb} and his foot resistance to the actual ground R_g' will be in series to the same ground as the l.v. circuit ground. Therefore, the current flowing through his body is

$$I_{hb} = V_{TH}/(R_g + Z_{TH} + R_{hb} + R_g') \qquad (7.46)$$

If this exceeds the let go current then a shock will result.

Review Questions and Problems

1. What is the effect of high electrostatic fields on human beings under a line? Why does a normal human being not experience a shock when walking underneath a line? Why do birds survive even though they come into contact with e.h.v. lines?
2. Describe the difference between primary shock currents and secondary shock current. What is the meaning of 'let-go' current?
3. A 400-kV horizontal line has 2 conductors of 3.18 cm diameter in a bundle with $B = 0.4572$ metre spacing. The line height and phase spacing are $H = 15$ m and $S = 12$ m. A D/C 230–kV line runs parallel to it with its centre 25 metres from the centre of the 400 kV line. The heights of the conductors are 18, 14, and 10 metres with equal horizontal spacings of 10 metres between conductors. Calculate the voltages induced in the 3 conductors closest to the 400–kV

line if the voltage is 420 kV, line-to-line. Assume both lines to be fully transposed.

4. A 1150 kV Δ line has conductors at heights 26m, and 44 m with 24 m spacing between the lowest conductors. Each phase is equipped with 8×46 mm diameter conductor on a circle of 1.2 metre diameter.

At 1200 kV, calculate the e.s. field at ground level at distances from the line centre $d = 26, 39,$ and 52 metres.

CHAPTER 8

Theory of Travelling Waves and Standing Waves

8.1 Travelling Waves and Standing Waves at Power Frequency

On an electrical transmission line, the voltages, currents, power and energy flow from the source to a load located at a distance L, propagating as electro-magnetic waves with a finite velocity. Hence it takes a short time for the load to receive the power. This gives rise to the concept of a wave travelling on the line which has distributed line parameters r, l, g, c per unit length. The current flow is governed mainly by the load impedance, the line-charging current at power frequency and the voltage. If the load impedance is not matched with the line impedance, which will be explained later on, some of the energy transmitted by the source is not absorbed by the load and is reflected back to the source which is a wasteful procedure. However, since the load can vary from no load (infinite impedance) to rated value, the load impedance is not equal to the line impedance always; therefore there always exist transmitted waves from the source and reflected waves from the load end. At every point on the intervening line, these two waves are present and the resulting voltage or current are equal to the sum of the transmitted and reflected voltages. The polarity of voltage is the same for both but the directions of current are opposite so that the ratio of voltage to current will be positive for the transmitted wave and negative for the reflected wave. These can be explained mathematically and have great significance for determining the characteristics of load flow along a distributed-parameter line.

The same phenomenon can be visualized through standing waves. For example, consider an open-ended line on which the voltage must exist with maximum amplitude at the open end while it must equal the source voltage at the sending end which may have a different amplitude and phase. For 50 Hz, at light velocity of 300×10^3 km/sec, the wavelength is 6000 km, so that a line of length L corresponds to an angle of $(L \times 360°/6000)$. With a load current present, an additional voltage caused by the voltage drop in the characteristic impedance is also present which will stand on the line. These concepts will be explained in detail by first considering a loss-less line ($r = g = 0$) and then for a general case of a line with losses present.

8.1.1   Differential Equations and Their Solutions

Consider a section of line Δx in length situated at a distance x from the load end. The line length is L and has distributed series inductance l and shunt capacitance c per unit length, which are calculated as discussed in Chapter 3. The inductance is $(l.\Delta x)$ and capacitance $(c.\Delta x)$ of the differential length Δx. The voltages on the two sides of l are $(V_x + \Delta V_x)$ and V_x, while the currents on the two sides of c are $(I_x + \Delta I_x)$ and I_x. From Fig. 8.1, the following equations can be written down:

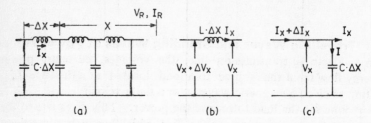

Fig. 8.1.   Transmission line with distributed inductance and capacitance.

$$(V_x + \Delta V_x) - V_x = \Delta V_x = (jwl.\Delta x)\, I_x \tag{8.1}$$

and
$$(I_x + \Delta I_x) - I_x = \Delta I_x = (jwc.\Delta x)\, V_x \tag{8.2}$$

But making Δx infinitesimal, the change in voltage and current along the line are expressed as differential equations thus:

$$dV_x/dx = z.I_x \text{ and } dI_x/dx = y.V_x \tag{8.3}$$

where $z = jwl$ and $y = jwc$, the series impedance and shunt capacitive admittance at power frequency. Since all quantities are varying sinusoidally in time at frequency $f = w/2\pi$, the time-dependence is not written, but is implicit.

By differentiating (8.3) with respect to x and substituting the expressions for dV_x/dx and dI_x/dx, we obtain independent differential equations for V_x and I_x as follows:

$$d^2V_x/dx^2 = z.y.V_x = p^2\, V_x \tag{8.4}$$

and
$$d^2 I_x/dx^2 = z.y.I_x = p^2\, I_x \tag{8.5}$$

where
$$p = \sqrt{z.y} = jw\sqrt{lc} = jw/v = j2\pi/\lambda \tag{8.6}$$

which is the propagation constant, $v =$ velocity of propagation, and $\lambda =$ wavelength. Equations (8.4) and (8 5) are wave equations with solutions

$$V_x = A\, e^{px} + B\, e^{-px} \tag{8.7}$$

and
$$I_x = (1/z)dV_x/dx = \left(\frac{p}{z}\right).(A\, e^{px} - B\, e^{-px})$$

$$= \sqrt{\frac{y}{z}}\, (A\, e^{px} - B\, e^{-px}) \tag{8.8}$$

The two constants A and B are not functions of x but could be possible functions of t since all voltages and currents are varying sinusoidally in time. Two boundary conditions are now necessary to determine A and B. We will assume that at $x = 0$, the voltage and current are V_R and I_R.

Then, $A + B = V_R$ and $A - B = Z_0 I_R$, where $Z_0 = \sqrt{z/y} = \sqrt{l/c} = $ the characteristic impedance of the line.

Then, $\qquad V_x = \tfrac{1}{2}(V_R + Z_0 I_R)\, e^{j2\pi x/\lambda} + \tfrac{1}{2}(V_R - Z_0 I_R)\, e^{-j2\pi x/\lambda} \qquad (8.9)$

and $\qquad I_x = \tfrac{1}{2}(V_R/Z_0 + I_R)\, e^{j2\pi x/\lambda} - \tfrac{1}{2}(V_R/Z_0 - I_R)\, e^{-j2\pi x/\lambda} \qquad (8.10)$

We can also write them as

$$V_x = V_R \cdot \cos(2\pi x/\lambda) + j\, Z_0 I_R \sin(2\pi x/\lambda) \qquad (8.11)$$

and $\qquad I_x = I_R \cdot \cos(2\pi x/\lambda) + j\,(V_R/Z_0)\cdot \sin(2\pi x/\lambda) \qquad (8.12)$

Now, both V_R and I_R are phasors at power frequency and can be written as $V_R = V_R \cdot e^{(jwt+\phi)}$, and $I_R = I_R \cdot \exp\{(jwt + \phi - \phi_L)\}$ where $\phi_L = $ internal angle of the load impedance $Z_L \angle \phi_L$.

We can interpret the above equations (8.11) and (8.12) in terms of standing waves as follows:

(1) The voltage V_x at any point x on the line from the load end consists of two parts: $V_R \cos(2\pi x/\lambda)$ and $j\, Z_0 I_R \sin(2\pi x/\lambda)$. The term $V_R \cos(2\pi x/\lambda)$ has the value V_R at $x = 0$ (the load end), and stands on the line as a cosine wave of decreasing amplitude as x increases towards the sending or source end. At $x = L$, it has the value $V_R \cdot \cos(2\pi L/\lambda)$. This is also equal to the no-load voltage when $I_R = 0$. Fig. 8.2(a) shows these voltages. The second term in equation (8.11) is a voltage contributed by the load current which is zero at $x = 0$ and $Z_0 I_R \sin(2\pi L/\lambda)$ at the source end $x = L$, and adds vectorially at right angles to I_R as shown in Fig. 8.2.

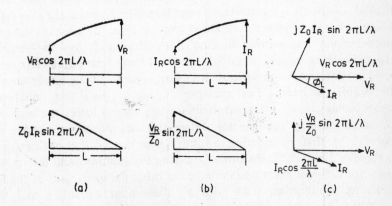

(a)   (b)   (c)

Fig. 8.2. Standing waves of (a) voltage, and (b) current at power frequency.

(2) The current in equation (8.12) also consists of two parts. $I_R \cos(2\pi x/\lambda)$ and $j(V_R/Z_0) \sin(2\pi x/\lambda)$. At no load, $I_R = 0$ and the current supplied by the source is $j(V_R/Z_0) \sin(2\pi L/\lambda)$ which is a pure charging current leading V_R by 90°. These are shown in Fig. 8.2(b) and the vector diagram of 8.2(c).

A second interpretation of equations (8.9) and (8.10) which are equivalent to (8.11) and (8.12) is through the travelling-wave concept. The first term in (8.9) is $\frac{1}{2}(V_R + Z_0 I_R) e^{jw(t+x/v)}$ after introducing the time variation. At $x = 0$, the voltage is $(V_R + Z_0 I_R)/2$ which increases as x does, i.e. as we move towards the source. The phase of the wave is $e^{jw(t+x/v)}$. For a constant value of $(t + x/v)$, the velocity is $dx/dt = -v$. This is a wave that travels from the source to the load and therefore is called the forward travelling wave. The second term is $\frac{1}{2}(V_R - Z_0 I_R)$ $e^{jw(t-x/v)}$ with the phase velocity $dx/dt = +v$, which is a forward wave from the load to the source or a backward wave from the source to the load.

The current also consists of forward and backward travelling components. However, for the backward wave, the ratio of voltage component and current component is $(-Z_0)$ while for the forward waves the ratio is $(+Z_0)$ as seen by the equations (8.9) and (8.10) in which the backward current has a negative sign before it.

8.2 Differential Equations and Solutions for General Case

In section 8.1, the behaviour of electrical quantities at power frequencies on a distributed-parameter line was described through travelling-wave and standing-wave concepts. The time variation of all voltages and currents was sinusoidal at a fixed frequency. In the remaining portions of this chapter, the properties and behaviour of transmission lines under any type of excitation will be discussed. These can be applied specifically for lightning impulses and switching surges. Also, different types of lines will be analyzed which are all categorized by the four fundamental distributed parameters, namely, series resistance, series inductance, shunt capacitance, and shunt conductance. Depending upon the nature of the transmitting medium (line and ground) and the nature of the engineering results required, some of these four parameters can be used in different combinations with due regard to their importance for the problem at hand. For example, for an overhead line, omission of shunt conductance g is permissible when corona losses are neglected. We shall develop the general differential equations for voltage and current by first considering all four quantities (r, l, g, c) through the method of Laplace Transforms, studying the solutions and interpreting them for different cases when one or the other parameter out of the four loses its significance. As with all differential equations and their solutions, boundary conditions in space and initial or final conditions in time play a vital role.

8.2.1 General Method of Laplace Transforms

According to the method of Laplace Transform, the general series impedance operator per unit length of line is $z(s) = r + ls$, and the shunt admittance operator per unit length is $y(s) = g + cs$, where $s =$ the Laplace-Transform operator.

Consider a line of length L energized by a source whose time function is $e(t)$ and Laplace Transform $E(s)$, as shown in Fig. 8.3. Let the line be terminated in a general impedance $Z_t(s)$. We will neglect any lumped series impedance of the source for the present but include it later on. Also, $x = 0$ at the terminal end as in Section 8.1.2. Let the Laplace Transforms of voltage and current at any point x be $V(x, s)$ and $I(x, s)$.

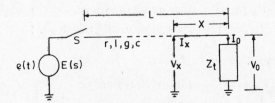

Fig. 8.3. Distributed-parameter transmission line with source $E(s)$ and terminating impedance $Z_t(s)$.

Then, the following two basic differential equations will hold as for the steady-state excitation discussed earlier:

$$\partial V(x, s)/\partial x = z(s).I(x, s) \text{ and } \partial I(x, s)/\partial x = y(s).V(x, s) \qquad (8.13)$$

For simplicity in writing, we may omit s in all terms but only remember that we are discussing the properties of the Laplace-Transforms of all quantities. The solutions for voltage and current in equation (8.13) will be

$$V(x) = A.e^{px} + B.e^{-px} \qquad (8.14)$$

and
$$I(x) = (p/z).(Ae^{px} - Be^{-px}) \qquad (8.15)$$

where again, $p =$ the propagation constant $= \sqrt{z.y} \qquad (8.16)$

Also, $(p/z) = \sqrt{y/z} = Y_0$, and $(z/p) = \sqrt{z/y} = Z_0$, the characteristic or surge impedance.

For this problem, the boundary conditions are:

(1) At $x = L$, $V(L) =$ source voltage $= E(s)$; and

(2) At $x = 0$, $V(0) = Z_t.I(0)$. Using them in (8.14) and (8.15) yields

$$A = (Z_t + Z_0) E(s)/[(Z_t + Z_0) e^{pL} + (Z_t - Z_0) e^{-pL}] \qquad (8.17)$$

and
$$B = (Z_t - Z_0) A/(Z_t + Z_0) \qquad (8.18)$$

$$\therefore \quad V(x) = \frac{\cosh px + (Z_0/Z_t) \sinh px}{\cosh pL + (Z_0/Z_t) \sinh pL} E(s) \qquad (8.19)$$

and $\quad I(x) = \dfrac{1}{z}\dfrac{\partial V}{\partial x} = \sqrt{\dfrac{y}{z}} \cdot \dfrac{\sinh px + (Z_0/Z_t)\cosh px}{\cosh pL + (Z_0/Z_t)\sinh pL} E(s) \qquad (8.20)$

These are the general equations for voltage V and current I at any point x on the line in operational forms which can be applied to particular cases as discussed below.

Line Terminations

For three important cases of termination of an open-circuit, a short circuit and $Z_t = Z_0$, the special expressions of equations (8.19) and (8.20) can be written.

Case 1. Open Circuit. $Z_t = \infty$

$$\left.\begin{array}{l} V_{oc}(x, s) = \cosh px \cdot E(s)/\cosh pL \\ I_{oc}(x, s) = \sqrt{y/z} \cdot \sinh px \cdot E(s)/\cosh pL \end{array}\right\} \qquad (8.21)$$

Case 2. Short-Circuit. $Z_t = 0$

$$\left.\begin{array}{l} V_{sc}(x, s) = \sinh px \cdot E(s)/\sinh pL \\ I_{sc}(x, s) = \sqrt{y/z} \cosh px \cdot E(s)/\sinh pL \end{array}\right\} \qquad (8.22)$$

Case 3. Matched Line. $Z_t = Z_0$

$$\left.\begin{array}{l} V_m(x, s) = (\cosh px + \sinh px) \cdot E(s)/(\cosh pL + \sinh pL) \\ I_m(x, s) = \sqrt{y/z}\,(\cosh px + \sinh px) \cdot E(s)/(\cosh pL + \sinh pL) \end{array}\right\} (8.23)$$

The ratio of voltage to current at every point on the line is

$$Z_0 = \sqrt{z/y}.$$

$$[\sqrt{y/z} = Y_0 = \sqrt{(g + cs)/(r + ls)} = 1/Z_0.]$$

Source of Excitation

The time-domain solution of all these operational expressions are obtained through their Inverse Laplace Transforms. This is possible only if the nature of source of excitation and its Laplace Transform are known. Three types of excitation are important:

(1) *Step Function.* $e(t) = V$. $E(s) = V/s$, where $V =$ magnitude of step.

(2) *Double-Exponential Function.* Standard waveshapes of lightning impulse and switching impulse used for testing line and equipment have a shape which is the difference between two exponentials. Thus $e(t) = E_0 (e^{-\alpha t} - e^{-\beta t})$ where E_0, α and β depend on the timings of important quantities of the wave and the peak or crest value. The Laplace Transform is

$$E(s) = E_0(\beta - \alpha)/(s + \alpha)(s + \beta) \qquad (8.24)$$

(3) *Sinusoidal Excitation.* When a source at power frequency suddenly energises a transmission line through a circuit breaker, considering only a single-phase at present, at any point on the voltage wave, the time function is $e(t) = V_m \sin(wt + \phi)$, where $w = 2\pi f$, $f =$ the power frequency, and $\phi =$ angle from a zero of the wave at which the circuit breaker closes on the positively-growing portion of the sine wave. Its Laplace Transform is

$$E(s) = V_m(w.\cos\phi + s.\sin\phi)/(s^2 + w^2) \tag{8.25}$$

Propagation Factor. For the general case, the propagation factor is

$$p = \sqrt{z.y} = \sqrt{(r + ls)(g + cs)} \tag{8.16}$$

In the sections to follow, we will consider particular cases for the value of p, by omitting one or the other of the four quantities, or considering all four of them.

Voltage at Open-End. Before taking up a detailed discussion of the theory and properties of the voltage and currents, we might remark here that for the voltage at the open end, equation (8.21) gives

$$V(o, s) = E(s)/\cosh pL = 2.E(s)/(e^{pL} + e^{-pL})$$

$$= 2E(s)(e^{-pL} - e^{-3pL} + e^{-5pL} - ...) \tag{8.26}$$

Under a proper choice of p (especially by the omission of g) this turns out to be a train of travelling waves reflecting from the open end and the source, as will be discussed later. Equation (8.26) also gives standing waves consisting of an infinite number of terms of fundamental frequency and all its harmonics.

8.2.2 The Open-Circuited Line: Open-End Voltage

This is a simple case to start with to illustrate the procedure for obtaining travelling waves and standing waves.   At the same time, it is a very important case from the standpoint of designing insulation required for the line and equipment since it gives the worst or highest magnitude of overvoltage under switching-surge conditions when a long line is energized by a sinusoidal source at its peak value. It also applies to energizing a line suddenly by lightning. The step response is first considered since by the Method of Convolution, it is sometimes convenient to obtain the response to other types of excitation by using the Digital Computer (see any book on Operational Calculus).

Travelling-Wave Concept: Step Response

Case 1. First omit all losses so that $r = g = 0$, Then,

$$p = s\sqrt{lc} = s/v \tag{8.27}$$

where $v =$ the velocity of e.m. wave propagation.

$$\therefore \quad V_0(s) = 2V/s(e^{sL/v} - e^{-sL/v}) = 2V\,(e^{-sL/v} - e^{-3sL/v} + ...)/s \qquad (8.28)$$

Now, by the time-shifting theorem, the inverse transform of

$$2V.e^{-ksL/v}/s = 2V.U(t - kL/v) \qquad (8.29)$$

where $\qquad U(t - kL/v) = 0$ for $t < kL/v$

$$= 1 \text{ for } t > kL/v.$$

We observe that the time function of open-end voltage obtained from equations (8.28) and (8.29) will be

$$V_0(t) = 2V\,[U(t - L/v) - U(t - 3L/v) + U(t - 5L/v) - ...] \qquad (8.30)$$

This represents an infinite train of travelling waves. Let $L/v = T$, the time of travel of the wave from the source to the open end. Then, the following sequence of voltages are obtained at the open end.

t	T	$2T$	$3T$	$4T$	$5T$	$6T$	$7T$	$8T$	$9T...$
V_0	$2V$	$2V$	0	0	$2V$	$2V$	0	0	$2V...$

A plot of the open-end voltage is shown in Fig. 8.4(a) from which it is observed that

(1) the voltage reaches a maximum value of twice the magnitude of the input step,

(2) it alternates between $2V$ and 0, and

(3) the periodic time is $4T$, giving a frequency of $f_0 = 1/4T$.

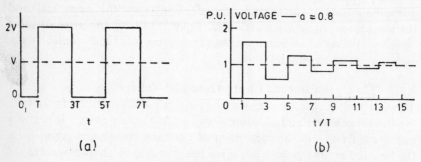

Fig. 8.4.   Step response of transmission line.
(a) Losses neglected.   (b) Losses and attenuation included.

Since all losses have been omitted, attenuation is absent and the amplitude between successive maxima does not decrease. The open-end voltage can never stabilize itself to a value equal to the excitation voltage.

Case 2.   Omit g

For this case, $p = \sqrt{(r + ls)cs}$ or $p^2 = lcs^2 + rcs$.

Then, $\qquad p^2 = (\sqrt{lc})^2\,(s + r/2l)^2 - r^2c/4l \qquad (8.31)$

The last term $(r^2c/4l)$ is usually negligible compared to the first. For example, consider $r = 1$ ohm/km, $l = 1.1$ mH/km and $c = 10$ nF/km.

Then, $r^2c/4l = 2.25 \times 10^{-6}$. Therefore, we take $p = s/v + r/2lv = s/v + \alpha$, where $\alpha = r/2lv =$ attenuation constant in Nepers per unit length.

Now, equation (8.26) becomes, with $E(s) = V/s$ for step,

$$V_0(s) = 2V(e^{pL} + e^{-pL})/s = 2V(e^{-pL} - e^{-3pL} + e^{-5pL} - \ldots)/s$$

$$= 2V/s \,(e^{-\alpha L}.e^{-sL/v} - e^{-3\alpha L}.e^{-3sL/v} + e^{-5\alpha L}.e^{-5sL/v} - \ldots) \qquad (8.32)$$

The time response contains the attenuation factor $\exp(-k\alpha L)$ and the time shift factor $\exp(-ksL/v) = \exp(-kTs)$. The inverse transform of any general term is $2V.\exp(-k\alpha L).U(t - kL/v)$. Let $a = \exp(-\alpha L)$, the attenuation over one line length which the wave suffers. Then the open end voltage of (8.32) becomes, in the time domain,

$$V_0(t) = 2V[a \cdot U(t - T) - a^3 \cdot U(t - 3T) + a^5 \cdot U(t - 5T)\ldots) \qquad (8.33)$$

With the arrival of each successive wave at intervals of $2T = 2L/v$, the maximum value also decreases while the minimum value increases, as sketched in Fig. 8.4(b) for $a = 0.96$ and 0.8. The voltage finally settles down to the value V, the step, in practice after a finite time. But theoretically, it takes infinite time, and because we have neglected the term $(r^2c/4l)$, the final value is a little less than V as shown below. The values of attenuation factor chosen above $(0.96$ and $0.8)$ are typical for a line when only the conductor resistance is responsible for energy loss of the wave and when a value of 1 ohm/km for ground-return resistance is also taken into account.

The following tabular form gives a convenient way of keeping track of the open-end voltage and Fig. 8.5 shows a graphical sketch which is due to Bewley. It is known as the Bewley Lattice Diagram. Note the

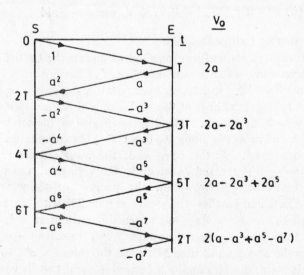

Fig. 8.5. The Bewley Lattice Diagram.

systematic method employed in finding the voltage after any number of reflections.

time t/T	$V_0(t)$		$a = 0.96$	$a = 0.8$
1	$V_1 = 2V.a$		1.92 V	1.6 V
2	$V_2 = 2V.a$			
3	$V_3 = 2V(a - a^3) = V_1 - a^2V_1$		0.15 V	0.576 V
4	$V_4 = V_3$			
5	$V_5 = 2V(a - a^3 + a^5) = V_1 - a^2V_3$		1.78 V	1.23 V
6	$V_6 = V_5$			
7	$V_7 = 2V(a - a^3 + a^5 - a^7) = V_1 - a^2V_5$	0.28 V	0.81 V
8	$V_8 = V_7$			
9	$V_9 = 2V(a - a^3 + a^5 - a^7 + a^9) = V_1 - a^2V_7$	1.66 V	1.08 V
10	$V_{10} = V_9$			
⋮				
N	$V_N = V_1 - a^2V_{N-2}$			
$N+1$	$V_{N+1} = V_N$			

According to this procedure, the final value is

$$V_\infty = 2V a(1 - a^2 + a^4 - a^6 + \ldots) = 2V a/(1 + a^2) \qquad (8.34)$$

Example 8.1: For $a = 0.96$ and 0.8, find the final value of open-end voltage and the % error caused by omitting the term $(r^2c/4l)$ in equation (8.31).

Solution. (a) $a = 0.96$, $V_\infty = 0.9992$ V, error $= 0.08\%$.

(b) $a = 0.8$, $V_\infty = 0.9756$ V, error $= 2.44\%$.

8.2.3 The Bewley Lattice Diagram

Before we consider another form for p, let us discuss the Lattice Diagram for keeping account of the infinite number of reflections on a line when suddenly energized by a source. A horizontal line is drawn to represent the line and two vertical lines at the ends on which equal intervals are marked at time T, as shown. The diagram begins at the top left corner at the source and proceeds along the line OT. The attenuation a is also shown for one travel. At the open end, the wave reflects completely as a. To the right is marked $2a$ which is the voltage at the open end after one travel time. Arrows show the progress of the wave. At T, the wave reflects and reaches the source at time $2T$ with an amplitude a^2, and since the source voltage has to remain constant at $+1$, the wave arriving at the source is reflected negatively and shown as $-a^2$. When this reaches the open end at time $3T$, it has the value $(-a^3)$. Again, at the end, this wave doubles and a reflection $(-a^3)$ travels back to the source. The total voltage at the open end is now $(a + a - a^3 - a^3) =$ $2(a - a^3)$ which is the sum of all voltages marked on the inclined lines up to time $3T$. Proceeding in this manner, we observe that in order to keep the source voltage at $+1$, continued reflections have to take place with negative signs at the source and positive signs at the open end.

8.2.4 $r/l = g/c = \beta$

If the above condition is satisfied, then

$$p = \sqrt{(r + ls)(g + cs)} = \sqrt{lc}\,[(s + r/l)(s + g/c)]^{1/2} = (s + \beta)/v \quad (8.35)$$
$$= s/v + r/lv.$$

The step response is the same as before in Section 8.2.2 and Fig. 8.4(b), but the attenuation factor is now $b = \exp(-\beta L/v) = \exp(-2\alpha L/v) = a^2$. The voltage attenuates more rapidly than when $g = 0$.

Example 8.2: For $a = 0.96$ and 0.8, find the resulting attenuation factor when $r/l = g/c$, assuming line length L to be equal in both cases. Also calculate the maximum values of the surge in the two cases in p.u.

Solution. (1) $a = e^{-\alpha L} = 0.96$ $\therefore$ $b = a^2 = 0.9216$.

Maximum value of surge is the first peak $= 1.8432$ p.u.

(2) $a = e^{-\alpha L} = 0.8$. $\therefore$ $b = a^2 = 0.64$.

$\therefore$ First peak $= 1.28$ p.u.

The relation $r/l = g/c$ is also known as the distortionless condition in which any waveshape of voltage applied at one end will travel without distortion of waveshape but will be attenuated. This will be shown later on. The relation is very useful in telephone and telegraph work where artificial loading coils are used to increase the value of inductance in cable transmission for which g and c are high while l is low.

The case of including all four parameters (r, l, g, c) in the propagation constant p leads to complicated expressions for the Inverse Laplace Transform involving Bessel Functions but, as will be shown in the next section, will yield easier solutions when the standing wave concept is resorted to.

8.3 Standing Waves and Natural Frequencies

In this method, the inverse Laplace Transform is evaluated by the Method of Residues and instead of an infinite number of reflections caused by both ends, we obtain the solution as the sum of an infinite number of frequencies which consist of a fundamental frequency and its harmonics. The infinite series of terms can be truncated after any suitable number of harmonics to yield engineering results. We will first show that the method yields the same result as the travelling-wave concept for a simple case, and then proceed with the Standing-wave Method for more complicated cases for $p = \sqrt{(r + ls)(g + cs)}$.

8.3.1   Case 1—Losses Neglected $r = g = 0$

The Laplace Transform of the open-end voltage, equation (8.26), can be written as (with $r = g = 0$),

$$V_0 = V/s \cdot \cosh pL = V/s \cdot \cosh (s\sqrt{lc}L) \qquad (8.36)$$

The location of the poles of the expression where the denominator becomes zero are first determined. Then the residues at these poles are evaluated as described below, and the time response of the open-end voltage is the sum of residues at all the poles.

Location of Poles. The denominator is zero whenever $s = 0$ and $\cosh s\sqrt{lc}L = 0$. Now, $\cosh j\theta = \cos \theta$ and this is zero whenever $\theta = \pm (2n + 1)\pi/2$, $n = 0, 1, 2,\ldots$. Thus, there are an infinite number of poles where $\cosh (s\sqrt{lc}L)$ becomes zero. They are located at

$$\sqrt{lc}Ls = \pm j(2n + 1)\pi/2.$$

$$\therefore \quad s = \pm j(2n + 1)\pi/2\sqrt{lc}L = \pm jw_n \text{ and } pL = \pm j(2n + 1)\pi/2 \qquad (8.37)$$

where $w_n = (2n + 1)\pi/2\sqrt{lc}L = (2n + 1)\pi v/2L = (2n + 1)\pi/2T$ (8.38)

upon using the property that $v = 1/\sqrt{lc}$ and $T = L/v$. We also observe that (lL) and (cL) are the total inductance and capacitance of the line of length L, and $\pi/2T = 2\pi/4T = 2\pi f_0$, with $f_0 =$ fundamental frequency. This was the same as was obtained in the travelling-wave concept.

Residues

In order to calculate the residue at a simple pole, we remove the pole from the denominator, multiply by e^{st} and evaluate the value of the resulting expression at the value of s given at the pole.

(1)   *At $s = 0$:*

$$\text{Residue} = \frac{V \cdot e^{st}}{\cosh (\sqrt{lc}Ls)} \bigg|_{s=0} = V, \text{ the input step} \qquad (8.39)$$

(2)   At the infinite number of poles, the procedure is to take the derivative of $(\cosh pL)$ with respect to s and evaluate the value of the resulting expression at each value of s at which $\cosh pL = 0$. Thus, the residues at the poles of $(\cosh pL)$ are found by the operation

$$V_{(n+)} = \frac{V}{s} \frac{e^{st}}{\dfrac{d}{ds}(\cosh pL)} \Bigg|_{pL = j(2n+1)\pi/2} \qquad s = jw_n \qquad (8.39)$$

Now, $\dfrac{d}{ds} \cosh \sqrt{lc}Ls = \sqrt{lc} \cdot L \cdot \sinh (\sqrt{lc} \cdot Ls) = \sqrt{lc} \cdot L \sinh pL$. But at

$pL = j(2n + 1)\pi/2$, $\sinh pL = j \sin (2n + 1)\pi/2 = j(-1)^n$.

$$\therefore \quad V_{(n+)} = \frac{V}{jw_n} \cdot \frac{\exp (jw_n t)}{j(-1)^n \cdot \sqrt{lc} \cdot L} = -(-1)^n \cdot V \cdot 2 \exp (jw_n t)/(2n + 1)\pi$$

$$(8.40)$$

Similarly, at $s = -jw_n$, the residue will be the complex conjugate of equation (8.41). Now using $e^{j\theta} + e^{-j\theta} = 2 \cos \theta$, the sum of all residues and therefore the time response is

$$V_0(t) = V\left[1 - \sum_{n=0}^{\infty}(-1)^n \cdot \frac{4}{(2n+1)\pi} \cdot \cos \frac{(2n+1)\pi}{2L\sqrt{lc}} \cdot t \right] \qquad (8.41)$$

The first or fundamental frequency for $n = 0$ is $f_0 = \pi/2L\sqrt{lc} = 1/4T$ and its amplitude is $(4/\pi)$. Let $\theta = \pi t/2L\sqrt{lc}$. Then equation (8.41) can be written as

$$V_0(t) = V\left[1 - \frac{4}{\pi}\left(\cos \theta - \frac{1}{3} \cos 3\theta + \frac{1}{5} \cos 5\theta - ...\right)\right] \qquad (8.42)$$

Now, the Fourier series of a rectangular wave of amplitude V is

$$\frac{4V}{\pi}\left(\cos \theta - \frac{1}{3} \cos 3\theta + \frac{1}{5} \cos 5\theta - ...\right).$$

Therefore, the open-end voltage response for a step input when line losses are neglected is the sum of the step input and a rectangular wave of amplitude V and fundamental frequency $f_0 = 1/4T$. This is the same as equation (8.30) and Fig. 8.4(a). Therefore, the travelling-wave concept and standing-wave method yield the same result.

8.3.2 General Case (r, l, g, c)

Instead of deriving the open-end voltage response for every combination of (r, l, g, c), we will develop equations when all four parameters are considered in the propagation constant, and then apply the resulting equation for particular cases. Now,

$$p = \sqrt{(r + ls)(g + cs)}$$

The poles of $(\cosh pL)$ are located as before when $pL = \pm j(2n + 1)\pi/2$, $n = 0, 1, 2,..., \infty$. The values of s at these poles will be obtained by solving the quadratic equation

$$(r + ls)(g + cs)L^2 = -(2n+1)^2\pi^2/4 \qquad (8.43)$$

This gives $s = -\left(\frac{r}{l} + \frac{g}{c}\right) \pm j\sqrt{\frac{rg}{lc} + (2n+1)^2 \frac{\pi^2}{4L^2lc} - \frac{1}{4}\left(\frac{r}{l} + \frac{g}{c}\right)^2}$

$$\qquad (8.44)$$

or $\qquad s = -a \pm jw_n$

where $\qquad a = (r/l + g/c)$ $\qquad (8.45)$

and $\qquad w_n^2 = rg/lc - \frac{1}{4}(r/l + g/c)^2 + (2n+1)^2\pi^2/4L^2lc$ $\qquad (8.46)$

Now $\qquad \dfrac{d}{ds} \cosh pL = L \cdot \sinh (pL) \cdot (dp/ds)$

where $\qquad \sinh pL = (-1)^n$ at the pole, and

$$\frac{dp}{ds} = \frac{1}{2} \frac{2lcs + (rc + gl)}{\sqrt{(r + ls)(g + cs)}}\bigg|_{s=-a+jw_n} = \frac{1}{2} \frac{2lc(-a + jw_n) + (rc + gl)}{j(2n+1)\pi/2L}$$

$$\qquad (8.47)$$

$$\therefore \quad \frac{d}{ds} \cosh pL \bigg|_{s=-a+jw_n} = j(-1)^n.2L^2lc\, w_n/(2n+1)\pi \qquad (8.48)$$

Thus, the residues at the poles $s = 0$, $s = -a + jw_n$, $s = -a - jw_n$ are found to be

(1) $At\ s = 0$. $V_0 = \dfrac{Ve^{st}}{\cosh\sqrt{(r+ls)(g+cs)}}\bigg|_{s=0} = V/\cosh L\sqrt{rg}$

$$(8.49)$$

(2) $At\ s = -a + jw_n$.

$$V(n+) = \frac{V.e^{st}}{s\,(d.\cosh pL/ds)}\bigg|_{s=-a+jw_n} = \frac{-(-1)^n.e^{-at}(2n+1)\pi}{2L^2lc\,w_n}\frac{e^{jw_nt}}{w_n+ja}$$

$$(8.50)$$

(3) $At\ s = -a - jw_n$.

$$V(n-) = \frac{-(-1)^n\,e^{-at}(2n+1)\pi}{2L^2lc\,w_n}\cdot\frac{e^{-jw_nt}}{w_n-ja} \qquad (8.51)$$

The sum of all residues and therefore the open-end voltage for step input will be, with $\tan\phi = a/w_n$,

$$V_0(t) = V\left[\frac{1}{\cosh L\sqrt{rg}} - \sum_{n=0}^{\infty}\frac{(-1)^n(2n+1)\pi.e^{-at}}{L^2.lc.w_n.\sqrt{a^2+w_n^2}}\cdot\cos(w_nt-\phi)\right]$$

$$(8.52)$$

Particular cases

(1) When $r=g=0$, the lossless condition, $a = 0$, $\phi = 0$, $\cosh L\sqrt{rg} = 1$, and $L^2lcw_n^2 = (2n+1)^2\pi^2/4$. The resulting open-end voltage reduces to equation (8.41).

(2) When $g = 0$, $a = -r/2l$, $w_n = [(2n+1)^2\pi^2/4L^2lc - (r/2l)^2]^{1/2}$.

$\cosh L\sqrt{rg} = 1$. $\quad\therefore\quad \sqrt{a^2+w_n^2} = (2n+1)\pi/2L\sqrt{lc}$.

(3) When $r/l = g/c$, the distortionless condition, $r/l = g/c = b$,

$$a = \tfrac{1}{2}(r/l + g/c) = r/l,\ rg/lc - \tfrac{1}{4}(r/l + g/c)^2 = 0,\ w_n$$
$$= (2n+1)\pi/2L\sqrt{lc}.$$

The natural frequency becomes equal to that when losses are omitted, equation (8.38).

At the velocity of light, for different lengths of line L, the following values of fundamental frequencies and travel time are found from the expressions $T = L/300$ ms, and $f_0 = 1/4T$.

L, km	100	200	300	400	500	750	1000
T, ms	0.333	0.667	1.0	1.333	1.667	2.5	3.333
f_0, Hz	750	375	250	187.5	150	100	75

These frequencies are important for calculating ground-return resistance and inductance according to Carson's Formulae.

8.4 Open-Ended Line: Double-Exponential Response

The response of an open-ended line when energized by a step input was described in Section 8.3 where the Laplace Transform of the step was $E(s) = V/s$. In considering lightning and switching surges, the excitation function is double exponential with the equation, as shown in Fig. 8.6(a),

$$e(t) = E_0(e^{-\alpha t} - e^{-\beta t}) \qquad (8.53)$$

Its Laplace Transform is

$$E(s) = E_0[1/(s + \alpha) - 1/(s + \beta)] = E_0 (\beta - \alpha)/(s + \alpha) (s + \beta) \quad (8.54)$$

Now, the operational expression of the open-end voltage is

$$V_0(s) = E(s)/\cosh pL = E_0(\beta - \alpha)/(s + \alpha)(s + \beta) \cosh pL \quad (8.55)$$

where $p = \sqrt{(r + ls)(g + cs)}$, as before.

The poles of (8.55) are located at $s = \alpha, -\beta$, and $-a \pm jw_n$, where by equations (8.45) and (8.46),

$$a = \frac{1}{2} (r/l + g/c) \text{ and } w_n = \sqrt{(2n + 1)^2\pi^2/4L^2lc + rg/lc - a^2}.$$

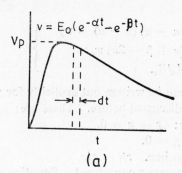

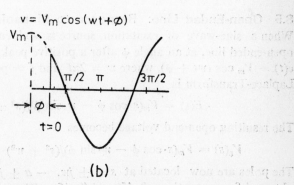

Fig. 8.6. (a) Double-exponential wave: $e(t) = (E_0 e^{-\alpha t} - e^{-\beta t})$.
(b) Sine wave of excitation switched at any point on the wave.

Residues

(1) At $s = -\alpha$. $\quad V(-\alpha) = \dfrac{E_0(\beta - \alpha)\, e^{-\alpha t}}{(\beta - \alpha)\cdot \cosh L\sqrt{(r - l\alpha)(g - c\alpha)}}$ (8.56)

(2) At $s = -\beta$. $\quad V(-\beta) = \dfrac{E_0(\beta - \alpha)\, e^{-\beta t}}{(\alpha - \beta)\cosh L\sqrt{(r - l\beta)(g - c\beta)}}$ (8.57)

(3) At $s = -a + jw_n$.

$$\frac{d}{ds}\cosh pL\,\big|_{\,s=-a+jw_n} = j(-1)^n.2L^2\, lc\, w_n/(2n + 1)\,\pi \quad (8.58)$$

$$\therefore \quad V(n+) = \frac{E_0\,(\beta - \alpha).e^{(-a+jw_n)\,t}.(2n + 1)\,\pi}{(\alpha - a + jw_n\,(\beta - a + jw_n)\, j\, 2L^2. lcw_n} \quad (8.59)$$

(4) At $s = -a - jw_n$. $\quad V(n -) = $ complex conjugate of $V(n +)$.

The sum of all residues will become

$$V_0(t) = \frac{E_0\, e^{-\alpha t}}{\cosh L \sqrt{(r - l\alpha)(g - c\alpha)}} - \frac{E_0\, e^{-\beta t}}{\cosh L \sqrt{(r - l\beta)(g - c\beta)}}$$

$$- E_0 \sum_{n=0}^{\infty} \frac{(-1)^n\,(\beta - \alpha)\,(2n + 1)\,\pi\, e^{-at}}{L^2\, lc\, w_n} \frac{1}{\sqrt{A_n^2 + B_n^2}} \cos(w_n\, t + \phi_n)$$

$$(8.60)$$

where $\qquad A_n = (\alpha - a)(\beta - a) - w_n^2$

$\qquad\qquad B_n = (\alpha + \beta - 2a)\, w_n \qquad \Big\}$ (8.61)

and $\qquad \tan \phi_n = A_n/B_n.$

Equation (8.60) can be written out explicitly for various combinations of r, l, g, c such as discussed before. These are:

(1) Lossless line: $\quad r = g = 0$.

(2) Neglect g. $\quad g = 0$.

(3) Distortionless line. $\quad r/l = g/c$.

(4) All four parameters considered. Equation (8.60).

8.5 Open-Ended Line: Response to Sinusoidal Excitation

When a sine-wave of excitation source is suddenly switched on to an open-ended line, at an angle ϕ after a positive peak, its time function is $e(t) = V_m \cos(wt + \phi)$, where $w = 2\pi f$ and $f =$ power frequency. Its Laplace-Transform is

$$E(s) = V_m(s\cdot\cos\phi - w\cdot\sin\phi)/(s^2 + w^2) \quad (8.62)$$

The resulting open-end voltage becomes

$$V_0(s) = V_m(s\cdot\cos\phi - w\cdot\sin\phi)/(s^2 + w^2)\cdot\cosh pL \quad (8.63)$$

The poles are now located at $s = \pm jw$, $-a \pm jw_n$, with a and w_n obtained from equations (8.45) and (8.46). The residue at each pole is evaluated as before and the resulting open-end voltage is as follows:

Let $J = (rg - w^2lc)L^2$, $K = w(rc + lg)L^2$

$a = \frac{1}{2}(r/l + g/c)$, $w_n = [(2n + 1)^2 \pi^2/L^2lc + rg/lc - a^2]^{1/2}$

$p_1 = 0.5(\sqrt{J^2 + K^2} + J)^{0.5}$, $q_1 = 0.5(\sqrt{J^2 + K^2} - J)^{0.5}$

$F = \cosh p_1 \cdot \cos q_1$, $G = \sinh p_1 \cdot \sin q_1$, $\tan \theta = G/F$,

$p_2 = a^2 + w^2 - w_n^2$.

Then, $V_0(t) = \dfrac{V_m \cdot \cos(wt + \phi - \psi)}{[F^2 + G^2]^{1/2}} + V_m \sum\limits_{n=0}^{\infty} \dfrac{(-1)^n (2n + 1)\pi e^{-at}}{L^2lc\, w_n\, (p_2^2 + 4a^2w_n^2)}$

$\times [\cos \phi \{w_n(p_2 - 2a^2) \cos w_n t - a(p_2 + 2w_n^2) \sin w_n t\}$

$+ w \sin \phi \cdot (p_2 \cdot \cos w_n t - 2aw_n \cdot \sin w_n t)]$ (8.64)

It consists of a steady-state response term, and the transient response with an infinite number of frequency components which decay with the time constant $\tau = 1/a$.

8.6 Line Energization with Trapped-Charge Voltage

Hitherto, we have been considering a line with no initial voltage trapped in it at the time of performing the switching or excitation operation with a voltage source such as the step, double-exponential or sinusoidal. When some equipment is connected between line and ground, such as a shunt-compensating reactor or a power transformer or an inductive potential transformer or during rain, the trapped charge is drained to ground in about half-cycle (10 ms on 50 Hz base). Therefore, when a switching operation is performed, the line is "dead". But there are many situations where the line is re-energized after a de-energization operation with a voltage trapped in it. This voltage in a 3-phase line has a value equal to the peak value of voltage with sinusoidal excitation or very near the peak. The resulting open-end voltage is higher than when trapped charge is neglected. The response or behaviour of the open-end voltage when there is an initial voltage V_t will now be discussed.

The basic differential equations for the line, Fig. 8.7, are

$$\partial V_x/\partial x = (r + l \cdot \partial/\partial t)i_x, \quad \text{and} \quad \partial i_x/\partial x = (g + c \cdot \partial/\partial t)V_x \quad (8.65)$$

As before, let $z = r + ls$ and $y = g + cs$. Then, taking the Laplace

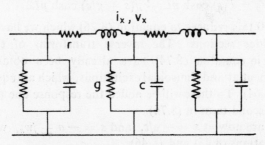

Fig. 8.7. Distributed-parameter line with all four parameters (r, l, g, c).

Transforms with the conditions that at $t = 0$, $i = 0$ and $v = V_t$, the trapped voltage, there results for the Laplace-Transform of voltage at any point x,

$$\partial^2 V_x(s)/\partial x^2 - p^2 V_x(s) + zc\, V_t = 0. \tag{8.66}$$

The complimentary function is obtained with $V_t = 0$ which is $V_{xc} = A\,e^{px} + B\,e^{-px}$. The particular integral will be (cV_t/y), and the complete solution is

$$V_x(s) = A\,e^{px} + B\,e^{-px} + c\, V_t/y \tag{8.67}$$

This can be verified by substituting in equation (8.66).

The boundary conditions are (1) at $x = L$, $V_L = E(s)$ and (2) at $x=0$, $V_0 = Z_t I_0$. Then, A and B have the values, with $Z_0 = \sqrt{z/y} = z/p$, the characteristic impedance,

$$A = (Z_t + Z_0)\, E(s)/D + (V_t\, c/y)(Z_0 \exp{(-pL)} - Z_t - Z_0)/D \tag{8.68}$$

and $B = (Z_t - Z_0)\, E(s)/D - (V_t\, c/y)(Z_0 \exp{(-pL)} + Z_t - Z_0)/D \tag{8.69}$

with $D = (Z_t + Z_0) \exp{(pL)} + (Z_t - Z_0) \exp{(-pL)} \tag{8.70}$

When $V_t = 0$, these equations reduce to equations (8.17) and (8.18).

The current, even with trapped charge, is

$$I_x = (I/z).\partial V_x/\partial x = Y_0.(A \exp{(px)} - B \exp{(-px)}) \tag{8.71}$$

which is the same as equation (8.15).

Combining equations (8.68), (8.69) and (8.67), since A and B also contain the trapped-voltage term,

$$V_x = E(s). \frac{\cosh px + (Z_0/Z_t) \sinh px}{\cosh pL + (Z_0/Z_t) \sinh pL} - \frac{V_t c}{y}$$
$$\cdot \frac{\cosh px + (1 - \exp{(-pL)})(Z_0/Z_t) \sinh px}{\cosh pL + (Z_0/Z_t) \sinh pL} \tag{8.72}$$

For an open-ended line, $Z_0/Z_t = 0$, and the voltage at any point x on the line from the open end is

$$V_{x0} = E(s).\cosh px/\cosh pL - (V_t\, c/y) \cosh px/\cosh pL \tag{8.73}$$

In particular, at the open end,

$$V_0 = E(s)/\cosh pL - V_t/(s + g/c) \cosh pL \tag{8.74}$$

When $V_t = 0$ this reduces to equation (8.26) which we have been dealing with in previous sections. The inverse transforms of the first term $E(s)/\cosh pL$ in equation (8.74) have already been obtained for step, double-exponential and sinusoidal excitations, which are equations (8.52), (8.60) and (8.64). To these will be added the response due to the trapped charge, the second term in (8.74).

The poles are now at $s = -g/c$, and $s = -a \pm jw_n$, with a and w_n given by equations (8.45) and (8.46).

(1) At the simple pole $s = -g/c$, the residue is

$$V_1 = V_t \cdot e^{st}/\cosh L\sqrt{(r+ls)(g+cs)}\,\big|_{s=-g/c} = V_t \cdot \exp\left[-(g/c)t\right] (8.75)$$

(2) The residue at $s = -a + jw_n$ is

$$V(n+) = \frac{-(-1)^n V_t \cdot \exp(-at) \cdot (2n+1)\pi}{2L^2\, lc\, w_n} \cdot \frac{\exp(jw_n t)}{w_n + j(a-g/c)} \qquad (8.76)$$

(3) At the pole $s = -a - jw_n$, the residue is equal to the complex conjugate of $V(n+)$, thus $V(n-) = [V(n+)]^*$.

The sum of all residues and the resulting contribution to the open-end voltage due to the trapped voltage finally becomes

$$V_t \cdot \exp\left[-(g/c)t\right] - V_t \sum_{n=0}^{\infty} (-1)^n \cdot \frac{\exp(-at) \cdot (2n+1)\pi}{L^2 \cdot lc \cdot w_n \{w_n^2 + (a-g/c)^2\}^{1/2}} \cos(w_n t - \psi)$$

$$(8.77)$$

where $\qquad \tan \psi = (a - g/c)/w_n \qquad (8.78)$

Both terms decay with their own time constants. Note that when $g=0$, the trapped-charge voltage is always present in the open-end voltage in spite of a line resistance r being present.

8.7 Corona Loss and Effective Shunt Conductance

When surges propagate on transmission lines they suffer attenuation or decrease in amplitude due to energy lost in the conductor resistance, ground resistance, and corona. This is particularly beneficial in the case of lightning and switching surges. In the previous discussions, a shunt conductance g has been included which represents the corona loss element. We will now derive an approximate expression for this quantity assuming the voltage to be of the double-exponential form. It was shown in Chapter 5 that for unidirectional surges the corona loss in Joules per unit length is given by the expression

$$W_e = \tfrac{1}{2}KC(V_m^2 - V_0^2) \qquad (8.79)$$

where $C = $ capacitance of conductor per unit length, $K = $ the increase in capacitance when corona is present, $V_m = $ peak value of voltage attained and $V_0 = $ corona-inception voltage, peak value.

This was obtained from the q-V relation approximated to a trapezoidal form, Fig. 5.3.

Let the waveshape of the double-exponential be

$$e(t) = E_0(e^{-\alpha t} - e^{-\beta t}) \qquad (8.80)$$

where E_0, α and β depend on the timings of front and 50% value on tail, and the crest voltage. Typical values of α, β, E_0 are given below:

	α	β	E_0/V_m
Lighting Impulse 1.2/50 μs	14.5×10^3	2.45×10^6	1.035
Switching Surge 250/2500 μs	320	12×10^3	1.13

Denoting by g_e the "effective conductance" per unit length, the differential energy loss is $(g_e.e^2.dt)$ per unit length, and the total energy loss in the full wave is

$$W_e = \int_0^\infty g_e.e^2.dt = g_e.E_0^2 \int_0^\infty (e^{-\alpha t} - e^{-\beta t})^2 = \frac{\beta - \alpha}{2\alpha\beta} g_e.E_0^2 \qquad (8.81)$$

Equating this to $\frac{1}{2} KC(V_m^2 - V_0^2)$, the effective conductance per unit length is

$$g_e = \{\alpha\beta/(\beta - \alpha)\} KC.(V_m^2 - V_0^2)/E_0^2 \qquad (8.82)$$

In practice, the crest value of voltage V_m for both lighting and switching surges is 2.5 to 3 times the crest value of corona-inception voltage V_0, and E_0 is approximately equal to V_m, Therefore the factor $(V_m^2 - V_0^2)/E_0^2$ is nearly unity. An approximation for g_e will then be

$$g_e \approx \alpha\beta \, KC/(\beta - \alpha) \text{ and } g_e/C = \alpha\beta \, K/(\beta - \alpha) \qquad (8.83)$$

The value of K is about 0.7.

Example 8.3: A 400 kV line has $c = 10$ nF/km. For lightning and switching surge type of voltage, calculate the effective conductance per unit length assuming $(V_m^2 - V_0^2)/E_0^2 = 1$.

Solution:

(a) *For lightning:* $g_e = \alpha\beta \, KC/(\beta - \alpha)$
$$= 14.5 \times 2.45 \times 0.7 \times 10^{-5}/2.436 = 102 \text{ } \mu\text{v/km}$$

(b) *For switching:* $g_e = (320 \times 12 \times 0.7/11.68)10^{-8} = 2.3 \text{ } \mu\text{v/km}$

Example 8.4: A 300 km line is to be represented by a model consisting of 12 Pi-sections for the above 400-kV line. Find the values of resistances to be connected in shunt for each section.

Solution: Let line length be L and number of Pi-sections be N. Then each π-section is equivalent to a length of line of (L/N) km. The total conductance is $G_e = g_e L/N$ mhos and the corresponding resistance is $R_e = N/g_e L$.

(a) *For lightning-impulse:* $R_e = 12/(102 \times 10^{-6} \times 300) = 392$ ohms

(b) *For switching-surge:* $R_e = 12/(2.3 \times 10^{-6} \times 300) = 17.4$ kilohms

Note that because of the lower resistance to be connected in shunt for the lightning case, the voltage loses energy faster than for the switching surge and the wave attenuation is higher. In general, lightning surges attenuate to 50% value of the incident surge in only 10 km whereas it may only be 80% for switching surges after a travel of even as far as 400 km.

8.8 The Method of Fourier Transforms

The method of Fourier Transforms when applied to propagation characteristics of lightning and switching surges is comparatively of recent origin and because of the availability of powerful Digital Computers offers a very useful tool for evaluating transient performance of systems, especially when distributed parameter lines occur in combination with lumped system elements. These are in the form of series source impedance, resistors in circuit breakers, shunt reactors for line compensation, transformers, bus bars, bushing capacitances, and entire sub-stations at the receiving end.

In previous sections, a solution for voltage at the open end of a line was derived in closed form by both the travelling-wave and standing-wave methods using the time-shifting theorem and residues by using the Laplace Transform. The Fourier Transform parallels the Laplace Transform with (1) the substitution of $s = a + jw$ in all quantities in their operational form, (2) separating real and j-parts in the resulting expression for a voltage or current under investigation, and (3) finally calculating the Inverse Fourier Transform (IFT) by performing an indicated numerical integration, preferably using the Digital Computer. We will illustrate the method for obtaining the open-end voltage considered before and then extend them to include other types of line termination and series and shunt impedances. Specific cases useful for design of insulation of lines will be taken up in Chapter 10 on Switching Surges.

The open-end voltage was found to be, equation (8.26),

$$V_0(s) = E(s)/\cosh pL \text{ where } p = f(r, l, g, c)$$

For a step input, $E(s) = V/s$, and in general $p = \sqrt{(r + ls)(g + cs)}$. Substitute $s = a + jw$. Then,

$$V_0(a + jw) = V/(a + jw).\cosh L \sqrt{\{r + l(a + jw)\}\{g + c(a + jw)\}} \tag{8.84}$$

We can separate the real and j-parts easily. Let $\cosh pL$ be written as the complex number $(M + jN)$. Then,

$$V_0 = V/(a + jw)(M + jN) = V/\{(aM - wN) + j(aN + wM)\}$$

$$= V.\frac{(aM - wN) - j(aN + wM)}{(aM - wN)^2 + (aN + wM)^2} = P + jQ \tag{8.85}$$

The real and j-parts of the required open-end voltage are

$$P = V(aM - wN)/D, \text{ and } Q = -V(aN + wM)/D \tag{8.86}$$

where $\quad D = (aM - wN)^2 + (aN + wM)^2$, the denominator $\tag{8.87}$

The inverse transform is then given by the integral

$$V_0(t) = F^{-1}[V_0(a + jw)] = \frac{2}{\pi}\int_0^\infty P.e^{at}.\cos wt.\sigma_f.dw, \tag{8.88}$$

where σ_f is called the "sigma factor" which helps in the convergence of the integral. In a practical situation, because of the upper limits being 0 and ∞, division by zero occurs. In order to obviate this, a lower limit for $w = W_i$ is assumed (could be a value of 10) and the integration is terminated at a final value $w = W_F$. The sigma factor is then written as

$$\sigma_f = \sin\,(\pi w/W_F)/(\pi w/W_F) \tag{8.89}$$

and the desired integral becomes

$$V_0\,(t) = \frac{2e^{at}}{\pi}\int_{W_i}^{W_F} P.\cos\,wt.\,\frac{\sin\,(\pi w/W_F)}{(\pi w/W_F)}.dw \tag{8.90}$$

The j-part could also be used and is then written as

$$V_0\,(t) = -\,\frac{2}{\pi}\,e^{at}\int_{W_i}^{W_F} Q.\sin\,wt.\,\frac{\sin\,(\pi w/W_F)}{(\pi w/W_F)}.dw \tag{8.91}$$

The numerical integration can be carried out by utilizing any of the several methods available in books on Numerical Methods, such as (a) the trapezoidal rule, (b) Simpson's Rule, (c) Gauss's method, and finally by the application of (d) the Fast-Fourier Transform. These are considered beyond the scope of this book.

There are several factors which are now discussed on the choice of values for several quantities such as a, W_i and W_F, which are very critical from the point of view of application of this method. First, the values of M and N are found as follows:

Let
$$pL = L\sqrt{\{r + l\,(a + jw)\}\,\{g + c\,(a + jw)\}} = m + jn \tag{8.92}$$

Then, $\quad (m + jn)^2 = J + jK$ where $\tag{8.93}$

$$\left. \begin{array}{l} J = L^2\,\{(r + la)\,(g + ca) - lc\,w^2\} \\ K = L^2\,\{(r + la)\,cw + (g + ca)\,lw\} \end{array}\right\} \tag{8.94}$$

$\therefore \quad m^2 - n^2 = J$ and $2\,mn = K$ with $K > 0 \tag{8.95}$

$\therefore \quad$ Solving for m and n in terms of J and K, there is

$$m = [\tfrac{1}{2}(\sqrt{J^2\quad K^2} + J)]^{1/2} \text{ and } n = [\tfrac{1}{2}(\sqrt{J^2 + K^2} - J)]^{1/2} \tag{8.96}$$

Then, $\cosh\,(m + jn) = \cos\,m.\cos\,n + j\,\sinh\,m.\sin\,n = M + jN$ or

$$M = \cosh\,m.\cosh\,n \quad \text{and} \quad N = \sinh\,m.\sin\,n \tag{8.97}$$

The use of the Fourier-Transform Method for very fast-rising input voltages such as the step function is very important and any choice of values for a, W_i and W_F used for this type of excitation will hold for other types. From experience, the following rules are formulated which should be tried out for each case and the final values chosen:

(1) The choice of the converging factor "a" which is the real part of the Fourier-Transform operator $s = (a + jw)$ is very crucial. If $T_f = $ final value of time up to which the response is to be evaluated, then a rule is to choose $(a\,T_f) = 1$ to 4. For example, for a switching-surge study carried out to one cycle on 50 Hz base, that is

20 ms, the value of $a = 50$ to 200. If the calculation is to proceed up to 40 ms, two cycles, then choose $a = 50$ to 200 for the first 20 ms, and 25 to 100 for the next 20 ms.

(2) The choice of final value of $w = W_F$ at which the integration of (8.90) or (8.91) is to be truncated is governed by the rise time of the phenomenon. The shortest rise time occurs for a step function (theoretically zero) and W_F may be large as 10^6, but for other wave-shapes it may be 10^5.

(3) The choice of interval (Δw) depends on the accuracy required. The number of ordinates chosen for integration is $N_0 = (W_F - W_i)/\Delta w$. This can be 500 to 1000 so that the choice of W_F will determine the frequency step to be used.

Chapter 10 will describe some calculations using this method.

8.9 Reflection and Refraction of Travelling Waves

The methods described earlier can be very usefully applied to simple but important system configurations and the Fourier Transform method can handle entire systems given sufficient computer time. In many situations, several components are connected in series and a wave travelling on one of these propagates in a different component with a different value. This is caused by the discontinuity at the junction which give rise to reflected and refracted (or transmitted) waves. They are described by reflection and refraction factors which are derived as follows:

The voltage and current at any point on a line was found to be, equations (8.19) and (8.20), and Fig. 8.3.

$$V = \frac{\cosh px + (Z_0/Z_t) \sinh px}{\cosh pL + (Z_0/Z_t) \sinh pL} E(s) \qquad (8.19)$$

$$I = \frac{1}{Z_0} \frac{\sinh px + (Z_0/Z_t) \cosh px}{\cosh pL + (Z_0/Z_t) \sinh pL} E(s) \qquad (8.20)$$

These resulted from the general solutions, equations (8.14) and (8.15),

$$V(x) = Ae^{px} + Be^{-px} \text{ and } I(x) = \frac{1}{Z_0} (Ae^{px} - Be^{-px})$$

The voltage and current consist of two parts:

$$V(x) = V_1 + V_2 \text{ and } I(x) = I_1 + I_2 = \frac{1}{Z_0} V_1 - \frac{1}{Z_0} V_2 \qquad (8.98)$$

We counted $x = 0$ from the terminal, Fig. 8.3, (at the impedance Z_t) while $x = L$ at the source end. The term e^{px} increases as we move from Z_t to the source which is an unnatural condition. But the concept of a wave decreasing in magnitude as one moves from the source end into the line is a natural behaviour of any phenomenon in nature. Consequently, this is a forward travelling wave from the source. When losses are neglected, $p = \sqrt{lc} = 1/v$ where $v =$ velocity of e.m. propagation. The current also consists of two parts: (1) $I_1 = Ae^{px}/Z_0 = V_1/Z_0$ which is the forward travelling component. The ratio of voltage to current is $+Z_0$, the characteristic impedance of line. (2) $I_2 = -Be^{-px}/Z_0 = -V_2/Z_0$, which is the backward-travelling component from the terminal end to the source. The ratio of voltage to current is $(-Z_0)$.

Let us re-write the expression in (8.19) as

$$V = \frac{e^{px}(1 + Z_0/Z_t) + e^{-px}(1 - Z_0/Z_t)}{e^{pL}(1 + Z_0/Z_t) + e^{-pL}(1 - Z_0/Z_t)} E(s) \qquad (8.99)$$

At $x = 0$, the surge is incident on Z_t where the incident voltage has the magnitude $(1 + Z_0/Z_t) E(s)/D$, where $D =$ the denominator of equation (8.99).

The total voltage across Z_t is

$$V = V_i + V_r$$

where $V_i =$ the incident voltage and $V_r =$ reflected voltage. The voltage V is also called the refracted voltage or transmitted voltage. At the junction of line and Z_t, from equation (8.99), we have the relation

$$K_r = \frac{\text{Reflected Voltage}}{\text{Incident Voltage}} = \frac{1 - Z_0/Z_t}{1 + Z_0/Z_t} = \frac{Z_t - Z_0}{Z_t + Z_0} = \left(\begin{array}{c}\text{Reflection}\\\text{Coefficient}\end{array}\right) \qquad (8.100)$$

The refraction or transmission coefficient is defined as

$$K_t = \frac{\text{Total Voltage at junction}}{\text{Incident Voltage at junction}}$$

$$= \frac{(1 + Z_0/Z_t) + (1 - Z_0/Z_t)}{1 + Z_0/Z_t} = 2Z_t/(Z_t + Z_0) \qquad (8.101)$$

We also note that

$$K_t = 1 + K_r \qquad (8.102)$$

Similarly, for the current. The total current can be written at $x = 0$, or at the junction of line and Z_t, as

$$I = \frac{1}{Z_0} \frac{(1 + Z_0/Z_t) + (Z_0/Z_t - 1)}{(1 + Z_0/Z_t) e^{pL} + (1 - Z_0/Z_t) e^{-pL}} E(s) = I_i + I_r \qquad (8.103)$$

The ratio of reflected component of current to the incident current at the junction, or the reflection coefficient, is

$$J_r = I_r/I_i = (Z_0/Z_t - 1)/(Z_0/Z_t + 1)$$

$$= -(Z_t - Z_0)/(Z_t + Z_0) = -K_r \qquad (8.104)$$

while the transmission coefficient is

$$J_t = I/I_i = (2Z_0/Z_t)/(1 + Z_0/Z_t) = 2Z_0/(Z_t + Z_0) = (Z_0/Z_t)K_t \quad (8.105)$$

The reflection coefficients for voltage and current, K_r and J_r, are of opposite sign since they are backward travelling components. The refraction coefficients for voltage and current are of the same sign.

These coefficients are usually derived in a simpler manner (without proving the relations between incident, reflected and refracted waves) as follows:

Consider the equations

$$V_t = V_i + V_r \text{ and } I_t = I_i + I_r$$

where $\qquad V_t = Z_t I_t, \; V_i = Z_0 I_i, \text{ and } V_r = - Z_0 I_r$ $\qquad$ (8.106)

Then, $\qquad Z_t I_t = Z_0 I_i - Z_0 I_r \text{ giving } I_i - I_r = Z_t I_t / Z_0$ $\qquad$ (8.107)

Since $I_i + I_r = I_t$, we obtain

$$2I_i = (1 + Z_t/Z_0)I_t \text{ giving } I_t = 2Z_0 I_i/(Z_t + Z_0) = J_t I_i \quad (8.108)$$

and

$$2I_r = (1 - Z_t/Z_0)I_t \text{ giving } I_r = I_i(Z_0 - Z_t)/(Z_0 + Z_t) = J_r \quad (8.109)$$

Similarly, $V_t/Z_t = (V_i - V_r)/Z_0$ or $V_i - V_r = (Z_0/Z_t)V_t$ and $V_i + V_r = V_t$. These yield

$$V_t = [2Z_t/(Z_t + Z_0)]V_i = K_t V_i \qquad (8.110)$$

and

$$V_r = V_t - V_i = (K_t - 1)V_i = V_i(Z_t - Z_0)/(Z_t + Z_0) = K_r V_i \quad (8.111)$$

These coefficients can be used very effectively in order to find junction voltages in simple cases which are experienced in practice. Let us consider certain special cases.

	Case	Z_t	K_t	K_r	J_t	J_r
(1)	Open circuit	∞	$+2$	$+1$	0	-1
(2)	Short circuit	0	0	-1	$+2$	-1
(3)	Matched	Z_0	$+1$	0	$+1$	0
(4)	Two outgoing lines	$Z_0/2$	$+2/3$	$-1/3$	$+4/3$	$+1/3$
(5)	n outgoing lines	Z_0/n	$2/(n+1)$	$\dfrac{-(n-1)}{(n+1)}$	$\dfrac{2n}{(n+1)}$	$\dfrac{(n-1)}{(n+1)}$

Example 8.5: An overhead line with $Z_0 = 400$ ohms continues into a cable with $Z_c = 100$ ohms. A surge with a crest value of 1000 kV is coming towards the junction from the overhead line. Calculate the voltage in the cable.

Solution: $K_t = 2Z_c/(Z_c + Z_0) = 200/500 = 0.4$.

Therefore Cable voltage $= 400$ kV, crest.

Example 8.6: In the above example, the end of the cable is connected to a transformer whose impedance is practically infinite to a surge, when the bushing capacitance is omitted. Calculate the transformer voltage.

Solution: When the 400 kV voltage reaches the junction of cable and transformer, it reflects positively with $K_r = +1$ and $K_t = +2$.
Therefore

Transformer voltage $= 800$ kV.

8.10 Transient Response of Systems with Series and Shunt Lumped Parameters and Distributed Lines

Up to Section 8.8 we considered the equations and solutions of a transmission line only when energized by different types of sources. In practice, lumped elements are connected to lines. In this section we will consider the development of equations suitable for solution based on transform methods or finite difference methods using the Digital Computer. The actual procedure for solution is left to Chapter 10 where some examples will be given.

Figure 8.8 shows a general single-line diagram of a source $E(s)$ energizing the line with distributed-parameters (r, l, g, c per unit length). There

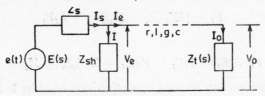

Fig. 8.8.   Transmission system with source, series impedance, shunt impedance, distributed-parameter line, and terminating impedance.

is an impedance Z_s in between the source and line which normally is composed of the transient reactance x' of the synchronous machine and any resistance that can be included in the circuit breaker during the switching operation. A shunt impedance Z_{sh} is also in the circuit which can represent the shunt-compensating reactor. The line is terminated with an impedance Z_t which consists of a transformer, shunt reactor, or an entire substation. For the present, only single-phase representation is used which we can extend to a complete 3-phase system as will be done in Chapter 10.

Let $\quad z = r + ls$ and $y = g + cs$ as before $\qquad$ (8.112)

Then, at any point x from the termination Z_t, the equations are

$$\partial V/\partial x = zI \text{ and } \partial I/\partial x = yV \qquad (8.113)$$

with solutions $\qquad V = A\,e^{px} + B\,e^{-px}$

and $\qquad\qquad I = (p/z)(A\,e^{px} - B\,e^{-px})$ $\qquad\qquad$ (8.114)

The boundary conditions are: (1) at $x = L$, $V = V_e$, the voltage at the line entrance, and (2) at $x = 0$, $V(0) = I(0) \cdot Z_t$. Using

$$Z_0 = \sqrt{z/y} = z/p,$$

we obtained the solutions, equations (8.19) and (8.20) for the voltage and current at any point on the line as follows:

$$V(x) = \frac{\cosh px + (Z_0/Z_t)\sinh px}{\cosh pL + (Z_0/Z_t)\sinh pL}\cdot V_e \qquad (8.115)$$

and
$$I(x) = \frac{1}{Z_0}\cdot\frac{\sinh px + (Z_0/Z_t)\cosh px}{\cosh pL + (Z_0/Z_t)\sinh pL}\cdot V_e \qquad (8.116)$$

At the end of the line, $x = 0$, the voltage and current are

$$V(0) = V_e/[\cosh pL + (Z_0/Z_t)\sinh pL] \qquad (8.117)$$

and
$$I(0) = V(0)/Z_t \qquad (8.118)$$

At the entrance to the line, $x = L$, they are

$$V(L) = V_e \text{ and } I(L) = \frac{1}{Z_0}\cdot\frac{\sinh pL + (Z_0/Z_t)\cosh pL}{\cosh pL + (Z_0/Z_t)\sinh pL}\cdot V_e \qquad (8.119)$$

Therefore, if the voltage V_e is found in terms of the known excitation voltage of source (step, double exponential, or sinusoidal) then all quantities in equations (8.115) to (8.119) are determined in operational form. By using the Fourier Transform method, the time variation can be realized.

Referring to Fig. 8.8, the following equations can be written down:

$$I_{sh} = V_e/Z_{sh}, \; I_s = I_{sh} + I(L), \; V_e = E(s)-Z_sI_s \qquad (8.120)$$

$$\therefore \quad V_e = E(s) - Z_s(I_{sh} + I(L))$$

$$= E(s) - \frac{Z_s}{Z_{sh}}V_e - \frac{Z_s}{Z_0}\frac{\sinh pL + (Z_0/Z_t)\cosh pL}{\cosh pL + (Z_0/Z_t)\sinh pL}V_e \qquad (8.121)$$

Solving for V_e, the voltage at the line entrance, we obtain

$$V_e = E(s)\frac{(1+Z_s/Z_{sh}+Z_s/Z_t)\cosh pL + (Z_0/Z_t+Z_sZ_0/Z_{sh}Z_t+Z_s/Z_0)}{\cosh pL + (Z_0/Z_t)\sinh pL}\sinh pL$$

$$\qquad (8.122)$$

When $Z_s = 0$ and $Z_{sh} = \infty$, the entrance voltage equals the source voltage $E(s)$.

Review Questions and Problems

1. A transmission line is 300 km long and open at the far end. The attenuation of surge is 0.9 over one length of travel at light velocity. It is energized by (a) a step of 1000 kV, and (b) a sine wave of 325 kV peak when the wave is passing through its peak. Calculate and plot the open-end voltage up to 20 ms.

2. For the step input in problem 1, draw the Bewley Lattice Diagram. Calculate the final value of open-end voltage.

3. A transmission line has $l = 1$ mH/km and $c = 11.11$ nF/km.

The conductor plus ground resistance amounts to 0.4 ohm/km. Taking only a single-phase representation, calculate (a) the velocity of propagation, (b) the surge impedance, (c) the attenuation factor for 400 km in Nepers and dB, (d) the maximum value of open-end voltage.

4. The above line is terminated in a load of 500 ohms resistance. Calculate the reflection and refraction factors at the load.

CHAPTER 9

Lightning and Lightning Protection

9.1 Lightning Strokes to Lines

Lightning is a major source of danger and damage to e.h.v. transmission lines, resulting in loss of transmission up to a few hours to complete destruction of a line. This entails a lot of expense to power utilities and consumers. Lightning-protection methods are based on sound scientific and engineering principles and practice; however, extensive damages do occur in power systems in spite of this. Thus, no transmission line design can be considered lightning-proof, nor do designers aim for this goal. Line outages, or a line being taken out of service for a short time, and line designs against these are based on statistical procedures. An acceptable design is to allow a certain number N_t outages per year per 100 km of line, or other time durations and other line lengths. The possibility of an outage depends on so many factors which are statistical in nature that a worst-case design is neither practical nor economical.

It is evident that the number of strokes contacting a tower or ground wire along the span can only depend on the number of thunderstorm days in a year called the 'keraunic level' or also called 'isokeraunic level', and denoted by I_{kl}. On the basis of a vast amount of experience from all over the world, it is estimated that the number of lightning strokes occurring over 1 sq. km. per year at any location is fairly well given by

$$n_s = \text{strokes to earth/km}^2 - \text{year} = (0.15 \text{ to } 0.2)\, I_{kl} \qquad (9.1)$$

The actual factor must therefore be determined from observational data in any given region. Also, from experience, the area intercepted by a line with its metallic structures is taken to be proportional to

(a) (height of tower, $h_t + 2 \times$ height of ground wire at mid span, h_g) and

(b) the distance s_g between ground wires if there is more than 1 ground wire on the tower. Combining all these factors, it is estimated that the number of strokes intercepted by 100 km of line per year is given by

$$N_S = (0.15 \text{ to } 0.2)\, I_{kl} \{0.0133\, (h_t + 2\, h_g) + 0.1\, s_g\} \qquad (9.2)$$

with h_t, h_g and s_g in metres.

Example 9.1: A region has 100 thunderstorm days in a year. A line with a single ground wire has tower height $h_t = 30$ m with ground wire height at midspan $h_g = 24$ m. Calculate the probable number of strokes contacting 100 km of line per year anywhere on the line.

Solution: $s_g = 0$ for 1 ground wire.
Therefore
$$N_s = (0.15 \text{ to } 0.2) \times 100 \times 0.0133 \ (30 + 48)$$
$$= 15.6 \text{ to } 20.8.$$

Example 9.2: If the above line is protected by 2 ground wires with $s_g = 12$m, and other dimensions remain same, calculate N_S.

Solution: $N_S = (0.15 \text{ to } 0.2) \ 100 \ \{0.0133 \times 78 + 1.2\} = 33.6 \text{ to } 44.8.$

We observe that providing 2 ground wires has nearly doubled the number of strokes contacting the line. A larger area of the line is taking part in attracting strokes to the line hardware which otherwise would have contacted ground. However, the better shielding provided by 2 ground wires will generally keep the risk of line outage lower than with only 1 ground wire by reducing the possibility of a line conductor being hit by lightning. It is the job of the line designer to see that proper shielding procedure is adopted such that all strokes contact the tower and ground wire, and none to the line conductors. This is based on probabilistic aspects.

For lines with longer spans, the number of strokes contacting the ground wires between towers is higher than a line with shorter spans. Since e.h.v. lines have higher tower heights, higher ground wire heights at midspan, larger separation between ground wires and in addition longer spans than lower voltage lines, the mechanism of an overvoltage being developed when ground wires are hit assume more prominence. It is therefore essential to have a knowledge of the number of strokes out of N_S which will contact the tower and the rest to the ground wires. This is necessary because at the tower, the lightning stroke is grounded immediately through a low resistance and the insulator voltage is consequently lower as compared to a stroke hitting the ground wire away from the tower. This makes the insulator voltage rise nearly to (stroke current × ground wire surge impedance).

Before we take up a discussion of the problem of design requirements of line insulation based upon lightning voltage and current, we shall briefly examine the mechanism of a lightning stroke when it reaches the vicinity of the transmission line with its metallic towers, ground wires, and phase conductors.

9.2 Lightning-Stroke Mechanism

As transmission voltages increased in 1950 after the second world war to 345 kV and higher, tower height and bulk increased which was reflected in more flashovers than were experienced by the earlier 132 kV and 230 kV lines. The old outage formulas developed for these lower voltage lines were found to yield lower outages than experience indicated. This necessitated a critical examination of the lightning mechanism as it pertains to conditions when contacting the line. In addition to the tower height, the width and breadth also increased because of wider phase spacing and increased weight of hardware the tower had to support as well as the tension. Back flashover mechanism became important in which the tower top potential at the location of outer phases increased to such a high value so as to cause an insulator flashover due to high pre-discharge currents induced by high fields at the pointed tip of the cross-arm. Due to travelling-wave effects inside the long framework of the tower itself, the tower-top potential is high and it takes time for all reflections to die down sufficiently to consider the tower as acquiring ground potential even though it is grounded well at the tower footing.

Prior to these very serious observations, all description of lightning mechanism began and ended with understanding the formation of cloud-charge centres, the physics of propagation of leaders and streamers from the cloud down to earth and the return stroke. This is of limited use for e.h.v. line designs although the physics of these mechanisms might find their own academic or other uses. But the advent of tall towers changed this emphasis and we will commence the discussion with the more important phenomenon of pre-discharge currents caused by the ground plane being elevated to the metallic parts of the transmission line at the towers, and at the same time the infinitely flat ground surface being replaced by metallic parts of smaller curvatures yielding very high stress concentrations which the lightning stroke looks from above.

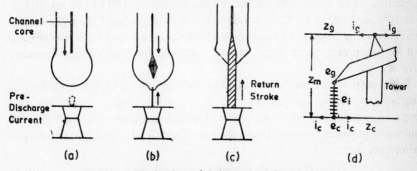

Fig. 9.1.   Mechanism of lightning stroke to tower.
(a) Channel and corona descending from cloud.
(b) Pre-discharge current from tower meeting the core.
(c) Return stroke.
(d) Build-up of insulator voltage.

It is now accepted that there are pre-discharge currents through the tower flowing from ground which are due to the flow of charges of opposite polarity to that coming down in the lightning stroke to contact a tower or a ground wire. The description of the entire work has taken a very long time commencing with Schonland, Komelkov, Bruce, Golde,

Griscom, and finally by Wagner and Hileman in recent years. The downward propagating stroke usually has a channel of about 1 to 3 mm diameter, but is accompanied by a very large corona sheath which can be considered cylindrical up to the channel tip and hemispherical ahead of it, as depicted in Fig. 9.1(a). The leader from the cloud approaches in steps and as it nears the earth the steps become shorter in length and the velocity increases to about 300 km/sec (1/1000-th of light velocity). The actual values vary over wide limits, but this is crucial for the stroke current. All probabilistic aspects of stroke current magnitudes depend on the vagaries of this velocity. The potential of the channel at the commencement of the pre-discharge current is of the order of 50 MV with respect to earth. At the high electric fields present, a corona discharge develops at the tip of the tower and this also propagates upwards to meet the down-coming leader and its corona envelope, Fig. 9.1(b). They meet at heights ranging from 60 to 100 metres above earth. The average potential gradient in the corona sheath is $50 \times 10^3/(60 \text{ to } 100)$ = 833 to 500 kV/m = 8.33 to 5 kV/cm or an average of 6 kV/cm which is necessary for breakdown of moist air.

The charge along the stroke has been estimated as about 5μC/cm to give this voltage gradient and the velocity of the upward channel at the instant and after contact with the downward leader is of the order of 10% light velocity. The formation of front time of the current waveform and the crest value depend on these factors. The stroke current is then $i = qv = 5 \times 10^{-6} \times 0.1 \times 3 \times 10^{10} = 15,000$ Amperes. If the velocity is higher, say 30% light velocity, the current increases to 45000 Amperes. The front time of the stroke is about $(60 \text{ to } 100 \text{ m})/(0.1 \times 3 \times 10^8) = 2$ μs, at 10% light velocity. It is now clear how the statistical variations observed in lightning-stroke currents and their wavefronts can be explained since they depend entirely on the charge content, velocity, etc.

When a stroke contacts a tower, the voltage stress experienced by the insulator is found as follows: If Z_c and Z_g, Fig. 9.1(d), are the surge impedances of the line conductor and ground wire, and Z_m the mutual impedance between them, then the ground wire and line-conductor potentials are

$$e_g = Z_g i_g + Z_m i_c, \quad \text{and} \quad e_c = Z_m i_g + Z_c i_c \qquad (9.3)$$

where i_g and i_c = currents flowing from each side in the two conductors. The conductor voltage is then

$$e_c = (Z_m/Z_g) e_g + (Z_c - Z_m^2/Z_g) i_c = K_f \cdot e_g + (Z_c - K_f Z_m) i_c$$
$$= K_f \cdot e_g + (1 - K_f Z_m/Z_c) Z_c i_c \qquad (9.4)$$

where K_f = coupling factor between ground wire and conductor for a voltage applied to the ground wire.

If the surge impedances are equal, $Z_g = Z_c$, then

$$e_c = K_f e_g + (1 - K_f^2) Z_c i_c.$$

The voltage across the insulator string is

$$e_i = e_g - e_c = (1 - K_f)\, e_g - (1 - K_f.Z_m/Z_c)\, Z_c\, i_c \qquad (9.5)$$

The second term is caused by the presence of pre-discharge current contributed by the line conductor, and if this can be artificially increased, the insulator voltage is lowered. We observe that $i_c = 0.5\, i_{pd}$, where $i_{pd} =$ the pre-discharge current flowing in the conductor. The second term gives an effective impedance of

$$Z_e = \tfrac{1}{2} Z_c\, (1 - K_f\, Z_m/Z_c) \qquad (9.6)$$

The reduction in insulator voltage in some designs by providing long pipes along the conductor up to 4 metres on each side has been estimated to lower the insulator voltage by 15%, and the pre-discharge current measured from laboratory experiments in about 1000 Amperes.

Stroke Contacting Midspan, Fig. 9.2 (a).

In this case,

$$e_g = Z_g\, i_g + Z_m\, i_c, \text{ and } e_c = Z_m\, i_g + Z_c\, i_c \qquad (9.7)$$

Also, $i_g + i_c = \tfrac{1}{2} I_s$, where $I_s =$ stroke current $\qquad (9.8)$

$$\therefore \quad I_s = 2(e_g - e_c)/(Z_g - Z_m) + 2(Z_c + Z_g - 2Z_m)\, i_c/(Z_g - Z_m) \quad (9.9)$$

Example 9.3: Taking an insulator voltage of 8000 kV, surge impedances $Z_g = 500, Z_c = 300$, and $Z_m = 150$ ohms, and the pre-discharge current $i_c = 15000$ Amperes, calculate the stroke current.

Solution: $I_s = 2 \times 8 \times 10^6/350 + 2 \times 15000 \times 500/350$

$$= 46000 + 21,400 = 67,400 \text{ Amperes}$$

We observe that the pre-discharge current has been responsible for contributing almost 33% to the intense stroke current. Photographs taken in laboratories on conductor-to-conductor gaps and on outdoor lines have shown the intense sheath of corona existing between the conductors inside which streamers are observed to propagate giving rise to such high pre-discharge currents, as shown in Fig. 9.2 (b).

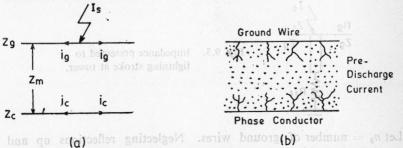

Fig. 9.2. Stroke to midspan. (a) conductor and stroke configuration (b) pre-discharge current and corona sheaths between ground wire and phase conductor.

9.3 General Principles of the Lightning Protection Problem

This section is a preliminary to subsequent material and will discuss the general principles which affect the number of tripouts per 100 km per year. This will give an idea to a designer on the type of information that is required to be gathered, and for a research worker on what types of observations must be carried out in a lightning investigation. Any work of this kind and the information obtained is done over a long period of time, and continuously during days of thunderstorm activity in any region. Much laboratory investigation can be carried out during the remaining portion of the year inside the laboratory using impulse generators. These will be covered in Chapter 13.

Since the lightning-protection problem is based on statistical procedures, we shall first outline the procedure here.

Ist Step. A knowledge of the number of strokes contacting 100 km of line per year is essential as given by equation (9.2). This will depend upon (a) the isokeraunic level I_{kl} which needs field observation, along the line route, (b) the tower height h_t, ground wire height h_g at midspan and the separation between ground wires, s_g. Thus, assumptions must be made for (b) for alternate designs. Since a stroke can contact anywhere on a line, the fraction that might hit a tower top has to be ascertained. For preliminary estimates, a stroke contacting a ground wire within $\frac{1}{4}$ span from the tower and those that hit the tower directly are considered as strokes to a tower. From observations, this is estimated as $0.6\,N_S$. The remaining fraction contacting the middle $\frac{1}{2}$ span can be assumed to cause danger to the clearance between ground wire and phase conductor.

IInd Step. The strokes contacting a tower, Fig. 9.3, will see an impedance to ground governed by (a) the tower-footing resistance R_{tf}, (b) the surge impedance of the ground wire, Z_g, (c) the coupling to line conductors, K_f, and (d) the surge impedance Z_s of the lightning stroke channel itself. This is about 400 ohms.

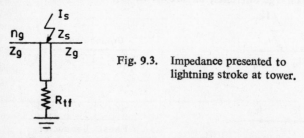

Fig. 9.3. Impedance presented to lightning stroke at tower.

Let n_g = number of ground wires. Neglecting reflections up and down the tower which will not be important after 4 reflections, or about 0.5 μs for a 40-m tower height, and the coupling to the phase conductor, the tower-top potential is approximately,

$$V_t = I_s/(1/R_{tf} + 2n_g/Z_g + 1/Z_s) \tag{9.10}$$

where the denominator is the total admittance, and I_s = stroke current delivered to a zero-impedance ground.

Example 9.4: A tower has a 40-ohm footing resistance and two ground wires each with $Z_g = 500$ ohms. The lightning stroke surge impedance is $Z_s = 400$ ohms. For $I_s = 50$ kA, crest, calculate the tower top potential (a) considering all impedances, (b) neglecting the ground wire and stroke surge impedances, and (c) considering only one ground wire and stroke surge impedance.

Solution: (a) $Z_e = 1/(1/R_{tf} + 4/Z_g + 1/Z_s) = 39.97$ ohms

Therefore tower top potential $V_t = 39.97 \times 50 = 1998.4$ kV, crest.

(b) $Z_e = R_{tf} = 40$ ohms, $\therefore$ $V_t = 2000$ kV, crest

(c) $Z_e = 1/(1/R_{tf} + 2/Z_g + 1/Z_s) = 31.75$ ohms

Therefore $V_t = 31.75 \times 50 = 1587.3$ kV, crest.

IIIrd Step. A knowledge of tower-footing resistance is therefore essential and methods for calculation of this quantity for various types of footing arrangements such as driven rods, horizontally buried wires called counterpoises, etc., will be given in a later section.

IVth Step. When the tower-top experiences the above voltage, the insulator strings supporting the conductors will experience a maximum voltage of

$$V_t = V_t(1 - K_f) + E_m \tag{9.11}$$

where E_m = crest value of line-to-ground power-frequency voltage, and K_f = coupling factor between ground wire(s) and the phase conductor.

The coupling factor can be calculated from the capacitance matrix of the multi-conductor system as shown in chapter 5, even when corona is present. The capacitance matrix is calculated from Maxwell's Potential coefficient matrix, $[c] = 2\pi e_0 [P]^{-1}$ as shown in chapter 3. A value for $K_f = 0.2$ to 0.3 is an average. Equation (9.11) is based on worst case and one must evaluate the probability that at the instant the stroke hits the tower top the power-frequency voltage is passing through its peak value but of opposite sign to the polarity of the lightning voltage at tower top. The trip-out rate will depend on this observed probability, which is again a field investigation on an energized line in the region under consideration.

Example 9.5: A 735-kV line ($735\sqrt{2/3} = 600$ kV, crest, line-to-ground) has a coupling factor between line and ground conductors $K_f = 0.2$. The tower-footing resistance is 40 ohms. Using $V_t = 2000$ kV crest for 50 KA lightning-stroke current, find the voltage experienced by the insulator string.

Solution: $V_i = 2000(1 - 0.2) + 600 = 1600 + 600 = 2200$ kV. We observe that the power frequency voltage is $600/2200 = 0.2727$ or 27.27% of the insulator voltage for this example. If the stroke current had been 100 KA, the reduction due to coupling to the phase conductor will be 800 kV whereas the power-frequency voltage increases the insulator voltage to an extent of 600 kV, crest. Thus, there always exist counter-balancing effects in such a game.

Vth Step. When the voltage V_i across the insulator has been estimated, we now enquire if this will exceed the flashover voltage of the string. For no flashover, V_i must equal the withstand voltage of the string. Therefore, flashover and withstand voltages of insulator strings when subjected to lightning impulses must be available from laboratory studies under all types of atmospheric conditions. Until reflections arrive from the low tower-footing resistances at adjacent towers to the tower hit by lightning, the insulator will be subjected to the full lightning voltage. For span length of 300 metres, a reflection arrives after 2 µs. Thus, it is essential to have the 2-µs flashover data.

For standard $5\frac{3}{4}'' \times 10''$ (14.6 cm $\times$ 25.4 cm) discs the average 2 µs 50% flashover values are 125 kV per disc under dry conditions and 80 kV under rain. It is evident that the lightning voltage must be kept below the impulse withstand voltage by a suitable margin, usually 10%.

VIth Step. The voltage experienced by the insulator is a function of the crest value of lightning-stroke current feeding into a zero-resistance ground. Therefore, the probability that this crest value of current will be achieved in the region where the transmission line will run must be known in order to calculate the number of trip-outs. This is a very extensive investigation and some examples of stroke currents and their statistical occurrence will be discussed in a later section. From a large amount of data available in published literature, the author has found that the following expression is valid for crest currents from 25 to 70 KA with I_c in Kiloamperes:

$$p_i = 1.175 - 0.015 I_c \qquad (9.12)$$

where p_i = probability (fraction) of strokes having a current of crest value I_c kiloamperes. This will also depend upon the isokeraunik level I_{kl}, so that a designer and research worker must obtain all the data for their region.

VIIth Step. We can now calculate the number of times in a year per 100 km of line which will give a voltage in excess of the flashover voltage of the insulator. This is

$$N_t = p_i.p_t.N_s \tag{9.13}$$

where p_i = probability that the crest value of lightning stroke current yielding the flashover voltage of insulator will be found,

p_t = the fraction of strokes N_s which will contact the tower,

N_s = number of strokes contacting the line per year per 100 km length,

and N_t = probable number of tripouts for those strokes contacting the tower.

To this must be added the number of strokes among N_s that cause mid-span flashover.

Example 9.6: A 400-kV horizontal line has 22 discs in the insulator and two ground wires spaced 15 metres apart at 20 m height at mid-span and 26 m at the tower. The tower-footing resistance is 40 ohms. The surge impedances are: Ground wire: 500 ohms, stroke: 400 ohms. Assume 60% of strokes to contact within $\frac{1}{4}$ span of line from the tower and at the tower top. The coupling factor between ground and phase conductor is 0.2 and the factor in N_s is 0.2. The isokeraunik level is 60 thunderstorm days per year. Calculate the number of tripouts per year per 100 km of line. See Fig. 9.3.

Solution:
(1) From equation (9.2),
   $N_s = 0.2 \times 60 \{0.0133 (26 + 2 \times 20) + 0.1 \times 15\}$
   $= 29.6$ strokes/100 km-year.
(2) From equation (9.10),
   $V_t = I_s/(1/R_{tf} + 4/Z_g + 1/Z_s) = 28.2 \, I_s$, kV, with the stroke current I_s in KA.
(3) From equation (9.11),
   $V_i = (1 - 0.2) V_t + 400 \sqrt{2/3} = 22.56 \, I_s + 326$, kV, crest
(4) Flashover voltage of 22 discs $= 22 \times 125 = 2750$ kV in fair weather. In rain, flashover voltage $= 1760$ kV.
(5) Therefore $2750 = 22.56 \, I_s + 326$, which gives the stroke current to be
   $I_s = (2750 - 326)/22.56 = 107.4$ KA.
   In rainy weather, $I_s = (1760 - 326)/22.56 = 63.6$ KA.
(6) The probability of reaching these crest currents are.

   In rain: $p_i = 1.175 - 0.015 \times 63.6 = 0.221$.

   In fair weather: The value of 107.4 KA is beyond the range of equation (9.12). But from Fig. 9.5, it is $p_i = 0.04$.
(7) Using values under rainy conditions, the number of probable tripouts is, from equation (9.13),

$$N_t = 0.22 \times 0.6 \times 29.6 = 3.925 \text{ per 100 km/year}$$

Using dry-weather values,

$$N_t = 0.04 \times 0.6 \times 29.6 = 0.7 \text{ per } 100 \text{ km/year}$$

We shall now discuss all the factors involved in the estimation of tripouts in some detail in the following sections.

9.4 Tower-Footing Resistance

The tower-footing resistance R_{tf} will depend on (a) the type of electrode configuration employed, and (b) the soil resistivity. The most common types of electrode shapes are (1) the hemisphere, (2) long slender rods of about 2 to 5 cm diameter and 10 to 15 metres in length driven vertically down into the soil and connected to tower legs, and (3) buried horizontal wires called the 'counterpoise' of 50 to 150 metres in length in soils where vertical rods cannot be driven. In all cases, sufficient area must be exposed between the electrode and soil in order for the current to spread over a large area.

Soil resistivities, ρ_s, have the following ranges:

Sea water	Moist soil	Loose Soil and Clay	Rock
10^0	10^1	10^2	$> 10^3$ ohm-metre

The electrode shapes and their dimensions are shown in Fig. 9.4 and the formulae for resistances of various types of electrodes are given in the following table.

Electrode Shape	Resistance
(1) Hemisphere, R	$\rho_s/2\pi R$
(2) Vertical driven rod (radius a, length $2l$)	$\dfrac{\rho_s}{2\pi l} \ln(2l/a)$, or, $\dfrac{\rho_s}{2\pi l}\left(\ln\dfrac{4L}{a} - 1\right)$ (Rudenberg) $= \dfrac{\rho_s}{2\pi l} \ln(1.472\, l/a)$ (Dwight).
(3) Horizontal wire. (radius a, length $2l$, depth y)	$\dfrac{\rho_s}{2\pi l} \ln(2l/a) \cdot \left[1 + \dfrac{\ln(l/y)}{\ln(2l/a)}\right]$ $= (\rho_s/2\pi l) \ln(2l^2/ay)$.

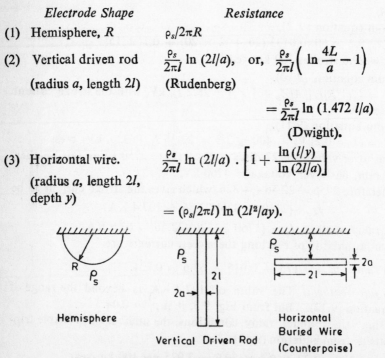

Fig. 9.4. Tower-footing electrodes. Hemisphere, driven rod and buried counterpoise.

Example 9.7: It is necessary to obtain a tower-footing resistance of 20 ohms in a soil of resistivity $\rho_s = 100$ ohm-m using the three different types of electrodes shown above. Take $a = 1.25$ cm for rods and counterpoises and a depth $y = 0.5$ m for the counterpoise. Calculate the required dimensions.

Solution:

(1) *Hemisphere.* $R = \rho_s/2\pi \ R_{tf} = 100/2\pi \times 20 = 0.8$ metre.

(2) *Vertical Driven Rod.* The equation to be solved for l is

$$\ln (2l/0.0125) = 2\pi l \times 20/100 = 0.4\pi l.$$

Trial and error gives $l = 5.4$ metres.

(3) *Horizontal Counterpoise.* The equation is

$$\ln (2l^2/0.0125 \times 0.5) = \ln (320l^2) = 0.4\pi l.$$

This gives $l = 8$ metres.

In the case of driven rods and counterpoises, the resistance is distributed along their lengths and the current entering them will exhibit travelling-wave effect with the result that initially the surge current will see the surge impedance but after a few reflections the resistance assumes the values given above. Usually 4 reflections are sufficient to change the resistance from the surge impedance to the steady value.

9.5 Insulator Flashover and Withstand Voltages

Under positive polarity lightning impulses, a standard $5\frac{3}{4}'' \times 10''$ disc (14.6 cm $\times$ 25.4 cm) shows a highly linear characteristic between sparkover voltage and number of discs. In fair or dry weather conditions, they can be expressed by the following values:

Time to Breakdown, μs	0.5	1.0	2.0	3.0	4.0	6.0	8.0
kV/disc, crest		188	150	125	110	105	97.5	92.5

1.20/50 μs wave, kV/disc $= 87.5$ kV crest

Power Frequency. In fair weather the flashover voltage is 75 kV/disc, crest value, or 53 kV/disc, r.m.s. value. A standard disc has a leakage distance of 31.8 cm (12.5 inches) over its surface. The usual creepage strength used is 1 kV/cm of leakage distance, r.m.s. value.

Strings in Parallel. When lightning hits a line, many insulators are stressed in parallel. It has been found in practice from outdoor experiments that under lightning-type of voltage with extremely small wavefront $(1 - 2 \ \mu s)$, this point is not important as it definitely is under longer surges such as the switching surge. Therefore, single-string values for voltage can be used for flashover and withstand strength.

Conductor-conductor Flashover. Under positive lightning wave the following empirical relation based upon experimental work can be taken:

Flashover voltage $V_{cc} = 590$ kV/metre, crest

9.6 Probability of Occurrence of Lightning Stroke Currents

The probability or the fraction of strokes reaching the ground whose magnitude is above a given anticipated value must be ascertained for the region in which the line will be run. Such experiments use 'magnetic links' connected near tower tops and conductors whose magnetization intensity will depend upon the crest value of current. A typical probability curve is presented in Fig. 9.5.

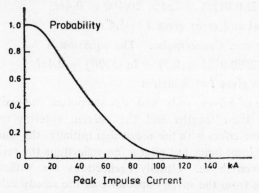

Fig. 9.5. Probability of occurrence of stroke-current magnitudes.

9.7 Lightning Arresters and Protective Characteristics

Lightning arresters, also called 'surge absorbers' because they are also meant for switching-surge protection, protect primarily major equipment such as transformers, rotating machines, shunt reactors, and even entire substations. Less expensive protective devices such as rod gaps can be used for protection of transformer and circuit-breaker bushings and open contacts. When substations have to be protected, they are located at the entrance of the incoming and outgoing lines. Modern surge absorbers for e.h.v. levels are designed to offer protection to equipment and lines for both lightning overvoltages and currents as well as switching surge overvoltages where the energy involved is much higher. Their characteristics will be described later. Arresters are of three important types and classified according to their internal structure. They are

(1) Gap type arrester without current-limiting functions;

(2) Gap type arrester with current-limiting capability; and

(3) Gapless Metal Oxide Varistors.

The first is commonly known by trade names such as Thyrite, Magnavalve, Autovalve, Miurite, etc., each one being associated with its manufacturer. The non-linear resistance material is usually sintered Silicon-Carbide (SiC) and is designed to dissipate the energy in short duration lightning-stroke current and the current at power frequency that will follow this current when the series gap conducts. The current is finally interrupted at a power-frequency zero. The second, or the current-limiting gap type, is of North-American and European design in which a magnetic action on the arc between the gap creates a lengthening of the arc with consequent large resistance capable of limiting the current. In such a design, the power-frequency current can be extinguished prior to reaching a current zero. Such an arrester can perform switching-surge duty also. The non-linear resistance is still SiC. The last one, the gapless MOV, is of recent origin, having been patented only in 1968 by a Japanese firm for low-voltage low-current electronic circuitry but now is sufficiently well developed to handle e.h.v. requirements.

Protective Ratio
The most important property of a surge absorber is the 'protective ratio' which is defined as

$$N_p = \frac{\text{Peak Impulse Insulation Level of Protected Equipment}}{\text{Rated Arrester Power-frequency Voltage, R.M.S. value}}.$$

Typical examples of lightning arrester protective ratios are given in Table 9.1.

The selection of an arrester with a specified voltage rating is governed by the value of 'earthing coefficient' or the 'earth-fault factor'. These are defined as follows and are based on a single line to ground fault condition.

$$EC = \frac{\text{RMS Value of healthy phase voltage at arrester locatio}}{\text{Line-to-line voltage at arrester location}} \quad (9.14)$$

$$\text{The earth-fault factor, EFF} = \sqrt{3} \times EC \quad (9.15)$$

and uses the line-to-ground voltage in the denominator of equation (9.14). The Indian standards and the I.E.C. use the EFF but arresters are still known by the earthing coefficient value. Thus, on a 400-kV line with a maximum operating voltage of 420 kV, an 80% arrester has a rating of 336 kV, r.m.s. with $EC = 0.8$ and EFF $= 0.8\sqrt{3}$.

Under a single line to ground fault, the voltage of the faulted phase is zero at the fault while the voltages of the remaining two healthy phases will rise above normal working voltage. The lightning arrester must withstand this higher voltage without sparkover of the series gap nor conduct current in a gapless arrester. The calculation of earthing coefficient is discussed in section 9.8. From Table 9.1, column 5 shows values of EC used in various countries which have ranged from a low value of 0.72 in the U.S.A. with solidly-earthed neutrals of equipment to a high of 1.0 in Finland using isolated-neutral system (tuned coil).

Discharge Current
The second important quantity for the arrester selection is the lightning-type impulse current which the arrester material has to discharge without damage to itself. Representative values of standard 8/20 μs surge

Table 9.1. Protective Characteristics of Lightning Arresters

Sl. No.	Country	Max. Service Voltage V_m, kV, r.m.s.	Arrester Rating V_a Ph-Gr. kV, r.m.s.	Earthing Coefficient* $C_e = V_a/V_m$	Impulse Protective Level, V_p, kV, Peak	Protective Level Ratio $N_p = V_p/V_a$	Station Equipment Impulse Withstand Level, V_s, kV, Peak	Protective Ratio* $C_p = V_{sl}/V_p$	Impulse withstand Ratio $C_i = V_{sl}/V_m$
(1)	(2)	(3)	(4)	(5)	(6)	(7)	(8)	(9)	(10)
1.	U.S.A.	360	258	0.72	750	2.91	1050	1.4	2.92
2.	India	420	336	0.8	1100	3.27	1550	1.41	3.7
3.	Finland	420	420	1.0	1350	3.22	1650	1.22	3.93
4.	France	420	375	0.89	1200	3.2	1450	1.21	3.45
5.	Germany	420	390	0.93	1156	2.97	1450	1.25	3.45
6.	Sweden	400	360	0.9	1100	3.75	1425	1.29	3.56
7.	Switzerland	420	345	0.82	1100	3.19	1550	1.41	3.7
8.	Canada	525	440–480	0.84–0.91	1260–1350	2.85	1675 (1800)	1.33	3.19 (3.43)
9.	U.S.A.	550	396	0.72	1130	2.85	1425 (1550)	1.27 (1.37)	2.59 (2.82)
10.	U.S.S.R.	525	395	0.75	1260	3.19	1500	1.19	2.86
11.	Canada	735	636–686	0.87–0.93	1700	2.67	2050 (2200)	1.21 (1.3)	2.79 (2.99)

*EFF = $\sqrt{3}\, C_e$

*IEC Recommends 1.2

currents taken from *IS* 3070 (Part I)—1974 are given below:

Crest 8/20 μs current, Amps	5000	10000	15000	20000
System voltage, kV	up to 230	345–400	500	750 (735–765)

Protective Level

The third important characteristic of an arrester is the protective level offered by it to the connected equipment. Having selected the arrester rating based on the system earthing coefficient for a bus fault, the manufacturer of the arresters can provide a resistance material which has a specified *I-V* characteristic. In ideal cases, the voltage across the arrester material should be constant for all current values which flow through it. In a gap-type arrester there is the additional quantity, the sparkover voltage of the gap. When this sparks over, the lightning current is drained to ground through the arrester material. This will hold an *IR*-voltage which depends on the non-linear resistance characteristic of the SiC material. The protective level offered by a lightning arrester is the higher of the following two voltage values:

(1) Sparkover voltage of the series gap under a standard 1.2/50 μs impulse; or

(2) Residual voltage (the *IR*-drop) when discharging the specified test impulse current of 8/20μs waveshape. For gapless arresters only item (2) applies.

With the constant improvements taking place in arrester material and voltage-control (grading) circuits incorporated in the arrester chamber, the above two values are made nearly equal. Table 9.2 gives representative values. In older types (before 1970) the 250/2500 μs switching-surge sparkover voltage was taken as 83% of the short-duration steep-front lightning-impulse breakdown. But in modern arresters, the two are made nearly equal by proper voltage control circuitry.

Table 9.2. Lightning Arrester Sparkover—Discharge Voltage Characteristics

System Maximum Continuous Operating Voltage, kV	Arrester Rating, kV, RMS	1.2/50μs Sparkover kV, crest	Discharge Voltage 8/20μs, KA		Switching Surge Sparkover kV, crest
362	258 (72%)	625	585		610
	276 (76%)	670	625	10	650
	312 (86%)	770	705		735
550	372 (68%)	895	905		870
	396 (72%)	955	965	15	925
	420 (76%)	1010	1020		980
800	540 (67.5%)	1300	1390		1200
	576 (72%)	1400	1500	20	1285
	612 (76%)	1465	1580		1370

When current-limiting gap-type arresters were not used, the power-follow current flowing due to 50Hz voltage when the gap has sparked over had to wait until a current zero, in order that the current through the arrester resistor material could be interrupted, as it does in a circuit breaker. However, in modern current-limiting arresters, the arc can be elongated to any desired extent in a suitably-designed arcing chamber to interrupt the current very quickly without waiting for a current zero. This reduces the energy dissipated by the SiC material and the performance is improved considerably even under switching-surge conditions when the energy discharged by a long transmission line could be considerable.

The lower the protective level offered by the arrester, it is evident that lower can be the insulation level of the equipment it protects. This will bring down the cost of major equipment such as transformers. Examples are already provided in Table 9.1. According to IS 2026. Part III (1981) examples of equipment insulation levels are given in Table 9.3 below.

Table 9.3.   Lightning Arrester Protective Levels and Equipment Insulation (Transformers)

Nominal System kV, RMS	Highest Equipment Voltage, kV, RMS	Rated Switching Impulse withstand (Phase-Neutral) kV, crest	Rated Lightning Impulse withstand Voltage, kV crest	L.A. Protective Level, kV, crest
275	300	750	850 and 950	
		850	950 and 1050	
345	362	850	950 and 1050	625–770
		950	1050 and 1175	
400	420	950	1050 and 1175	
		1050	1175, 1300 and 1425	
500	525	1050	1175, 1300 and 1425	895–1010
		1175	1425 and 1550	1010
750	765	1425	1550 and 1800	1300–1465
		1550	1800 and 1950	

The ideas discussed above can be represented pictorially as shown in Fig. 9.6.

9.8 Dynamic Voltage Rise and Arrester Rating

The selection of arrester voltage rating was shown to depend upon the voltage rise of the healthy phases at the arrester location when a single line to ground fault takes place. Three cases are shown in Fig. 9.7: (1) the isolated-neutral system with a bus fault on phase C, (2) the solidly-grounded-neutral system with a bus fault; and (3) a general system with a fault beyond connected equipment.

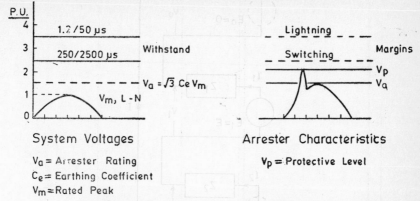

System Voltages

V_a = Arrester Rating
C_e = Earthing Coefficient
V_m = Rated Peak

Arrester Characteristics

V_p = Protective Level

Fig. 9.6. Pictorial representation of arrester voltage rating, earthing coefficient C_e, overvoltages under lightning and switching impulses, protective level of arrester, and margins.

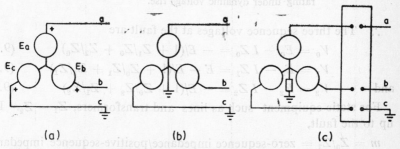

(a)   (b)   (c)

Fig. 9.7. System configurations. (a) Isolated neutral, (b) Solidly grounded, and (c) general configuration for illustrating dynamic voltage rise under 1-phase to ground fault.

For the isolated-neutral system, the voltage of the healthy phases to ground will rise to line-to-line voltage. Thus, Earthing Coefficient EC = Healthy Phase Voltage/line-line voltage = 1.0. The corresponding earth fault factor $EFF = \sqrt{3}\, EC$.

For the solidly-earthed system, the healthy phases do not experience any rise in voltage from the normal operating condition. Therefore, $EC = 1/\sqrt{3}$ and $EFF = 1$.

For the general system, the earthing coefficient lies between $1/\sqrt{3}$ and 1. The value is derived in terms of the ratio of zero-sequence impedance Z_0 to the positive-sequence impedance Z_1 up to the fault. When single line to ground fault occurs, according to symmetrical component theory, the three sequence networks are connected in series, as shown in Fig. 9.8. Only the positive-sequence network contains a driving voltage E. The current is

$$I = E/(Z_0 + Z_1 + Z_2) \tag{9.16}$$

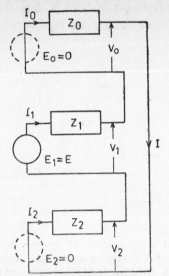

Fig. 9.8.   Connection of sequence networks and calculation of arrester rating under dynamic voltage rise.

∴ The three sequence voltages at the fault are

$$V_0 = E_0 - I Z_0 = -E/(1 + Z_1/Z_0 + Z_2/Z_0) \qquad (9.17)$$
$$V_1 = E_1 - I Z_1 = E - E/(1 + Z_0/Z_1 + Z_2/Z_1) \qquad (9.18)$$

and
$$V_2 = E_2 - I Z_2 = -E/(1 + Z_0/Z_2 + Z_1/Z_2) \qquad (9.19)$$

For static equipment such as lines and transformers, $Z_1 = Z_2$. Let, up to the fault,

$m = Z_0/Z_1 = $ zero-sequence impedance/positive-sequence impedance.

Then, $V_0/E = -m/(m + 2)$, $V_1/E = (m + 1)/(m + 2)$

and $V_2/E = -1/(m + 2)$ $\qquad (9.20)$

The healthy phase voltage such as V_b is with $a = 1\angle 120°$,

$$\frac{V_b}{E} = \frac{V_0 + a^2 V_1 + aV_2}{E}$$

$$= \frac{-m + (-0.5 - j0.866)(m + 1) - (-.5 + j0.866)}{m + 2}$$

and its p. u. value is

$$\left| \frac{V_b}{E} \right| = \sqrt{3} \left| \frac{\sqrt{m^2 + m + 1}}{m + 2} \right|, \text{ when } m \text{ is a real quantity. } (9.21)$$

This will apply when $R_0/x_0 = R_1/x_1$, where R and x are the resistance and reactance components of Z. The earthing coefficient is therefore

$$EC = \left| \frac{V_b}{\sqrt{3}E} \right| = \left| \frac{\sqrt{m^2 + m + 1}}{m + 2} \right|, m \neq 0.$$

When m varies from 2 to ∞, the earthing coefficient and earth fault factors have the following values.

$m = Z_0/Z_1$	1	2	2.5	3	3.5	4	5	7.5	10	∞
EC		0.578	0.66	0.7	0.721	0.744	0.764	0.795	0.849	0.88	1
EFF		1	1.14	1.20	1.25	1.29	1.323	1.38	1.47	1.52	$\sqrt{3}$

9.9 Operating Characteristics of Lightning Arresters

In this section we will describe the three types of arresters used in e.h.v. systems in detail. These are (1) the gap-type SiC arrester, (2) the current-limiting gap-type SiC arrester, and (3) the gapless metal oxide varistor or the zinc-oxide (ZnO) arrester.

9.9.1 Gap Type SiC Arresters

In both the non current-limiting and current-limiting types, the material is sintered Silicon Carbide which is made for a voltage rating of 6 kV, r.m s., per disc. As many discs as are necessary for the arrester rated voltage are stacked in series and provided with voltage-grading circuits. These may consist of high voltage resistors, capacitors, or a combination of both. The I-V characteristics of both types are shown in Fig. 9.9. They function in two stages: (a) Upon occurrence of an over-voltage, the gaps sparkover and provide a low-impedance path to ground. (b) The series resistor reduces power-frequency follow current so that the arc across the gap is able to re-seal either before or at the following voltage and current zero. In the magnetically-blown out surge arrester, Fig. 9.9 (b), the arc is lengthened so that the resulting back e.m.f. helps to reduce the power follow current. The energy loading is about 7 Kw/kV during limiting switching surges with a protection level of 2.2 p.u. being attained.

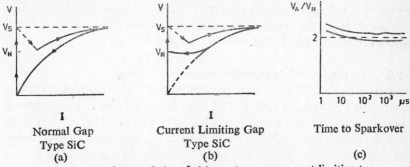

Fig. 9.9. I-V characteristics of (a) gap-type non current-limiting type, (b) gap-type current-limiting type, arresters. (c) Time lag of breakdown of series gap with voltage-grading circuits.

The resistance of a valve element is non-linear with 8.5 kV crest $(6\sqrt{2})$ for 1250 Amperes and only 12 kV crest for 10,000 Amperes surge current. These values give an I-V relation.

$$V = 2604 \, I^{0.166} \text{ (or } I \propto V^6) \quad (9.21)$$

where V and I are in volts and amperes. The resulting resistance is

$$R = V/I = 2604 \, I^{-0.834} \text{ ohms.} \quad (9.22)$$

Ideally, if the material is to hold a constant value of voltage at all currents, the resistance of the material must vary as $R_{id} = K_{id}.I^{-1}$. This is nearly fulfilled in Metal Oxide Varistors, but not in a SiC arrester.

In a non current-limiting arrester, the gap breaks down on the incidence of a lightning overvoltage, which is set at a power-frequency voltage of 1.5 to 2.0 p. u. and at a lightning voltage equal to the protective level. In addition to the lightning current of standard waveshape of 8/20 μs, the material conducts the power-follow current with a very large dissipation of heat in the resistor. In addition, under a switching surge discharge, currents are high and the duration can be as much as 2000 μs. Therefore, lightning arresters are subjected to 3 types of tests:

(1) The power-frequency flashover of the gap. This should not take place for 1 minute at less than 1.5 p. u. voltage [With an earthing coefficient of 0.8 say, the arrester rated voltage is $0.8 \sqrt{3} = 1.3856$ times the maximum line-to-neutral operating voltage].

(2) An 8/20 μs standard lightning impulse current test. The resulting voltage must not exceed the protective level offered by the arrester at rated currents of 5, 10, 15 or 20 Kiloamperes, crest. The protective level is given in Table 9.3.

(3) A long-duration current test of 100-150 Amperes of 1000 μs duration. This simulates switching-surge duty when a line is energized from a source.

By using a suitably-designed grading network to control the distribution of voltage amongst the series-connected gaps, the voltage-time characteristics of the arrester is controlled much better than without them as shown in Fig. 9.9(c).

9.9.2 The Gapless Metal Oxide Arrester

Recent advances in solid-state physics in electronics have been applied to develop a material which is ideal for surge arrester. SiC resistors have a characteristic $I = KV^{\alpha}$, with $\alpha = 4$ to 6. This is not high enough for arresters without spark gaps. A ceramic material based on oxides of Zn, Bi, and Co has $\alpha > 20$ and can handle a very large current range. The I-V characteristic is nearly ideal with a constant protective level from a few milliamperes to thousands of amperes over a 5 decade range. The base material is ZnO (n-material) grains sintered in a flux of various insulating oxides such as Bi_2O_3. Other constituents are CoO, MnO, and Cr_2O_3. These change energy levels and hence the conduction and insulating properties. These oxides coat the high-conductivity ZnO grains with a thin insulating layer of 100-200 Å as shown in Fig. 9.10. The resulting I-V characteristic is nearly $I \propto V^{35-50}$.

Example 9.8: Two arrester materials have the characteristics $I \propto V^6$ and $I \propto V^{40}$. For current variations from 10 Amperes to 10000 Amperes, determine the ratio of voltages at these currents.

Solution: (a) $10 = KV_1^6$ and $10^4 = KV_2^6$

$\therefore \quad V_2/V_1 = (10^4)^{1/6} = 10^{0.667} = 4.645$

(b) $10 = KV_1^{40}$ and $10^4 = KV_2^{40}$

$\therefore \quad V_2/V_1 = (10^4)^{1/40} = 10^{0.1} = 1.26$.

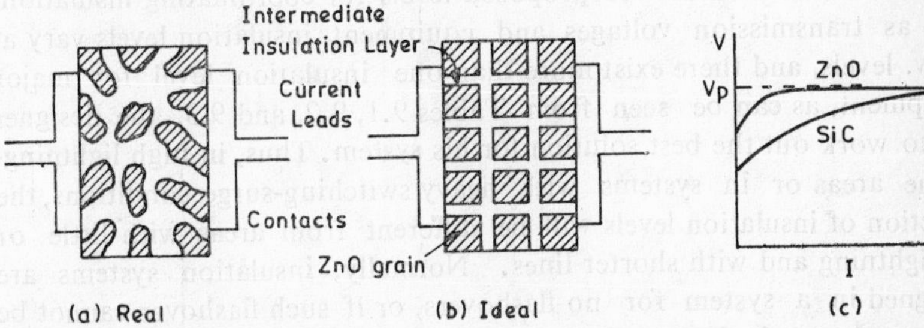

Fig. 9.10. Internal structure and external I-V characteristics of ZnO gapless arrester. Comparison with SiC arrester.

Therefore, higher the index of V, the less is the rise in voltage and more it approaches the ideal. A comparison of gap-type SiC arrester and gapless ZnO arrester is shown in Fig. 9.10(c). ZnO arresters are usually made in discs of 80 mm diameter and 32 mm thickness. At the reference voltage the current conducted is only 5 mA at about 150°C. At lower temperatures (60°C) it is well below 1 mA. Thus, even though the material is always connected in the circuit, it conducts negligible amount of current. This is because of the insulating oxides surrounding the ZnO. When an overvoltage occurs, the energy band levels change and the current will rise by a continuous transition from insulating to conduction state. At the termination of the voltage transient, either lightning or switching, the current is reduced and there is no power-follow current. The rated voltage or the reference voltage is controlled by the earthing coefficient EC.

The advantages of ZnO technology are evidently their simple construction, absence of spark gaps which gives a shock to the system when the gap breaks down. But the disadvantages are a continuous flow of power frequency current with the theoretical possibility of thermal runaway present in all solid-state materials. The earlier arresters demonstrated instability effects with watts loss under normal voltage increasing with time and with the number of discharges. But modern developments have eliminated this danger practically completely. In order to prove the reliability, a line-discharge test is specified and the temperature of the material must regain its normal value after two or three

quick discharges in succession. The absence of spark gaps also eliminates the need for voltage-grading system which in turn eliminates the
volt/time-lag property present in gap-type arresters.

9.10 Insulation Coordination Based on Lightning

Insulation coordination consists of selecting insulation of various lines
and equipment that have to be interconnected into a system for desired
operational requirement. The system must be reliable and economical.
The I.E.C. and Indian standards (or other standards) have only recommended certain values or proposed levels for coordinating insulation.
But as transmission voltages and equipment insulation levels vary at
e.h.v. levels, and there exist more than one insulation level for major
equipment, as can be seen from Tables 9.1, 9.2, and 9.3, the designer
has to work out the best solution for his system. Thus, in high lightning-
prone areas or in systems with heavy switching-surge conditions, the
selection of insulation levels will be different from areas with little or
no lightning and with shorter lines. Normally, insulation systems are
designed in a system for no flashovers, or if such flashover cannot be
prevented such flashovers should be restricted to places where damage is
not done, such as air gaps or in gap-type arresters. The flashover should
not disturb normal system operation and must occur in re-sealable
insulation structures. The overvoltages that can cause damage are due
to external origin, namely lightning, and operation of the system itself
which are at power frequency, earth faults, and switching operations. We
will consider here the insulation co-ordination principles based on
lightning. These insulation levels are known as Basic Impulse Insulation
Level or BIL. Those based on switching-surge requirements are known
as Switching Impulse Levels or SIL.

The lightning arrester is the foundation of protection in e.h.v. ranges,
which is selected for both lightning and switching-surge duty. It is
usually of the magnetic blow-out (current-limiting) gap type, or in recent
years she gapless ZnO type. For atmospheric overvoltages, the duty or
task of the arrester is to limit these overvoltages to the protective level
V_p given in column 6 of Table 9.1. This is the peak value of impulse
voltage as determined by the higher of the 1.2/50 µs spark-over value of
the gap or residual voltage for standard 8/20 µs surge current in the
10 KA to 20 KA range. The latter applies to gapless type while both
the voltages apply to gap type arresters. The lightning current passing
through the arrester material is calculated as follows.

Consider a travelling wave of voltage V_w, crest, which is accompanied
by a current wave I_w on a line with surge impedance Z, Fig. 9.11. They
strike an arrester whose duty is to hold the voltage across it constant at
the protective level V_p. Now, by using Thevenin's theorem, with the
arrester terminals open, the incident travelling wave will give a voltage
$2V_w$ due to total reflection. The Thevenin impedance looking through
the open arrester terminals is equal to the surge impedance of the line Z.

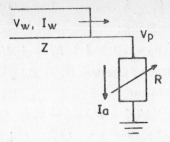

Fig. 9.11. Calculation of arrester current.

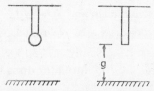

Fig. 9.12. Sphere-to-plane and rod-to-plane protective gaps for insulation coordination of equipment.

Therefore, with the arrester connected, the current through it will be

$$I_a = (2V_w - V_p)/Z \qquad (9.23)$$

The maximum value of travelling-wave voltage V_w can reach is the flashover voltage of the line insulation. Also, it is assumed that V_p stays fairly constant at all current values discharged by the arrester.

Example 9.9: For a 750 kV line, take $V_w = 3000$ kV, crest, travelling on the line and $V_p = 1700$ kV. The line surge impedance is $Z = 300$ ohms. Calculate and discuss (a) the current flowing in the line before reaching the arrester, (b) the current through the arrester, and (c) the value of arrester resistance for this condition and verify the reflection and refraction coefficients giving rise to the voltage and current conditions.

Solution:

  (a) $I_w = V_w/Z = 3000/300 = 10$ Kiloamperes.

  (b) $I_a = (2V_w - V_p)/Z = (6000 - 1700)/300 = 14.33$ KA.

  (c) The reflected current in the line is $+ 4.33$ KA.

This gives rise to a reflected voltage of $- 4.33 \times 300 = - 1300$ kV. Under these conditions, the arrester resistance is

$$R_a = V_p/I_a = 1700 \text{ kV}/14.33 \text{ KA} \approx 120 \text{ ohms. (118.6 ohms)}.$$

With the line surge impedance $Z = 300$ ohms, the following reflection and refraction coefficients are found.

  (1) Voltage reflection factor: $K_r = \dfrac{R_a - Z}{R_a + Z} = \dfrac{120 - 300}{120 + 300} = -\dfrac{3}{7}$

∴ Reflected voltage $V_r = K_r \cdot V_w = -3 \times 3000/7 = -1300$ kV.

(2) Voltage refraction factor:

$$K_t = 2R_a/(R_a + Z) = 1 + K_r = +4/7.$$

∴ Junction voltage at arrester, $V_p = K_t V_w = 1700$ kV.

(3) Current reflection factor:

$$J_r = (Z - R_a)/(Z + R_a) = -K_r = +3/7.$$

∴ Reflected current $I_r = J_r I_w = +4.33$ KA

(4) Current transmission factor: $J_t = 2Z/(Z + R_a) = +10/7$

∴ Arrester discharge current $I_a = J_t I_w = 14.33$ KA.

This example shows that for such a 750 kV line an arrester rated for 15 KA would be necessary. Usually a 750-kV line will be equipped with about 35 standard $5\frac{3}{4}'' \times 10''$ insulator discs whose withstand value is about 3000 kV.

The protective ratio can be calculated if the rated voltage of the arrester is known from the system conditions.

Example 9.10: For the above example, if an 80% arrester is used, calculate the protective ratio $N_P = V_P/V_a$.

Solution: For rated line-to-line voltage of 750 kV, arrester rating is

$$V_a = 0.8 \times 750 = 600 \text{ kV (R.M.S.)}.$$

∴ Protective ratio $N_p = 1700/600 = 2.83$.

Voltage Across Equipment Protected By Arrester

In the ideal case, the arrester must be located adjacent to the equipment which is usually a large transformer or shunt reactor. In practice, however, there may be a length of line between the two extending to 20 to 40 metres. This results in a slightly higher voltage across the equipment due to repeated reflections. The high inductance of a transformer or reactor represents nearly an open-circuit to a surge. The excess voltage experienced is given by an empirical equation and depends on the line length and the rate of rise of the voltage, thus:

$$\Delta V = (dV_w/dt) \cdot (l/150) \text{ kV}, \tag{9.24}$$

where l = length of line in metres

and dV_w/dt = steepness of wave front in kV/μs of the incoming wave. This can be taken as approximately 500 kV/μs for lines with overhead ground wires and 1000 kV/μs when a line conductor is hit (A line without earth wires).

Example 9.11: A transformer is connected by a length of 20 metres of line to an arrester. The rate of rise of voltage is 700 kV/μs. The arrester voltage is 1700 kV. Calculate the voltage across the transformer.

Solution: $\Delta V = (700) \times (20/150) = 93$ kV.

∴   Transformer voltage $= 1793$ kV, impulse crest.

Note that from Table 9.3, for 750 kV system voltage, the transformer lightning-impulse withstand voltages have levels ranging from 1550 to 1950 kV.

The transformer insulation level is kept higher than the arrester protective level by a safe margin as shown by column 8 of Table 9.1. The I.E.C. suggests a value of $1.2\,V_p$ as the equipment insulation level. This is the protective ratio $C_p = V_s/V_p$, where $V_s =$ voltage level of station equipment insulation. It depends on the earthing coefficient, C_e, the impulse protective level ratio of arrester, C_p, and the impulse protective ratio of equipment, C_i. Thus, the equipment insulation level is

$$V_s = C_e.C_p.C_i.V \qquad (9.25)$$

where $V =$ rated r.m.s. value of power-frequency line-line voltage of the system.

Example 9.12: For a 750-kV system with maximum operating voltage of 765 kV at the receiving end substation, the earthing coefficient is $C_e = 0.84$, the protective level ratio of the arrester is 2.83, and the equipment insulation level ratio is 1.3. Calculate the impulse withstand voltage of the equipment insulation.

Solution: $V_s = 0.84 \times 2.83 \times 1.3\,V = 3.094 \times 765$

$$= 2370 \text{ kV, crest.}$$

[*Note*. I.E.C. suggests a 2400 kV level. But this has not yet found acceptance as a standard].

Example 9.13: For a 400-kV system (420 kV maximum) the impulse level of a transformer is 1425 kV, crest or peak. Calculate the ratio of impulse withstand level of transformer insulation to the maximum service voltage.

Solution: $C_t = C_e.C_p.C_i = V_s/V = 1425/420 = 3.4$

Rod-Plane Spark Gap

Thus far, the requirement and protection afforded by a lightning arrester were discussed. For providing further safety to major equipment insulation in transformers, reactors and circuit breakers as well as their bushings, rod-plane and rod-rod gaps will normally be used in parallel, which are variously known as protective gaps or spill gaps. These have a time lag of sparkover ranging from 2-10 μs depending upon the gap length between electrodes so that the protected equipment must be capable of withstanding the flashover voltage of the gaps for this length of time.

Average 50% flashover voltage values of rod-plane and rod-rod gaps are given by well-known formulas, which are on the average as follows, with d in metres:

Electrode Geometry	Power Frequency KV crest	Lightning Impulse kV crest
Rod-Plane	$V_{50} = 652.d^{0.576}$	$500\ d$
Rod-Rod	$V_{50} = 850\ d^{0.576}$	$650\ d$

The withstand value for these gaps is normally 85% of the 50% flashover voltage. Characteristics of 50% flashover and withstand voltages of long air gaps will be discussed in Chapter 11.

Example 9.14: A 750 kV bushing is protected by gaps which withstand 2 p.u. power-frequency voltage. Determine their 50% flashover value under 50 Hz and lightning-impulse voltages, if (a) rod-plane gap is used, and (b) rod-rod gap is used.

Solution: The calculation will be based on the power-frequency voltage of 750 kV, r.m.s., line-line. The 1 p.u. line-to-ground crest value is

$$750\sqrt{2}/\sqrt{3} = 612 \text{ kV}.$$

∴ 2 p.u. = 1224 kV, which is the withstand voltage, V_w.
The flashover value is taken to be $V_{50} = V_w/0.85 = 1440$ kV.

(a) For a rod-plane gap, $d = (V_{50}/652)^{1/0.576} = 2.2086^{1.736}$

$$= 3.958 \text{ metres.}$$

For this gap length, the lightning-impulse 50% flashover value is $V_{50} = 500\ d = 1980$ kV, crest. The withstand voltage will be approximately 85% of this value, or $V_{wi} = 0.85 \times 1980 = 1683$ kV, crest.

(b) For a rod-rod gap, $d = (1440/850)^{1.736} = 2.5$ metres. The 50% flash over under lightning impulse voltage is

$$V_{50} = 650\ d = 1625 \text{ kV, crest}$$

and the impulse withstand is $V_{wi} = 0.85 \times 1625 = 1381$ kV, crest.

For adequate protection and proper insulation coordination, the protective gap flashover values must be higher than the lightning arrester protective voltage level and lower than the transformer or bushing insulation levels.

Review Questions and Problems

1. A region has 75 thunderstorm days in a year. A 400-kV line has a tower height of 35 m with two ground wires at 25 metre height at midspan and separated by 20 metres. What is the probable number of strokes contacting 400 km of line per year?

2. Describe with neat sketches the mechanism of lightning stroke contacting (a) a tower, and (b) midspan.

3. A 750-kV horizontal line has 35 discs in the insulator. The two ground wires are spaced 30 m apart at heights of 30 metres at midspan and 40 metres at the tower. The tower-footing resistance is 20 ohms. The coupling factor between a ground wire and phase conductor is 0.15 and the factor in N_s is 0.2 for calculating number of strokes contacting the line per 100 Km in a year. The isokeraunik level is 100 days per year. Assume 50% of strokes to contact the tower. Calculate the stroke current to flashover the insulator string if the surge impedance of stroke is 400 ohms, and ground-wire surge impedance is 500 ohms. Take the flashover value of one disc as 125 kV, peak, for lightning impulse.

4. Define the terms (a) earthing coefficient, (b) earth fault factor, (c) residual voltage, (d) arrester rating, and (e) insulation co-ordination.

5. Compare the performance characteristics of silicon carbide arrester with a Zinc Oxide arrester. What are the advantages and disadvantages of each?

6. A voltage wave of 2500 kV is travelling on a line of surge impedance 275 ohms. The arrester connected to the line has a protective level of 1500 kV. Calculate (a) the current in the wave, (b) the current through the arrester, and (c) the arrester resistance at this current.

LIGHTNING AND LIGHTNING PROTECTION 253

Review Questions and Problems

1. A region has 75 thunderstorm days in a year. A 400-kV line has a
tower...

2. Describe...

3. ...ground wires are spaced...
...span and 40 metres at the lower...
20 ohms...
...conductor...

6. A voltage wave of 2500 kV...

CHAPTER 10

Overvoltages in EHV Systems Caused by Switching Operations

10.1 Origin of Overvoltages and Their Types

Overvoltages due to the release of internally trapped electromagnetic and electrostatic energy in an e.h.v. system cause serious damages to equipment insulation. These could, under many circumstances, be more severe than lightning damage which we considered in the previous chapter. Surge diverters and resistances included purposely while making switching operations as well as other schemes reduce the danger to a considerable extent. This chapter will describe all these schemes and discuss the equations that indicate overvoltages and the background for suggested remedies. Investigation of switching overvoltages has assumed very great importance as transmission voltages are on the increase and line lengths and capacity of generating stations are also increased. The short-circuit capacity of sources are responsible for a large amount of damage to insulation.

Overcurrents are generated by short circuits and lightning, but they do not form the subject matter of this chapter. The problems of calculating short-circuit currents from symmetrical-component theory is well covered in a large number of treatises. Here, some of the overvoltage conditions caused by interruption of short-circuit currents will only be discussed with due regard to the duty imposed on the circuit breaker through the restriking and recovery voltages. They bring in the concept of terminal fault, short-line fault, two-and four-parameter representations of the recovery voltage. In addition, interruption by the circuit breaker of low inductive currents such as dropping transformers and shunt reactors cause overvoltages because of sudden collapse of current described by the phenomenon of 'current chopping'. Interruption of small capacitive currents caused by dropping an unloaded line bring overvoltages because of possibility or re-striking in the arcing chamber of the breaker. Ferro-resonance conditions exist when the circuit-breaker poles do not close simultaneously as is usually the case with poorly-maintained circuit breakers. However, the most important operation caused by the circuit breaker is to energize a long e.h.v. transmission line at desired intervals. The line may be carrying no trapped voltage or it could be re-energized while a voltage is trapped in it. These energizing and re-energizing transients will be discussed at great length and will form the bulk of the material of this chapter. Some of the methods of handling such problems have already been described in Chapter 8 where travelling-wave and standing-wave methods were explained which result when a switching operation is performed. Line re-energization with trapped-charge voltage was also considered.

10.2 Short-Circuit Current and the Circuit Breaker

Consider a simple system with one generating station G connected through 2 lines and feeding a load, Fig. 10.1. If a short circuit occurs

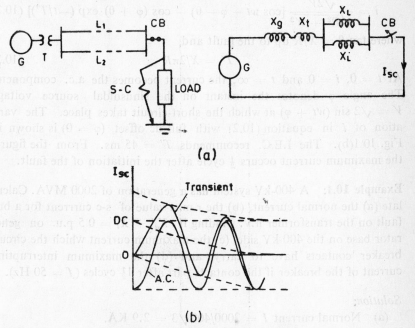

Fig. 10.1.   (a) Short circuit across load bus in a system.
  (b) Short-circuit current components. Steady state ac component, transient dc component and total s-c current.

at the load bus necessitating breaker CB to open and isolate the faulted bus, the question is what stresses come on the breaker.

When all resistances are neglected, the a.c. component of short circuit fed by the source is

$$i = V/(X_g + X_t + 0.5 X_L) \qquad (10.1)$$

If there are other generating areas connected to the load, their currents add to i.

In modern 2-cycle high-speed breakers, the contacts separate in about 30 milliseconds or $1\frac{1}{2}$ cycles on 50 Hz base after the initiation of the short circuit. This is governed by the operating time of the protective system and the pneumatic or mechanical operating system. This time is equal to or longer than the subtransient time constant of large alternators. Therefore the value of X_g to use is the transient reactance X_d' which is nearly 0.3 p.u. for 500 MVA sets and 0.3 to 0.45 p.u. for 1000 MVA sets. For smaller sets they may range from 0.15 to 0.2 p.u.

For transformers, X_t is approximately 0.1 to 0.15 p.u. For lines, approximate values are $X_L = 0.32$ ohm/Km for 400 kV and 0.275 for 750 kV. In practice, $X = X_d' + X_t \approx 0.5$ p.u. so that for a bus fault not including lines, the short-circuit current is twice the rated current of the generating station. When the d-c component is also included, total short-circuit current with full offset will be

$$i = \frac{\sqrt{2V}}{\sqrt{R^2 + X^2}} \left[\cos wt - \varphi - \theta \right) - \cos (\varphi + \theta) \cdot \exp (-t/T') \right] \quad (10.2)$$

where $\tan \theta = X/R$ up to the fault and

$$T' = X/2\pi R \quad (10.3)$$

At $t = 0$, $i = 0$ and $t = \infty$, the current becomes the a.c. component. The angle φ denotes the instant on the sinusoidal source voltage $V = \sqrt{2} \sin (wt + \varphi)$ at which the short-circuit takes place. The variation of i in equation (10.2) with full dc offset ($\varphi = 0$) is shown in Fig. 10.1(b). The I.E.C. recommends $T' = 45$ ms. From the figure, the maximum current occurs $\frac{1}{2}$ cycle after the initiation of the fault.

Example 10.1: A 400-kV system has a generation of 2000 MVA. Calculate (a) the normal current, (b) the r.m.s. value of s-c currrent for a bus fault on the transformer h.v. winding if $X_d' + X_t = 0.5$ p.u. on generator base on the 400-kV side, (c) the maximum current which the circuit breaker contacts have to carry, and (d) the maximum interrupting current of the breaker if the contacts part after $1\frac{1}{2}$ cycles ($f = 50$ Hz).

Solution:
- (a) Normal current $I = 2000/400\sqrt{3} = 2.9$ KA.
- (b) RMS value of s-c current $I_{rms} = 2 \times 2.9 = 5.8$ KA.
- (c) Maximum current through breaker contacts at 10 ms is

$$I_m = \sqrt{2}I_{rms} (1 + e^{-10/45}) = 1.8\sqrt{2} \, I_{rms} = 14.8 \text{ KA}$$

- (d) At 30 ms, ($1\frac{1}{2}$ cycles after fault initiation) a peak value occurs.

$$\therefore \quad I_t = \sqrt{2} \, I_{rms} (1 + e^{-30/45}) = 1.513\sqrt{2} \, I_{rms} = 12.4 \text{ KA}$$

The final interruption of the circuit is at 2 cycles after fault initiation when the current passes through a zero if the decrement of the dc component is rapid.

In this example, the short-circuit current from one 2000 MVA source at 400 kV was 5.8 KA, r.m.s. In a large interconnected system with several generating stations, the s-c current level will be very high. Presently, air-blast breakers are available for 80 KA and SF_6 breakers for 90 KA. This shows that a system engineer must keep s-c levels down to what currently-available circuit breakers can handle. In 400-kV networks the maximum specified is 40 KA.

It is evident that the d-c component must decay fast in order that interruption might take place at the first current zero after the contacts part. At 40 ms, the d.c. component has a value $e^{-40/45} = 0.41$ of the a.c. component and usually a current zero can occur. The above discussion has assumed a 3-phase bus fault.

Single-Phase Short Circuit

Nearly 80% of all faults in a system involve only a single phase and the s-c current magnitude is lower than for a 3-phase fault, which occur only in 10% of cases. However, the most severe duty of a breaker occurs under a 3-phase fault and this governs the breaker design. As described in Chapter 9, under a single phase to ground fault, the three sequence networks are connected in series and the fault current is

$$I_{1ph} = 3E/(Z_0 + 2Z_1) \qquad (10.4)$$

But it is $I_{3ph} = E/Z_1$ under a 3-phase fault since only the positive-sequence network is involved. Thus, the ratio of currents for these two types of fault is

$$I_{1ph}/I_{3ph} = 3/(2 + Z_0/Z_1) = 3/(2 + X_0/X_1) \qquad (10.5)$$

when resistances are neglected. For a solidly-grounded neutral, $X_0/X_1 \cong 2$ so that the single-phase s-c current is 75% of that for a 3-phase bus fault.

Delayed Current Zero Condition

The attainment of a current zero depends on the rate of decay of the d-c component, which is governed by the resistance up to the fault. When faults occur very close to large power stations of higher than 1000 MVA capacity and line reactance and resistance are not present, it is difficult for the current to pass through zero quickly. In such cases, the arc resistance of the circuit breaker must be increased by providing multiple interrupters. For low-voltage generator breakers the s-c current is even higher and in order to interrupt 100 KA, air-blast breakers are preferred because of higher arc resistance than SF_6 breakers in order to effect a current zero.

10.3   Recovery Voltage and the Circuit Breaker

When the contacts have separated and the arc has been finally quenched, the contacts have to withstand the recovery voltage. The final value of this voltage equals the source voltage while the initial value is equal to the low arc voltage which may be practically zero. Thus, an oscillatory condition exists which may be of single frequency or contain multiple frequencies, depending upon the connected network. For a single frequency it is

$$V_R = \sqrt{2}\,V.(\cos wt - \cos w_0\,t).K \qquad (10.6)$$

where $w_0 = 2\pi f_0 =$ the natural frequency, and

K = a constant which depends on the type of fault.

The rate of rise of this recovery voltage (RRRV) determines the ability of the quenching medium to interrupt the arc, since the rate of rise of dielectric strength must exceed the RRRV. For systems with low natural frequency, which occur when long lines are involved with their high inductance and capacitance, oil circuit breakers were found adequate, although air blast and SF_6 breakers can do as well. But in systems with high natural frequency, the rate of rise of recovery voltage is very high and air-blast and SF_6 breakers are necessary.

These are complicated further when a short line of 1 or 2 km is interposed between the circuit breaker and the fault location. Two types of fault are distinguished when the severity of RRRV is assessed: (1) the terminal fault, TF, and (2) the short-line fault, SLF.

Terminal Faults (TF)

Such faults involve maximum s-c currents. 3-phase terminal faults yield highest s-c currents and result in most severe conditions for the recovery voltage. This is further enhanced when one pole of the circuit breaker clears ahead of others. The first pole to clear experiences the highest recovery voltage since the transient component is higher than when the second or subsequent poles clear. This is between 1.5 to 2 times the phase voltage appearing after final interruption.

Short-Line Fault (SLF)

In this type of fault, reflections arriving on the line of 1 or 2 km length between the breaker and fault are superimposed on the source voltage, as shown in Fig. 10.2. These give the highest rates of recovery voltage, and in many breakers the interrupting capability will be governed by SLF rather than by TF. Time delays between contacts opening cannot be avoided and in this type of fault the last pole to open experiences highest recovery voltage because of induction from the cleared phases.

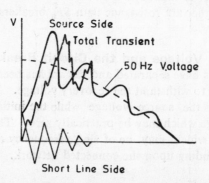

Fig. 10.2.   Transient recovery voltage across poles of circuit breaker under Short Line Fault.

These stresses can be reduced in circuit breakers by connecting resistances or capacitances in parallel to absorb the shock of high-frequency transients under very high s-c currents. Resistances are preferred for air-blast breakers and capacitances for SF_6 breakers since the difference in arc resistance controls the effectiveness of these remedial measures.

Definition of Transient Component of Recovery Voltage

(a) 2-Parameter Definition
For single-frequency circuits, the recovery voltage is defined through two parameters: (1) the magnitude, V_1, and (2) the rate of rise, V_1/t_1, Fig. 10.3.

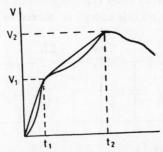

Fig. 10.3.   2- and 4-parameter representation of recovery voltage across circuit breaker.

(b) 4-Parameter Definition
With large interconnection in a system, the 2-parameter definition has been superseded by the 4-parameter definition because of the presence of multiple frequencies, as shown in Fig. 10.3. These are V_1, V_P, t_1, t_2 (or V_1/t_1). The initial time t_1 is clearly equal to twice the travel time of the wave on the shortest connected line. In a short-circuit to ground, the capacitance of the grounding system is involved which may help to keep the steepness of the recovery voltage or *RRRV* down.

Under these two types of fault the following values of *RRRV* are recommended by the I.E.C. for circuit-breaker designs.

Terminal Fault   5 kV/μs   (This is being revised to 5.5 to 12.6 kV/μs).

Short-Line Fault   9 kV/μs

These are maximum values and will depend upon the current and the degree of assymmetry. The first peak of recovery voltage is usually 2.25 p. u.

10.4 Overvoltages Caused by Interruption of Low Inductive Current

When disconnecting transformers or reactors, the current is low but highly inductive. When a circuit breaker designed normally to interrupt very high short-circuit currents interrupts such low currents on the high voltage side, overvoltages occur on the equipment by premature reduction of current to zero prior to reaching a normal current zero. Current magnitudes in such cases are

Transformer on no load—2 to 5 Amperes,

Reactor-loaded transformer—up to 400 Amperes,

High-voltage reactors—100 to 200 Amps.

Fig. 10.4 shows a current magnitude i_c when chopping occurs with the system voltage v_c across the inductive load. The stored energy is $\frac{1}{2} i_c^2 L_2 + \frac{1}{2} v_c^2 C_2$ and oscillates at the natural frequency $f_2 = 1/2\pi\sqrt{L_2 C_2}$, which in e.h.v. systems is 200 to 400 Hz for a transformer on no load and may be as high as 1000 Hz for a shunt reactor. The maximum voltage appears when all this energy is stored in the capacitance.

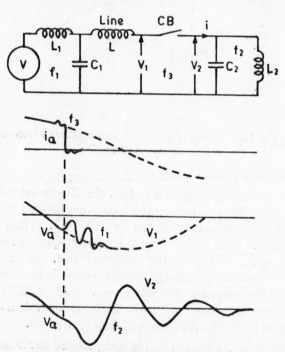

Fig. 10.4. Conditions of voltage and current during 'current chopping' when interrupting low inductive currents.

Thus, $V_{\max}^2 C_2 = v_c^2 C_2 + i_c^2 L_2$ which yields

$$V_{\max} = (v_c^2 + i_c^2 L_2/C_2)^{1/2} \tag{10.6}$$

The quantity $\sqrt{L_2/C_2}$ equals the characteristic or natural impedance of the equipment.

If i_c is low, the voltage is at its peak and no overvoltage occurs. However, if the current at the instant of chopping is at its peak value and $v_c = 0$, then $V_{\max} = i_c \sqrt{L_2/C_2}$. The following values are typical overvoltages which may be expected based on experience.

220 kV — 2.5 p.u. 400 kV — 1.8 p.u. 750 kV — 1.2 p.u.

This type of overvoltage can be reduced by surge arresters and series resistances used in circuit breakers.

10.5 Interruption of Capacitive Currents

When transmission lines are dropped or de-energized or capacitor banks switched off, overvoltages are generated. Consider Fig. 10.5 where a line is represented by a lumped capacitance C_2. Before interruption, $V_1 = V_2$. After the current is interrupted C_2 remains charged to $V_2 = V_s\sqrt{2/3}$, which is the crest value of the source voltage at power frequency. However, the source voltage V_1 changes polarity and the breaker voltage is $V_b = V_s\sqrt{2/3}\,(1+K)$ where $K \approx 1$, giving $2V_s\sqrt{2/3}$. If the insulating medium in the breaker has not gained sufficient dielectric strength to withstand V_b, the arc may restrike and connect the line to the source. A current flows which is 90° leading. When the circuit is interrupted again at a current zero, the voltage is at its peak value and the line holds a negative voltage. There may be repeated restrikes such as this and breaker failure may occur.

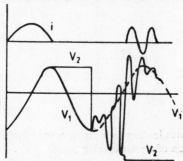

Fig. 10.5. Voltages during re-striking of breaker when interrupting small capacitive currents (line dropping).

Therefore, restrike-free breakers are essential. Modern SF_6 and airblast breakers meet these requirements but in all cases proper maintenance is absolutely necessary. Lines equipped with shunt reactors help to drain the trapped charge of the line and aid in proper interruption. Capacitor banks cause the same type of stresses on the circuit breaker.

10.6 Ferro-Resonance Overvoltages

Partial resonance conditions occur in power systems when unbalanced configuration occurs so as to place capacitances in series with inductances. When a transformer is connected to a long transmission line and both are switched together, such a condition might occur as shown in Fig. 10.6. Under normal operating conditions, the line capacitance to ground is energized by the phase voltage. However, suppose during a switching operation, one pole opens or closes non-simultaneously with the others. The equivalent circuit, Fig. 10.6(b), shows that the line capacitance is in series with the transformer inductance in the open phase. The condition when two poles are open is also shown in Fig. 10.6(b). The problem involves very difficult analysis and must take into account the distributed capacitance of line and the non-linear magnetization curve of the transformer. This is a highly specialized topic and will not be considered further. The problem is also very important in urban distribution where long cables are used underground

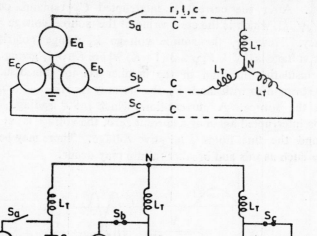

Fig. 10.6. Conditions leading to ferro-resonance during non-simultaneous operation of circuit breaker poles.

and failures to both cable and transformer insulations have resulted in great expense and inconvenience to consumers and power companies, especially since load-shedding and frequent switching has become common.

10.7 Calculation of Switching Surges—Single Phase Equivalents

Switching operations in e.h.v. systems give rise to overvoltages due to the interaction between electrostatic and electromagnetic energies trapped in the long lines which release in connected equipment. Insulation designs are governed by these overvoltages and the remaining parts of this chapter will be devoted to the calculation of overvoltages and their reduction in practice.

We will commence the detailed discussion on formulating and solving equations controlling switching surges with (a) lumped-parameter networks, then (b) distributed-parameter lines, and finally (c) a combination of both lumped-parameter elements connected to distributed-parameter lines. Some of these ideas have already been developed in Chapter 8 where the theory was presented. A very good engineering idea of the magnitudes of switching surges experienced in system (c) can be obtained if systems (a) and (b) are first analyzed and their results interpreted.

10.7.1 Single-Frequency Lumped-Parameter Circuit

Consider Fig. 10.7 showing a series L-R-C circuit energized by a source $e(t)$ by closing the switch S. The capacitor has an initial trapped-charge voltage V_0 with the polarity shown and there is no initial current. The operational equation for the current using Laplace-Transform is

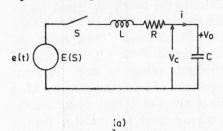
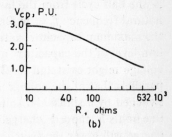

Fig. 10.7. (a) Lumped-parameter L-R-C circuit with initial voltage on capacitor.
(b) Effect of increase of R on peak value of voltage across capacitor.

$$(Ls + R + 1/Cs)\, I(s) + V_0/s = E(s) \tag{10.7}$$

giving
$$I(s) = \frac{s \cdot E(s) - V_0}{L[(s + \alpha)^2 + w_0^2]} \tag{10.8}$$

where $\alpha = R/2L$ and $w_0^2 = 1/LC - \alpha^2$ $\qquad$ (10.9)

The resulting capacitor voltage in operational form is

$$V_c(s) = I(s)/Cs + V_0/s = \frac{sE(s) - V_0}{LCs[(s + \alpha)^2 + w_0^2]} + \frac{V_0}{s} \tag{10.10}$$

The source of excitation will be considered to be (a) a step function, and then (b) a sinusoidal function which occurs in practice.

10.7.2 Step Response of L-R-C Circuit

For this case, $E(s) = E/s$, where $E =$ the step magnitude. Now, the Inverse Laplace Transform of the first expression in equation (10.10) is

$$L^{-1}\left[\frac{1}{s[(s+\alpha)^2 + w_0^2]}\right] = \frac{1}{\alpha^2 + w_0^2}\left[1 - e^{-\alpha t}\left(\cos w_0 t + \frac{\alpha}{w_0}\sin w_0 t\right)\right] \tag{10.11}$$

Consequently, the capacitor voltage is

$$v_c(t) = (E - V_0)\,[1 - e^{-\alpha t}\sqrt{1 + \alpha^2/w_0^2}\,\cos(w_0 t - \varphi_0)] + V_0 \tag{10.12}$$

where $\tan \varphi_0 = \alpha/w_0$. Also $\sqrt{1 + \alpha^2/w_0^2} = 1/w_0\sqrt{LC}$ $\qquad$ (10.13)

This expression applies when damping is low and $R < 2\sqrt{L/C}$. As a check we note that at $t = 0$, equation (10.12) yields $v_c(0) = V_0$ and at $t = \infty$, $v_c(\infty) = E$, the step to which the capacitor will be finally charged.

For the design of insulation level of the capacitor, the worst case occurs when at the instant of switching the initial voltage on the capacitor is $V_0 = -E$. For this case

$$v_c(t) = E\left[1 - \frac{2}{w_0\sqrt{LC}} e^{-\alpha t}.\cos(w_0 t - \varphi_0)\right] \tag{10.14}$$

For very lightly damped circuits, $\alpha \ll w_0$, $w_0^2 = 1/LC$, and $\phi_0 \approx 0$.

$$\therefore \qquad v_c(t) = E(1 - 2e^{-\alpha t}\cos w_0 t) \tag{10.15}$$

It reaches a maximum value of nearly $3E$ or 3 p.u. when $w_0 t = \pi$, which is one half cycle from the instant of closing the switch S, based on the natural frequency $f_0 = w_0/2\pi$. When the trapped-charge voltage is zero, the maximum capacitor voltage reaches $2E$, or 2 p.u. Thus, the insulation level of the capacitor must be designed for 3 p.u. when trapped voltage might exist and for 2 p.u. when this voltage is zero. This case parallels that of the voltage at the open end of a transmission line when switched by a sinusoidal source while it passes at its peak value and the line holds a trapped charge. Any scheme used for draining the line charge will lower the switching surge from 3 p.u. to 2 p.u. We also note that the maximum amplitude during the oscillatory period is $3E$ which is written as

$$\text{Amplitude of oscillation} = 2 \times \text{Final value} - \text{initial value} \tag{10.16}$$

Example 10.2: A 400-kV 400-km line has the distributed parameters $r = 0.031$ ohm/km, $l = 1$ mH/km, and $c = 10$ nF/km. The equivalent lumped parameter for the circuit of Fig. 10.7 are assumed as $R = 12.4$ ohms, $L = 400$ mH, and $C = 4$ μF. It is excited by an equivalent step voltage of magnitude $E = 420\sqrt{2/3} = 343$ kV. Calculate

(a) the attenuation factor α,

(b) the natural angular frequency w_0 and $f_0 = w_0/2\pi$, and

(c) the peak value of voltage across C with the line holding an initial trapped-charge voltage equal to -343 kV.

Solution:

(a) $\alpha = R/2L = 12.4/2 \times 0.4 = 15.5$ sec^{-1}

(b) $w_0^2 = 1/LC - \alpha^2 = 0.625 \times 10^6$. $\therefore$ $w_0 = 790.6$ and $f_0 = 125$ Hz.

Time for $\frac{1}{2}$ cycle $= 4$ ms when the peak occurs.

(c) Peak value of capacitor voltage is, by equation (10.15),

$$v_{cp} = E[1 - 2.e^{-15.5 \times 4 \times 10^{-3}}.\cos \pi] = 2.88\, E = 988 \text{ kV}.$$

Equation (10.14) can be maximized as follows:

$$\frac{d}{dt}\left[\frac{v_c(t)}{E}\right] = 0 = \frac{2e^{-\alpha t}}{w_0\sqrt{LC}}[\alpha \cos(w_0 t - \varphi_0) + w_0 \sin(w_0 t - \varphi_0)]$$

giving $\tan(w_0 t - \varphi_0) = -\alpha/w_0 = -\tan \varphi_0 = \tan(-\varphi_0)$ or $\tan(\pi - \varphi_0)$.

Therefore $w_0 t - \varphi_0 = \pi - \varphi_0$, for the first peak.

Subsequent peaks occur at $(2n + 1) \pi - \varphi_0$, $n = 1, 2, 3, \ldots$

The peak value occurs at $\qquad t_m = \pi/w_0$. $\qquad$ (10.17)

The maximum p.u. value from equation (10.14) is

$$v_{cp}/E = 1 + \frac{2}{w_0 \sqrt{LC}} \cdot e^{-\pi \alpha/w_0} \cdot \cos \varphi_0 = 1 + 2e^{-\pi \alpha/2w_0} \qquad (10.18)$$

since $\cos \varphi_0 = w_0/\sqrt{\alpha^2 + w_0^2} = w_0 LC$ from equation (10.13).

In example 10.2, $\alpha = 15.5$ and $w_0 = 790.6$, giving $e^{-\pi \alpha/w_0} = 0.94$, and therefore $v_{cp} = 1 + 2 \times 0.94 = 2.88$ p.u.

We also note that the ratio of adjacent peaks above the input step can be written as

$$K_p = \frac{v_p(2n-1) - E}{v_p(2n+1) - E} = \frac{\exp(-\pi(2n-1)\alpha/w_0)}{\exp(-\pi(2n+1)\alpha/w_0)} = \exp(2\pi\alpha/w_0) \quad (10.19)$$

Therefore $2\pi\alpha/w_0 = \ln(K_p)$ giving $\alpha = \dfrac{w_0}{2\pi} \ln(K_p) = f_0 \cdot \ln(K_p)$ $\quad$ (10.2)

This gives a very convenient method of measuring the attenuation factor from an oscilloscopic record. The frequency of the oscillatory voltage is easily measured while the amplitudes of successive peaks can also be found. Such a method is useful in all transient studies where the attenuation is governed by the ac value of resistance at the frequency of oscillation while the d.c. resistance can be measured on a bridge.

10.7.3 Reduction of Switching Surge Overvoltage

On long e.h.v. lines, introduction of a series resistance with the circuit breaker at the instant of switching reduces the open-end voltage. After nearly 10 ms the series resistance is removed from the circuit by having another circuit breaker closing across the terminals of the resistor. We can examine the effect of increasing R in Fig. 10.7. We observe that $w_0 = 0$ when $R = 2\sqrt{L/C}$, and the circuit is 'critically damped'. For this condition, equation (10.10) becomes

$$V_c(s) = \frac{E - V_0}{LC} \frac{1}{s(s+\alpha)^2} + \frac{V_0}{s} \text{ where } \alpha^2 = R^2/4L^2 = 1/CL \quad (10.21)$$

The time response is

$$v_c(t) = (E - V_0)[1 - e^{-\alpha t}(1 + \alpha t)] + V_0 \qquad (10.22)$$

For the case $\qquad V_0 = -E,$

$$v_c(t) = E[1 - 2 e^{-\alpha t}(1 + \alpha t)] \qquad (10.23)$$

Differentiating and taking $dv_c/dt = 0$, there results $dv_c/dt = 2\alpha^2 E \cdot e^{-\alpha t} \cdot t$ which is zero at $t = 0$. The resulting maximum value of capacitor voltage is $v_{cp} = -E = V_0$, from (10.23). Consequently, the voltage across the capacitor will never exceed the step or 1 p.u.

Example 10.3: In example 10.2, with $L = 0.4$ Henry, $C = 4\mu F$ calculate the maximum voltage across capacitor as the resistance R is changed from 10 ohms to 632 ohms ($= 2\sqrt{L/C}$). Assume the trapped voltage to be -1 p.u.

Solution: The maximum value is $1 + 2\exp(-\pi\alpha/w_0)$ p.u., equation (10.18). The resistance values may be selected on a logarithmic scale.

$R =$	10	20	30	40	60	80	100	200	300	400	600	632
$\alpha =$	12.5	25	37.5	50	75	100	125	250	375	500	750	790
$w_0 =$	790.6	790.3	790	789	787	784	781	750	696	612	250	0
$v_{cp} =$	2.9	2.81	2.723	2.64	2.48	2.34	2.21	1.7	1.37	1.15	1.003	1

Fig. 10.7(b) shows the p.u. values of capacitor voltage as the series resistance is increased.

10.7.4 Sinusoidal Excitation—Lumped Parameter Circuit

Equation (10.10) is the general expression for the capacitor voltage in operational form in the L–R–C circuit shown in Fig. 10.7(a). The forcing function is $e(t)$ whose Laplace Transform is $E(s)$. For a sine-wave of excitation with $e(t) = V_m \cos(wt + \varphi)$, the Laplace-Transform is

$$E(s) = V_m(s \cdot \cos\varphi - w \cdot \sin\varphi)/(s^2 + w^2) \tag{10.24}$$

In solving for the capacitor voltage for this type of excitation, it is easy if the responses due to the forcing function and the initial voltage V_0 on the capacitor are separated. Also, instead of evaluating the residues, we will use an intuitive method, although the Inverse Laplace Transform can be worked out in the usual manner.

Response due to Initial Trapped Voltage V_0
For this voltage, the operational expression for the capacitor voltage in equation (10.10) is

$$V_0(s) = \frac{V_0}{s} - \frac{V_0}{LC} \cdot \frac{1}{s \cdot [(s + \alpha)^2 + w_0^2]} \tag{10.25}$$

The inverse transform is, using equation (10.11),

$$V_0(t) = V_0[1 - \{1 - e^{-\alpha t} \cdot \sqrt{1 + \alpha^2/w_0^2} \cdot \cos(w_0 t - \varphi_0)\}]$$

$$= \frac{V_0}{w_0 LC} e^{-\alpha t} \cdot \cos(w_0 t - \varphi_0) \tag{10.26}$$

where the values of α, w_0 and φ_0 are taken from equations (10.9) and (10.13). The above solution is valid for $R < 2\sqrt{L/C}$.

For the critically-damped case, $R = 2\sqrt{L/C}$, $w_0 = 0$, it becomes

$$v_0(t) = V_0[1 - (1/\alpha^2 LC)\{1 - e^{-\alpha t}(1 + \alpha t)\}]$$

$$= V_0[1 - (4L/R^2 C)\{1 - e^{-\alpha t}(1 + R t/2L)\}] \tag{10.27}$$

The quantity $\sqrt{L/C}$ can be called the natural impedance Z_0 of the L–R–C circuit, giving $4L/R^2C = 4Z_0^2/R^2$. Thus, the magnitude of the second quantity in (10.27) depends on the ratio of characteristic impedance to the actual resistance included in the network.

Response due to Forcing Function $e(t) = V_m \cos(wt + \varphi)$

The transient resulting from sudden energization of the circuit will consist of two parts which are (1) the steady state term or particular integral surviving when all transients have vanished, and (2) the exponentially decaying transient term or the complimentary function.

(1) Steady-State Term

Using phasor algebra, the circuit current phasor leads the applied voltage by the angle ψ and the magnitude is V_m/Z. Thus, its value is

$$I_{ss} = \frac{V_m}{Z} \cos(wt + \varphi + \psi) \tag{10.28}$$

where

$$Z = \sqrt{R^2 + (1/wC - wL)^2} \tag{10.29}$$

and

$$\psi = \arctan[(1/wC - wL)/R]$$

The capacitor voltage in the steady state lags 90° behind the current and has the value

$$V_{cs} = \frac{1}{wC} I_{ss} \angle -90° = \frac{V_m}{wCZ} \sin(wt + \varphi + \psi) \tag{10.30}$$

(2) Complimentary Function or Transient Term

The transient term will be of the same form as equation (10.11) and is written as

$$V_{cc} = (K_1 \cos w_0 t + K_2 \sin w_0 t)\, e^{-\alpha t}, \tag{10.31}$$

where K_1 and K_2 are constants to be determined from initial and final conditions. The total capacitor voltage will then be the sum of equations (10.30) and (10.31).

Total Response

The capacitor voltage is

$$V_{co} = \frac{V_m}{wCZ} \sin(wt + \varphi + \psi) + e^{-\alpha t} (K_1 \cos w_0 t + K_2 \sin w_0 t) \tag{10.32}$$

and the current is

$$I_{co} = C\, \frac{dV_{co}}{dt} = \frac{V_m}{Z} \cos(wt + \varphi + \psi) + C\, e^{-\alpha t} [(w_0 K_2 - \alpha K_1) \cos w_0 t$$

$$- (w_0 K_1 + \alpha K_2) \sin w_0 t] \tag{10.33}$$

Because of the presence of inductance, at $t = 0$, $I_{co} = 0$. Also, since the initially-trapped voltage V_0 has been considered separately, the response in equation (10.32) applies for $V_{co} = 0$ at $t = 0$. Using these two initial conditions in equations (10.32) and (10.33) yields

$$K_1 = \frac{-V_m}{wCZ} \sin(\varphi + \psi)$$

and
$$K_2 = \frac{\alpha}{w_0} K_1 - \frac{1}{w_0 CZ} V_m \cos(\varphi + \psi) \Bigg\} \quad (10.34)$$

Let
$$\tan \delta = \alpha/w = R/2wL = R/2X$$

and
$$\tan \eta = w_0 \cdot \sin(\varphi + \psi)/\sqrt{\alpha^2 + w^2} \cos(\varphi + \psi - \delta) \Bigg\} \quad (10.35)$$

Then, the capacitor voltage will be

$$V_{co} = \frac{V_m}{wCZ} \Bigg[\sin(wt + \varphi + \psi) - e^{\alpha t} \cdot \sin(w_0 t + \eta) \cdot \Bigg[\sin^2(\varphi + \psi)$$

$$+ \frac{\alpha^2 + w^2}{w_0^2} \cos^2(\varphi + \psi - \delta) \Bigg]^{1/2} \Bigg] \quad (10.36)$$

We also observe that

$$w_0^2/(\alpha^2 + w^2) = 4w_0^2 L^2/(R^2 + 4w^2 L^2) \quad (10.37)$$

For $R \ll 4w^2 L^2$, this becomes simply $(w_0/w)^2$, where $w = 314$ for $f = 50$ Hz.

Finally, the capacitor voltage in the presence of an initial voltage V_0 is obtained as the sum of equations (10.36) and (10.26) for low damping, $R < 2\sqrt{L/C}$. For this case, we can also write approximately, $w_0 = 1/\sqrt{LC}$, $\alpha^2 + w^2 \approx w^2$, $\delta \approx 0$, $\psi \approx 90°$, $\varphi_o = 0°$, $\tan \eta = (w_0/w) \times$ $\tan(\varphi + \psi)$, and $Z = (1/wc - wL)$. Then, for $R \ll 2\sqrt{L/C}$,

$$V_c = \frac{V_m}{1 - w^2 LC} [\cos(wt + \varphi) - e^{-\alpha t} \cdot \sin(w_0 t + \eta) \cdot$$

$$\sqrt{\cos^2 \varphi + (w/w_0) \sin^2 \varphi} + V_0 e^{-t} \cdot \cos w_0 t \quad (10.38)$$

Critical Damping. $R = 2\sqrt{L/C}$.

For this case, the response function with V_0 is

$$V_c = \frac{V_m}{wCZ} [\sin(wt + \varphi + \psi) - e^{-\alpha t} \{\sin(\varphi + \psi) + t \cdot \sqrt{\alpha^2 + w^2} \cdot$$

$$\cos(\varphi + \psi - \delta)] + V_0[1 - 4L/R^2 C] \{1 - e^{-\alpha t}(1 + Rt/2L)\}] \quad (10.39)$$

Short-circuit Power

In usual sources, the inductance is considerable. If in the circuit of Fig. 10.7(a), the inductance L is considered as contributed by the sinusoidal source only, we can define a short-circuit power of the source as

$$P_{sc} = V_m^2/2wL \quad (10.40)$$

This will be delivered when a short-circuit occurs at the terminals of the source whose series reactance at power frequency is $X = wL$. Then, the p.u. value of capacitor voltage for $R < 2\sqrt{L/C}$ can be written in terms of this s-c power as follows from equations (10.26) and (10.36):

$$\frac{V_c}{V_m} = \frac{P_{sc}}{V_m^2} \frac{2L}{CZ} \left[\sin(wt + \varphi + \psi) - e^{-\alpha t} \cdot \sin(w_0 t + \eta) \cdot \right.$$

$$\left. \left[\sin^2(\varphi + \psi) + \frac{R^2 + 4w^2 L^2}{4w_0^2 L^2} \cos^2(\varphi + \psi - \delta) \right]^{1/2} \right.$$

$$+ \frac{P_{sc} V_0}{V_m^3} \frac{2wL}{w_0^2 \sqrt{LC}} \cos(w_0 t - \varphi_0) \qquad (10.41)$$

The term $P_{sc}L = V_m^2/2w$ is a constant for given values of forcing function V_m and power frequency $f = w/2\pi$.

10.8 Distributed-Parameter Line Energized by Source

We have already analyzed the travelling-wave and standing-wave responses of a distributed-parameter line with open end when energized by a step function in Chapter 8. The operational expressions for the voltage and current at any point on an open-ended line were, neglecting the initial trapped voltage,

$$\left. \begin{array}{l} V(x, s) = E(s) \cdot \cosh px / \cosh pL \\[2mm] I(x, s) = E(s) \cdot \sinh px / Z_0 \cdot \cosh pL \end{array} \right\} \qquad (10.42)$$

and

where $p = \sqrt{(r + ls)(g + cs)}$, the propagation factor

and $Z_0 = \sqrt{(r + ls)/(g + cs)}$, the surge impedance of line $\approx \sqrt{l/c}$.

Also, $v = 1/\sqrt{lc}$, the velocity of propagation.

At the open end, the voltage is

$$V_0 = E(s)/\cosh pL. \qquad (10.43)$$

Step Response. Denoting $A_0 = e^{-\alpha L}$ where $\alpha = r/2lv = r/2Z_0$, the step response was obtained as shown in Fig. 8.4 and the Bewley Lattice Diagram, Fig. 8.5. The general expression for the voltage at time $t = NT$, where $T =$ time taken for travel of surge over one line length $(T = L/v)$ was $V_{NT} = 2A_0 - A_0^2 V_{(N-2)T}$, N odd. The maximum voltage is $2A_0$ at the open end when trapped charge is neglected. With a trapped voltage of 1 p.u., the open-end voltage reaches a maximum of $(2A_0 + 1)$ p.u. In usual transmission lines, the trapped voltage is 0.8 p.u. on a reclosing operation. We may note here that the value of first peak is very nearly the same as was obtained from the lumped-parameter network.

The attenuation factor must be properly calculated by considering the resistance of not only the conductor but also the ground return which exceeds the conductor resistance by a factor of as much as 10. The following example shows the effect of ground-return inductance on the velocity of propagation and the resistance on the attenuation.

Example. 10.4: The 400–km 400–kV line considered in example 10.2 has the following details:

Conductor resistance $r_c = 0.031$ ohm/km

Ground return resistance $r_g = 0.329$ ohm/km at 125 Hz

Series inductance of line $l_s = 1$ mH/km

Ground-return inductance $l_g = 0.5$ mH/km

Shunt capacitance $c = 11.1$ nF/Km

Calculate

(a) velocity of propagation v and travel time T;

(b) surge impedance;

(c) attenuation factor A_0;

(d) maximum p.u. value of open-end voltage without trapped voltage.

Compare all these quantities by neglecting ground and considering the ground-return parameters.

Solution:

	Neglecting r_g and l_g	*Considering r_g and l_g*
Total resistance r	0.031 ohm/km	0.36 ohm/km
Total inductance l	1 mH/km	1.5 mH/km
capacitance c	11.1 nF/km	11.1 nF/km
velocity v, km/sec	3×10^5	$10^6/\sqrt{1.5 \times 11.1}$
		$= 2.45 \times 10^5 = 82\%$ light velocity
Time for l travel, T, ms	1.33	1.633
Surge impedance Z_0, ohms	300	$\sqrt{1.5/11.1} \times 10^3 = 367$
Attenuation factor, A_0	$e^{-12.4/600} = 0.98$	$e^{-144/734} = 0.822$
Maximum open-end voltage, $2A_0$	1.96 p.u.	1.644 p.u.

We note that the effect of ground-return resistance and inductance on the switching surge response (step response) is to

(a) decrease the velocity of propagation of the surge,

(b) increase the time of travel,

(c) increase the surge impedance,

(d) increase the attenuation of surge over one travel, and

(e) lower the maximum value of open-end voltage.

Sine-Wave Response

Following the same procedure as for the step response, the travelling-wave concept yields the following sequence of voltages at the open end when the source voltage has the form $e(t) = V_m \cos (wt + \varphi)$.

At $t = 0$, the switch is closed and the source voltage will appear at the open end at time $t = T$ with the magnitude $A_0 \cos \varphi$. Due to total reflection at the open end, the voltage will be $V_1 = 2A_0 \cos \varphi$, p.u. At $t = 2T$, the voltage $V_m \cos (wT + \varphi)$ will reach the open end and become double. $\therefore V_2 = 2A_0 \cos (wT + \varphi)$, p.u At $t = 3T$, the reflected voltage at $t = T$ would have gone back to the source and returned with a negative sign after attenuating by A_0^2. Also, the voltage $V_m \cos (2wT + \varphi)$ would reach the open end at the same time. $\therefore$ At $t = 3T$, the p.u. value of open-end voltage is $V_3 = 2A_0 \cos (2wT + \varphi) - A_0^2 V_1$. The sequence of voltages are therefore as follows:

At $\quad t = 0 \qquad : V_0 = 0$

$\qquad t = T \qquad : V_1 = 2 A_0 \cos \varphi$

$\qquad t = 2T \qquad : V_2 = 2 A_0 \cos (wT + \varphi)$

$\qquad t = 3T \qquad : V_3 = 2 A_0 \cos (2wT + \varphi) - A_0^2 V_1 \qquad (10.44)$

$\qquad t = 4T \qquad : V_4 = 2 A_0 \cos (3wT + \varphi) - A_0^2 V_2$

$\qquad \vdots$

$\qquad t = NT \qquad : V_N = 2 A_0 \cos (\overline{N-1}\, wt + \varphi) - A_0^2 V_{N-2}$

When the source is switched on to the line at the instant it is passing through its peak value, $\varphi = 0°$. This usually gives the maximum stress on line insulation.

Example 10.5: For the 400-kV 400-km line of previous example, when the source of excitation is sinusoidal with $f = 50$ Hz, and is switched on to the line at its peak ($\varphi = 0$), considering ground-return resistance and inductance, calculate and plot the open-end voltage in per unit up to 20 ms (1 cycle).

Solution: $\quad T = 1.633$ ms, $\qquad v = 2.45 \times 10^5$ km/s, $A_0 = 0.822$

$\therefore \quad N = 20/1.633 = 12.25$ travel times. We will go up to 14 travel times ($V = 0$ up to $t = T$). ($A_0^2 = 0.6757$)

$\quad t = 0 \qquad V_0 = 0$

$\quad t = T \qquad V_1 = 2A_0 = 1.644$

$\quad t = 2T \qquad V_2 = 2A_0 \cos 1.633 \times 18° = 2A_0 \cos 29.4° = 1.452$

$\quad t = 3T \qquad V_3 = 2A_0 \cos 2 \times 29.4° - A_0^2 V_1 = -0.259$

$\quad t = 4T \qquad V_4 = 2A_0 \cos 3 \times 29.4 - A_0^2 V_2 = -0.9362$

Continuing in this manner, we obtain

$V_5 = -0.587$, $V_6 = -0.746$, $V_7 = -1.244$, $V_8 = -0.976$, $V_9 = -0.098$,

$V_{10} = 0.5037$, $V_{11} = 0.735$, $V_{12} = 0.98$, $V_{13} = 1.134$,

and $\quad V_{14} = 2A_0 \cos 13 \times 29.4 - A_0^2 V_{12} = 0.86$.

The oscillations have nearly vanished and the open-end voltage follows the source voltage as shown in Fig. 10.8.

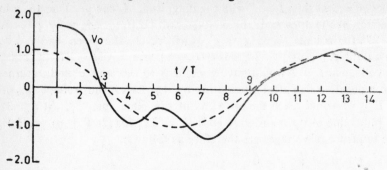

Fig. 10.8. Open-end voltage on a line excited by a sinusoidal source.

10.9 Generalized Equations for Single-Phase Representation

A more general case with series and shunt impedance lumped elements connected to distributed-parameter line was considered in chapter 8, last section. It was shown that the voltage at the entrance to the line was, with a little re-writing of equation (8.122),

$$V_e = E(s) \cdot \frac{\cosh pL + (Z_0/Z_t) \sinh pL}{\left(1 + \dfrac{Z_s}{Z_{sh}} + \dfrac{Z_s}{Z_t}\right) \cosh pL + \left(\dfrac{Z_0}{Z_t} + \dfrac{Z_s Z_0}{Z_{sh} Z_t} + \dfrac{Z_s}{Z_0}\right) \sinh pL} \qquad (10.45)$$

Having found V_e, all other voltages and currents can be obtained. In particular, the voltage at the far end or at the termination is

$$V_0 = E(s) \bigg/ \left[\left(1 + \frac{Z_s}{Z_{sh}} + \frac{Z_s}{Z_t}\right) \cosh pL + \left(\frac{Z_0}{Z_t} + \frac{Z_s Z_0}{Z_{sh} Z_t} + \frac{Z_s}{Z_0}\right) \sinh pL\right] \qquad (10.46)$$

These form the generalized equations for a single-phase representation of the system. These equations also yield the more complicated set of performance equations of a 3-phase system if all quantities are suitably replaced by matrices. This will be taken up in the next section. The matrices can be diagonalized to yield three mutually-decoupled single-phase systems or modes of propagation for the surge. Obtaining the inverse transform to yield the corresponding time variation in closed form is not easy for a general case and must be attempted through the Fourier Transform or other methods suitable for the Digital Computer.

Particular Cases of the General Equations
Equations (10.45) and (10.46) for the entrance and termination voltages can be used for particular cases which occur in practice. Some typical cases are shown in the preceding table. 7 cases are considered which are shown in Fig. 10.9.

Numerical Examples
The inverse transform has been evaluated by using the Fourier Transform Method for three cases which are shown in Fig. 10.10. These are:

S.No.	System Configuration	Impedances	Open-end Voltage V_0	Entrance Voltage V_e
1	Fig. 10.9(a)	$Z_s=0, Z_{sh}=\infty$ $Z_t=\infty$	$E(s)/\cosh pL$	$E(s)$
2	Fig. 10.9(b)	$Z_s=0, Z_{sh},$ $Z_t=Z_{sh}$	$\dfrac{E(s)}{\cosh pL + \dfrac{Z_0}{Z_{sh}}\sinh pL}$	$E(s)$
3	Fig. 10.9(c)	$Z_s=0, Z_t,$ $Z_{sh}=\infty$	$\dfrac{E(s)}{\cosh pL + \dfrac{Z_0}{Z_t}\sinh pL}$	$E(s)$
4	Fig. 10.9(d)	$Z_s, Z_{sh}=\infty$ $Z_t=\infty$	$\dfrac{E(s)}{\cosh pL + \dfrac{Z_s}{Z_0}\sinh pL}$	$V_0 \cdot \cosh pL$
5	Fig. 10.9(e)	Z_s, Z_{sh} $Z_t=\infty$	$\dfrac{E(s)}{\left(1+\dfrac{Z_s}{Z_{sh}}\right)\cosh pL + \dfrac{Z_s}{Z_0}\sinh pL}$	$V_0 \cdot \cosh pL$
6	Fig. 10.9(f)	$Z_s, Z_{sh},$ $Z_t=Z_{sh}$	$\dfrac{E(s)}{\left[\left(1+\dfrac{2Z_s}{Z_{sh}}\right)\cosh pL + \left(\dfrac{Z_0}{Z_{sh}}+\dfrac{Z_s}{Z_0}+\dfrac{Z_0 Z_s}{Z_{sh}^2}\right)\sinh pL\right]}$	
7	Fig. 10.9(g)	$Z_s, Z_{sh},$ Z_t	Equation 10.46	$V_0\left(\cosh pL + \dfrac{Z_0}{Z_{sh}}\sinh pL\right)$ Equation 10.45

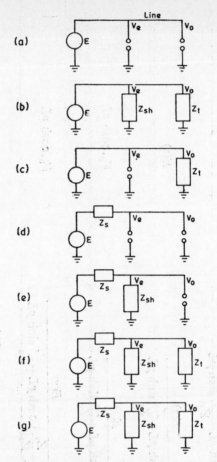

Fig. 10.9. 7 particular cases of system configuration for illustrating application of generalized equations.

1. 250 km line between Dehar Power Station and Panipat Receiving Station with line open. Line parameters are $r = 0.0274$ ohm/km, $l = 0.992$ mH/km, $c = 7.4$ nF/km, $T = 916$ μs, $Z_s = 0.075$ ohms and 0.2184 Henry which represent the resistance and transient reactance of power station. Switching at peak of sine wave. Both entrance voltage and open-end voltage are shown.

2. 160 km line, $Z_s = 0.384$ ohm and 0.049 Henry, $T = 545$ μs, $r = 0.02$ ohm/km, $l = 0.89$ mH/km, $c = 13$ nF/km. Initial $w_i = 10$, final $w_F = 10^5$, $\Delta w = 100$, number of ordinates used in numerical integration $n = 1000$, $a = 1/10T$ from 0 to $10T$, $a = 1/20T$ from 10T to 20T. Both entrance voltage V_e and open-end voltage are plotted for sinusoidal excitation at peak value.

3. Step response of a transformer-terminated line. 160 km line, $r = 0.29$ ohm/km, $l = 2$ mH/km, $c = 14.1$ nF/km, transformer

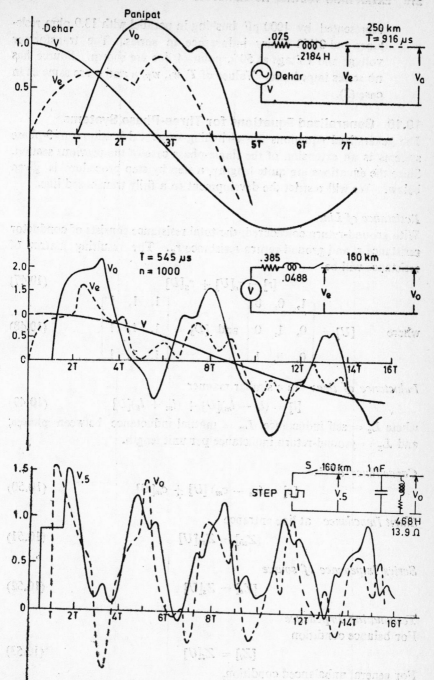

Fig. 10 10.   Overvoltages during switching operations on systems calculated by
Fourier Transforms.
(a) 250 km line energized at peak of sine wave at generating station with series
impedance. (b) 160 km line energized at peak of sine wave, with source having
small inductance (sub-transient). (c) Step response of a receiving-station transformer termination.   No source series impedance.

represented by 1000 pF bushing in parallel with 13.9 ohm resistance and 0.468 Henry inductance in series. The transformer voltage and voltage at 50% point of line are shown. Source has no series impedance. Values of T, w_l, w_F, n and a are same as in case (2).

10.10 Generalized Equations for Three-Phase Systems

The generalized equations for switching-surge calculation on 3-phase systems is an extension of the single-phase case of the previous section. Since the equations are quite lengthy, a step by step procedure is given below. We will restrict the development to a fully transposed line.

Resistance of Line

With ground-return considered, the total resistance consists of conductor resistance r_c and ground-return resistance r_g. The resulting matrix of resistance will be

$$[r] = r_c[U] + r_g[D] \tag{10.47}$$

where
$$[U] = \begin{bmatrix} 1, & 0, & 0 \\ 0, & 1, & 0 \\ 0, & 0, & 1 \end{bmatrix} \text{ and } [D] = \begin{bmatrix} 1, & 1, & 1 \\ 1, & 1, & 1 \\ 1, & 1, & 1 \end{bmatrix} \tag{10.48}$$

Inductance of line: In a similar manner
$$[l] = (l_s - l_m)[U] + (l_m + l_g)[D] \tag{10.49}$$
where L_s = self inductance, L_m = mutual inductance between phases, and L_g = ground-return inductance per unit length.

Capacitance of line
$$[c] = (c_s - c_m)[U] + c_m[D] \tag{10.50}$$

Shunt Impedance at line entrance
$$[Z_{sh}] = Z_{sh}[U] \tag{10.51}$$

Series Impedance of source
$$[Z_s] = Z_s[U] \tag{10.52}$$

Terminating Impedance
For balance condition
$$[Z_t] = Z_t[U] \tag{10.53}$$
For general unbalanced condition,

$$[Z_t] = \begin{bmatrix} Z_{t1}, & & \\ & Z_{t2}, & \\ & & Z_{t3} \end{bmatrix} \tag{10.54}$$

Source Voltage

$$[e(t)] = V_m \begin{bmatrix} \cos{(wt + \varphi_1)} \\ \cos{(wt + \varphi_2)} \\ \cos{(wt + \varphi_3)} \end{bmatrix} \qquad (10.55)$$

For synchronous closing,

$$\varphi_2 = \varphi_1 - 120°, \quad \varphi_3 = \varphi_1 + 120°.$$

The angles $\varphi_1, \varphi_2, \varphi_3$ denote the points on the wave at which switching takes place from the positive peaks of the respective phase voltages.

Voltage Equations
Commencing with the output end, we have the following sequence of equations, as shown from Fig. 10.11

$$[V_0] = [Z_t][I_0], [V_0] = [F][V_e] \qquad (10.56)$$

where $[F]$ is a function which is to be determined from the differential equations. For a single-phase representation, F is given by the ratio of equations (10.46) and (10.45), which is $[\cosh pL + (Z_0/Z_t) \sinh pL]$. $[I_e] = [G][V_e]$ where $[G]$ is another function still to be evaluated. Refer

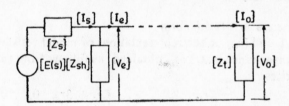

Fig. 10.11.   General representation of a system.

to Chapter 8 for the single-phase case.   A similar value in matrix form can be derived

$$[I_s] = [I_e] + [Z_{sh}]^{-1} [V_e] \qquad (10.57)$$

$[E] =$ Laplace transform of source voltages

$$= [V_e] + [Z_s][I_s] = [V_e] + [Z_s] \{[I_e] + [Z_{sh}]^{-1}[V_e]\}$$

$$= [[U] + [Z_s] \{[G] + [Z_{sh}]^{-1}\}][V_e] \qquad (10.58)$$

∴   The entrance voltage is

$$[V_e] = [[U] + [Z_s] \{[G] + [Z_{sh}]^{-1}\} [E] \qquad (10.59)$$

Having obtained the entrance voltage in terms of the source voltage $[E]$, the output voltage $[V_0]$ and the currents can be obtained in operational form.   The time-domain solutions have to be evaluated by the Inverse Transform.

10.10.1   Resolution into Component Modes of Propagation

In equations (10.47) to (10.50), the line parameters involve the special matrix $[D]$. If this is diagonalized, then the voltages and currents can all be resolved into components which will show no mutual interaction. They are resolved into three 'modes of propagation.' At any desired time, they can be re-combined to yield the phase quantities. This can be carried out by the Digital computer after every calculation of the inverse transform.

It has been shown in Chapter 3 that the modified Clarke Transformation matrix will diagonalize $[D]$. The matrices are

$$[T] = \frac{1}{\sqrt{6}} \begin{bmatrix} \sqrt{2}, & \sqrt{3}, & 1 \\ \sqrt{2}, & 0, & -2 \\ \sqrt{2}, & -\sqrt{3}, & 1 \end{bmatrix} \text{ and } [T]^{-1} = \frac{1}{\sqrt{6}} \begin{bmatrix} \sqrt{2}, & \sqrt{2}, & \sqrt{2} \\ \sqrt{3}, & 0, & -\sqrt{3} \\ 1, & -2, & 1 \end{bmatrix} \tag{10.60}$$

Then
$$[T]^{-1}[D][T] = \begin{bmatrix} 3, & 0, & 0 \\ 0, & 0, & 0 \\ 0, & 0, & 0 \end{bmatrix} = [\lambda], \tag{10.61}$$

the eigenvalue matrix.

Line Parameters in Modal Form

Let $[r_m]$, $[l_m]$, and $[c_m]$ be the representations of modal resistance, inductance and capacitance matrices, obtained as follows from the line parameter matrices.

$$[r_m] = [T]^{-1}[r][T] = \begin{bmatrix} r_1, & & \\ & r_2, & \\ & & r_3 \end{bmatrix} = \begin{bmatrix} r_c + 3r_g, & 0, & 0 \\ 0, & r_c, & 0 \\ 0, & 0, & r_c \end{bmatrix} \tag{10.62}$$

$$[l_m] = [T]^{-1}[l][T] = \begin{bmatrix} l_1 & & \\ & l_2 & \\ & & l_3 \end{bmatrix} = \begin{bmatrix} l_s + 2l_m + 3l_g & & \\ & l_s - l_m & \\ & & l_s - l_m \end{bmatrix} \tag{10.63}$$

$$[c_m] = [T]^{-1}[c][T] = \begin{bmatrix} c_1, & & \\ & c_2, & \\ & & c_3 \end{bmatrix} = \begin{bmatrix} c_s + 2c_m, & & \\ & c_s - c_m, & \\ & & c_s - c_m \end{bmatrix} \tag{10.64}$$

Lumped Impedances in Modal Form

$$[Z_{shm}] = [T]^{-1}[Z_{sh}][T] = Z_{sh}[U] \tag{10.65}$$

$$[Z_{sm}] = [T]^{-1}[Z_s][T] = Z_s[U] \qquad (10.66)$$

The terminating impedance needs special attention. For balanced impedances, equation (10.53),

$$[Z_{tm}] = [T]^{-1}[Z_t][T] = Z_t[U] \qquad (10.67)$$

But when the impedances connected to ground in the three phases are unequal as in equation (10.54), there will be mutual interaction between the phases since

$$[T]^{-1}[Z_t][T] = \begin{bmatrix} (Z_{t1} + Z_{t2} + Z_{t3})/3, & (Z_{t1} - Z_{t3})/\sqrt{6}, \\ (Z_{t1} - Z_{t3})/\sqrt{6}, & (Z_{t1} + Z_{t3})/2, \\ (Z_{t1} - 2Z_{t2} + Z_{t3})/2\sqrt{3}, & (Z_{t1} - Z_{t3})/2\sqrt{3}, \end{bmatrix}$$

$$\begin{bmatrix} (Z_{t1} - 2Z_{t2} + Z_{t3})/2\sqrt{3} \\ (Z_{t1} - Z_{t3})/2\sqrt{3} \\ (Z_{t1} + 4Z_{t2} + Z_{t3})/6 \end{bmatrix} \qquad (10.68)$$

For $Z_{t1} = Z_{t2} = Z_{t3}$, this reduces to equation (10.67).

Special methods are available for handling unbalanced terminal conditions which have been developed by Uram, Miller and Feero, and by Dommel and others. But a very extensive investigation of the effect produced on switching surges carried out on a Transient Network Analyzer by Sujatha Subhash, Meera, Jyoti and Kanya Kumari have revealed the following important theorem:

'When unbalanced terminal conditions exist, the system response to switching is nearly the same as if the system consists of three balanced systems, each one having terminating impedances equal to those connected in the individual phases.' Using this property, the system can be analyzed three times by using balanced load of Z_{t1} in all phases and calculating the switching-surge response in phase 1. Then replace a balanced load of Z_{t2} and determine the response of phase 2. Finally, replace a balanced load Z_{t3} in the system and the resulting response of phase 3 will be the same as when the unbalanced load is used. We might also remark that since switching-surge calculations under unbalanced loads or when considering resistances in lines is very complicated and consumes a very long computer time, shorter methods, giving results acceptable for engineering designs must be continuously devised. Hedman has found that the line resistance can be omitted in solving the travelling-wave equations but can be included as a lumped series resistance with the load. Similarly, the problem of frequency-dependence of ground-return parameters have also caused grave concern and form very advanced topics which cannot be discussed in this introductory chapter to calculation of switching surges. The reader is referred to the extensive work listed in the bibliography.

Voltages and Currents in Modal Form: Differential Equations:
Source Voltage

$$[T]^{-1}[e(t)][T] = \frac{V_m}{\sqrt{6}} \begin{bmatrix} \sqrt{2}\{\cos{(wt+\varphi_1)} + \cos{(wt+\varphi_2)} + \cos{(wt+\varphi_3)}\} \\ \sqrt{3}\{\cos{(wt+\varphi_1)} - \cos{(wt+\varphi_3)}\} \\ \cos{(wt+\varphi_1)} - 2\cos{(wt+\varphi_2)} + \cos{(wt+\varphi_3)} \end{bmatrix}$$

$$(10.69)$$

The operational forms of these voltages can be evaluated suitably by using $\angle[\cos{(wt+\varphi)}] = (s\cdot\cos{\varphi} - w\cdot\sin{\varphi})/(s^2 + w^2)$.

The governing differential equations and their solutions for each of the three independent modes can now be derived. The basic differential equation for the phase quantities are

$$\frac{d[V_x]}{dx} = \{[r] + [l]s\}[I_x], \quad \text{and} \quad \frac{d[I_x]}{dx} = [c]s[V_x] \qquad (10.70)$$

These are resolved into modal form thus:

$$[T]^{-1}\frac{d}{dx}[V_x] = \frac{d}{dx}[T]^{-1}[V_x] = \frac{d}{dx}\begin{bmatrix} V_{1m} \\ V_{2m} \\ V_{3m} \end{bmatrix} = \frac{d}{dx}[V_m] \qquad (10.71)$$

But $\quad [T]^{-1}\{[r] + [l]s\}[T][T]^{-1}[I_x] = \{[r_m] + [l_m]s\}[I_m] \qquad (10.72)$

$$\therefore \quad \frac{d}{dx}[V_m] = \{[r_m] + [l_m]s\}[I_m] \qquad (10.73)$$

Similarly, $\qquad \dfrac{d}{dx}[I_m] = [c_m]s[V_m] \qquad (10.74)$

These constitute 6 independent equations which are as follows by using equations (10.62) to (10.64):

$$d V_{1m}/dx = \{(r_c + 3r_g) + (l_s + 2l_m + 3l_g)s\}I_{1m}$$
$$d V_{2m}/dx = \{r_c + (l_s - l_m)s\}I_{2m}$$
$$d V_{3m}/dx = \{r_c + (l_s - l_m)s\}I_{3m} \qquad (10.75)$$
$$d I_{1m}/dx = (c_s + 2c_m)s\cdot V_{1m}$$
$$d I_{2m}/dx = (c_s - c_m)s\cdot V_{2m}$$
$$d I_{3m}/dx = (c_s - c_m)s\cdot V_{3m}$$

We observe that these are equivalent to the following 3 sets of equations, one for each mode, as if each pertains to a single-phase quantity but in modal form,

$$d\cdot(V_{km})/dx = z_{km}\cdot I_{km}, \quad \text{and} \quad d(I_{km})/dx = y_{km}V_{km} \qquad (10.76)$$

where $k = 1, 2, 3$. Omitting the subscripts, for any one of the modes, the differential equations are of the form

$$dV/dx = zI, \quad \text{and} \quad dI/dx = yV \tag{10.77}$$

Their solutions are
$$\left. \begin{array}{l} V = Ae^{px} + Be^{-px} \\[2mm] I = \dfrac{1}{Z_0}\left(Ae^{px} - Be^{-px}\right) \end{array} \right\} \tag{10.78}$$
and

We have encountered these before in Chapter 8. In these equations, the propagation factor p and surge impedance Z_0 belong to the mode under consideration, and their values are follows:

Mode 1:

$$\begin{aligned} p_1 &= \sqrt{[\{(r_c + 3r_g) + (l_s + 2l_m + l_g)\,s\}(c_s + 2c_m)\,s]} \\ &= \left. \sqrt{(r_1 + l_1 s)\,c_1 s} \right\} \\ Z_{01} &= \sqrt{(r_1 + l_1 s)/c_1 s} \end{aligned} \tag{10.79}$$

Mode 2:

$$p_2 = \sqrt{(r_2 + l_2 s)\,c_2 s}, \quad Z_{02} = \sqrt{(r_2 + l_2 s)/c_2 s} \tag{10.80}$$

Mode 3:

$$p_3 = p_2 \quad \text{and} \quad Z_{03} = Z_{02} \tag{10.81}$$

The values of r_1, r_2, l_1, l_2, c_1, c_2 are given in equations (10.62) to (10.64). Equations (10.78) can be applied for each mode in turn which will yield expressions for the required voltages or currents in the system when the entrance voltage V_e to the line is found in terms of the source voltage as shown in equations (10.45) and (10.46) for a single phase case.

After having calculated the inverse transform at every instant of time by a suitable method on the Digital Computer, the modal voltages and currents can be re-converted to phase quantities by the inverse transformation to equations (10.71) to (10.74):

$$[V_{ph}] = [T]\,[V_m] \quad \text{and} \quad [I_{ph}] = [T]\,[I_m] \tag{10.82}$$

For example,

$$\begin{aligned} \begin{bmatrix} V_{ph1} \\ V_{ph2} \\ V_{ph3} \end{bmatrix} &= \frac{1}{\sqrt{6}} \begin{bmatrix} \sqrt{2}, & \sqrt{3}, & 1 \\ \sqrt{2}, & 0, & -2 \\ \sqrt{2}, & -\sqrt{3}, & 1 \end{bmatrix} \begin{bmatrix} V_{1m} \\ V_{2m} \\ V_{3m} \end{bmatrix} \\ &= \frac{1}{\sqrt{6}} \begin{bmatrix} \sqrt{2}V_{1m} + \sqrt{3}V_{2m} + V_{3m} \\ \sqrt{2}V_{1m} - 2V_{3m} \\ \sqrt{2}V_{1m} - \sqrt{3}V_{2m} + V_{3m} \end{bmatrix} \end{aligned} \tag{10.83}$$

In switching-transient calculations, the most difficult operation is obtaining the inverse transform. We will discuss this aspect next and only indicate the steps involved in evaluating the same by using a Digital

Computer for the Inverse Fourier Transform. This is direct and the main ideas have already been set forth in Chapter 8. Various other very powerful methods have been evolved by many eminent engineers to obtain the time response. Chief among these are (See Greenwood, Reference 11 under Books in Bibliography):

1. Dommel's Method based on Bergeron's analysis of Water Hammer in hydraulic pipe lines and the development of the Electromagnetic Transients Programmes (EMTP);

2. Uram, Miller, and Feero's Method using Laplace Transforms;

3. Ametani's Modified Refraction Coefficient Method;

4. Barthold and Carter's Reflection Coefficient Method which is an application and improvement of Bewley's Lattice Diagram for handling a large system using the Digital Computer.

5. Finally, the Fourier Transform Method developed and used by the British team of Day, Mullineux, Reed, Doeppel, and others.

10.11 Inverse Fourier Transform for the General Case

We will illustrate the steps involved in evaluating the Inverse Fourier Transform (IFT) for the far-end voltage given by equation (10.46) and for each mode in a 3-phase system after the phase voltages have been resolved into the 3 independent modes. Consider the equation

$$V_0 = E(s) \bigg/ \left[\left(1 + \frac{Z_s}{Z_{sh}} + \frac{Z_0}{Z_t} \right) \cosh pL + \left(\frac{Z_0}{Z_t} + \frac{Z_s Z_0}{Z_t Z_{sh}} + \frac{Z_s}{Z_0} \right) \sinh pL \right]$$

(10.46)

where $p = \sqrt{(r + ls)(g + cs)}$, $Z_o = \sqrt{(r + ls)/(g + cs)}$ (10.84)

In previous equations the shunt conductance g could be introduced as a matrix without losing the generality. For the sake of illustration the steps, let $r = g = 0$, and only shunt reactors be used at both ends so that $Z_t = Z_{sh}$. Then,

$$p = s\sqrt{lc} = s/v, \; Z_0 = \sqrt{l/c}, \; Z_t = Z_{sh} = L_{sh}s,$$

$$Z_s = R_s + L_s s$$ (10.85)

Here, v = velocity of propagation for the mode under consideration with l and c computed for each mode,

Z_0 = surge impedance of the line for the mode,

L_{sh} = inductance of the shunt-compensating reactor,

L_s = series inductance of source = transient reactance/$2\pi f$,

R_s = series resistance of source and any resistance incorporated in the circuit breaker during switching, and

$E(s)$ = source voltage in operational form for the mode obtained from equation (10.69).

Then,

$$V_0(s) = E(s) \bigg/ \left[\left(1 + \frac{2(R_s + L_s s)}{L_{sh} s} \right) \cosh s/v \right.$$

$$\left. + \left(\frac{Z_0}{L_{sh} s} + \frac{(R_s + L_s s)Z_0}{L_{sh}^2 s^2} + \frac{R_s + L_s s}{Z_0} \right) \sinh s/v \right] \qquad (10.86)$$

Next, we replace $s = a + jw$ where a and w are as yet unspecified but they enter in the integration as discussed in Chapter 8. When $s = a + jw$, the equation (10.86) will become a function of complex numbers. We then proceed to separate the real and j-parts, designating them P and Q respectively. Now,

$$1 + \frac{2R_s}{L_{sh}} \frac{1}{s} + \frac{2L_s}{L_{sh}} = \left(1 + \frac{2L_s}{L_{sh}} + \frac{2R_s a}{L_{sh}(a^2 + w^2)} \right) - j \frac{2R_s w}{L_{sh}(a^2 + w^2)} \qquad (10.87)$$

$$\frac{Z_0}{L_{sh} s} + \frac{R_s Z_0}{L_{sh}^2} \frac{1}{s^2} + \frac{L_s Z_0}{L_{sh}^2} \frac{1}{s} + \frac{R_s}{Z_0} \frac{1}{s}$$

$$= \left[\frac{Z_0}{L_{sh}} \frac{a}{a^2 + w^2} + \frac{R_s Z_0}{L_{sh}^2} \frac{a^2 - w^2}{(a^2 + w^2)^2} + \frac{R_s Z_0}{L_{sh}^2} \frac{a}{a^2 + w^2} + \frac{R_s}{Z_0} + \frac{L_s a}{Z_0} \right]$$

$$- j \left[\frac{Z_0 w}{L_{sh}(a^2 + w^2)} + \frac{2R_s Z_0 aw}{L_{sh}^2(a^2 + w^2)^2} + \frac{R_s Z_0 w}{L_{sh}^2(a^2 + w^2)} - \frac{L_s w}{Z_0} \right] \qquad (10.88)$$

$$\cosh (s/v) = \cosh (a/v) \cdot \cos (w/v) + j \sinh (a/v) \cdot \sin (w/v) \qquad (10.89)$$

$$\sinh (s/v) = \sinh (a/v) \cdot \cos (w/v) + j \cosh (a/v) \cdot \sin (w/v) \qquad (10.90)$$

By multiplying equation (10.87) with (10.89), and (10.88) with (10.90), the denominator of equation (10.86) can be written as a complex number

$$D = D_r + jD_i \qquad (10.91)$$

If the source of excitation $e(t)$ is sinusoidal and the modal resolution has been carried out according to equation (10.69) and its corresponding operational expression is evaluated with $s = a + jw$, it can be written as

$$E(s) = N_r + jN_i \qquad (10.92)$$

$$\therefore \quad V_0(a + jw) = (N_r + jN_i)/(D_r + jD_i)$$

$$= \frac{(N_r D_r + N_i D_i) + j(N_i D_r - N_r D_i)}{D_r^2 + D_i^2} \qquad (10.93)$$

The real part is

$$P = (N_r D_r + N_i D_i)/(D_r^2 + D_i^2) \qquad (10.94)$$

and the j-part is

$$Q = (N_i D_r - N_r D_i)/(D_r^2 + D_i^2) \qquad (10.95)$$

The inverse transform of $V_0(a + jw)$ can be obtained by performing a numerical integration. Thus,

$$V_0(t) = \frac{e^{at}}{2\pi} \int_{w_l}^{w_F} P \cdot \cos wt \cdot \frac{\sin (\pi w/w_F)}{(\pi w/w_F)} \cdot dw \qquad (10.96)$$

or $\qquad = -\frac{e^{at}}{2\pi} \int_{w_l}^{w_F} Q \cdot \sin wt \cdot \frac{\sin (\pi w/w_f)}{(\pi w/w_F)} \cdot dw \qquad (10.97)$

10.12 Reduction of Switching Surges on EHV Systems

In practice, the most severe overvoltages on lines which the insulation must withstand are caused under the following conditions:

(a) energizing an open-ended line;
(b) re-energizing an open-ended line;
(c) re-energizing a line after clearing a single-phase to earth fault.

Energizing refers to switching a line without trapped charge while re-energizing is done with a trapped charge left from a previous energization or re-energization. In single-pole reclosing schemes, there is no voltage on the faulted phase when it is cleared from both ends, but due to mutual coupling from the other phases the circuit-breaker re-energizes the faulted phase with a trapped charge. If resistance is inserted in the breaker during re-energization, a maximum value of 1.5 p.u. can be expected at the open end.

In e.h.v. lines, energization and re-energization causes a maximum switching-surge amplitude of not higher than 2.5 p.u. at the open end to ground and the designer has to make every effort to keep it below 2 p.u. by taking measures which will be described later. While we discussed re-energization for lumped-parameter network in section 10.7, in e.h.v. lines trapped charge is very rarely present because of shunt-connected equipment which drain the trapped charge. Thus, all closing operations may be considered more or less equivalent to energizing a line without trapped charge.

The following measures are adopted in e.h.v systems to reduce overvoltage magnitudes.

1. Draining of Trapped Charge of Line

Shunt reactors are invariably used at both ends of an e.h.v. line as shown in Fig. 10.12 (a). The schemes used resemble Fig. 10.12 which is known as the '4-legged reactor'. The reactor in the common neutral connection serves to quench secondary arc produced under single-pole reclosing which is not discussed in this book. The shunt reactors are designed with a very low resistance (high Q at power frequency of the order of 200). These provide compensating VARs at nearly zero power factor during normal steady-state operation. If one of the purposes of using shunt reactors is also to drain the trapped charge after a de-energizing or line-dropping operation, the reactor in the neutral will be replaced by a resistor. In such schemes, the time constant is low and the line discharges completely in 5 time constants which usually is set at 5 ms or

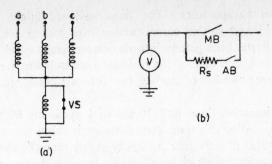

Fig. 10.12. (a) Four-legged reactor for draining trapped charge and quenching
secondary arc during single-pole reclosing.
(b) Switching arrangement of series resistance in circuit breaker.
MB—Main breaker. AB—Auxiliary breaker.

1/4 cycle on 50 Hz basis. The resistor is short-circuited by a vacuum
switch VS rated for 15 kV. It is interlocked with the main circuit breaker
such that VS opens at the same instant as the circuit breaker and closes
just prior to the main circuit breaker does.

Instrument transformers such as the inductive type potential transformers (IPT) can also discharge the trapped line charge in contrast to
capacitive voltage transformers (CVT). Power transformers help to
drain the trapped charge in about 20 ms or 1 cycle, if they are still left
connected. But switching a transformer-terminated line is not looked
with favour because of the possibility of ferro-resonance conditions. The
Hydro-Quebec Company of Canada relies solely on shunt reactors to keep
the switching overvoltage to 2.1 p.u. on their 735 kV line and have not
equipped the circuit breakers with series resistances.

2. Series Resistance Switching

For lines of 400 kV and higher (or on some very long 220 kV lines
also) reduction of switching surges to 2 p.u. or less can be attained by inser-
ting a resistance R in series with the line. At the time of energization,
Fig. 10.12 (b), the main breaker is open while the auxiliary breaker
closes. The voltage impressed at the line entrance is thus $V_e =$
$e(t) \cdot Z_0/(R + Z_0)$. If $R = Z_0$, only 50% of the source voltage is impressed
on the line giving 1 p.u. at the open end due to total reflection. Because
the line is matched at the source end, the voltage settles down to the
source voltage very quickly. However, when re-energization with trapped charge occurs a maximum of 2 p.u. will be attained. Thus, with
series-resistance switching the overvoltage is never higher than 2 p.u.
This has been verified by a large number of switching surge studies using
the Transient Network Analyzer and Digital Computer.

The value of resistance R in general depends on a large number of
factors as follows:

(a) The value of R is selected to achieve optimum results for the system.
(b) The surge impedance of connected lines when there is a single

line or multiple lines. The lines switched might not all be of equal length so that complications arise due to reflections from the shorter lines getting into the longer ones and vice-versa.

(c) The insertion time of the resistance controls the overvoltage.

From a large number of studies, the following recommendations are made:

1. The insertion time is 8–10 ms or $\frac{1}{2}$ cycle on 60 Hz or 50 Hz basis. After this time, the resistance is shorted.

2. The value of resistance is slightly higher than the surge impedance of a single line which is switched. In older designs a value of the order of 1000 ohms was used, but modern practice is nearly 400 ohms.

3. The closing span of the circuit-breaker poles must be controlled within 60°.

The last item is very important under 3-phase reclosing operations. Poorly-maintained breakers can have a 180° lag between the first and last pole to close which result in high overvoltages since the last phase has a trapped voltage induced in it by the other phases which have already been energized. On the other hand, because of the non-synchronous or non-simultaneous closure of the poles with resulting unbalanced conditions, ground-return currents are present which help to attenuate the surges. However, each case must be studied carefully on models and the worst case guarded against.

3. Electronic Sensing of Voltage Polarities

In the resistance-insertion scheme the maximum overvoltage condition exists when the main breaker closes to short-circuit R in the auxiliary breaker, and at this instant the polarity of line-side voltage is opposite to that of the source. Very sophisticated electronic circuitry using sensors and logic elements to sense the polarities of the two voltages and to activate the closing mechanism of the main circuit breaker exist. This connects the line directly to the source while the polarities of voltages are the same. This applies when there is a trapped charge on the line. In such schemes the overvoltage is brought down as low as 1.5 p.u. at the open end. The scheme is improved further if the main breaker closes when the current in the line is zero when there are oscillations caused by the inductance and capacitance of the line itself. Such a scheme has been developed and used successfully in the U.S.A. by the Bonneville Power Administration.

4. Limiting Value of Minimum Switching Surge

While a designer or user of such sophisticated and expensive equipment aims at lowering the overvoltages to 1.5 p.u. or less, it has been observed that there is not much advantage in lowering the overvoltage to less than 1.5 p.u. The main reason for this is that under a single line to ground fault, the dynamic voltage rise is 1.5 p.u. or very near this value.

For an 80% arrester of the conventional type, the overvoltage under a fault reaches $0.8\sqrt{3} = 1.4$ p.u. Therefore there is not much advantage gained in lowering only the switching overvoltage.

However, with new gapless Metal Oxide arresters, the voltage rating of the arrester can be as low as 60 to 65% of line-to-line voltage which permits a lowering of equipment insulation levels. These arresters are meant for switching-surge duty so that e.h.v. insulation levels can be brought down further.

10.13 Experimental and Calculated Results of Switching-Surge Studies

We conclude this chapter with a discussion of switching-surge overvoltages as observed on e.h.v. systems of typical configurations. These studies are carried out by three methods:

(a) Mathematical modelling using the Digital Computer,
(b) Physical modelling using the Transient Network Analyzer (TNA), and
(c) Field Tests.

Every designer involved in the selection of air-gap clearances or other types of insulation must have exhaustive data on these two points:

(1) The expected maximum amplitudes of overvoltages at all points of all types in the system network, and
(2) The flashover or withstand voltages of the insulation structure to be selected which will insulate against the overvoltages.

Both these aspects are statistical in nature and involve application of probability theory. Some of these concerning insulation strength of long airgaps will be discussed in the next chapter, Chapter 11.

The main factors to be investigated on a TNA or with a Digital Computer prior to line design are listed below. These vary with the system but the list covers the most important items used in e.h.v. systems.

(1) *Line Length L and Line Constants r, l, g, c:* These quantities may not be truly constant because of the effect of frequency especially the ground-return parameters. In general, a frequency of $f = 1/4T = v/4L$ can be used where v = velocity and T = travel time $= L/v$.

(2) *High Voltage Reactors:* When these are depended upon to reduce the switching-surge magnitudes by draining the trapped charge, the loss of one leg or as many legs as are anticipated must be studied. The loss of shunt reactor also raises the power-frequency open-end voltage due to Ferranti Effect. In lowering the switching-surge magnitudes, both the steady-state and transient components of voltage must be kept down.

(3) *Series Capacitors:* The presence of series capacitors up to 50% line-inductance compensation normally do not affect switching surges, but will affect the steady-state component in the total transient.

(4) *Short-Circuit Capacity of Source*: This is reflected as the value of series reactance of source. In general, an infinite source (zero series impedance) yields higher switching surges at the open end of a line, but there are cases where this depends upon the length of line switched as well as their number in parallel.

(5) *Damping Factor of Supply Network*: The presence of ground currents on the switching overvoltages plays an important role. The frequency-dependence of the zero-sequence resistance and inductance must be properly taken into account in any study.

(6) *Presence of Transformers or Autotransformers*: When a line is connected to a source through a transformer at the generating station, as is usual in unit-connected schemes, the line can be switched either on the high-voltage side or the low-voltage side. In general, when switched from the l.v. side the series-impedance of the transformer lowers the entrance voltage of the line and the switching over-voltage at the open end is lower than when switched from the h.v. side. The presence or absence of tertiaries has some effect on the switching surges.

(7) *Saturation of Transformers and Reactors*: These give rise to harmonic oscillations with the line capacitance.

(8) *Load Transmitted*: A line-dropping operation followed by a fault due to the overvoltage gives very high surges if the circuit breakers at both ends do not open simultaneously. The power factor of load has considerable effect on the switching surges.

(9) *Single-Phase Reclosing*: When a fault is cleared on a single line to ground fault by opening circuit breakers, due to mutual induction from the healthy phases, a current of 20–30 Amperes keeps flowing in the arc at the fault on the grounded line. The dead time, or the time interval between opening of the faulted line and its reclosure has to be determined. With shunt reactors using 4-legged construction, the dead time is of order of 5 seconds or 250–300 cycles.

(10) *Frequency of Source*: This has only a secondary effect.

(11) *Switching Resistors*: These clearly have very important effect in reducing switching overvoltages. The optimum values of resistance and the time of their insertion in series with the circuit breaker has been the subject of intense study.

(12) *Order of Phases to Clear*: Where circuit breaker poles have a time lag in closing or opening, the switching-surge magnitude will have different value from simultaneous closure called synchronous closing. The last phase to close has a higher overvoltage because of coupling of voltage from the other phases.

(13) *Restriking in Circuit Breaker*: When breaker arcing chambers are unable to interrupt properly, there is a restrike which essentially connects a line to the source with trapped charge on the line. This has been considered in section 10.5. Long lines cause

maximum trouble due to line dropping with restrike in circuit breaker causing lightning-arrester failures or circuit-breaker explosions due to frequent operation.

(14) *Lightning Arrester Sparkover Characteristics*: These are normally set for 1.5 p.u. with gap type SiO arresters. If magnetic blow-out or current-limiting gaps fail, then they cannot handle switching-surge duty. Their effect on insulation levels must be investigated.

(15) *Line-to-line Voltages*: In order to design conductor-to-conductor clearances, a knowledge of phase-to-phase overvoltage magnitude is necessary. It is insufficient to obtain the maximum expected line-to-ground voltage on one phase and multiply by $\sqrt{3}$. The maximum is lower than this value because of the waveshapes not being sinusoidal.

Out of an enormous amount of studies carried out all over the world, the following types of conditions leading to overvoltages are considered very important. Against each item, a list of parameters which affect the overvoltage are given.

(1) *Interruption of Line at No Load*: Length of line, line constants, shunt reactors, series capacitors, reactance of source, and switching resistors.

(2) *Load Shedding*: All of the above factors and in addition saturation effects of transformers and reactors, load characteristics and damping factors.

(3) *Load Shedding at End of Line Followed by Disconnection of Unloaded Line*: All of the above factors plus natural frequency of supply network as it affects the restriking in the breaker.

(4) *Clearing a Short Circuit*: All the above factors.

(5) *Closing a Transformer-Terminated Line with Load*: All the above factors plus order of phases to close which causes possible ferro-resonance condition.

(6) *Energizing or Re-energizing an Unloaded Line*: All the above factors minus the transmitted load.

In order to give an idea of p.u. values of overvoltages that may be realized, Table 10.1 has been prepared summarizing the findings of a 500 kV/230 kV system which were obtained from model studies on a

Table 10.1. Switching-Surge Voltages from Model Studies on T.N.A.

A. Switching an Open-ended Line

1. *Switching from High Voltage Side (500 kV)*

    (a) With trapped charge on Line

2,260 MVA source short circuit power	3.3
13,600 MVA	3.8
Infinite Bus at source	4.1

(Contd.)

    (b)   Without trapped charge on line:

           2,260 MVA source short circuit power   2.1

           13,600 MVA   2.6

    (c)   De-energizing without fault: 1.p.u.

           No restrike in breaker   trapped on all lines

                                           $(+1, -1, -1)$

    (d)   De-energizing line with a line to ground fault

           (270 Kms)   1.3

2.   *Switching from Low-Voltage Side (220 kV)*

    Transformer has no tertiary, Line has no trapped charge

    since transformer drains the charge in 5 to 10 cycles   2.0

B.   **Switching a Transformer Terminated Line**

    High-voltage side closing; source transformer has

    no tertiary   2.2

C.   **Series Capacitor Compensation**

    Up to 50%   No changes in over-

                                         voltage magnitudes

D.   **Source Transformer Tertiary Effect**

1.   *Switching Open Ended Line*

    (a)   High-side energizing with trapped charge

           (i)   With transformer tertiary   3.0

           (ii) Without tertiary   3.3

    (b)   High-side energizing a dead line without

           trapped charge:

           (i)   With transformer tertiary   2.0

           (ii) Without tertiary   2.1

    (c)   Low-side Energizing of transformer and line:

           (i)   With transformer tertiary   2.8

           (ii) Without tertiary   2.0

           (iii) With shunt reactor connected to tertiary for

                50% compensation   2.6

E.   **Lightning Arrester**

    Arrester at sending end set to spark at:

    (a)   2.0 p.u.   3.0

    (b)   1.5 p.u.   2.6

    Arrester included at receiving end, and set to

    spark at 1.5 p.u.   2.2

F.   **Surge Suppressing Resistor in Breaker**

    In all cases where overvoltages exceeded 2.3 p.u.,

    1200 ohms was included   2.0

(Contd.)

G. **High-Speed Reclosing After Fault Clearing**

(with trapped charge)   3.6

H. **Effect of Non-Synchronous Operation of Circuit Breaker Poles**

(a) Breaker poles close within 1/2 cycle   2.4
(b) Breaker poles close on 1 cycle   3.5

Transient Network Analyzer and reported in an I.E.E.E. Paper (No. 50 in list of Bibliography).

Review Questions and Problems

1. A 750 kV system has a generation of 4000 MVA. The transient reactance of the source is 0.25 p.u. on 750 kV 4000 MVA base and the unit-connected transformer has a leakage reactance of 0.15 p.u. For a 3-phase bus fault on the high-voltage side of the transformer, calculate the short-circuit current in kiloamperes and in p.u.

2. Explain the terms (a) sub-transient reactance, (b) transient reactance, (c) synchronous reactance of a source, and (d) a.c. component, (e) d.c. component, and (f) the interrupting current capacity of a circuit breaker.

3. Explain the terms (a) terminal fault, (b) short-line fault, (c) 2-parameter definition of recovery voltage, and (d) 4-parameter definition of recovery voltage.

4. Explain clearly how overvoltages are generated when interrupting (a) low inductive currents and (b) low capacitive current. Draw a figure showing ferro-resonance condition in a network when two poles of a circuit breaker are open and one pole is closed.

5. A series L-R-C circuit has a $L = 0.3$ H, $R = 9$ ohm, and $C = 3.3 \ \mu$F. It is excited by a step input of magnitude 600 kV with the initial value of capacitor voltage equal to -600 kV, in Fig. 10.7. Calculate the following—
(a) attenuation factor $\alpha = R/2L$, (b) the natural frequency of oscillation w_0 and f_0, (c) the first peak of voltage across the capacitor, (d) the ratio of adjacent positive peaks of voltage from zero value.

6. Repeat problem 4 if the resistance is increased to the value $R = 2\sqrt{L/C}$. Only parts (a) and (c) apply.

CHAPTER 11

Insulation Characteristics of Long Air Gaps

11.1 Types of Electrode Geometries Used in EHV Systems

In e.h.v. transmission systems, other than the internal insulation of equipment such as transformers, circuit breakers, etc., all external insulation is air at atmospheric pressure and electronegative gases such as SF_6. Air gaps used for insulation have now reached enormous lengths—up to 20 metres and more—and will increase with further increase in transmission voltage. As of this date, the highest commercially-used voltage is 1200 kV, ac. But experimental projects at 1500 kV ac have been in progress since 1980. It is very important and vital to know the breakdown and withstand properties of such long gaps and in this chapter we will discuss only air gaps.

The insulation characteristics of long air gap lengths namely the flashover and withstand voltages are governed mainly by the geometry of the electrodes insulated by the gap. The most usual electrode geometries are shown in Fig. 11.1. Representative air-gap lengths used in conductor-to-tower gaps are:

400 kV	500 kV	750 kV	1000 kV	1150–1200 kV
2.134 m	2.59 m	5.6 m	7 m	8 m (U.S.A.)
(84″)	(102″)			10 m (U.S.S.R.)

These gaps are subjected to several types of voltage waveshapes so that breakdown and withstand voltages have to be ascertained for all of them. These are:

(a) Power Frequency A.C.; (b) Lightning Impulses—positive and negative; (c) Switching Surges—positive and negative; (d) D.C.—positive and negative; and (e) A.C. with a D.C. offset.

In both indoor and outdoor high-voltage laboratories, experiments at great expense have been and are being conducted in order to determine the breakdown voltages of air-gap geometries. Along with these important experiments, a very large amount of theoretical work concerning the basic mechanisms of breakdown have also been handed down in technical literature. Once these mechanisms are thoroughly understood, the breakdown characteristics of very long air gaps have been known to follow the same pattern as small gaps. An understanding of basic breakdown mechanisms helps to reduce the expense involved in large-scale experiments and proper interpretation of such experimental results.

All breakdown phenomena depend on the electric field distribution in the gap which initiate the primary electron-avalanche mechanism, leading to secondary mechanisms, as these depend on the volocities or energies of the resulting ions. In very long gaps, especially under switching surges, the secondary mechanisms lead to very dangerous conditions which are controlled by the environment leading to the well-known saturation phenomenon and sometimes to anomalous flashover.

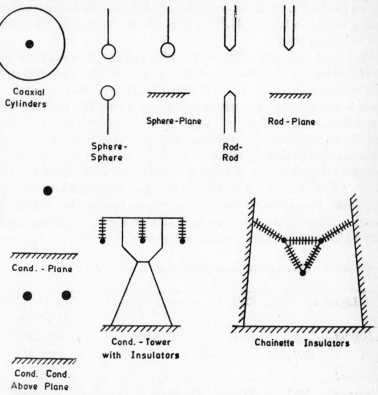

Fig. 11.1. Types of electrode geometries encountered in e.h.v. transmission lines with long air gaps.

It is not possible to discuss at any great length the electric field distribution of all types of geometries and the reader should consult books on electrostatics and highly specialized articles on such field distribution. It is one of the single most important items that must be carefully examined. Some field patterns will be given as and when the discussion proceeds in the later sections of this chapter. Some problems have already been described in Chapter 4 under Voltage Gradients.

11.2 Breakdown Characteristics of Long Air Gaps

The calculation of field distribution in the long gap will not lead to as clear an understanding of the breakdown characteristics as they do in very small gaps because of the importance of secondary phenomena becoming the controlling factors. We are therefore constrained to resort to experimental results conducted from full-scale mock-ups in laboratories using properly-designed electrodes. These models should represent as faithfully as possible the situations encountered in the field, for example, a tower must be full-scale with tubular members and ground can be represented by wire mesh, and so on. Artificial rain apparatus must be mounted properly to wet insulators evenly. The electrode geometries shown in Fig. 11.1 must be capable of being tested under the types of voltages listed in Section 11.1, namely, power frequency, lightning, dc, and switching surges.

The most important factors by which engineers describe the breakdown voltage is the critical flashover voltage or 50% flashover voltage which is the voltage that yields flashover in 50% of the number of shots given to the air gap. The withstand voltage is based on probability basis and will be explained later on. The following empirical relations between the CFO (50% flashover voltage) and gap length are valid for design purposes for positive-polarity voltages, except for ac excitation. All voltages are peak values of voltage, including power frequency. Negative polarity flashover, except under rain under switching surges, will always be higher than positive polarity or become equal to it. Under exceptional circumstances, such as in heavy rain under switching surges, it may be lower than positive-polarity flashover.

Rod-Plane: *Gap Length in metres*, V_{50} *in kV crest*

Lightning	D.C.	Power Frequency	Switching Surge ($120/4000$ μs)
$1.2/50$ $\mu s - 550d$	$500d$	$652d^{0.576}$	$535.5d^{0.552}$
$1/50 \quad - 667d$			$(3 < d < 9)$
Rod-Rod			$500d^{0.6}$
$1.2/50$ $\mu s - 580d$	$555d$	$500d$	$872d^{0.429}$
			$687d^{0.6}$

For positive rod-negative plane electrodes, the following additional formulas are also in use.

Feser's Formulas

Static voltage:	$500d + 12$, kV (d in metres)	(11.1)
Power Frequency:	$455d + 25$	(11.2)
$1.2/50$ μs Lightning:	$540d$	(11.3)
Switching Surge:	$100\sqrt{50d + 1} - 250$	(11.4)

Leroy and Gallet Formula

Switching Surge	$3400/(1 + 8/d)$	(11.5)

Paris et al.   Switching surge—$500d^{0.6}$ or $535d^{0.552}$ (11.6)

All the above formulas are based on experimental results and are compared in Fig. 11.2. There are however, many attempts made from time to time to derive a theoretical expression for the sparkover voltage

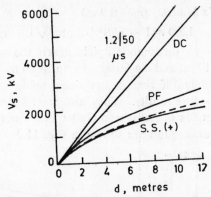

Fig. 11.2.   Comparison of flashover characteristics of long air gaps under lightning impulse (1.2/50 μs), dc, power frequency (PF), and positive switching surges.

in terms of the basic physical processes that might be occurring in long gaps. They are based on the following broad assumptions:

(1) Prior to breakdown, there exists an intense corona envelope on the highly-stressed electrode. This is particularly prominent on the positive rod in a rod-plane arrangement and on both rods in a rod-rod air gap. This spherical corona envelope is assumed to have a diameter ranging from 20 cm to 1 metre.

(2) The corona gives rise to leader channels of about 1–3 mm in diameter such that the radial electric field is 20 kV/cm with a charge of about 0.5 to 1 μc/cm. These leaders fork out in all directions, but one that eventually causes the sparkover propagates along the axis of the gap and is preceded by a leader-corona at its tip. The axial electric field in the leader column is about 5 kV/cm.

(3) When the leader corona has reached a length of 4.5 to 5 metres in a 7 m gap, or about 65–70% length, the influence of the plane cathode is so intense that there results a jump-phase where the acceleration of the leader corona is high enough to bridge the remaining portion of the gap. In a rod-rod gap, a negative-leader propagates from the negative rod also to meet the positive leader and sparkover occurs.

Without going through the derivation of the following equations which are based on differing assumptions, three expressions are given for the 50% sparkover voltage, V_s, for positive switching surges in terms of the gap length, d.

I. *Lemke's Model.*   ($d < 10$ m)

$$V_s = 450 \, [1 + 1.33 \ln (d - \ln d)], \, kV \qquad (11.7)$$

II. Waters' Model

$$V_s = (1.5 \times 10^6 + 3.2 \times 10^5 \, d)^{0.5} - 350, \, kV \qquad (11.8)$$

III. Alexandrov's Model. $(r = 0.9 \, m)$

$$V_s = 1260 \, r \, (1 - r/d)^{0.5} \cdot \tanh^{-1}\sqrt{1 - r/d}, \, kV \qquad (11.9)$$

All of the above three models yield nearly the same value of V_s for gap length up to 10 m, and the latter two up to 20 m. It is therefore very surprising how widely differing assumptions made by the three investigators can lead to close results. They also correspond very closely to the empirical formulas of Leroy and Gallet, and Paris et al. Results of comparison of these models are shown in Fig. 11.3.

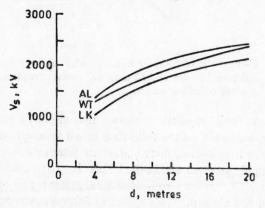

Fig. 11.3. Comparison of CFO vs gap length according to the models of Lemke (LK), Waters (WT), and Alexandrov (AL).

11.3 Breakdown Mechanisms of Short and Long Air Gaps

Literature describing the physical mechanism of sparkover of gaps is immense and varied. We will only give the broad outline here of several basic and important theories which have been put forward by eminent physicists and engineers to give the design engineer an indication of some of the physical processes in understandable terms. It is recognized by all that a comprehension of these basic mechanisms of sparkover will help to design experiments of breakdown involving less expense and time.

We limit ourselves to the mechanism of gaps near atmospheric pressure. The two classic theories of breakdown of small gaps with uniform fields are the well-known: (1) Townsend and Raether mechanisms called the avalanche theory, and (2) the streamer theory of Loeb and Meek. After discussing these two theories, we will briefly outline the several theories or models of breakdown of long gaps put forward by many workers to explain the dependence of the 50% breakdown voltage with change in gap length, which in one way or another are based on the avalanche and streamer theories of uniform field gap breakdown.

11.3.1 Townsend's Theory and Criterion of Sparkover

Fig. 11.4(a) shows a uniform-field gap between an anode and a cathode with $E = V/d$. An initial free electron in the gap acquires sufficient energy to release further electrons from the intervening gas molecules to cause an electron avalanche. It is evident that if the electric field is high the electrons cause increased ionization within limits while with increase in gas pressure p, the number of molecules or density of gas is increased so that the free path of an electron is decreased before another collision can occur. This results in an electron acquiring low velocity and kinetic energy between collisions, and the probability that it will release another electron from a gas molecule is reduced. The effect of electric field intensity E and the gas pressure p on the probability of creating new electrons by an incident electron on gas molecules is therefore proportional to E and $1/p$. This is expressed through Townsend's first ionization coefficient, α. Thus,

$$\alpha = \text{mean number of collisions taking place in 1 cm of drift by } 1 \text{ electron} \qquad (11.10)$$

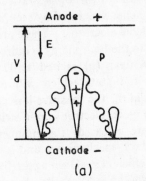
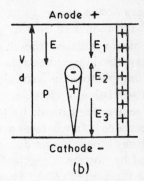

Fig. 11.4. (a) Theory of avalanche breakdown of uniform-field gap
(Townsend and Raether)
(b) Theory of streamer breakdown (Loeb and Meek).

The value of α is determined experimentally and has the following characteristics in Nitrogen, where $n =$ number density of molecules at the given temperature and pressure [$n = 2.69 \times 10^{19}$/cm^3 at 0°C and 760 mm].

E/n	100	200	300	400	500	600
α/n	10^{-19}	4×10^{-18}	1.5×10^{-17}	3×10^{-17}	4.5×10^{-17}	5×10^{-17}.

The best fit for this variation (carried out by the author) gives a formula of the type

$$\log_{10}(\alpha/n) = 15.82/[16.74/(E/n) - 1] \qquad (11.11)$$

The ion density can be converted to pressure of gas by the relation

$$n = 2.69 \times 10^{19}\,(p_T/760)\cdot(273/T) \qquad (11.12)$$

where p_T = pressure of gas in mm Hg at temperature $T°K$. Then, at 20°C ($T = 293\,°K$), $n = 3.295 \times 10^{16}\,p_{20}$. This gives

$$\alpha = 5.p_{20}.10^{\theta} \qquad (11.13)$$

where $\qquad \theta = 0.8725 \times 10^{19}/(E/p_{20} - 5.516 \times 10^{17}) \qquad (11.14)$

This shows that $\alpha = p.f_1\,(E/p)$ where $f_1\,(E/p)$ is a function of (E/p) only.

A single avalanche is attracted by the positive electrode because of the intense negative charge at its head, at a velocity of about 10^7 cm/sec, whereas the drift velocity of positive ions left behind and travelling towards the cathode is about 10^5 cm/sec since they are much heavier than electrons. Thus, the cloud of positive ions is almost stationary as compared to the drift of the electron avalanche. The positive ions, however, disturb the electric field due to the space charge.

The single electron avalanche reaching the anode does not constitute sparkover since the gap is not completely bridged by electrons to constitute a heavy enough current flow. The breakdown or sparkover of the gap occurs due to a succession of avalanches generated by secondary mechanisms caused by the primary-avalanche electrons. These are (a) thermionic emission from the cathode due to the high energy acquired by the positive ions or unstable neutral molecules; (b) excited molecules emitting photons on returning to the gronnd state which will cause photo-emission of electrons; (c) the potential of the positive-ion space charge near the cathode being sufficient to eject electrons upon striking the cathode. In whatever manner these are released, the number of electrons now increases due to the secondary mechanisms and is described by Townsend's second ionization coefficient γ. Once again (like α) γ is a function of (E/p). From available experimental results, the author has found that the functional relationship between γ and (E/p) is of the type

$$\gamma = 1.52 \times 10^{-6}\,(E/n)^{1.66} = 5.782 \times 10^{-34}\,(E/p)^{1.66} \qquad (11.15)$$

The current intensification due both the primary and secondary mechanisms is, when the avalanches cross the gap distance d,

$$\alpha\text{-process only:} \quad I = I_0\cdot e^{\alpha d} \qquad (11.16)$$
$$\gamma\text{-process also:} \quad I = I_0\cdot e^{\alpha d}/[1 - \gamma(e^{\alpha d} - 1)] \qquad (11.17)$$

The Townsend criterion for spark breakdown is taken as $I \to \infty$, or when the denominator of equation (11.17) becomes zero, that is,

$$\gamma(e^{\alpha d} - 1) = 1 \qquad (11.18)$$

Since the functional relationship of α and γ with gas pressure p and electric field intensity E is known, the value of (E/p) is calculated by trial and error from equation (11.18). In the uniform field, $E = V_s/d$ so that the value of (V_s/pd) is found, and the sparkover for given values of p and d can be calculated.

11.3.2 Paschen's Law

Using $\alpha = p \cdot f_1(V/pd)$ and $\gamma = f_2(V/pd)$ for a uniform-field gap, the Townsend breakdown criterion of equation (11.18) can be re-written as

$$f_2(V_s/pd) \cdot [\exp\{f_1(V_s/pd) \cdot pd\} - 1] = 1 \qquad (11.19)$$

Thus, the sparkover voltage V_s is a function of (pd), which is Paschen's Law. In air, the formula for sparkover voltage is found to be

$$V_s = 24.22 \cdot \frac{273 + 20}{273 + t} \cdot \frac{pd}{760} + 6.08 \sqrt{\frac{273 + 20}{273 + t} \cdot \frac{pd}{760}}, \text{kV}$$

$$= 9.34 \, pd/T + 3.775 \, pd/T, \text{kV} \qquad (11.20)$$

where p is in mm Hg, d in cm, and t in °C.

The above relation holds even for gases with electron attachment such as SF_6, whose relative electric strength compared to air is 2.5/1.

At standard temperature and pressure, equation (11.20) gives the uniform-field sparkover for 1 cm gap as 24.22 kV.

In non-uniform fields such as co-axial cylindrical geometry and cylinder-cylinder (conductor-conductor on a transmission line) gaps, F.W. Peek, Jr., has determined the following expressions for onset of partial discharge (corona) in the highly divergent field distribution for air-density factors close to unity.

Concentric Cylinders. $\quad E_0 = 31 \, \delta(1 + 0.308/\sqrt{r\delta})$, kV/cm, peak $\quad$ (11.21)

Parallel Cylinders. $\quad E_0 = 30 \, \delta(1 + 0.301/\sqrt{r\delta})$, kV/cm, peak $\quad$ (11.22)

Here, r = radius of inner cylinder or conductor in cm,

and $\quad \delta$ = air-density factor $= \dfrac{273 + 20}{273 + t} \cdot \dfrac{p}{760} = 0.3856 \, p/T$, $\quad$ (11.23)

with p = pressure in mm Hg.

The pressure varies with temperature and altitude as described in Chapter 4.

11.3.3 Streamer Breakdown Theory of Loeb and Meek

In the Townsend mechanism, the time lag of breakdown, which is the time interval between application of potential and the breakdown of the gap, is equal to the transit time of the electron avalanche. However, it has been observed with long air gaps in uniform fields, in practice, the time lags are shorter than those postulated by electron-transit times. This has led to the Streamer Theory of breakdown. It uses the property that a very high concentration of the electric field occurs in front of the electron-avalanche head, as well as in the space between the positive-charge cloud and the cathode. These two are directed in the same sense as the applied electric field as shown in Fig. 11.4(b). In the space between the electron-avalanche head and the positive space-charge cloud the field is lower than the applied uniform field. The photo-electrons emitted in the augmented field initiate auxiliary avalanches, which are intense along the axis of the main avalanche, will be directed towards the anode to be absorbed there. The positive ions then form a self-propagating streamer extending from the anode to cathode forming a conducting plasma. This constitutes the breakdown of the gap.

Loeb and Meek considered the transition from electron avalanche to streamer to occur when the field produced between the positive ions and the avalanche head is equal to the applied field $E = V/d$. The breakdown criterion obtained by them, with a spherical avalanche head, can be written as

$$V_s = 5.27 \times 10^{-7} \, (\alpha d) . e^{(\alpha d)}/\sqrt{d/p} \tag{11.24}$$

Since $\alpha = p.f_1(V_s/pd)$, the sparkover voltage is once again a function of (pd). The equation can be solved for V_s for given values of p and d by trial and error. The functional relationship between α and (V_s/pd) is of the type of equation (11.13).

11.4 Breakdown Models of Long Gaps with Non-Uniform Fields

As mentioned earlier, several models for explaining the observed characteristics of the 50% breakdown of long air gaps have been put forward from time to time. The main characteristics of gap sparkover which need to be explained by any of the models are listed below:

(1) The functional relationship between sparkover voltage and gap length for a given electrode geometry. Most theories confine themselves to a rod-to-plane gap, and try to evolve a factor called the "gap factor" to extend the equation to apply to other types of gap geometries found in practical engineering situations, usually obtained as a result of experimental observations.

(2) The volt-time characteristics which produce the time lag for breakdown after application of the voltage to the gap.

(3) The relation between switching-surge front time and the spark-over voltage called the U-characteristic.

(4) The average electric field in the gap at the instant of break-down.

(5) The charge in the leader column and its velocity of propagation, as well as the size of corona envelope around the highly-stressed electrode (rod).

(6) Some other characteristics which are governed by the convergent or divergent electric fields.

None of the models put forward so far in technical and scientific literature consider all the above factors in one model nor the complete mechanism of distribution of the charge, the temperature, the velocity, number density, column radius, leader-corona size, microscopic voltage gradient, etc. Some theories or models use experimental data to obtain numerical estimates for the salient governing factors, as will be outlined below. We will first list the common factors to all theories and then point out where they differ, and yet in the final result they all show agreement among each other within limits.

The mechanism of breakdown depends on a leader or streamer propagating from the rod anode towards the plane cathode in a rod-plane gap with the rod positive. The onset of this "positive leader" is preceded by an initial corona discharge, and also accompanied by corona at the leader tip. The leader carries a current of less than 1 Ampere and propagates at a velocity varying from 1 to 10 cm/μs. When the leader corona approaches the plane-cathode, secondary processes give rise to a streamer which meets the leader in most cases, that is, there is a "jump-phase" in which conditions are ripe for the latter portion of the leader to reach the anode at velocities exceeding 100 cm/μs. Once the gap is bridged, a return current stroke from the plane cathode follows in the ionized channel whose velocity can reach 300 cm/μs. The potential gradient in the return stroke is of the order of 1 kV/cm or more.

Pre-breakdown streamer pulses in the corona are followed by breakdown streamers. The charge in the leader channel varies from 0.5 μC/cm to 4 μC/cm depending on the resistance connected in series with the gap to the source, and the waveshape of the applied impulse voltage. The corona envelope at the rod gives rise to 50 or more leader branches out of which only one causes the streamer breakdown. The rest give rise to leader-tip coronas which cause photo-ionization of the gap and help secondary processes. The temperature of the leader has been known to exceed 5000 °K and the longitudinal electric field approaches 2–3 kV/cm. The corona envelope has been known to extend to diameters ranging from 20 cm to 2 metres depending upon the overvoltage. For small gaps, the space charge in the corona envelope lowers the electric field in the gap to such an extent which require large voltages for breakdown. This effect is less pronounced as the gap length increases so that breakdown voltages less than suggested by a linear increase are indicated. The entire phenomenon is therefore non-linear with the breakdown voltage showing a lower increase with gap spacing for switching surges ($V_s \propto d^{0.5-0.6}$). Since most of these conditions are governed by the waveshape of the applied voltage, several factors governing the sparkover voltage have to be measured under actual conditions in the laboratory from smaller gaps in order to predict the breakdown characteristics of very long gaps.

We will now consider a few important models.

(a) Lemke's Model

The simplest and a very straight forward model is due to Lemke, which is based on certain contradictory assumptions as pointed out by Waters (see Meek and Craggs, (ed), Electrical Breakdown of Gases, Chapter 5, 1978). It takes into account the properties of two components in the leader channel: (i) a leader length L_1, and (ii) a leader-corona discharge tip of length L_t with potential gradient E_t. Sparkover condition is assumed to reach when $L_1 + L_t = d$, the gap length. The sparkover voltage is

$$V_s = E_1 \cdot L_1 + E_t \cdot L_t \qquad (11.25)$$

It is the determination of the four controlling quantities that give rise to some problems.

The potential drop along the leader turns out to be

$$V_1 = E_1 \cdot L_1 = E_0 \cdot L_0 \cdot \ln (1 + L_1/L_0) \qquad (11.26)$$

where $E_0 = 1.5$ kV/cm chosen by Lemke which is felt low,

and $L_0 = 1$ m from photographs taken of the leader propagation in gaps up to 2 m.

Turning to the leader-corona tip, the potential gradient E_t in the streamers of the leader-corona is assumed to be 4.5 kV/cm based on observations made with 1 m gap. The length of the streamers, L_t, is also obtained from measurements and is

$$L_t = 1 + \ln d \qquad (11.27)$$

Finally, the critical sparkover voltage comes out to be

$$V_s = E_t[1 + (1 + E_0/E_t) \cdot \ln (d - \ln d)], \ d \text{ in cm,}$$
$$= 450[1 + 1.33 \ln (d - \ln d)], \text{ kV.} \qquad (11.28)$$

This formula yields values of V_s which agree with experimental results for switching-surge flashovers up to gap lengths of 10 m, but lower values for larger lengths. Therefore, gap clearances required beyond 10 m will be larger than required when using this formula and will give uneconomical designs. However, it is worthwhile mentioning at this stage that up to 1150 kV transmission, external insulation using air gaps on towers are not in excess of 9 metres and this simple model is adequate.

(b) Waters' and Jones' Mechanism

In this model, a critical length for the leader propagation is postulated which if attained, a large increase in current takes place and sparkover results. The calculation of the critical leader length is attempted. In a rod-plane gap, the average potential gradient in the leader channel is governed by the degree of ionization given by Saha's equation through the channel temperature which is caused by i^2R heating. Because of this heat and increased ionization, the average potential gradient in the leader channel is increased from 3–4 kV/cm to a value of about 10 kV/cm when the leader reaches 24 cm length in a 2-metre rod-plane gap. If the leader propagates beyond this length, due to still higher ionization (and possibly due to intense secondary mechanisms) the average voltage gradient falls to 1–2 kV/cm very quickly and sparkover occurs.

The model does not give a specific relation between sparkover voltage and gap length, but gives the mean potential gradient in the leader during the decreasing gradient process to be

$$E_1 = 1.25\,(i)^{-1/2} \cdot (1 + 5.25\,e^{-t/85}),\ kV/cm \tag{11.29}$$

where t is in μs. The current i in most cases is equal to 0.75 A but is related to the temperature.

(c) Alexandrov's Model

In this theory, three quantities are assumed to govern the breakdown: (1) the leader-tip potential, (2) electric field strength at the head of the developing discharge, and (3) the radius of curvature of the tip. The relation among them is found by Alexandrov to be

$$V_t = E_t \cdot r \cdot \sqrt{1 - r/d} \cdot \tanh^{-1} \sqrt{1 - r/d} \tag{11.30}$$

where V_t = leader-tip potential.

E_t = electric field gradient at the head of the developing discharge,

and r = radius of curvature of the tip = 0.9 metre.

The value of E_t was obtained experimentally, while r was calculated to be 0.91 m for a 2-m gap and 0.98 for a 30-m gap. For sparkover condition, $E_t = 12.6$ kV/cm = 1260 kV/m. Therefore, equation (11.30) for the sparkover voltage becomes

$$V_s = 1260\,r\ \sqrt{1 - r/d} \cdot \tanh^{-1} \sqrt{1 - r/d} \tag{11.31}$$

(d) Waters' Model

Another model proposed by Waters assumes that the current flow in the positive leader column, i, depends upon the tip potential V_t in a parabolic manner, as derived by Townsend. The unbridged gap between the leader tip and plane cathode is assumed to break down when the leader-tip potential is $(0.5\,V + 175)$ kV, where V = anode potential. From this, and other experimental observations that the leader length is 20 cm in a 2-m gap at 1000 kV, the sparkover voltage derived by Waters is

$$V_s = [1.5 \times 10^6 + 3.5 \times 10^5 d]^{1/2} - 350\ kV \tag{11.32}$$

Another expression derived by Waters from certain observations made of the basic processes by the team from Electricité de France is, with d in metres,

$$V_s = 563 \cdot d^{0.5},\ kV \tag{11.33}$$

With all these models based on differing assumptions, it must be clear that a unified model can be attempted by a research worker even today.

11.5 Positive Switching-Surge Flashover—Saturation Problem

In selecting the required air-gap insulation clearance for e.h.v. towers between the conductor and tower structure, its ability to withstand positive switching surges with conductor positive is of paramount importance. The shortest air gap occurs in the tower window and has been the subject of intensive investigation. The problem is complicated since the critical flashover voltage depends upon (a) the wave-front time of the switching surge, (b) the gap length, (c) the width of the tower structure, and (d) the presence of insulator and other hardware. Therefore, no two laboratories agree on the CFO voltage of a given gap length in a conductor-to-tower insulation structure. From a very large number of experimental results available, some important observations and properties are worth noting.

(1) The CFO voltage V_{50} of tower windows with insulator strings inside them varies with the front-time of the switching surge. Typical flashover voltages are shown in Fig. 11.5.

(2) The CFO voltage also varies with the percentage of space filled by porcelain of the insulator string which has a higher permittivity than air.

(3) The CFO voltage depends upon the ratio of the shortest distance in air between conductor and tower (the "strike distance") to the length of the porcelain or glass insulator string. A ratio between 0.85 and 1 is normally used in tower windows.

(4) The first switching-surge tests performed by J.W. Kalb, Jr., of the Ohio Brass Co., U.S.A., established the important property that the minimum value of CFO occurred at a certain front time. His experiments for 525 kV towers with 24 insulators in the string gave the minimum CFO at a wave-front time of 250 µs. The CFO

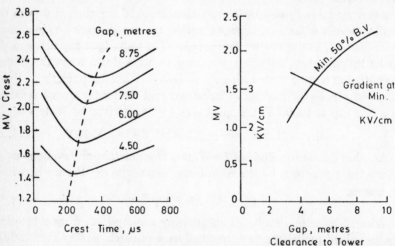

Fig. 11.5. (a) CFO of long air gaps between conductor and tower showing time to crest of switching surge against gap distance.

(b) Minimum breakdown voltage and gradient at minimum CFO. Variation with gap length.

was higher at shorter and longer wave-front times than this value.

(5) From results obtained in laboratories all over the world since Kalb's experiments in the early 1960's, it has been established that the minimum CFO occurs for switching-surge wave-front times varying from 55 to 300 μs, Fig. 11.5, depending upon the mock-up the investigators have used.

(6) The values quoted by different investigators for the wave-front time have not been based on a uniform definition of the wave-front. Some have used the time to actual crest value of voltage, while others have used the I.E.C. standard definition of $t_f = 1.67$ $(t_{90} - t_{30})$ where t_{90} and t_{30} are times from virtual zero to 90% and 30% crest values of voltage on the rising portion of the wave.

(7) Since Kalb's pioneering experiments, the standard waveshape for a switching surge has been adopted as 250/2500 μs.

(8) This does not necessarily imply that tower insulation structures tested under the standard waveshape will yield the worst switching-surge strength. Each air-gap clearance or gap length for a given electrode geometry must be tested individually with varying front times to ascertain the minimum CFO. This necessarily requires a switching-surge generator capable of delivering surges with varying front times. The problem of design of such impulse generators is discussed in detail in Chapter 13.

(9) The width of tower structure used in mock-ups in laboratory investigation have some effect on the CFO voltage, as also the length of conductor used on either side of the tower. This has been clearly pointed out by Dr. Alexandrov who recommends conductor lengths not less than 20 metres on either side. The conductor profile and size must be exact replicas of what will be used on actual lines. This is because the pre-breakdown mechanisms and therefore the breakdown characteristics of long gaps under positive switching surges are very sensitive to secondary mechanisms which are controlled by metallic parts in the environment of the air gap undergoing test.

(10) The CFO of long gaps under positive switching-surge voltage does not increase linearly with the gap length, as shown in Figs. 11.2 and 11.3. It bends towards the gap-length axis showing that a saturation might be occurring.

Possible Ceiling Voltage for AC Transmission

The most serious of the above listed problems is the non-linear breakdown characteristic with gap length and saturation. It shows that increase in gap length does not bring about a corresponding increase in insulation strength or CFO voltage. The group of engineers of E. de F. in France, while testing gaps up to 8 metres between rod and plane found that the CFO for positive switching surge can be described by the equation

$$V_{50} = 3400/(1 + 8/d), \text{ kV } (d \text{ in metres})\tag{11.34}$$

This was interpreted at the Les Renardiéres Laboratory to imply that as the gap length d increased to very large values, the highest voltage that can be supported is 3400 kV, crest. Beyond this, no air gap length however large ($d = \infty$) can sustain a switching surge in excess of 3400 kV. This might imply that there is a limit to the transmission voltage that can be used for ac lines as shown below. This ceiling voltage has been variously taken to be 1850 kV, r.m.s., line-to-line, to 2800 kV.

Example 11.1: If the crest value of s.s. supported by an air gap has a ceiling of 3400 kV, using a safety factor of 1.5 and switching overvoltage of 1.5 p.u., calculate the maximum permissible r.m.s. value of line-to-line voltage for ac transmission.

Solution: With a safety factor (withstand voltage $= 2/3 \times$ CFO voltage) of 1.5, the withstand voltage is $3400/1.5 = 2267$ kV crest. Since the p.u. value of s.s. is 1.5, the crest or peak value of line-to-neutral voltage is $2267/1.5 = 1510$ kV. The corresponding line-to-line r.m.s. value of voltage will be $1510 \times \sqrt{3/2} = 1850$ kV.

If the p.u. value of s.s. can be reduced or the safety factor decreased, the ceiling voltage can be increased. The relation between the 50% flashover voltage (CFO voltage) and the withstand voltage depends upon the statistical properties of breakdown which will be discussed in a later section. The flashover probability of a given gap length with voltage follows a nearly Gaussian or normal distribution. The standard deviation σ has to be determined for each gap length. Then the accepted rule up to date is that

$$\text{Withstand voltage} = (1 - 3\sigma) \times 50\% \text{ flashover voltage}\tag{11.35}$$

The existence of a ceiling voltage for ac transmission is now being hotly debated. Dr. Alexandrov working with very long gaps has come to the opinion that there is a slow but perceptible increase in CFO for very long air gaps and there is no such value as a ceiling voltage which an air gap can exhibit, i.e. there is no ultimate saturation. This can be compared to the breakdown under lightning wave shapes on the basis of what we observe in nature. If there be such a phenomenon as saturation, voltages of the order of 100 to 1000 MV observed in lightning might not be present in order to cause breakdown of air gaps from 500 to 5000 metres from cloud to ground. Alexandrov's equation (11.31) gives this slow rise.

Example 11.2: Using equation (11.31), $V_s = 1260 \, r\sqrt{1 - r/d} \cdot \tanh^{-1} \sqrt{1 - r/d}$, calculate V_s for $r = 0.9$ and $d = 25$, 50, and 100 metres.

Solution. Let $x = \tanh^{-1}\sqrt{1 - r/d}$. Then $\sqrt{1 - r/d} = \tanh x$.

$\therefore (e^x - e^{-x})/(e^x + e^{-x}) = \sqrt{1 - r/d}$,

giving
$$x = \tfrac{1}{2} \ln \frac{1 + \sqrt{1 - r/d}}{1 - \sqrt{1 - r/d}}.$$

(a) For $d = 25$, $x = \tfrac{1}{2} \ln (1.81835/0.18165) = 2.69731$

$\therefore \ V_s = 1260 \times 0.9 \times 0.81835 \times 2.69731 = 2503$ kV, crest.

(b) $d = 50: \sqrt{1 - r/d} = 0.990959$

$$x = \tfrac{1}{2} \ln \frac{1 + .990959}{1 - .990959} = 2.6773$$

$\therefore \ V_s = 1260 \times 0.9 \times 0.990959 \times 2.6773 = 3008.6$ kV, crest.

(c) $d = 100: \sqrt{1 - r/d} = 0.99549$.

$$x = \tfrac{1}{2} \ln \frac{1 + .99549}{1 - .99549} = 3.046155$$

$\therefore \ V_s = 1260 \times 0.9 \times 0.99549 \times 3.046155 = 3438.8$ kV, crest.

The equation $3400/(1 + 8/d)$ gives the following values:

(a) $d = 25$, $V_s = 2576$ kV;

(b) $d = 50$, $V_s = 2931$;

(c) $d = 100$, $V_s = 3148$ kV.

These observations are meant to give the reader the idea that the investigation of basic mechanisms leading to breakdown of very long gaps is still as necessary as when they were first attempted to explain the formation of charges in clouds, their propagation as leaders and streamers, and the mechanism of induced pre-discharge currents from transmission-line structures and ground plane. This must be investigated under different types of excitation voltages. Such investigations are being undertaken at some expense of money, time, and effort in e.h.v. laboratories and by engineers dealing with e.h.v. transmission in conjunction with physicists. Dr. K. Berger of the E.T.H. in Zurich, Switzerland, has pointed out in one of his discussions that his studies and that of his students have revealed the fact that there is quite a lot to understand about the basic characteristics of breakdown of a 60-cm gap and a 6-metre gap even though they show similar external characteristics between V_s and d.

11.6 CFO and Withstand Voltages of Long Air Gaps—Statistical Procedure

Insulation desigin of e.h.v. lines based on the use of long air gaps tends towards a statistical procedure instead of deterministic methods based on worst-case situation. This is primarily because of the large number of variables involved in the problem each of which has its own characteristic probability of occurrence either alone or in conjunction with other variables. For example, the shortest gap between the tower structure and conductor supported from a vertical or I-string of insulators occurs under maximum conductor swing in high winds and large oscillatory conditions. This distance must withstand the highest switching-surge voltage normally encountered in the system under the worst-case design. However, the above is true only if the probability of maximum conductor swing coincides with the probability that the maximum switching surge will also occur at the same time. Such a probability of both events occurring simultaneously is very remote and a design bsaed on worst case gives uneconomically large air-gap lengths necessitating extremely heavy towers. In actual practice, the U.S.S.R. 1150 kV line does not consider both events to occur simultaneously during the life of the line, and the gap clearance is designed for withstanding the maximum switching surge with the insulator vertical in an I-string or the conductor occupying its normal rest position for a double 90°-V string.

In adopting the probabilistic philosophy of design, it is also evident that even though the probability of a flashover under worst condition is neglected, there exists always the danger that such a condition could occur. Therefore, a flashover once in so many switching operations should be allowed. This depends entirely on the experience of the designer acquired from existing lines, if data of such information have been properly logged. The most usual case is to allow 1 flashover in 1000 switching operations and it is the determination of this probability that has formed the entire basis for design of air gaps insulation for switching surge and lighting-impulse voltages based on statistical considerations. Some designers use a 0.2% probability of flashover (1 in 500 operations). It has however been recognized that a flashover under these two types of impulse voltages is not catastrophic on the system in that it is not any more serious than initiation of a single line to ground fault or a phase-to-phase fault. Surge absorbers are also improved a lot to handle the severity imposed by switching-surge duty.

From what has been described before, it is evident that many probabilities have to be determined, chief among which are the following:

(a) Magnitude of switching surge experienced in the system during all possible types of switching operations and system conditions during switching.

(b) Environmental conditions such as rate of rainfall, humidity (relative and absolute).

(c) Wind conditions which give rise to aeolian vibrations, wake-induced oscillations, etc., that determine the swing and clearances under service conditions.

(d) Snow, ice, and such other conditions which will affect the insulation strength.

Although the task of correlating all these factors in order to evolve an economical or optimal design is a formidable task and may sound impossible, in practice, digital computer programmes such as the METIFOR, attempt to include the simultaneity of the probability of occurrence of the above factors in helping to evolve a suitable design for the insulation structure of towers and conductors. Needless to say, costs can be brought down considerably when all factors are determined to a high degree of confidence level.

As one example, we will quote one instance of a 735–kV line from Churchill Falls in Labrador (New Foundland) in Canada as designed by the Bechtel Corporation. With an air gap in the window of 15.4 feet (4.7 m) and 27 insulators, the cost per kilometre of line was $ 61,000 (1970 price quoted by Mr. Price), while it could be brought down to $ 48,000/km by using 13.2 feet (4.024 m) in the window, that is a reduction of 2.2 feet or 14.3%. The CFO voltages in the two cases are 1307 kV crest and 1062 kV crest. The phase spacing could be reduced from 50 feet to 45 feet (15.44 m to 13.72 m) with a grand saving of $ 5 million out of $ 65 million for transmitting 5000 MW which costs $ 13/kW in transmission line towers and foundations.

The data necessary for line insulation design as it is practiced for the present are therefore the following.

(a) The flashover voltage of an insulator string on a tower expressed through the 50% flashover level, V_{50}. This is the relation between probability of flashover and the surge amplitude.

(b) The relation between flashover voltage and waveshape, the weather or atmospheric conditions.

(c) The standard deviation, σ_f, for flashover when using an assumed statistical variation for the flashover voltage.

(d) The statistical distribution of amplitudes and waveshapes of the surges occurring on the system during all possible switching conditions which can be defined through a mean value μ and standard deviation σ_s.

(e) The number, n, of insulator strings stressed simultaneously on the line.

(f) The statistical relation between atmospheric conditions and the probability of occurrence of a certain level of the switching surges.

Items (a) to (c) are evaluated in outdoor and indoor e.h.v. and u.h.v. laboratories, while item (d) on evaluating the switching-surge magnitudes on electrical networks is carried out from model studies on a Transient Network Analyzer (TNA) or by field tests on existing lines or calculation on Digital Computers. Very extensive literature exists on case-by-case as well as general studies on this topic and the reader is referred to the bibliography on this important topic.

Even though much stress has been laid on the V_{50} or CFO voltage, the designer is really interested in the withstand voltage of an insulation structure at the design stages. This value has been pointed out to be the voltage at which 1 flashover in 500 or 1000 switching operations result, and cannot be determined from experiments with available time in a laboratory. Thus, the low probability region of flashover, 0.1% or 0.2%, must be obtained from flashover probabilities of higher values.

Using a procedure paralleling insulation breakdown values of solid insulation structures used in transformers, the time required to evaluate the 0.1% or low probability of flashover is obtained by assuming the most popular and widely-used Normal or Gaussian distribution of the type

$$p(V) = \frac{1}{\sqrt{2\pi}} \int_{-\infty}^{V} (1/\sigma_f) \cdot \exp \{- (x - V_{50})^2/2\sigma_f^2 \} \cdot dx \qquad (11.36)$$

where $p(V)$ = probability of flashover at voltage V,

V_{50} = 50% flashover voltage, or the mean,

and σ_f = standard deviation as percentage of V_{50}.

The type of relation in equation (11.36) is used chiefly for obtaining the standard deviation σ_f from a set of experimental results obtained in a laboratory. The probability of flashover $p(V)$ at a certain voltage level as well as the V_{50}-level are determined, as shown in Example 11.3 below.

Example 11.3: A flashover test on an insulation structure of a tower gave the following results.

No. of shots	24	25	24	24	25	20	20
Voltage kV	1600	1620	1650	1690	1730	1770	1800
No. of flashovers	4	5	8	12	15	16	18
% flashover	16.67	20	33.3	50	60	80	90

(a) Plot the % flashover against voltage on linear and probability graph papers.

(b) Give the value of 50% flashover, V_{50}.

(c) Assuming the standard deviation σ_f to be the difference between V_{50} and $V_{16 \cdot 7}$, and V_{50} and $V_{83 \cdot 3}$, calculate the average standard deviation in kV.

(d) Determine σ_f/V_{50} in percentage.

Solution:

(a) Fig. 11.6 shows the two graphs.

(b) V_{50} = 1690 kV, crest, from the table of data given.

(c) $V_{50} - V_{16 \cdot 7}$ = 1690 − 1600 = 90 kV

$V_{83 \cdot 3} - V_{50}$ = 1780 − 1690 = 90 kV.

$\therefore$ σ_f = 90 kV.

(d) % σ_f/V_{50} = 90 × 100/1690 = 5.3%.

The following properties should be noted regarding the relation between flashover probability, 50% flashover voltage, and standard deviation when the distribution is Gaussian.

(1) As we deviate from V_{50} by one standard deviation σ_f, either below

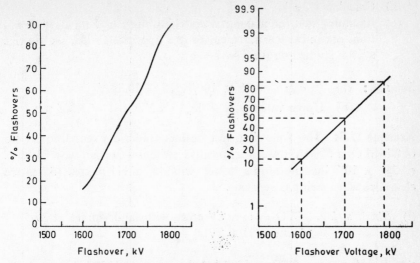

Fig. 11.6.   Plot of number or % flashovers with voltage on (a) linear graph,
(b) probability paper, to illustrate median (50%) flashover voltage,
and standard deviation.   Example 11.3.

or above it, the probability of flashover changes by 33.3%, and nearly 67% of all flashovers lie within one standard deviation.

(2) It can pe shown that at $2\sigma_f$ from V_{50}, the flashover probabilities will be 2% at $(V_{50} - 2\sigma_f)$ and 98% at $(V_{50} + 2\sigma_f)$. Since we are interested in the low-probability region of flashover, we only consider voltages less than the mean value V_{50}.

(3) At 3 standard deviations from V_{50}, the flashover probabilities will be 0.1% at $(V_{50} - 3\sigma_f)$ and 99.9% at $(V_{50} + 3\sigma_f)$.

(4) We now have established the withstand voltage if this is taken to give 0.1% probability of flashover. For the present state of the art of design followed in most countries, this is the accepted value for withstand voltage.

(5) However, some designers use $(V_{50} - 4\sigma_f)$ as the withstand voltage which gives a probability of flashover of 0.0033% or 1 flashover in 30,000 switching operations.

We can now appreciate the full significance of controlled experiments in the laboratory which should yield accurate values for the V_{50} value and the standard deviation σ_f. In a design, the withstand voltage must equal or exceed the highest crest value of anticipated switching surge in the system. This also must be ascertained on a probability basis.

Example 11.4: The results of a large number of switching operations performed on a system using models on a TNA gave the following results when pre-insertion resistors were not used in the circuit breaker.

Mean value of switching surge $\mu = 2.325$ p.u.

Standard deviation in s.s. overvoltage $\sigma_0 = 0.25$ p u.

(a) Calculate % σ_0/μ.

(b) Assuming that the maximum switching surge is 4 standard deviations above the mean, calculate the upper value for s.s. on this system studied on the TNA.

Solution: (a)   % σ_0/μ = 0.25 × 100/2.325 = 10.8%.

(b)   Upper value = $\mu + 4\sigma_0$ = 2.325 + 1 = 3.325 p.u.

Example 11.5: The following test results under dry conditions were obtained on a 500-kV tower for positive switching surges with 23 units of $5\frac{3}{4}'' \times 10''$ insulators in a tower window of 11 metres (5.5 metre clearance when string is vertical).

(i) *Single string.*   CFO = 1340 kV crest, withstand voltage at $3\sigma_f$
                          down = 1100 kV.

(ii) *Double string.*   CFO = 1390 kV, $3\sigma_f$ down = 1250 kV.
For the two cases, calculate
(a) % σ_f/CFO; (b) The maximum allowable p.u. switching surge based on operating voltage of 525 kV.

Solution: 1 p.u. switching surge = $525\sqrt{2/3}$ = 428·66 kV, crest, line-to-ground.

(a) *For single string.* σ_f = (1340 − 1100)/3 = 80 kV.

∴   % σ_f/V_{50} = 80 × 100/1340 = 5.95%.
Fur double string σ_f = (1390 − 1250)/3 = 47 kV.

∴   % σ_f/V_{50} = 47 × 100/1390 = 3.4%.
These are the percentage standard deviations.

(b) Switching surge per unit values:
Single string   1100/428.66 = 2.566 allowable
Double string   1250/428.66 = 2.916 allowable.

Example 11.6: In a test setup for a 1100-kV tower, the CFO was 1640 kV on an insulator of 5 m length inside the window. The standard deviation obtained from a graph on normal probability paper of all flashover data was σ_f = 6.2%. Calculate the maximum allowable p.u. value of switching surge on this insulation structure.

Solution: Withstand voltage will be taken as $3\sigma_f$ down from CFO, which will be V_W = 1640 (1 − 3 × 0.062) = 1330 kV. Peak value of line-to-ground voltage, 1 p.u. = $1100\sqrt{2/3}$ = 898 kV.
∴   Allowable p.u. value of s.s. = 1330/898 = 1.481 p.u.
From the above examples, it should be evident that

(a) as the system voltage increases (from 500 kV to 1100 kV) the p.u. value of maximum switching surge is decreased (from 2.5 — 3 to 1.5).

(b) Pre-insertion resistors are necessary in order to lower the s.s. amplitude in practice. Also, shunt-compensating reactors are used as the system voltage is increased.

(c) The standard deviation under dry conditions for air gaps in towers with insulators is in the neighbourhood of 5% CFO. With this value, the withstand voltage is 85% of CFO voltage. The value of $\sigma_f = 5\%$ has been recommended by the I.E.E.E. even though actual test results may show variations.

Further Properties of CFO and σ_f

In Section 11.5, mention was made of the properties of the Critical Flashover voltage of insulation structures encountered in e.h.v. transmission lines. We will consider some further properties here, chiefly the variation of CFO with waveshape and atmospheric conditions. It must be emphasized beforehand that very little data is available to date regarding the exact waveshape of switching surge obtained on actual systems during switching operations. It is also evident that no two systems will encounter the same waveshape because of reflections in the connected equipment of the system and the system layout. Therefore, a waveshape of 250/2500 μs has been standardized for testing purposes. But this does not yield the critical or lowest value of CFO for all gap lengths as shown in Fig. 11.5. With increasing gap length, the wavefront time at which minimum flashover occurs also increases. For example, for a 3-metre gap it is 150 μs whereas for 15 metres and longer the minimum CFO occurs at a wavefront times 500 μs and longer. In e.h.v. transmission, gap lengths range between 3 m and 10 m for transmission voltages from 400 kV to 1200 kV.

An additional property is that the standard deviation is also not constant. It is also a function of time to crest of the applied impulse voltage. However, the minimum value of σ_f generally occurs for waveshapes of switching surges having the critical wavefront.

Atmospheric conditions play a large part in influencing the CFO of air gps, as can be expected. Standard atmospheric conditions are defined according to ANSI and I.E.C. specifications as follows:

	Pressure	Temperature	Absolute Humidity	Vapour Pressure
ANSI	760 mm	25°C	15 grams/m³	15.5 mm Hg
I.E.C.	760 mm	20°C	11 grams/m³	11.37 mm Hg

For a particular set of weather conditions, the CFO is given by the equation

$$V_{50} = K_{rad} \cdot K_h \cdot K_p \cdot \text{(CFO under standard conditions)}. \qquad (11.37)$$

The three correction factors are as follows:

K_{rad} = correction for relation air density = $(RAD)^{0.5-0.7}$,

K_h = correction for humidity $\approx 0.82 + 0.656\,v_p$,

v_p = vapour pressure in mm Hg,

and K_p = correction for precipitation = 1 to 0.96 depending upon intensity of precipitation.

Equation (11.37) gives a quantity defined as "relative insulation strength," defined as

RIS = $(V_{50}/V_{50}$ under standard conditions) = $K_{rad} \cdot K_h \cdot K_p$. (11.38)

It has also been observed that for long wavefront times, breakdown occurs prior to the crest time and for shorter than critical front, breakdown occurs after the crest value. The time duration during which the surge remains higher than 90% crest value has also been found important for breakdown.

11.7 CFO Voltage of Long Air Gaps—Paris' Theory

In Section 11.3 all available formulae relating to the breakdown of air gaps were summarized under the usual four types of voltage waveshapes encountered in e.h.v. transmission and dc. Section 11.4 dealt with several mechanisms of breakdown and formulae were given for air-gap flashover voltages based on physical models of the processes. In most cases, the breakdown for positive-polarity switching surges of the highly-stressed electrode region govern the insulation clearance, d, required. The CFO voltage varies approximately as $d^{0.5-0.6}$. This relation was found from a very large amount of experimental data gathered all over the world, particularly by Paris and his co-workers in Italy. The formulae are known as Paris' formulae after the principal investigator. Because of this work, it may generally be said that the design of insulation of e.h.v. systems using long air gaps has been placed on a scientific footing, but requires investigations of a more accurate nature on a case-by-case basis. The design of insulation clearances will be taken up in Chapter 14.

Basically, the air-gap clearance required from a conductor to tower bears a relation to the clearance required for a rod-plane gap of the same length. The rod-plane geometry has the lowest CFO under positive polarity switching impulse of any gap geometry encountered in practice. For the rod-plane gap, Paris' formula for V_{50} is, with d in metres,

$$V_{50(r-p)} = 500 \cdot d^{0.6}, \text{ kV, crest} \tag{11.39}$$

The flashover voltages of other gaps also vary as $d^{0.6}$, but are different from the rod-plane gap V_{50} by a factor called the "gap factor" and denoted by kg, which is greater than unity. Therefore, the general CFO voltage is given by

$$V_{50} = 500 \cdot \text{kg} \cdot d^{0.6}, \text{ kV}. \tag{11.40}$$

The following table gives values of kg generally used.

Values of Gap Factors, kg

kg	Electrode Configuration	kg	Electrode Configuration
1.00	Rod-Plane	1.55	Conductor-Cross Arm
1.05	Rod-Tower Structure	1.30	Rod-Rod
1.15	Conductor-Plane		
1.20	Conductor-Tower Window		

In general, the above values of gap factors kg are valid when insulator strings are absent, and nearly valid when insulator strings exist in the tower window. They are not valid for negative-polarity switching surges. Some experimental work carried out at the EHV-UHV Project of E.P.R.I. by the General Electric Co., U.S.A., have indicated that the ratio (insulator height/shortest air-gap clearance to tower) (H/d ratio) has a bearing on the value of the gap factor. Also it is affected by the insulator string, being different for I-strings and V-strings. The exponent of d in the presence of insulators is also higher than 0.6.

Application of Paris' Formula for Design

The chief merit of Paris' formulae lies in its having been adopted in many countries for design of air-gap clearances, as will be discussed below. The equation $V_{50} = 3400/(1+8/d)$ kV obtained by the Les Renardières Group of Electricité de France is also used for rod-plane gaps under positive switching surges for gap lengths up to 10 metres and above 2 metres.

The steps used in design of air-gap clearances making use of gap factors are as follows:

(1) From network studies using TNA or Digital Computer ascertain the expected maximum switching-surge magnitudes, and decide on the safety factors. For example, these were found to be between 2.5 p.u. to 1.5 p.u. for transmission voltages from 400 kV to 1150 kV. [1 p.u. = crest line-ground power-frequency voltage at maximum operating voltage. 362 kV for 345 kV system to 1200 kV for 1150 kV system]. Because of Ferranti effect 1 p.u. voltage is also based on the maximum expected power-frequency voltage where the equipment is connected in the system.

(2) Assuming $\sigma_f = 5\%$ and allowing a further 5% to cover for differences between laboratory measurements and field conditions, the withstand voltage is

$$V_w = (V_{50} - 3\sigma_f)/1.05 = V_{50} \times 0.85/1.05 = 0.8095 V_{50}. \qquad (11.41)$$

The factor $3\sigma_f$ is suggested as adequate for most design purposes but some designers use $4\sigma_f$ below V_{50} as the withstand voltage.

(3) The value of withstand voltage V_w is known since it is equal to the expected value of switching-surge voltage V_{ss}.

(4) The CFO is then, with $\sigma_f = 0.05\ V_{50}$,

$$V_{50} = 1.05\ V_{ss} + 3\sigma_f = V_{ss}/0.8095 = 1.235V_{ss}. \tag{11.42}$$

(5) This must equal $500.\text{kg}.d^{0.6}$ $\qquad\qquad$ (11.43)

$\therefore\quad d = (1.235\ V_{ss}/500\ \text{kg})^{1/0.6}$ $\qquad\qquad$ (11.44)

Although the I.E.E.E. suggests $\sigma_f = 5\%$ of CFO, the I.E.C. recommends 6% which is used in European designs.

(6) Once the line-to-tower clearance is fixed from (11.44), the line-to-line clearance is taken to be $1.7\ d$.

(7) The line-to-ground clearance from switching-surge consideration is also fixed as $h = 4.3 + 1.4\ d$ metres.

Example 11.7: A 400 kV line (420 kV maximum) has a 2.5 p.u. switching surge when resistor switching is used and trapped charge is neglected. Design the clearance between (a) conductor to tower, (b) conductor to conductor, and (c) the conductor to ground. Assume a factor of safety of 1.2.

Solution: 1 p.u. crest voltage $= 420\sqrt{2/3} = 343$ kV.

$\therefore$ S.S. crest voltage $V_{ss} = 1.2 \times 2.5 \times 343 = 1029$ kV.

The value of $kg = 1.2$ from the table of values of kg for conductor-window.

(a) $\therefore$ $d = (1.235 \times 1029/1.2 \times 500)^{1.667} = (2.118)^{1.667} = 3.493$ metres.

(b) Then, line-to-line clearance $= 1.7\ d = 5.938$ metres.

(c) The conductor-to-ground clearance is $1.4\ d + 4.3 = 9.19$ metres.

[The minimum clearance to ground used for 400 kV lines in India is 8.9 metres and conductor-to-tower is $84'' = 2.134$ metres].

Example 11.8: Using a conductor-to-tower clearance to be $84''$ (2.134 m) as per NESC recommendations, determine the p.u. value of allowable switching surge level for the 400-kV line in Example 11.7. Neglect the safety factor.

Solution: $d = 2.134.$ $\therefore$ $V_{50} = 1.2 \times 500 \times 2.134^{0.6} = 945.5$ kV.

With $\quad \sigma_f = 5\%$ of V_{50} and $V_W = V_{50} - 3\sigma_f = 0.85\ V_{50}$,

$\qquad V_{ss} = V_W = 0.85 \times 945.5 = 803.7$ kV.

$\therefore$ p.u. switching surge allowed $= 803.7/343 = \mathbf{2.343}$.

Example 11.9: For a 735-kV line (maximum operating voltage 750 kV) the anticipated switching surge is 2.1 p.u. without resistors in the breakers but only shunt reactors draining the trapped charge during reclosing. Design the line-to-tower and line-to-line clearance on the assumption of (a) no safety factor, and (b) a safety factor of 1.1 on the anticipated switching surge $(2.1 \times 1.1 = 2.31$ p.u.). Take $\sigma_f = 6\%$ of CFO.

Solution. Crest line-to-ground voltage $= 750\sqrt{2/3} = 612.5$ kV.

(a) $\therefore$ $V_{ss} = 2.1 \times 612.5 = 1286$ kV, crest

$\qquad V_W = (1 - 3 \times 0.06)\,V_{50} = 0.82\,V_{50}.$

$\therefore$ $V_{50} = V_W/0.82 = 1286/0.82 = 1568.3 = 600 \cdot d^{0.6}.$

and $\quad d = (1568.3/600)^{1.667} = 4.96$ metres.

$\therefore$ Line-to-line clearance $= 1.7 \times 4.96 = 8.432$ metres.

(b) $V_{ss} = 1286 \times 1.1 = 1415$ kV, crest

$\qquad V_{50} = 1415/0.82 = 1725.1$ kV $= 600 \cdot d^{0.6}$

$\therefore$ $d = (1725.1/600)^{1.667} = 5.814$ metres.

Line-to-line clearance $= 1.7\,d = 9.883$ metres.

Review Questions and Problems

1. The 50% flashover voltage under positive switching surge are given by the three formulae (1) Leroy and Gallet, (2) Paris, and (3) Feser. They are, respectively,
   $3400/(1 + 8/d)$, $500\,d^{0.6}$, and $100(\sqrt{50d + 1} - 2.5)$ kV,
   with d in metres. On a 750 kV line, the switching surge expected is 2.1 p.u. which the air gap has to withstand. Take $V_{50} = 1.15\,V_w$. Calculate the gap lengths required according to the three formulae.
2. Describe the mechanisms of breakdown of a long air gap by (a) Lemke's model, (b) Waters' model, and (c) Alexandrov's model.
3. Explain the differences between Townsend's avalanche mechanism and the streamer theory of breakdown of short gaps in uniform fields.
4. Comparing the two formulae $3400/(1 + 8/d)$ and $500d^{0.6}$ for a rod-plane gap breakdown voltage, discuss why one team of investigators claim that there is a ceiling voltage for ac transmission while the other does not indicate such a limit.
5. The withstand voltage of a gap is 2000 kV with a standard deviation of $\sigma = 6\%$. Calculate the 50% flashover voltage and 83.3% flashover value if withstand voltage is taken to be 3σ lower than the 50% flashover value.

CHAPTER 12

Power-Frequency Voltage Control and Overvoltages

12.1 Problems at Power Frequency

Power-frequency voltage is impressed on a system continuously as compared to transients caused by faults, lightning, and switching operations. Certain abnormal conditions arise when overvoltages of a sustained nature can exist in systems which have to be guarded against. Insulation levels will be governed by these, and it is very important to know all the factors which contribute to such overvoltages. E.H.V. lines are longer than and their surge impedance lower than lines at 345 kV and lower voltages. Also, e.h.v. lines are used more for point-to-point transmission so that when load is dropped, a large portion of the system is unloaded and voltage rise could be more severe than when there is a vast interconnected network. Due also to the high capacitance of e.h.v. lines, possibility of self-excitation of generators is quite serious. Shunt reactors are employed to compensate the high charging current, which not only prevent overvoltages during load dropping but also improve conditions for load flow, and the risk of self-excitation can also be counteracted. In order to improve conditions, variable static VAR systems can also be employed as well as switched capacitors which introduce harmonics into the system. Finally, the use of series capacitors to increase line loading in long lines might bring about the danger of subsynchronous resonance in which electrical conditions in generators can produce torques which correspond to the torsional frequencies of the shaft and result in mechanical damage.

The system at power frequency consists of lumped-parameter network elements connected to distributed-parameter transmission lines, and the calculations are best handled through generalized constants in matrix form.

12.2 Generalized Constants

We have already derived equations for voltage and current at any point on a distributed-parameter line in Chapter 10 in terms of the voltage at the entrance to the line for any general line termination impedance, Z_t. They are, for a line of length L,

$$E(x) = E_e \frac{\cosh px + (Z_0/Z_t) \sinh px}{\cosh pL + (Z_0/Z_t) \sinh pL} \tag{12.1}$$

$$I(x) = \frac{E_e}{Z_0} \cdot \frac{\sinh px + (Z_0/Z_t) \cosh px}{\cosh pL + (Z_0/Z_t) \sinh pL} \tag{12.2}$$

In the steady state, the propagation constant and surge impedance are

$$p = \sqrt{(r + jWl)(g + jWc)} \text{ and } Z_0 = \sqrt{(r + jWl)/(g + jWc)} \tag{12.3}$$

Also, at the load end, $x = 0$, $E_0 = Z_t I_0$ so that

$$E_0 = E_e/[\cosh pL + (Z_0/Z_t) \sinh pL] \tag{12.4}$$

This can be written for the entrance voltage as

$$E_e = (\cosh pL) E_0 + (Z_0 \cdot \sinh pL) \cdot (E_0/Z_t)$$
$$= E_0 \cdot \cosh pL + I_0 \cdot Z_0 \sinh pL \tag{12.5}$$

Similarly, from equation (12.2), at $x = 0$ and at $x = L$,

$$I_0 = \frac{E_e}{Z_0} \cdot \frac{Z_0}{Z_t} \bigg/ [\cosh pL + (Z_0/Z_t) \sinh pL] \tag{12.6}$$

and

$$I_e = \frac{E_e}{Z_0} [\sinh pL + (Z_0/Z_t) \cosh pL]/[\cosh pL$$
$$+ (Z_0/Z_t) \sinh pL] \tag{12.7}$$

$$= \left(\frac{1}{Z_0} \sinh pL\right) E_e/[\cosh pL + (Z_0/Z_t) \sinh pL]$$
$$+ (\cosh pL) \cdot (E_e/Z_t)/[\cosh pL + (Z_0/Z_t) \sinh pL]$$

$$= \left(\frac{1}{Z_0} \sinh pL\right) E_0 + (\cosh pL) I_0 \tag{12.8}$$

Equations (12.5) and (12.8) give expressions for the voltage and current at the entrance to the line in terms of the voltage and current at the output or load end.

These can be re-written as

$$\left.\begin{array}{l} E_e = A \cdot E_0 + B \cdot I_0 \\[2mm] I_e = C \cdot E_0 + D \cdot I_0 \end{array}\right\} \tag{12.9}$$

and

For steady-state conditions, we will designate the line entrance as the source end and the load end as the receiving end. The subscripts "s" and "r" will be used to indicate these. Thus,

$$E_s = AE_r + BI_r \text{ and } I_s = CE_r + DI_r \tag{12.10}$$

We observe that

$$A = D = \cosh pL = \cosh \sqrt{ZY} \tag{12.11}$$

$$B = Z_0 \cdot \sinh pL = \sqrt{\frac{Z}{Y}} \cdot \sinh \sqrt{ZY} \tag{12.12}$$

$$C = \frac{1}{Z_0} \cdot \sinh pL = \sqrt{Y/Z} \cdot \sinh \sqrt{ZY} \tag{12.13}$$

$$\left.\begin{array}{l} \text{where } Z = (r + jWl)L = \text{total series impedance of line} \\[2mm] \text{and } Y = (g + jWc)L = \text{total shunt admittance of line} \end{array}\right\} \tag{12.14}$$

In dealing with steady-state voltage and currents on an overhead line, we can take $g = 0$. For underground cables, this is not valid.

From (12.11) to (12.13) we note that

$$AD - BC = \cosh^2 pL - \sinh^2 pL = 1 \qquad (12.15)$$

The propagation constant p can be written in various forms.

$$p = \sqrt{(r + jWl)jWc} = \alpha + j\beta, \qquad (12.16)$$

where

$$\alpha^2 = \frac{W^2lc}{2} \left[\sqrt{1 + 4r^2/W^2l^2} - 1\right]$$

and

$$\beta^2 = \frac{W^2lc}{2} \left[\sqrt{1 + 4r^2/W^2l^2} + 1\right] \qquad (12.17)$$

For the case where the resistance is low enough such that

$$2r \ll Wl, \ \sqrt{1 + (2r/Wl)^2} \approx 1 + 2r^2/W^2l^2,$$

we have

$$\alpha^2 = r^2c/l \text{ and } \beta^2 = W^2lc(1 + r^2/W^2l^2) \qquad (12.18)$$

The real part α of p is the attenuation constant and is approximately

$$\alpha = r \sqrt{c/l} = r/Z_{00} \qquad (12.19)$$

where $\qquad Z_{00} =$ surge impedance of line when $r = 0$.

Also, if we define $v_0 =$ velocity of propagation of e.m. wave on a resistance-less line, $v_0 = 1/\sqrt{lc}$ and we have

$$Z_{00} = v_0 l = l/v_0 c \text{ and } \alpha = r/v_0 l = v_0 rc \qquad (12.20)$$

The j-part β of p is the phase-shift constant and it is approximately

$$\beta = \frac{W}{v_0} \sqrt{1 + r^2/W^2l^2} \approx \frac{W}{v_0} (1 + \tfrac{1}{2}r^2/W^2l^2) \qquad (12.21)$$

The wavelength is related to the frequency by $f\lambda = v_0$.

$$\therefore \ \beta = \frac{2\pi}{\lambda}(1 + r^2/2W^2l^2) \qquad (12.22)$$

For a line without losses,

$$\alpha = 0 \text{ and } \beta = 2\pi/\lambda \qquad (12.23)$$

The wavelength λ at 50 Hz is 6000 km based on light velocity of 300,000 km/sec. Therefore, for a line of 100 km, $2\pi/\lambda = 6°$. This usually indicates that on an uncompensated line at no load, and when line resistance

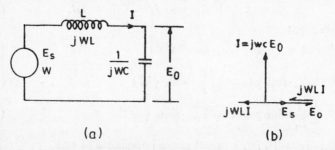

(a) (b)

Fig. 12.1. Ferranti Effect in lumped L-C circuit excited by sine wave of voltage, and phasor diagram.

is negligible, the phase difference between the source voltage and receiving-end voltage is 6° for every 100 km of line length at 50 Hz.

12.3. No-Load Voltage Conditions and Charging Current

When there is no-load at the receiving end, $I_r = 0$, the control of voltage at line ends poses certain problems due to overvoltage conditions. Unlike dc lines, ac lines have a charging current flowing through the series inductance of line and can cause a rise in the output voltage at the receiving end. This is the well-known "Ferranti Effect." Consider a simple series L-C circuit with lumped inductance and capacitance, as shown in Fig. 12.1(a) and the phasor diagram of voltages and currents in 12.1(b). By proper voltage division, the source voltage will be

$$E_s = (1/jWc + jWL) E_0/(1/jWc) = (1 - W^2LC)E_0 \quad (12.24)$$

This shows that E_s is less than E_0 and in phase with it. The voltage drop in the inductive reactance due to the charging current of the capacitor, $I = jWC\, E_0$, is subtracting from the load voltage to give the source voltage.

In a distributed-parameter line, neglecting series resistance for understanding the phenomenon, equation (12.5) gives at no load,

$$E_s = (\cosh jWL/v_0) E_0 = E_0 \cdot \cos (WL/v_0) = E_0 \cdot \cos (2\pi L/\lambda) \quad (12.25)$$

Since $\cos WL/v_0 \leqslant 1$, the source voltage is lower than the receiving-end voltage. Normally, the source voltage is held constant at the station bus so that the receiving-end bus voltage rises with line length.

Example 12.1: Calculate the expected p.u. value of load-end voltage for various line lengths from 100 km to 1000 km at no load. Neglect line resistance and assume source-end voltage to be held constant at 1 p.u.

Solution: The following table shows all values.

L, km	100	200	300	400	500	600	700	800	900	1000
$(2\pi L/\lambda)°$	6°	12°	18°	24°	30°	36°	42°	48°	54°	60°
cos $2\pi L/\lambda$	0.9945	0.978	0.951	0.9135	0.866	0.809	0.743	0.669	0.588	0.5
E_r/E_s	1.0055	1.022	1.05	1.095	1.155	1.236	1.346	1.494	1.7	2

If we continue this table, we soon find that the ratio E_0/E_s on a resistanceless line increases to infinity for a line whose length is equal to one quarter wavelength or 1500 km, and then will decrease.

Since the standard specifications state the maximum operating voltages for given nominal system voltages, some measures must be taken to control the voltage rise at no load at the receiving end. This requires shunt inductiv reactive vars to be provided. If the equipment is connected only at one end, usually the load end, then a synchronous condenser can be used operating in the under-excited condition. The use of shunt-compensating reactors either of the fixed type or variable type are used at both ends, and in recent years, static var systems of the switchable type using high-speed thyristor switches are coming into practice.

Example 12.2: Determine the limiting lengths for uncompensated lines if the voltages at the two ends must be held at the following sets of values. Neglect resistance and assume no-load condition.

(a) $E_s = 400$ kV, $E_0 = 420$ kV,

(b) $E_s = 380$ kV, $E_0 = 420$ kV,

(c) $E_s = 750$ kV, $E_0 = 765$ kV,

(d) $E_s = 735$ kV, $E_0 = 750$ kV,

(e) $E_s = 720$ kV, $E_0 = 750$ kV.

Note that in all cases, $E_s < E_0$.

Solution: Since $E_s = E_0 \cos(2\pi L/\lambda)$, the line length is

$$L = (\lambda/2\pi) [\cos^{-1} (E_s/E_0)] = \frac{100}{6} \cos^{-1} (E_s/E_0) \text{ km.}$$

For the given data, the line lengths will be
(a) 296 km, (b) 420 km, (c) 189 km, (d) 191 km, (e) 271 km.
Note that the larger the difference between E_s and E_0, the longer will be the line length allowed without shunt reactor compensation.

Charging Current and MVAR

The current supplied by the source into the line at no load $(I_r = 0)$ is, from equation (12.10)

$$I_s = CE_0 = \frac{1}{Z_0} \sinh pL \cdot E_r. \tag{12.26}$$

When resistance is neglected,

$$Z_{00} = \sqrt{l/c} \text{ and } p = jW/v_0 = j2\pi/\lambda.$$

$$\therefore \quad I_s = j \sqrt{\frac{c}{l}} \sin (WL/v_0) . E_r = jE_r . \sqrt{\frac{c}{l}} \cdot \sin (2\pi L/\lambda). \tag{12.27}$$

It leads the receiving-end voltage by 90°. An effective capacitance for the distributed line connected across the receiving-end voltage E_r and drawing the same current has the value

$$c_0 = \frac{1}{WZ_{00}} \sin (2\pi L/\lambda), \text{ Farad.} \tag{12.28}$$

The corresponding charging reactive power supplied by the source per phase will be

$$Q_0 = E_s \cdot I_0 = E_s \cdot E_r \sqrt{\frac{c}{l}} \cdot \sin(2\pi L/\lambda). \tag{12.29}$$

If E_s and E_r are line-to-line voltages in kV, r.m.s., then Q_0 will be the 3-phase charging MVAR. Also,

$$Q_0 = E_s^2 \cdot \sqrt{\frac{c}{l}} \cdot \tan(2\pi L/\lambda), \text{ since } E_r = E_s/\cos(2\pi L/\lambda)$$

$$\tag{12.30}$$

Note that for a resistanceless line, where $L = \lambda/2$, there is no charging current.

Example 12.3: For a 400 kV line, $l = 1$ mH/km and $c = 11.1$ nF/km, and $E_s = 400$ kV from the source, line-line, r.m.s. Calculate the charging MVAR for line lengths varying from 100 km to 1000 km. Neglect resistance.

Solution: $Z_{00} = \sqrt{l/c} = 300$ ohms.

L, km	100	200	300	400	500	600	700	800	900	1000
$\tan\left(\frac{6L}{100}\right)^\circ$	.105	.213	.325	.445	.577	.727	.9	1.11	1.38	1.732
Q_0, MVAR	56	113.4	173.3	237.5	308	387.5	480	592	734	924
Q_0/L	.56	.567	.578	.594	.616	.646	.686	.74	.816	·924

For normally-encountered distances from 300 km to 600 km, the charging MVAR per 100 km is 58 to 65 with an average of 60 MVAR for each 100 km length.

Example 12.4: Repeat example 12.3 for a 750 kV line with surge impedance of $Z_{00} = 250$ ohms. Take line lengths from 400 km to 1000 km.

Solution: $E_s^2/Z_{00} = 2250$ *MVAR*.

L, km	400	500	600	700	800	900	1000
Q_0, MVAR	1001	1298	1635	2025	2500	3096	3897

For lines between 400 and 1000 km, the charging MVAR is 250 to 390 per 100 km (With an average value of 300 MVAR/100) km).

Example 12.5: For the 400-kV and 750-kV lines, calculate the surge-impedance loading, SIL.

Solution: The surge-impedance load is equal to the MVA delivered to a load equal to Z_{00}. This is

$$\text{SIL} = E_s^2/Z_{00}.$$

For 400 kV line, SIL = $400^2/300$ = 533.3 MVA.

For 750 kV line, SIL = $750^2/250$ = 2250 MVA.

Example 12.6: For a 400 kV 400 km line, 50% of the line-charging MVAR is to be compensated by connecting shunt reactors. Calculate the approximate MVAR required in these.

Solution: From example 12.3, the charging MVAR supplied by the source is 237.5 MVAR. For 50% compensation, the shunt reactors have to provide approximately 120 MVAR. In practice, 60 MVAR reactors will be connected at each end of line.

More accurate shunt-reactor compensation will be calculated in later sections, including line resistance.

12.4. The Power Circle Diagram and Its Use

Equation (12.10) can also be written for E_r and I_r in terms of E_s and I_s as follows, with $AD - BC = 1$, and $A = D$

$$\begin{bmatrix} E_r \\ I_r \end{bmatrix} = \begin{bmatrix} A, & B \\ C, & D \end{bmatrix}^{-1} \begin{bmatrix} E_s \\ I_s \end{bmatrix} = \begin{bmatrix} A, & -B \\ -C, & D \end{bmatrix} \begin{bmatrix} E_s \\ I_s \end{bmatrix} \tag{12.31}$$

Also,
$$I_r = (E_s - AE_r)/B \tag{12.32}$$

All these quantities are complex numbers. The receiving-end power is

$$W_r = P_r + jQ_r = E_r I_r^* = E_r(E_s^* - A^*E_r^*)/B^* \tag{12.33}$$

If we consider E_r as reference giving $E_r = |E_r| \angle 0^\circ$, and the sending-end voltage $E_s = |E_s| \angle \delta$, the angle δ is called the power-angle. Also, let $A = |A| \angle \theta_a$ and $B = |B| \angle \theta_b$. Then

$$W_r = P_r + jQ_r = \frac{E_r \angle \theta_b}{|B|}(\; |E_s| \; \angle -\delta - A\angle -\theta_a \cdot E_r)$$

$$= \frac{E_r \, |E_s|}{|B|} \angle \theta_b - \delta - \frac{E_r^2}{|B|} \angle \theta_b - \theta_a \tag{12.34}$$

Separating the real and j-parts, there result

$$P_r = \frac{E_r \, |E_s|}{|B|} \cos(\theta_b - \delta) - \frac{E_r^2}{|B|} \cos(\theta_b - \theta_a) \tag{12.35}$$

and
$$Q_r = \frac{E_r \, |E_s|}{|B|} \sin(\theta_b - \delta) - \frac{E_r^2}{|B|} \sin(\theta_b - \theta_a) \tag{12.36}$$

These can be re-written as

$$P_r + \frac{E_r^2}{|B|} \cos(\theta_b - \theta_a) = \frac{E_r \, |E_s|}{|B|} \cos(\theta_b - \delta) \tag{12.36}$$

and $$Q_r + \frac{E_r^2}{|B|} \sin(\theta_b - \theta_a) = \frac{E_r |E_s|}{|B|} \sin(\theta_b - \delta) \qquad (12.37)$$

Then, eliminating δ by squaring and adding the two equations, we obtain the locus of P_r and Q_r to be a circle with given values of A and B, and for assumed values of E_r and $|E_s|$ as follows:

$$\left\{ P_r + \frac{E_r^2}{|B|} \cos(\theta_b - \theta_a) \right\}^2 + \left\{ Q_r + \frac{E_r^2}{|B|} \sin(\theta_b - \theta_a) \right\}^2$$
$$= \left(\frac{E_r |E_s|}{|B|} \right)^2 \qquad (12.38)$$

The coordinates of the centre of the receiving-end power-circle diagram are

$$x_c = - \frac{E_r^2}{|B|} \cdot \cos(\theta_b - \theta_a), \text{ MW, and}$$

$$y_c = - \frac{E_r^2}{|B|} \sin(\theta_b - \theta_a), \text{ MVAR} \qquad (12.39)$$

The radius of the circle is

$$R = E_r |E_s| / |B|, \text{ MVA} \qquad (12.40)$$

Figure 12.2 shows the receiving-end power-circle diagram. If the receiving-end voltage is held constant, the centre of the circle is fixed,

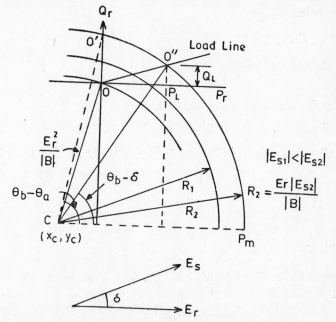

Fig. 12.2.   Receiving-end circle diagram for calculating reactive compensation for voltage control at buses.

but the radius will depend on the value of the sending-end voltage E_s. Therefore, for a chosen variation in $|E_s|$, a system of circles with centre at C and proper radius given by equation (12.40) can be drawn. The power-circle diagram and the geometrical relations resulting from it are extremely useful to a design engineer and an operating engineer to determine the status of power flow, reactive power flow, compensation requirements for voltage control and many properties concerning the system. These will be illustrated through typical examples.

Referring to Fig. 12.2, the angle $\angle OCY = \theta_b - \theta_a$ and from equations (12.35) and (12.36), when $P_r = 0$, that is when the circle intersects the vertical axis at O',

$$R \cdot \cos(\theta_b - \delta_0) = |x_c| \text{ giving } \angle O'CY = \theta_b - \delta_0 \qquad (12.41)$$

It is therefore a simple matter to determine the angle δ_0 for no-load condition.

For any other load, the load-line is drawn. Let the circle intersect the load line at O''. Then $\angle O''CY = \theta_b - \delta$, from equation (12.35). Because of the geometrical construction using instruments, there will be a certain amount of inherent error involved in the results but will serve engineering purposes. However, the circle diagram and the geometrical properties can be used to calculate all values with great accuracy.

Maximum Power

For a given set of E_r, E_s, A and B, the maximum power that can be transmitted, P_m, is

$$P_{max} = R - |x_c| = \frac{E_r |E_s|}{|B|} - \frac{E_r^2}{|B|} \cos(\theta_b - \theta_a) \qquad (12.42)$$

and the corresponding power angle is

$$\delta_{max} = \theta_b - \angle O'CP_m \qquad (12.43)$$

For the case when resistance is neglected,

$A \cosh pL = \cos WL/v_0$, giving $\theta_a = 0$, and

$B = \sqrt{l/c} \cdot \sinh jWL/v_0 = j\sqrt{l/c} \sin WL/v_0$, giving $\theta_b = 90°$.

$$\therefore \quad P_{max} = E_r |E_s|/|B| = E_r |E_s|/(Z_{00} \cdot \sin WL/v_0) \qquad (12.44)$$

From (12.35) and (12.36),

$P_r = E_r |E_s| \cdot \sin \delta/|B|$, which is maximum when $\delta = 90°$ and $P_{max} = E_r |E_s|/|B|$. We also observe from equation (12.44) that $|B| = Z_{00} \sin WL/v_0$.

Example 12.7: The following details are given for a 750-kV 3-phase line: Resistance $r = 0.014$ ohm/km, inductance $l = 0.866$ mH/km, reactance $x = 0.272$ ohm/km at 50 Hz, $c = 12.82$ nF/km, $y = 4.0275 \times 10^{-6}$ mho/km, $v_0 = 3 \times 10^5$ km/sec, line length = 500 km. Calculate items (a) and (b) below, and work parts (c) and (d). Give proper units for all quantities.

(a) $Z = L(r + jWl)$, $Y = jWcL$, $Z_{00} = \sqrt{l/c}$.

(b) The generalized constants A, B, C, D.

(c) For $E_r = 750$ kV and $|E_s| = 0.98\,E_r$, determine the coordinates of the centre of the receiving-end power-circle diagram and its radius.

(d) Find the power angle δ for transmitting a load of 2000 MW at 750 kV at the receiving-end at unity power factor.

Solution: All quantities are given in both rectangular and polar forms.

(a) $Z = 500\,(0.014 + j\,0.272) = 7 + j\,136 = 136.2\,\angle 87°$, ohms

$Y = j\,500 \times 4.0275 \times 10^{-6} = j\,2.014 \times 10^{-3}$

$$= 2.014 \times 10^{-3}\,\angle 90°,\ \text{mho}$$

$Z_{00} = \sqrt{0.866/12.82} \times 10^3 = 260$ ohms.

(b) $\sqrt{ZY} = (136.2 \times 2.014 \times 10^{-3})^{1/2}\,\angle 88°\!\cdot\!5 = 0.5237\,\angle 88°\!\cdot\!5$

$$= 0.0137 + j\,0.5235 \qquad (0.5235\ \text{radians} = 30°)$$

$\sqrt{Z/Y} = (136.2/2.014 \times 10^{-3})^{1/2}\,\angle -1°\!\cdot\!5$

$$= 260\,\angle -1°\!\cdot\!5 = 259.9 - j\,6.8\ \text{ohms}$$

$\therefore\quad A = D = \cosh\sqrt{ZY} = \cosh 0.0137 \cdot \cos 0.5235 +$

$j\sinh 0.0137 \cdot \sin 0.5235 = 0.86625 + j\,0.00685 = 0.8663\,\angle 0.45°$

[At no load, without compensation, for $E_r = 750$ kV, $E_s = AV_r = 650$ kV, line-to-line].

$$C = \sqrt{\frac{Y}{Z}}\,\sinh\sqrt{ZY} = \frac{1}{260\,\angle -1°\!\cdot\!5}\,\sinh\,(0.0137 + j\,0.5235)$$

$$= \frac{1}{260\,\angle -1°\!\cdot\!5}\,(\sinh 0.0137 \cdot \cos 0.5235$$

$$+\, j\cosh 0.0137 \cdot \sin 0.5235)$$

$$= 0.50024\,\angle 88°\!\cdot\!64/260\,\angle -1°\!\cdot\!5 = 1.924 \times 10^{-3}\,\angle 90°\!\cdot\!14$$

$$= (-0.0047 + j\,1.923994)\,10^{-3}\ \text{mho}.$$

$B = Z_0 \sinh\sqrt{ZY} = 0.50024\,\angle 88°\!\cdot\!64 \times 260\,\angle -1°\!\cdot\!5$

$$= 130\,\angle 87°\!\cdot\!14 = 6.486 + j\,129.84\ \text{ohms}.$$

(c) *Centre.* $x_c = -\dfrac{E_r^2}{|B|}\cos(\theta_b - \theta_a)$

$$= -\frac{750^2}{130}\cos(87.14 - 0.45°)$$

$$= -250\ \text{MW}.$$

$$y_c = -4327 \cdot \sin 86°\!\cdot\!69 = -4319\ \text{MVAR}.$$

Radius. $R = E_r\,|E_s|/|B| = 0.98 \times 750^2/130 = 4240\ \text{MVA}.$

(d) From the geometry of the circle diagram, Fig. 12.2, or, from equation (12.35),

$$2000 + 250 = 4240 \cos (87.14 - \delta),$$

giving $\delta = 87.14 - 57.95 = 29°\cdot2$.

[The surge impedance $Z_{00}=260$ ohms. The SIL is $750^2/260=2163$ MW. At this load with $6°/100$ km, the power angle would be $30°$ for 500 km length of line.]

Example 12.8: In the previous example, calculate the reactive compensation to be provided across the load at the receiving end for (a) no-load condition, and (b) at full rated load of 2000 MW at 0.95 power-factor lag. State the nature of reactive compensation in the two cases.

Solution:

(a) The no-load point is O' on the circle diagram, Fig. 12.2. By geometry, $(Q_0 + 4319)^2 + 250^2 = 4240^2$, which gives $Q_0 = -86.4$ MVAR showing that the compensation required is capacitive. The point O' falls below the origin O. See Fig. 12.3.

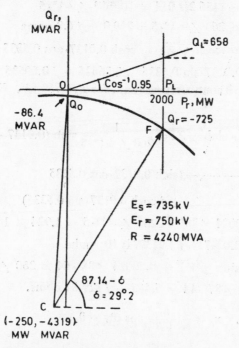

Fig. 12.3.   Circle diagram for Example 12.8.

(b) For locating the full-load point at 0.95 lag, a line OL is drawn from the origin at $\cos^{-1} 0.95 = 18°\cdot2$ in the 1st quadrant since the reactive power of load is positive, equal to $Q_L = 2000 \tan 18°\cdot2 = 657.6$ MVAR. From the circle diagram,

$$(\mid y_c \mid - PF)^2 + (P + \mid x_c \mid)^2 = R^2$$

or, $(4319 - PF)^2 + (2000 + 250)^2 = 4240^2$, giving

$PF = 725$ MVAR, capacitive. This is the total reactive power required at the receiving end for the given voltage and load conditions. To this must be added 657.6 MVAR of the load to obtain the rating of the compensating equipment, giving a total of $657.6 + 725 = 1383$ MVAR. From equation (12.35), the power angle δ is still $29°.2$.

From the above example, it may be observed that the sending-end voltage is low enough to require a capacitive compensation at no load. If this voltage is raised while keeping $E_r = 750$ kV, then shunt reactor compensation would be necessary at the load end, as shown in the example below.

Example 12.9: For raising E_s also to 750 kV, calculate the compensations to be provided (a) at no load, and (b) at full load, (c) calculate the power angles δ in the two cases.

Solution: Since E_r, A and B are the same, the coordinates of the centre of the receiving-end circle diagram stay the same at $(-250$ MW, -4319 MVAR). But the new radius is $R = 750^2/130 = 4327$ MVA.

(a) At no load: $(Q_0 + 4319)^2 + 250^2 = 4327^2$ giving $Q_0 = 0$. Thus no compensation is required at no load. [If $E_s > 750$ kV, then inductive compensation would be called for].

(b) At full load: $(4319 - PF)^2 + 2250^2 = 4327^2$ giving $PF = 623$ MVAR. Therefore, adding 658 MVAR in the load, the compensating equipment must provide a capacitive reactive power of $623 + 658 = 1281$ MVAR. Since the load can vary from no-load to full load, this will be in the nature of switched capacitors.

These examples have assumed that all compensation equipment is connected across the receiving end.

12.5. Voltage Control Using Synchronous Condensers

From the generalized constants (A, B, C, D) of a given input port and output port, the power-circle diagram or the corresponding geometrical relations can be utilized for deciding the proper compensating MVAR's to be provided at the receiving end when a set of magnitudes for E_r and E_s at the two ends of the line are specified. When the load has a lagging power factor or even a unity power factor, generally the control of voltage is achieved by providing leading power factor or capacitive compensation at the receiving end. This can take the form of switched capacitors, usually connected to the low-voltage tertiary of the sub-station transformer at the load end. At no load, usually inductive reactive compensation is required if the sending-end voltage is to be raised. This is provided by either switched type (regulated) or constant type (unregulated).

The synchronous condenser provides both types of MVAR's, lagging or leading, that is inductive or capacitive. This is achieved by control of dc field excitation. The motor runs without a shaft load and is enclosed in an explosion-proof casing which is filled with hydrogen at above atmospheric pressure in order to minimize rotational losses. The higher than atmospheric pressure is necessary in order to prevent oxygen from leaking into the casing and causing an explosive mixture with hydrogen. Because of limitations imposed on excitation, the synchronous phase modifier can provide only 60 to 70% of its rated capacity at lagging power factor (under-excited condition) and full rated leading reactive power (over-excited condition).

The design of the synchronous phase modifier (or condenser for short) and the voltage conditions are illustrated below. The general equation satisfied by the real and reactive powers at the load end at any point on the circle is

$$(P_r + \mid x_c \mid)^2 + (Q_r + \mid y_c \mid)^2 = R^2 \qquad (12.45)$$

At no load, point O', $P_r = 0$. The reactive power is

$$Q_0 = \sqrt{R^2 - x_c^2} - \mid y_c \mid \qquad (12.46)$$

At full load, point O'' in Fig. 12.2 or 12.3,

$$(P_L + \mid x_c \mid)^2 + (\mid y_c \mid - Q_r)^2 = R^2 \qquad (12.47)$$

∴ The total reactive power required at the receiving-end is

$$Q_r = \mid y_c \mid - \sqrt{R^2 - (P_L + \mid x_c \mid)^2} \qquad (12.48)$$

The capacitive MVAR required in the synchronous condenser is then

$$Q_c = Q_L + Q_r = Q_L + \mid y_c \mid - \sqrt{R^2 - (P_L + \mid x_c \mid)^2} \qquad (12.49)$$

If we use the relation $Q_0 = mQ_c$, $(m < 1)$, then the equation to be satisfied by the radius of the circle, which is the only quantity involving E_s, when E_r, A, B, P_L, Q_L are specified will be,

$$m\sqrt{R^2 - (P_L + \mid x_c \mid)^2} + \sqrt{R^2 - x_c^2} = (1 + m) \mid y_c \mid + mQ_L \qquad (12.50)$$

When all quantities except R are specified, equation (12.50) will yield the value of $R = E_r \mid E_s \mid / \mid B \mid$, from which the sending-end voltage is determined. This is shown by the following example.

Example 12.10: For example 12.9, taking $E_r = 750$ kV and $m = 0.7$ for the synchronous condenser, determine (a) the sending-end voltage, and (b) the proper rating of the synchronous condenser, (c) give your comments.

Solution: The data given are

$E_r = 750$, $A = 0.8663 \angle 0.45°$, $B = 130 \angle 87°.14$, $x_c = -250$,

$y_c = -4319$, $P_L = 2000$ MW, $Q_L = 658$ MVAR, inductive. $m = 0.7$.

(a) Solving equation (12.50) for R there results

$$R = 4822 = 750 \times E_s/130.$$

∴ The sending-end voltage has the magnitude

$$E_s = 4822 \times 130/750 = 836 \text{ kV, line-to-line.}$$

(b) The synchronous-condenser rating is $Q_r = Q_L + Q_r$, from equation (12.49).

$$Q_r = 4319 - \sqrt{4822^2 - (2000 + 250)^2}$$
$$= 54.12 \text{ MVAR, capacitive, at full load.}$$

∴ $Q_c = 54.12 + 657.6 = 711.7$ MVAR, over-excited.

As a check, the under-excited reactive power required at no load is given in equation (12.46) as

$$Q_0 = \sqrt{4822^2 - 250^2} - 4319 = 496.5 \text{ MVAR, Inductive.}$$

The ratio of

$$Q_0/Q_c = 0.698 \approx 0.7.$$

The rating of the synchronous condenser would be about 720 MVAR when delivering capacitive reactive power at over-excited condition of operation.

The circle diagram could also have been used by drawing circles with several values of E_s.

(c) We observe that a sending-end voltage of 836 kV, $l - l$, for a 750 kV system is very high and beyond that allowed by standard specifications. Therefore, use of synchronous condenser to provide compensating vars only at one end of the line is not a practical solution. We have to separate the functions of no-load compensation and full-load compensation by using shunt-compensating reactors for no-load, and switched capacitors for full-load operation. This will be discussed in the following sections.

12.6. Cascade Connection of Components—Shunt and Series Compensation

In the previous sections, the (A, B, C, D) constants of only the line were considered. It becomes evident that through the example of the 750 kV line parameters, it is impossible to control the voltages within limits specified by IS and IEC by providing compensation at one end only by synchronous condensers, or by switched capacitors if the voltages are to vary over wider limits than discussed. In practice, shunt-compensating reactors are provided for no-load conditions which are controlled by the line-charging current entirely, and by switched capacitors for full-load conditions when the load has a lagging power factor.

Generalized Equations

For no-load conditions, $Z_t = \infty$, and the equations for sending-end and receiving-end voltages are, Fig. 12.4,

$$E_r = E_s/[\cosh pL + (Z_0/Z_{sh}) \sinh pL] \tag{12.51}$$

For simplicity, let $r = 0$. Then,

$$p = jW/v_0 = j\,2\pi/\lambda,\ Z_0 = Z_{00} = \sqrt{l/c},\ Z_{sh} = jX_{sh},$$

so that

$$E_s/E_r = \cos 2\pi L/\lambda + (Z_{00}/X_{sh}) \sin\, 2\pi L/\lambda \tag{12.52}$$

When the ratio E_s/E_r is given for a system, the value of X_{sh} is easily determined. In particular, when $E_s = E_r$,

$$X_{sh} = Z_{00}/(\text{cosec}\ 2\pi L/\lambda - \cot 2\pi L/\lambda). \tag{12.53}$$

Example 12.11: For the 750-kV line of previous examples, $L = 500$ km, $\lambda = 6000$ km at 50 Hz and $Z_{00} = 260$ ohms. Assuming $E_s = E_r = 750$ kV, calculate the reactance and 3-phase MVAR required at each end in the shunt-compensating reactors. Neglect line resistance.

Solution: $2\pi L/\lambda = 30°$ at 6° per 100 km of length of line.

$\therefore$ $X_{sh} = 260/(\text{cosec}\ 30° - \cot 30°) = 3.73 \times 260 = 970$ ohms.

This is necessary at each end connected between line and ground so that there will be 6 such reactors for the 3-phases.

Current through each reactor $I_{sh} = 750/970\sqrt{3} = 0.4464$ kA.

$\therefore$ MVAR of each reactor per phase $= 750 \times 0.4464/\sqrt{3} = 193.3$.

Total 3-phase MVAR at each end $= 580$ **MVAR**.

Chain Rule

For the reactor only, the generalized constants are given by, see Fig. 12.4,

$$\begin{bmatrix} E_s \\ I_s \end{bmatrix} = \begin{bmatrix} 1, & 0 \\ -jB_{sh}, & 1 \end{bmatrix} \begin{bmatrix} E'_r \\ I'_r \end{bmatrix} \tag{12.54}$$

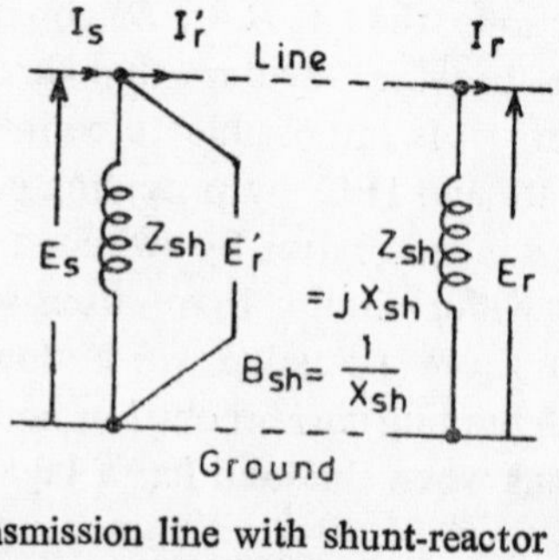

Fig. 12.4. Transmission line with shunt-reactor compensation for voltage control at no load.

where $B_{sh} = 1/X_{sh} =$ admittance of each reactor per phase.

If (E_s, I_s) refer to the input voltage and current from the source and (E_r, I_r) the output quantities at the load end, then the generalized constants (A_T, B_T, C_T, D_T) for the entire system is obtained by chain multiplication of the three matrices for the cascade-connected components. This is

$$
\begin{bmatrix} A_T, & B_T \\ C_T, & D_T \end{bmatrix} = \begin{bmatrix} 1, & 0 \\ -jB_{sh}, & 1 \end{bmatrix} \begin{bmatrix} A, & B \\ C, & D \end{bmatrix}_{Line} \begin{bmatrix} 1, & 0 \\ -jB_{sh}, & 1 \end{bmatrix}
$$

$$
= \begin{bmatrix} A, & B \\ C, & D \end{bmatrix}_{Line} -jB_{sh} \begin{bmatrix} B, & 0 \\ 2A - jB_{sh}B, & B \end{bmatrix} \quad (12.55)
$$

where A, B, C, D refer only to the line, equations (12.11) to (12.13). These are, in general, complex quantities whereas B_{sh} is a real number.

When once the total generalized constants (A_T, B_T, C_T, D_T) are calculated, the receiving-end power-circle diagram can be drawn and all requirements for compensation can be determined. The main requirement for the shunt reactors is control of voltage at no-load. For this case,

$$
E_s = A_T E_r = (A - jB_{sh}B) E_r
$$

$$
\therefore \qquad B_{sh} = (A - E_s/E_r) jB = j(E_s/E_r - A)/B \quad (12.56)
$$

For example 12.11,

$$
B_{sh} = j(1 - 0.866)/j\,130 = 1.0308 \times 10^{-3} \text{ mho}
$$

giving $X_{sh} = 970$ ohms.

12.6.1. Shunt Reactor Compensation of Very Long Line with Intermediate Switching Station

For very long lines, longer than 400 km at 400 kV, or at higher voltages, an intermediate station is sometimes preferable in lieu of series-capacitor compensation which will be discussed in the next section. Fig. 12.5 shows the arrangement where each line section has the generalized constants $(A, B, C, D)_{Line}$, and each of the four shunt reactors has an admittance of $(-jB_{sh})$ and reactance (jX_{sh}). Then, by using the chain rule of multiplication, the total generalized constants (A_T, B_T, C_T, D_T)

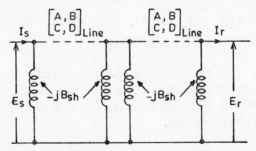

Fig. 12.5. Extra-long line with shunt reactors at ends and at an intermediate station.

relating (E_s, I_s) with (E_r, I_r) will be as follows, upon using equation (12.55):

$$\begin{bmatrix} A_T, & B_T \\ C_T, & D_T \end{bmatrix} = \begin{bmatrix} A - jB_{sh}\,B, & B \\ C - j2B_{sh}\,A - B_{sh}^2 B, & D - jB_{sh}\,B \end{bmatrix}^2$$

$$= \begin{bmatrix} A^2 + BC, & 2AB \\ 2AC, & A^2 + BC \end{bmatrix} - B_{sh}^2 B \begin{bmatrix} 2B, & 0 \\ 6A, & 2B \end{bmatrix} - jB_{sh}$$

$$\times \begin{bmatrix} 4AB & , & B^2 \\ 2BC + 4A^2 + 2B_{sh}^2 B^2, & 4AB \end{bmatrix} \quad (12.57)$$

For $B_{sh} = 0$, this reduces to

$$\begin{bmatrix} A_T, & B_T \\ C_T, & D_T \end{bmatrix}_{Bsh=0} = \begin{bmatrix} A^2 + BC, & 2AB \\ 2AC, & A^2 + BC \end{bmatrix} = \begin{bmatrix} A, & B \\ C, & D \end{bmatrix}^2 \quad (12.58)$$

The voltage and current at the intermediate station is

$$\begin{bmatrix} E_r' \\ I_r' \end{bmatrix} = \begin{bmatrix} A - jB_{sh}\,B, & B \\ C - j2A B_{sh} - B_{sh}^2 B, & D - jB_{sh}\,B \end{bmatrix} \begin{bmatrix} E_r \\ I_r \end{bmatrix} \quad (12.59)$$

Example 12.12: In example 12.11, shunt compensating reactors of 580 MVAR are provided at each end. Calculate the % compensation of charging MVAR provided by these shunt reactors. Neglect line resistance.

Solution: The generalized constants for the line are

$$A = 0.866, \ B = j130, \text{ and } C = \frac{1}{Z_{00}} \sin 2\pi L/\lambda = j\frac{1}{260} \sin 30°$$
$$= j1.923 \times 10^{-3} \text{ mho.}$$

$\therefore$ Charging MVAR of line without shunt reactors is

$$Q_0 = C E_s^2 = j1.923 \times 10^{-3} \times 750^2 = 1082 \text{ MVAR.}$$

With shunt reactors, with $B_{sh} = 1.0308 \times 10^{-3}$ mho, from equation (12.55),

$$C_T = C - j2A B_{sh} - B_{sh}^2\, B = j(1.923 - 1.7854 - 0.138)\,10^{-3} \approx 0.$$

$\therefore$ % compensation is total or 100%.

Example 12.13: For the examples, 12.11 and 12.12, instead of 580 MVAR, allow 60% shunt reactor compensation to maintain $E_s = 750$ kV. Calculate the MVAR of the shunt reactors at each end, and the voltage at the receiving end at no-load.

Solution: % compensation = $100 (C - C_T)/C$, giving

$$C_T = (1 - 0.6) C = 0.4 C = j0.4 \times 1.923 \times 10^{-3} \text{ mho}$$

$$= j0.7692 \times 10^{-3} \text{ mho.}$$

But $\quad C_T = C - j2A\,B_{sh} - B_{sh}^2\,B,$

or $\quad j0.7692 \times 10^{-3} = j1.923 \times 10^{-3} - j2 \times 0.866 \times B_{sh} - j130\,B_{sh}^2.$

Solving the quadratic equation for B_{sh} results

$$B_{sh} = 0.6358 \times 10^{-3} \text{ mho.}$$

∴ 3-phase MVAR of each shunt-reactor bank at 750 kV will be

$$Q_{sh} = B_{sh}\,E_s^2 = 0.6358 \times 10^{-3} \times 750^2 = 358 \text{ MVAR.}$$

[Note that this may be approximately 60% of 580 MVAR which was required for 100% compensation].

With these reactors connected, the receiving-end voltage will not be 750 kV. It is calculated as follows:

$$A_T = A - jB_{sh}\,B = 0.866 + 0.6358 \times 10^{-3} \times 130 = 0.9486.$$

∴ $\quad E_r = E_s/A_T = 790.6 \text{ kV.}$

The compensation is a bit low since the bus voltage at the receiving end is higher than 765 kV, which is specified by IS and IEC. If E_r is to be held at 765 kV, then E_s must be lowered. Since $A_T = 0.9486$, $E_s = 0.9486 \times 765 = 726$ kV. Then, the shunt-reactor rating will be

$$Q_{shs} = 726^2 \times 0.6358 \times 10^{-3} = 335 \text{ MVAR at the sending end,}$$

and $Q_{shr} = 765^2 \times 0.6358 \times 10^{-3} = 372$ MVAR at the receiving end. A compromise of 350 MVAR would be selected and the voltages at the two ends adjusted.

Example 12.14: A 400-kV line is 800 km long. Its inductance and capacitance per km are $l = 1$ mH/km and $c = 11.1$ nF/km ($Z_{00} = 300$ ohms). The voltages at the two ends are to be held at 400 kV at no load. Neglect resistance. Calculate

  (a) MVAR of shunt-reactors to be provided at the two ends and at an intermediate station midway with all four reactors having equal reactance.

  (b) The A, B, C, D constants for the entire line with the shunt reactors connected.

  (c) The voltage at the intermediate station. (Use 6°/100 km).

Solution: Refer to Fig. 12.5 and equations (12.57) and (12.59). For one 400-km section.

  (a) $A = D = \cos 24° = 0.9135$, $B = j300 \sin 24° = j122$ ohm and
  $C = (\sin 24°)/300 = j1.356 \times 10^{-3}$ mho.

  The total A_T from end to end for 800 km is

$$A_T = A^2 + BC - 2B_{sh}^2 B - j4AB\,B_{sh}$$
$$= 0.669 + 445.8B_{sh} + 29,768B_{sh}^2.$$

But since $E_s = E_r$, the value of $A_T = 1$.

By using this value of A_T and solving for B_{sh} yields $B_{sh} = 0.709 \times 10^{-3}$ mho. The MVAR of each of the reactors at the two ends will be

$$Q_e = 400^2 \times 0.709 \times 10^{-3} = 113.4 \text{ MVAR, 3-phase unit.}$$

(b) The A, B, C, D constants for the entire line is found as follows: From equation (12.55), for each 400-km section,

$$A - jB_{sh}B = 0.9135 + 0.709 \times 10^{-3} \times 122 = 1$$

$$C - jB_{sh}(2A - jB_{sh}B) = j1.356 \times 10^{-3} - j0.709 \times 10^{-3}$$
$$\times (1.827 + 0.709 \times .122)$$
$$= 0.$$

$$\therefore \begin{bmatrix} A_T, & B_T \\ C_T, & D_T \end{bmatrix} = \begin{bmatrix} 1, & j122 \\ 0, & 1 \end{bmatrix}^2 = \begin{bmatrix} 1, & j244 \\ 0, & 1 \end{bmatrix}$$

(c) At the intermediate station, $E_r' = 1.E_s = 400$ kV.

12.6.2. Series-Capacitor Compensation at Line Centre

In order to increase the power-handling capacity of a line, the magnitude of B must be reduced, as shown in equations (12.35) and (12.44). In normal practice, $\theta_b \approx 90°$ and $\theta_a \approx 0$ for very low series line resistance. Therefore, $P = E_r E_s \cdot \sin \delta / B$. We also observe that the value of B is very nearly equal to the series inductive reactance of the line so that by employing capacitors connected in series with the line, the power-handling capacity of a line can be increased for chosen values of E_s, E_r, and δ. All these three quantities are limited from considerations of highest equipment voltages and stability limits. Usually, the series capacitor is located at the line centre when one capacitor only is used, or at the one-third points if two installations are used. On a very long line with intermediate station, the series capacitor can be located here. In this section we will only consider the generalized constants when a series capacitor is located at the line centre without intermediate shunt-reactor compensation. Thus, the system considered consists of equal shunt reactor admittances B_{sh} at the two ends and a capacitor with reactance x_c at the line centre, as shown in Fig. 12.6.

For each half of the line section of length $L/2$,

$$A' = D' = \cosh \sqrt{ZY/4}, \quad B' = Z_0 \sinh \sqrt{ZY/4},$$

$$C' = \frac{1}{Z_0} \sinh \sqrt{ZY/4} \tag{12.60}$$

where Z and Y refer to the total line of length L. The surge impedance is not altered even though the line length is halved.

For the series capacitor, Fig. 12.6(b), the voltages and currents on the two sides are related by

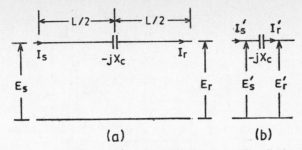

Fig. 12.6. Transmission line with series-capacitor compensation in middle of line.

$$\begin{bmatrix} E_s' \\ I_s' \end{bmatrix} = \begin{bmatrix} 1, & -jx_c \\ 0, & 1 \end{bmatrix} \begin{bmatrix} E_r' \\ I_r' \end{bmatrix} \tag{12.61}$$

Thus, for the two half-sections of line with the capacitor separating them, the total generalized constants will be, by the chain rule of multiplication,

$$\begin{bmatrix} A_T, & B_T \\ C_T, & D_T \end{bmatrix} = \begin{bmatrix} \cosh \sqrt{ZY/4}, & Z_0 \sinh \sqrt{ZY/4} \\ \dfrac{1}{Z_0} \sinh \sqrt{ZY/4}, & \cosh \sqrt{ZY/4} \end{bmatrix} \begin{bmatrix} 1, & -jx_c \\ 0, & 1 \end{bmatrix}$$

$$\begin{bmatrix} \cosh \sqrt{ZY/4}, & Z_0 \sinh \sqrt{ZY/4} \\ \dfrac{1}{Z_0} \sinh \sqrt{ZY/4}, & \cosh \sqrt{ZY/4} \end{bmatrix} = \begin{bmatrix} \cosh \sqrt{ZY}, & Z_0 \sinh \sqrt{ZY} \\ \dfrac{1}{Z_0} \sinh \sqrt{ZY}, & \cosh \sqrt{ZY} \end{bmatrix}$$

$$-j\frac{x_c}{2Z_0} \begin{bmatrix} \sinh \sqrt{ZY}, & Z_0 (\cosh \sqrt{ZY} + 1) \\ \dfrac{1}{Z_0} (\cosh \sqrt{ZY} - 1), & \sinh \sqrt{ZY} \end{bmatrix} \tag{12.62}$$

$$= \begin{bmatrix} A, & B \\ C, & D \end{bmatrix}_{\text{Line}} -j\frac{x_c}{2Z_0} \begin{bmatrix} \sinh pL, & Z_0 (\cosh pL + 1) \\ \dfrac{1}{Z_0} (\cosh pL - 1), & \sinh pL \end{bmatrix} \tag{12.63}$$

With no series capacitor, $x_c = 0$, equation (12.63) reduces to the generalized constants of the line.

Shunt Reactors at Both Ends and Series Capacitor in Middle of Line
If shunt-compensating reactors of admittance B_{sh} are located at both ends of line, the total generalized constants with series capacitor located in the line centre can be obtained in the usual way by using the chain rule, from Fig. 12.7. It is left to the reader to verify the result.

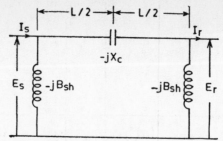

Fig. 12.7.   Transmission line with series capacitor in middle and shunt reactors at ends.

$$\begin{bmatrix} A_T, & B_T \\ C_T, & D_T \end{bmatrix} = \begin{bmatrix} A, & B \\ C, & D \end{bmatrix}_{\text{Line}} -j B_{sh} \begin{bmatrix} B, & 0 \\ 2A - j B_{sh} B, & B \end{bmatrix}$$

$$-j \frac{x_c}{Z_0} \begin{bmatrix} \sinh pL, & Z_0 \, (\cosh pL + 1) \\ \dfrac{1}{Z_0} (\cosh pL - 1), & \sinh pL \end{bmatrix}$$

$$-\frac{x_c B_{sh}}{2} \begin{bmatrix} \cosh pL + 1, & 0 \\ \dfrac{2}{Z_0} \sinh pL - j B_{sh} \, (\cosh pL + 1), & \cosh pL + 1 \end{bmatrix} \qquad (12.64)$$

We must point out that all quantities except x_c and B_{sh} are complex numbers with a magnitude and phase angle.

If shunt reactors are not used, $B_{sh} = 0$, then equation (12.64) reduces to (12.63). If no series capacitor is used, equation (12.55) results.

The value of capacitance is chosen such that its reactance amounts to a chosen percentage of the series inductive reactance of the line. For example, if 50% series capacitance is to be provided for a 800 km 400 kV line having a series inductive reactance of j 244 ohms, the capacitive reactance is $-j$ 122 ohms. The required capacitance will be $C = 1/W x_c$ = 26 µF at 50 Hz. The power-handling capacity of a line with 50% series compensation will increase to double that without series capacitor.

While the power-handling capacity can be increased, series capacitor compensation results in certain harmful properties in the system, chief among them are:

(a) Increased short-circuit current. Note that for a short-circuit beyond the capacitor location, the reactance is very low. At the load-side terminal of the capacitor, a short-circuit will result in infinite current for 50% compensation.

(b) Sub-harmonic or sub-synchronous resonance conditions during load changes and short circuits. This has resulted in unexpected failures to long shafts used in steam turbine-driven alternators and exciters when one or more of the resulting sub-harmonic currents due to series compensation can produce shaft torques that correspond to one of the several resonance frequencies of the shaft, called torsional modes of oscillation or critical speeds. This aspect will be discussed in some detail and counter measures used against possible failure will be described in the next section.

12.7. Sub-Synchronous Resonance in Series-Capacitor Compensated Lines

12.7.1. Natural Frequency and Short-Circuit Current

With series compensation used, the power-handling capacity of a single circuit is approximately

$$P = E_r E_s \cdot \sin \delta / X_s \qquad (12.65)$$

where
$$X_s = X_L - X_c = X_L (1 - m) \qquad (12.66)$$

with $m = X_c/X_L$ = degree of compensation, and the reactances are at power frequency f_0. The approximation occurs because we have considered the series inductive reactance X_L to be lumped. At any other frequency f,

$$X_L(f) = 2\pi f L_L = \text{total inductive series reactance of line}$$

and $X_c(f) = 1/2\pi f C =$ series capacitive reactance of capacitor.

$\therefore$ The resonance frequency occurs when $X_L(f_e) = X_c(f_e)$ giving

$$2\pi f_e L_L = 1/2\pi f_e C \qquad (12.67)$$

We can introduce the power frequency f_0 by re-writing equation (12.67) as $(2\pi f_0 L_L)(f_e/f_0) = 1/[2\pi f_0 C \cdot (f_e/f_0)]$ at resonance. Consequently,

$$(f_e/f_0)^2 = 1/(2\pi f_0 L_L) \cdot (2\pi f_0 C) = X_c/X_L = m. \qquad (12.68)$$

Thus, the electrical resonance frequency is $f_e = f_0 \sqrt{m}$ $\qquad$ (12.69)
The reduction of series reactance also reflects as decrease in effective length to $L_e = (1 - m) L$.

For $f_0 = 50$ Hz and 60 Hz, and for various degrees of series compensation, the following table gives the resonant frequencies and effective length which are also plotted in Fig. 12.8.

$m = X_c/X_L$	0.1	0.2	0.3	0.4	0.5	0.6	0.7	0.8	0.9	1.0
% compensation	10	20	30	40	50	60	70	80	90	100
f_e/f_0	0.3162	0.447	0.548	0.632	0.707	0.775	0.837	0.894	0.949	1
f_e for 50 Hz	15.8	22.4	27.4	31.6	35.36	38.7	41.8	44 7	47.4	50
f_e for 60 Hz	18.97	26.8	32.9	37.95	42.4	46.5	50.2	53.7	56.9	60
$L_e/L = 1-m$	0.9	0.8	0.7	0.6	0.5	0.4	0.3	0.2	0.1	0

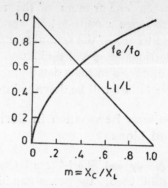

Fig. 12.8.   Variation of resonance frequency f_e and effective inductance L_e of line with degree of series compensation (X_C/X_L).

For a single line with series compensation between 40% and 70%, the line can resonate at frequencies between 63.25% and 83.7% of the power frequencies. Any disturbance in the line giving rise to currents at these frequencies will increase the voltage across the series capacitor and therefore the line drop to extremely high values which are limited only by the line resistance. The voltage across the capacitor during normal operation is $V_c = X_c I$, where $I =$ current flowing in line.

Example 12.15:   A 50-Hz 750 kV line with $l = 0.866$ mH/km is 500 km long. It is provided with 50% series compensation connected in the middle of line. The power delivered at 750 kV is 2000 MW 3-phase per circuit at unity power factor. Neglect shunt capacitance and line resistance and assume the line inductance to be lumped. Calculate

(a)   the reactance and capacitance of series capacitor,

(b)   the voltage drop across it at full load,

(c)   the current flowing through it and the voltage across it during a sustained short circuit occurring

  (i)   on the source-side terminal of the capacitor,

  (ii)   on the load-side terminal of the capacitor, and

  (iii)   across the load.

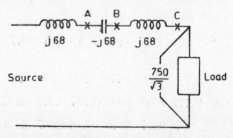

Fig. 12.9.   Illustrating dangers of increased short-circuit currents with series-capacitor compensation for faults at various important locations.

(d) the same as in (c) without the series capacitor.

Assume that no other generating area is supplying the load. See Fig. 12.9.

Solution: Total series reactance of 250 km of line at 50 Hz is

$$314 \times 0.866 \times 10^{-3} \times 250 = 68 \text{ ohms.}$$

(a) For 50% compensation, $X_c = 68$ ohms and $C = 46.8$ μF.

(b) At full load, $I = 2000/750\sqrt{3} = 1.54$ kA $= 1540$ Amps.

∴ Voltage across capacitor $V_c = 68 \times 1.54 = 104.7$ kV.

(c) (i) For a short-circuit to ground occurring at point A, the capacitor will discharge through the fault and its voltage will become zero. There will be an oscillatory stage consisting of a circuit with the capacitor, line inductance of $j68$ ohms and the load impedance.

(ii) For a short-circuit at point B, we observe that the steady-state component of current is infinity and dangerously high voltage will exist across the capacitor. Under such a condition, arrangements must be made to bypass the capacitor by a gap that will flashover at a preset overvoltage value. This is normally set at 2.2 p.u. where 1 p.u. voltage = voltage across capacitor during normal operation which was calculated as 104.7 kV. These protective schemes will be described later on.

(iii) When a short-circuit occurs at point C, the sustained r.m.s. value of current is

$$I_{sc} = 750/68\sqrt{3} = 6.368 \text{ kA} = 4.135 \times \text{full-load current}$$

The voltage across the capacitor will be $V_c = 750/\sqrt{3} = 433$ kV. This is the lowest value of voltage across the capacitor for a short-circuit occurring at any place on the line between points B and C.

(iv) Without the series capacitor, the currents during short circuit at points A or B and C will be 6.368 kA and 3.184 kA respectively. Consequently, using a series capacitor has increased the short-circuit level of the system.

Protective Schemes for Series Capacitor

Some of the methods used in practice for protecting series capacitors from damage due to overvoltages and overcurrents are described below and shown in Fig. 12.10. In some schemes, the capacitor banks are protected by a sparkover gap that will either short-circuit the capacitor or otherwise use a voltage-limiting surge arrester of the gap—SiC type or gapless ZnO type. When the short-circuit in the system is cleared, the capacitor is re-inserted in order to preserve the stability of

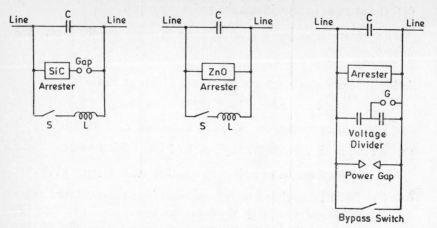

Fig. 12.10. Protective schemes for series capacitors during abnormal conditions.

the system. The re-insertion transient can be very high and cause gap flashover. This is prevented by either parallel resistor or SiC and ZnO arresters. The sparkover voltage of the gap G is between 2.2 and 3.5 p.u. of normal capacitor voltage at full rated current.

12.7.2. Sub-Synchronous Resonance Problem and Counter Measures

As pointed out before, electrical resonance frequencies lower than synchronous frequency f_0 exist when series-capacitor compensation is used. The frequency f_0 corresponds to the steady-state speed of the rotor in large power stations. With steam-turbine-driven generating units (designated T-G units), the shaft or shaft portions connecting the HP, IP, LP stages of the turbine and the generator-exciter are very long, with their own characteristic torsional mechanical resonant frequencies. When the electrical system operates in such a manner that the rotating fields in the generator due to sub-synchronous currents produce torques of the same frequency as one of the mechanical torsional frequencies of the shaft and of the correct phase, torques up to 10 times the break away or ultimate strength of the shaft can be reached resulting in shaft damage. This phenomenon of electro-mechanical interaction between electrical resonant circuits of the transmission system and the torsional natural frequencies of the T-G rotor is known as "Sub-Synchronous Resonance," and designated SSR.

The phenomenon of SSR has been studied very extensively since 1970 when a major transmission network in the U.S. experienced shaft failure to its T-G unit with series compensation in the 500 kV lines. This has now gone into technical literature as a classic problem and known as Project Navajo. The phenomenon, however, had been known to exist for a few years according to many experts who predicted such a phenomenon in series-compensated lines connected to T-G units. As a result of extensive study of Project Navajo, countermeasures to combat the SSR problem have been designed and are operating successfully. The SSR failure must therefore be considered as one of the governing factors in design of series-compensated lines when they are used for evacuating power from large thermal power stations. The combined cost of series-capacitor installation and the countermeasures is lower than the cost of additional transmission lines required when no series-compensation is used. We shall briefly describe the SSR problem and the counter-measures, since torque interactions in the generator belong to the realm of synchronous-machine theory which is outside the scope of this book.

Figure 12.11(a) shows the torque and speed conditions in the generator which cause failure of the shaft, while conditions when countermeasures are taken to suppress the breaking torque are shown in Fig. 12.11(b). Three distinct problems have been identified in SSR problem which are called

(1) Induction Generator Effect,
(2) Torsional Interaction, and
(3) Transient Torque Problem.

The first two are known as steady-state problems while the last one occurs when system conditions change due to short-circuits and switching operations.

12.7.2.1 Induction Generator Effect

The electrical resonance frequencies, f_e, produce sinusoidal perturbations in the generator which are superimposed on the synchronously-rotating field at f_0. The resulting torque variation produced on the rotor is also sinusoidal and occurs at the same speed as the perturbing field. When the frequency of perturbation is $f_m = f_0 - f_e$, the torque perturbation is in phase with the speed perturbation. This phenomenon is called "negative damping" and is the main cause of SSR oscillations. During the SSR condition, a large current flows in the stator circuit giving rise to large torques.

Thus, at sub-synchronous frequencies causing resonance, the reactance of the system viewed from the generator terminals is zero and the resis-tanu may be negative. This condition gives rise to self excitation of the oscillatory currents at this natural frequency and dangerous condi-tions will build up progressively. The shaft mechanical life is usually represented in graphical form as shown in Fig. 12.11 (c). If the mag-nitude of torque is below the level shown as ∞, no damage to the shaft occurs and its life can be counted as infinity. However, if the torque oscillation exceeds the level marked 0, the fatigue life of the shaft be-comes zero which indicates that the shaft will fail. This is also known as the "Once-in-a-Lifetime Torque Limit."

Normally, the long shaft of a T-G set consists of sections with varying

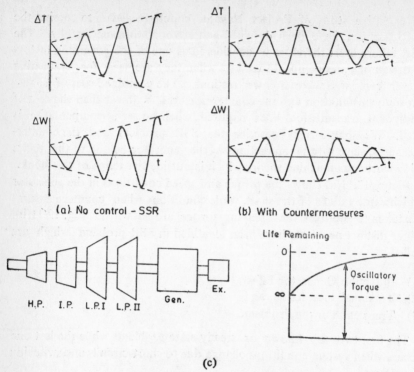

Fig. 12.11. Sub-synchronous Resonance condition.
(a) Variation of torque and angular velocity without counter measures.
(b) Same, with countermeasures.
(c) Life-strength of shafts of T-G units.

dimensions, resulting in different values of stiffness and damping giving rise to several modes of torsional resonance frequencies. An example of the Navajo Project electrical and mechanical torsional frequencies are given in the table below.

Electrical System

Natural Frequencies, f_e	44.5	43.5	30.7	28.3	25.5	10.5	10
Torsional Frequencies	44.2	39.8	34	26.7			6.8
Mode Number	1	2	3	4			5

Remedy for Countering Induction Generator Effect

The principles on which counter measures to combat the self-excitation problem and the resulting induction-generator effect are based on the following ideas:

(1) *Addition of pole-face or amortisseur windings to reduce the rotor resistance of generator at the sub-synchronous frequencies, which help to*

produce a net damping (positive resistance) at the slip frequencies. They are used in salient-pole generators to prevent hunting but are not normally used in cylindrical-rotor turbo-generator. But the SSR problem has shown the need for incorporating them in round-rotor generators also.

(2) *Addition of series reactance in stator circuit.* This will help to de-tune the resonant network as viewed from the generator terminals. The leakage inductance of transformer can also be increased. However, the drawback of increased series reactance is the reduction of stability margin, and it is also not easy to design the large reactors required for the high load currents they have to carry.

(3) *System switching and unit tripping.* During potentially dangerous SSR conditions, system switching can take the form of shorting the series capacitors as described in the previous section, or isolating the generator from the system by switching it to an uncompensated line if this is available in order that power flow is not interrupted. But this requires that the system must be designed for such a scheme, and withstanding the transients resulting from switching operations.

(4) *Armature-current relay protection.* This relay senses the sub-synchronous-frequency armature currents in the range of 25% to 75% of power frequency at a level of 5% of rated current, when these frequencies are known to cause *SSR* danger. The relay will trip the generator under sustained sub-synchronous oscillations. There must be a further discrimination for current level or torque changes that take place under low-magnitude sustained conditions and high level changes under fault conditions. These may vary from 1% of rated current under sustained conditions to 300% under faults.

12.7.2.2 *Torsional Interaction*

Dangerous conditions for a shaft exist when an electrical resonance frequency or minimum reactance occurs very near one of the natural torsional modes of the T-G shaft. These frequencies are the eigenvalues of the spring-mass-damping model of the electro-mechanical system equations. The analysis of this problem is considered beyond the scope of this book, since the mechanical damping inherent in the turbine increases when loaded and therefore the problem is non-linear. Torsional interaction takes place at an electrical resonance frequency which is near the complement of a torsional resonant frequency, that is, complement frequency = synchronous frequency f_0 − torsional resonance frequency f_m. When this condition exists, even a small voltage generated by the oscillating rotor can give rise to large SS current in the stator resulting in large undamped torques which keep growing.

Solution to Problem

(1) *Dynamic Stabilizer.* This is a shunt reactor controlled by thyristors

whose value is modulated by the oscillatory currents in the stator, or the rotor velocity deviations. The reactor introduces enough current in the armature at the critical sub-synchronous frequency so as to cancel the current resulting from the transmission system with its series capacitor.

(2) *Reduction in Series-Capacitor Compensation or Complete Removal.* The flashing gaps across the series-compensation capacitor prevents the dangerous condition.

(3) *Filters.* Filters in series with the generator increase the effective circuit resistance. These are known as Line Filters. They have to carry the full load current.

However, static blocking filter connected at the neutral end of the h.v. winding of the generator-transformer offers a better solution. The scheme is shown in Fig. 12.12. The N separate filters in series are each tuned in parallel resonance to block current at the offending frequency giving rise to the torsional modes. The resistances of the filters are in the range 200 to 600 ohms and have high Q.

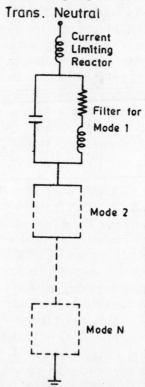

Fig. 12.12.   Filters used at transformer neutral for blocking sub-synchronous currents.

This can also be augmented by a device known as Excitation Damping

Control which acts on the Automatic Voltage Regulator by injecting a sinusoidal signal of proper phase which is obtained from a rotor-motion sensor.

(4) *Other Remedies.* Except for pole-face amortisseur windings, all countermeasures used for eliminating the induction-generator effect also help to reduce the torsional-interaction problem. In addition, a torsional-motion relay is used to detect excessive mechanical stresses in the shaft. This picks up signals from toothed wheels and proximity magnetic pick-ups at each end of the shaft. Filters are used to convert these signals proportional to the modal oscillations. It acts as a back-up protection as it is too slow in action. Its main function is to trip the generator in the event of excessive sustained torques.

12.7.2.3 *Transient Torque Problem*

Dangerous transient torques are obtained during faults in the system or when making switching operations to change system configuration. Two

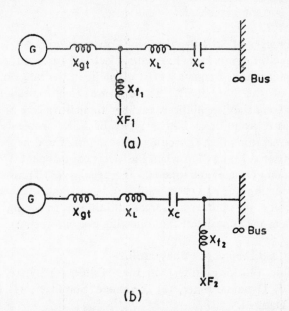

(a)

(b)

Fig. 12.13. Sub-synchronous conditions during faults on a system giving rise to transient torques of large amplitude.

types of fault conditions must be considered for their relative severity, as shown in Fig. 12.13.

In Figure 12.13(a) a fault occurs on the generator side of the transmission line. In this case the series capacitor will be charged to high voltage by the infinite bus. Before the fault occurs the natural frequency f_e is determined by X_{gt}, X_L and X_c of the generator-transformer, line, and the capacitor. This could be a complement to a rotor torsional mode frequency f_n, that is $f_e + f_n = f_0$ or $f_n = f_0 - f_e$, where f_0 = average synchronous frequency (50 or 60 Hz). When the fault is cleared, the capacitor will discharge partly through the generator at frequency f_e to produce an amplified torque.

On the other hand, referring to Fig. 12.13 (b), if a fault occurs at F_2 on the infinite-bus side, sub-synchronous current will flow into the fault and until the fault is cleared, torque amplification occurs. This case is worse than the previous case of fault at F_1. Nearly 10 times the once-in-a-lifetime torque level (0 life-line in Fig. 12.11 (c)) may be reached. In general, transient torques are large-amplitude problems and very sudden.

Remedy for Transient Torque Problem

Transient torques can be controlled by some of the schemes described before. These are: (a) Static Blocking Filter; (b) Series Line Filter; (c) Tripping the generator or unit tripping scheme; (d) Increased Series Reactance; and (e) Flashing the series capacitor which will also decrease the short-circuit current magnitude.

12.7.2.4 *Summary of SSR Problem and Countermeasures*

In series-capacitor compensated systems, electrical resonant frequencies below the synchronous frequency exist which lie in the range of torsional frequencies of the T-G shaft. These cause electro-mechanical sub-synchronous frequency oscillations leading to shaft failure in one of the several modes of excitation. Three types of SSR problems exist: (1) Induction Generator Effect caused by self-excitation and negative damping; (2) Torsional Interaction when the electrical resonant frequency is the complement of one of the torsional frequencies; (3) Transient Torque problems occurring during system faults and switching operations.

The remedies or countermeasures utilized in removing the shaft-failure problem may consist of the following five categories:

(1) *Filtering and Damping*: These utilize
   (a) Static Blocking Filter, (b) Line Filter, (c) Bypass Damping Filter, (d) Dynamic Filter, (e) Dynamic Stabilizer, (f) Excitation System Damper.

(2) *Relaying and Detecting*: These are
   (a) Torsional Motion Relay, (b) Armature Current Relay, (c) Torsional Monitor.

(3) *System Switching and Unit Tripping*

(4) *Modification to Generator and System*: These include
   (a) T-G modifications for new units altering stiffness, inertia and

damping of rotors, (b) Generator Series Reactance, (c) Pole-Face Amortisseur Windings.

(5) *Removal or Short-Circuiting the Series Capacitor*
This is already provided in the system during normal course for limiting short-circuit currents in the system.

12.8. Static Reactive Compensating Systems (Static VAR)

In the previous section, one type of reactor compensation for countering SSR was mentioned as a Dynamic Filter which uses thyristors to modulate the current through a parallel-connected reactor in response to rotor speed variations. The advent of high-speed high-current switching made possible by thyristors (silicon-controlled-rectifiers) has brought a new concept in providing reactive compensation for optimum system performance. Improvements obtained by the use of these static var compensators (SVC) or generators (SVG) or simply static var systems (SVS) are numerous, some of which are listed below:

(1) When used at intermediate buses on long lines, the steady-state power-handling capacity is improved.
(2) Transient stability is improved.
(3) Due to increased damping provided, dynamic system stability is improved.
(4) Steady-state and temporary voltages can be controlled.
(5) Load power factor can be improved thereby increasing efficiency of transmission and lowering of line losses.
(6) Damping is provided for SSR oscillations.
(7) Overall improvement is obtained in power-transfer capability and in increased economy.
(8) The fast dynamic response of SVC's have offered a replacement to synchronous condensers having fast excitation response.

In principle, when the voltage at a bus reduces from a reference value, capacitive Vars have to be provided, whereas inductive Vars are necessary to lower the bus voltage. Figure 12.14 shows schematically the control characteristics required and the connection scheme. The automatic voltage regulutor AVR provides the gate pulses to the proper thyristors to switch either capacitors or inductors. The range of voltage variation is not more than 5% between no load and full load.

Many schemes are in operation and some of them are shown in Fig. 12.15 and described below.

(1) *The TCR-FC System* (Fig. 12.15 (a))
This is the Thyristor Controlled Reactor-Fixed Capacitor system and provides leading vars from the capacitors to lagging vars from the thyristor-switched reactors. Because of the switching operation, harmonics

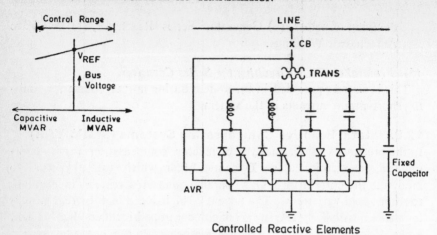

Fig. 12.14.   General circuit of Static Var Systems with controlled reactive
elements and control range exercised.

are generated.   Since there are 6 thyristors for the 3-phases, 6-pulse
harmonics are generated in addition to the 3rd.   These must be eliminated if they are not to affect normal system operation.

This can be improved by using a large number of small reactors to
reduce harmonics which normally depend on the reactor size. However,
it requires correspondingly larger number of thyristor switches. This is
known as segmented TCR-FC.

By using two transformers, one in Y/Δ and the other in Y/Y, and
dividing the fixed capacitors and controlled reactors into two groups,
12 thyristors can be used in a 12-pulse configuration. The lowest harmonic will be the 12th (600 Hz or 720 Hz in 50 or 60 Hz systems). It introduces complications in transformer design, and requires more thyristors.
The gate-firing angle control is made more difficult because of the 30°
phase shift between the secondary voltages of the Y/Δ and Y/Y transformers.

(2)   *The TCT Scheme* (Thyristor Controlled Transformer)
A transformer of special design with almost 100% leakage impedance is
controlled on the secondary side by thyristor switches.   By connecting
the 3 phases in Δ, 3rd harmonics are eliminated. This system has better
overload capacity and can withstand severe transient overvoltages.   But
it is more expensive than TCR scheme.   See Fig. 12.15(b).

(3)   *TSC-TCR Scheme* (Fig. 12.15(c))
This is Thyristor Switched Capacitors and Thyristor-Controlled Reactors.
It has TCR's and capacitance changed in discrete steps.   The capacitors
serve as filters for harmonics when only the reactor is switched.

(4)   *MSR-TCR Compensator Scheme*
This is the Mechanically Switched Capacitor-Thyristor Controlled Reac-

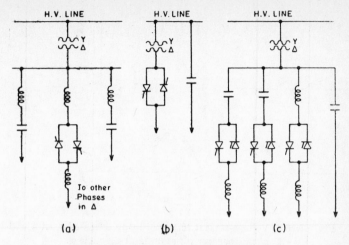

Fig. 12.15. Different types of Static Var Compensators.
(a) TCR-FC System, (b) TCT scheme, (c) TSC-TCR scheme

tors scheme. It utilizes conventional mechanical or SF_6 switches instead of thyristors to switch the capacitors. It proves more economical where there are a large number of capacitors to be switched than using TSC (thyristor-switched capacitors). The speed of switching is however longer and this may impair transient stability of the system.

(5) *SR Scheme* (Saturated Reactor)

In some schemes for compensation, saturated reactors are used. Fixed capacitors are provided as usual. A slope-correction capacitor is usually connected in series with the saturated reactor to alter the *B-H* characteristics and hence the reactance.

General Remarks

In all SVC schemes, harmonics generated by the switching operation are very important from the point of view of telephone and other types of interference, for example, signals. Filters are therefore an integral part of these Var compensating schemes. Fast-action of these SVC schemes with proper protection can offer great advantages in controlling voltages under load rejection. Most modern schemes utilize fibre optics for gate firing.

The current-voltage relationship as a function of firing angle are shown in Fig. 12.16(a) and (b) for switched reactors and capacitors respectively.

A complete 3-phase bank of switched reactor or capacitor connected in Δ is schematically shown in Fig. 12.16, which eliminates triplen harmonics.

12.9 High Phase Order Transmission

For e.h.v. transmission requirements, either ac 3-phase or bipolar dc is

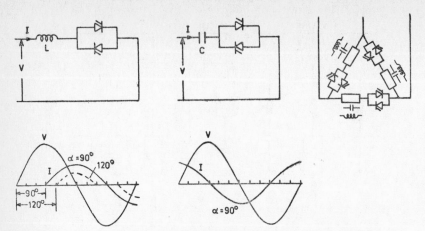

Fig. 12.16.   Voltage-current relationships in switched reactor and capacitor with variation in firing angle α.

used. Since such high voltage lines run mostly in the countryside, space for right-of-way does not pose as severe a problem as near metropolitan areas. However, wherever insulation clearances have to be optimized and lines have to be compacted, the use of higher than 3-phases on the same tower can be adopted. This is a recent development and studies are being carried out on a very extensive basis. One transmission line using 6 phases has already been commissioned in the U.S.A. and more are envisaged. We shall give here a brief discussion of a six-phase system only in so far as the transmission line is concerned and operating in the steady state. Transients and faults are not considered.

In a 3-phase system, the three voltages with respect to ground can be written as the phasors

$$V_a = V\angle 0°, \ V_b = V\angle -120°, \text{ and } V_c = V\angle 120° \qquad (12.70)$$

The magnitude of line-to-line voltages are also equal so that

$$|V_{ab}| = |V_{bc}| = |V_{ca}| = \sqrt{3}\,V.$$

When a double-circuit configuration is used, as shown in Fig. 12.17(a), we observe that the insulation clearance between the conductors are nearly

$$D_{ac'} = D_{bb'} = D_{ca'} = D, \ D_{ab} = D_{bc} = D_{b'c'} = D_{a'b'} = h,$$

$$D_{ac} = D_{a'c'} = 2h, \text{ and so on.}$$

On the other hand, the magnitude of voltages are

$$V_{ac'} = V_{ca'} = V_{ab} = V_{bc} = V_{b'c'} = V_{a'b'} = \sqrt{3}\,V;$$

$$V_{aa'} = V_{bb'} = V_{cc'} = 0 \text{ and so on.}$$

Comparing the insulation clearances with the voltages (neglecting the tower width) we observe that they bear no relation to each other. Thus, spatial distribution of insulation clearances do not match the voltages experienced by them. Therefore, the line is not optimized to this extent.

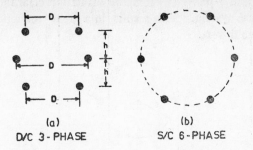

(a)   (b)

D/C 3 - PHASE   S/C 6 - PHASE

Fig. 12.17.   Double circuit 3-phase ac line and single-circuit 6-phase line.

Now consider the situation when the six conductors are energized by a 6-phase system of voltages, Fig. 12.17(b). These are

$$V_1 = V \angle 0°, \; V_2 = V \angle -60°, \; V_3 = V \angle -120°,$$

$$\left. \begin{array}{l} V_4 = V \angle -180°, \; V_5 = V \angle -240°, \; V_6 = V \angle -300° \\ V_4 = - V_1, V_5 = - V_2, V_6 = - V_3 \end{array} \right\} \quad (12.71)$$

Also

These are voltages to ground. The corresponding magnitudes of voltages between conductors are as follows;

$$\left. \begin{array}{l} V_{12} = V_{23} = V_{34} = V_{45} = V_{56} = V_{61} = V \\ V_{13} = V_{24} = V_{35} = V_{46} = V_{62} = \sqrt{3}\, V \\ V_{14} = V_{25} = V_{36} = 2V. \end{array} \right\} \quad (12.72)$$

From Fig. 12.17(b) it may be observed that the distances between the pairs of conductors denoted by the subscripts in equation (12.72) in the voltages are also equal to the corresponding voltage magnitudes. Thus, the spatial distribution of conductors correspond to the distribution of system voltages. Insulation clearances are optimized to a high degree and the lines can be compacted better than in a double-circuit 3-phase line.

For system voltages (line-to-ground) ranging from 80 kV to 442 kV, the voltages between adjacent phases in a 3-phase system and a 6-phase system are compared in the following table.

Phase-Ground Voltage, kV		80	133	199	230	289	442
Phase-Phase Voltage, kV	3-phase	138	230	345	400	500	765
(Between adjacent Conductors)	6-phase	80	133	199	230	289	442

We observe that it is better to define the phase-to-ground voltage as the system voltage since this is a common factor for both 3-phase and 6-phase systems. The voltage between conductors is in the ratio $1 : \sqrt{3}$

for the 6-phase and 3-phase systems and insulation clearances are smaller for the 6-phase system. This results in a great reduction in overall dimensions of the towers.

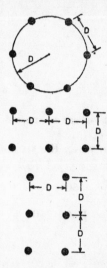

Fig. 12.18. Several alternative configurations for high phase order 6-phase line on tower.

Some of the proposed conductor configurations for 6-phase systems are depicted in Fig. 12.18.

Comparison of Power-Handling Capacity

For the sake of comparison, we assume that a double-circuit 3-phase system and a 6-phase system have the same conductor size and line-to-ground voltages. Therefore, the current per conductor is assumed the same for both systems. Then, the following power relations hold:

$$\left.\begin{array}{ll} 3\text{-}Phase\ D/C\ System & P_{3ph} = 6 \times V_{LG} \times I_C \\[2mm] 6\text{-}Phase\ S/C\ System & P_{6ph} = 6 \times V_{LG} \times I_C \end{array}\right\} \quad (12.73)$$

Therefore, cost for cost, there is no difference between the two systems as long as the line-to-ground voltages are equal and the current-carrying capacity of the conductors are also equal.

Other Factors

For medium-voltage lines up to 138 kV, 3-phase, (80 kV line-to-ground) and 6-phase 80 kV lines, with 1 metre clearance between adjacent conductors in the 6-phase system, the clearance between conductors in the 3-phase system will be 1.73 metres. As an example, with both having conductors of 2.3 cm diameter, the maximum surface voltage gradients on the conductors are:

3-phase system — 12 kV/cm (r.m.s)

6-phase system — 10.5 kV/cm (r.m.s)

Ground-level electrostatic fields (see Chapter 7) with 11-metre height of lowest conductor above ground are both less than 1 kV/m so that this is not a point of much concern.

Review Questions and Problems

1. Derive equations (12.1) and (12.2) for the voltage and current at any point on a transmission line in terms of the voltage at entrance to the line.

2. A 750 kV line has the distributed line constants $r = 0.025$ ohm/km, $l = 0.9$ mH/km, and $c = 12.3$ nF/km. At 50 Hz, calculate the following if the line is 600 km in length.

   (a) A, B, C, D constants.
   (b) The charging current and MVAR at a receiving end voltage of 750 kV, line-line, on no load.
   (c) The coordinates of the centre of the receiving end power-circle diagram.
   (d) The surge-impedance loading.

3. In Example 12.13, the shunt reactor compensation for maintaining

   $$E_s = 726 \text{ kV and } E_r = 765 \text{ kV, for a 750 kV}$$

   line, reactors had 3-phase ratings of 335 MVAR and 372 MVAR. With the sending-end voltage held constant at 726 kV, there is a total loss of reactor at the receiving end due to a fault. Calculate the receiving-end voltage resulting from taking the reactor out of service.

4. Repeat problem 3 if the reactors at both ends are taken out of service. Calculate the voltage at the receiving end with $E_s = 726$ kV.

5. Derive equations (12.63) and (12.64).

6. A very long line has series capacitance amounting to 50% of the series reactance. Calculate the natural electrical frequency.

7. List the dangers resulting from series capacitor compensation on long lines, and the remedies taken to counteract them.

8. What is the reason for the existence of SSR in the steady state and transient conditions in series-capacitor compensated lines?

9. List 3 dangerous conditions which give rise to SSR and all the countermeasures taken to guard against them. Use a tabular form.

CHAPTER 13

EHV Testing and Laboratory Equipment

13.1 Standard Specifications

All e.h.v. equipment voltage levels are governed by Standard Specifications adopted in each country. Equipment manufacturers produce goods not only for domestic use but also for export so that a design or testing engineer must be familiar with these specifications from all over the world. Some of the major specifications are I.E.C., I.S.I., V.D.E., S.A.E., A.N.S.I., B.S.S., C.S.A., U.S.S.R., Japan and others. In this chapter we shall discuss some of the requirements of e.h.v. insulation levels conforming to Indian Standards which parallel the I.E.C. specifications for most types of equipment and lines.

Standard Voltages

At and above 220 kV, the voltage levels that must not be exceeded in any part of the system are stipulated by IS 2026: Specifications for Power Transformers—Part III: Insulation levels and Dielectric Tests. Some typical values are quoted on next page.

Certain broad outlines can be discussed here regarding the voltage levels. With increasing thickness used for solid insulation structures in transformers immersed in oil, the danger of voids forming in the insulation is increased. For lower-voltage transformers, a test at twice the rated maximum voltage at double the power frequency for 60 seconds was specified without the need for partial-discharge measurement. However, at higher voltages the test voltage is 150% of maximum equipment voltage and the duration of voltage is increased to 30 minutes with the need for detection of partial discharges. From volt-time experiments carried out on such insulation structures, it has been ascertained that a transformer which does not show a failure under these test conditions can be expected to have reasonably long life if maintenance is excellent during service.

We also observe several levels of P.F., switching, and lightning-impulse withstand test voltages. These are necessary in view of the severity or otherwise of these voltages occurring in practice. For example, a transformer located in regions of high isokeraunik level should be tested at the higher test voltage level.

	220	275	345	400	500	750
1. Nominal System RMS kV (l – l)	220	275	345	400	500	750
2. Highest Equipment Voltage, kV RMS	245	300	362	420	525	765
3. Short Duration P.F. Withstand Voltage kV, RMS	325 360 395	395,460	460,510	570,630	790	1150
4. Rated Switching Impulse Withstand Voltage, kV crest		750,850	850,950	950,1050	1050,1175	1425,1550
5. Rated Lightning Impulse Withstand Voltage, kV crest	750 850 950	850,950 950,1050	950,1050 1050,1175	1050,1175 1175,1300 1425	1175,1425 1300,1550 1425	1550 1800 1800 1950

Note that there are several levels of P.F. voltages, lightning impulse voltages and switching-surge withstand voltages.

Typical values of interest are as follows:

Arrester Voltage Rating kV, RMS	Withstand Voltage up to 10 kA Arresters	Impulse Sparkover (1.2/50 μs), kV, Crest		kV, crest Residual Voltage
		10 KA	Front of the wave, kV/μs	
150	288	500/577 at	1080	500
174	334	570/660	1160	570
186	356	610/702	1180	610
198	380	649/746	1200	649
>198	1.9 × Rated Arrester Voltage			
To 225		3.28/3.78 Rated Voltage	1200	3.28 × Rated Voltage
To 396		3.26/3.76 Rated Voltage	1200	3.26 × Rated Voltage
>396		No values are agreed upon	1200	No values are specified

Lightning Arresters

Gapless type arresters are new and standard specifications are still being formulated. But for gap-type non-linear resistor type arresters, typical test values are given in IS 3070-Part I (1974), revised in 1982.

Three important items specified are (see page 358):

(1) Voltage withstand of the series gap;
(2) Maximum impulse sparkover of the gap;
(3) Maximum residual voltage of the resistor block.

The standard lightning-impulse waveshapes are (a) 1.2/50 μs for voltage, (b) 8/20 μs for current. These timings and their definitions will be described later.

In addition, a long-duration impulse current test is also required at 150 Ampere peak current for 2000 μs duration (of the peak) for 10 KA class of arresters. This simulates switching surges on lines.

A high-voltage laboratory should consist of sources or generators to provide the necessary test voltages, and in addition, be adequate to carry out research and development work. The remaining parts of this chapter will be devoted to description and design of these testing equipments.

13.2 Standard Waveshapes for Testing

The transient voltages impinging on equipment when connected in a system show a wide variation in magnitude and waveshape. It is only the power-frequency voltage which has a definite sinusoidal shape and is symmetric in both half cycles. However, from a vast experience and amount of experimental data collected, it has been ascertained that the severity of stress on the equipment insulation is properly represented during testing if these voltages and currents are standardized regarding their magnitudes and waveshapes. The magnitudes have already been given through examples from standard specifications. The waveshapes are discussed below. Their generation in a laboratory will be described later on.

13.2.1 Voltage Waveshapes

(1) *Lightning Impulse.* For testing all types of equipment including transformers under lightning impulses, the following definition for voltage will be used as per I.E.C. and Indian Standard Specifications as shown in Fig. 13.1. In this waveform, V_p = crest value, V_0 = overshoot, t_m = time to crest, t_t = time to 50% value on tail, t_1 = time to 90% value of crest, t_2 = time to 30% value of crest, 0 = actual zero, 0′ = virtual zero. All timings are measured from the virtual zero 0′, which is defined below.

The solid curve represents the actual desired waveshape generated

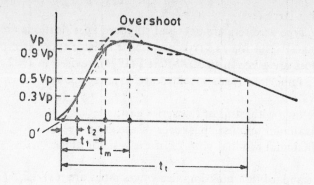

Fig. 13.1. Details of lightning-impulse voltage waveshape.

without oscillations on the front, and without the overshoot shown by broken line. The wave can be expressed as the difference between two exponentials with equation.

$$e(t) = E(e^{-\alpha t} - e^{-\beta t}) \tag{13.1}$$

There is an initial curvature at the wavetoe, followed by a nearly linear rise, then a very gradual rise to peak value, and finally a fall to zero value on the 'tail'. The important timings are:

(1) the time interval between 90% and 30% crest value, $t_1 - t_2$,
(2) the virtual zero of time (point 0′) obtained by joining the 90% and 30% voltages and extending it backwards to intersect the time axis (zero voltage),
(3) the time t_m from virtual zero to crest value,
(4) the time t_t from 0′ to 50% crest value on the tail,
(5) the time interval t_3 during which the voltage remains above 90% crest value.

On the same figure is shown an oscillatory portion with an overshoot V_0 since the presence of inductance in a two-element energy storage circuit with small damping gives rise to oscillations at the peak value.

The standard waveshape defined above has the following specifications for the timings and overshoot:

Front time: $t_f = 1.67 (t_1 - t_2) = 1.2\ \mu s \pm 30\%$ tolerance.

$$= 0.84\ \mu s\ \text{to}\ 1.56\ \mu s.$$

Tail Time: $t_t = 50\ \mu s \pm 20\%$ tolerance $= 40\ \mu s$ to $60\ \mu s$.
Overshoot: Not over 5% of crest value.

In normal laboratory-generated waves, the virtual zero occurs in less than 0.2 μs from the actual zero 0. The time to actual crest from zero, t_m, is in general twice the front time t_f. However, the actual value will depend on the waveshape timings with their tolerances and will be discussed later on.

(2) *Switching Surge for Testing Line Equipment*
For testing all equipment other than transformer under switching surge voltages, the front and tail timings for the double-exponential waveshape of Fig. 13.1 have the following values and tolerances:

Front Time: $t_f = 1.67(t_1 - t_2) = 250 \ \mu s \pm 30\% = 175$ to $325 \ \mu s$

Tail Time: $t_t = 2500 \ \mu s \pm 20\% = 2000$ to $3000 \ \mu s$.

For crest values longer than 100 μs, certain standard specifications permit the front time of 250 μs to equal the time to actual crest value of voltage. The test engineer must consult the standards for such variations allowed in definitions.

(3) *Switching Surge for Transformer Testing*
On account of the large inductance which an unloaded transformer presents to a high-voltage generator, oscillations result on the wavetail. For this reason, standards specify the following definitions for a switching surge for transformer testing: As per Indian Standards, the wavefront should be at least 20 μs, the length of surge from first zero to the occurrence of next zero must be at least 500 μs, and the magnitude of the surge must be in excess of 90% of crest value for at least 200 μs. When core saturation occurs the length of the surge may be shorter than 500 μs, but this is usually corrected by injecting a dc of proper polarity in special tests.

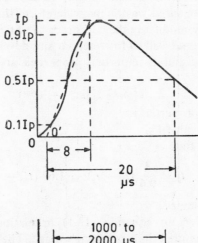

Fig. 13.2. Details of lightning-current waveshape.

13.2.2 Current Waveshapes

For type testing lightning arresters, the lightning current impulse wave-shapes have the following specifications, as shown in Fig. 13.2:

(1) *Short-Duration Impulse.* Time to crest = 8 μs on a straight line joining 10% and 90% crest values. Time to 50% value on tail = 20 μs.

(2) *Long-Duration Impulse.* Time duration of crest value is 1000 μs to 2000 μs depending upon the crest value of current. For 150 Amps, the duration is 2000 μs and for light-duty arresters for 100 Amps, it is 1000 μs.

Most of the voltage and current waveshapes of double-exponential types are ganerated by using capacitors, resistors, cascade-connected testing transformers. But the long-duration current wave is generated by a set of inductors and capacitors which represent a transmission line type of circuit.

13.3 Properties of Double-Exponential Waveshapes

A double-exponential voltage waveshape can be generated in principle by charging a capacitor and discharging it through properly-connected set of resistors and a load capacitance. The circuit description and design of such generators based on the principle of the Marx circuit will be discussed in the next section. Here, we shall describe the properties of a double-exponential waveshape which form the basis for designing the generators or modifying them in order to obtain suitable waveshapes for standard tests as well as for research and development purposes.

The waveshape and its important properties are shown in Fig. 13.1. The equation is

$$e(t) = E (e^{-\alpha t} - e^{-\beta t}) \tag{13.1}$$

The required properties are

(a) the crest value V_p,

(b) the front time

$$t_f = \frac{1}{0.6} (t_1 - t_2) = 1.67 (t_{90} - t_{30}) \tag{13.2}$$

(c) the tail time t_t.

The value of E in equation (13.1) in relation to the crest V_p will depend on the required timings for the voltage. The standard timings are

(a) for lightning, $t_f = 1.2 \pm 0.36$ μs, $t_t = 50 \pm 10$ μs.

(b) for switching, $t_f = 250 \pm 75$ μs, $t_t = 2500 \pm 500$ μs.

The values of α and β will be controlled by these timings, and vice versa.

Example 13.1: Examine the double-exponential waveshape for t_f, t_t, and V_p given that $E = 1.035$, $\alpha = 14.5 \times 10^3$, and $\beta = 2.45 \times 10^6$.

Solution: Normally, since t is in microseconds, we modify the values of α and β suitably. The given equation can be re-written as $e(t) = 1.035$ $(e^{-0.0145t} - e^{-2.45t})$ with t in μs.

For examining the wavefront, let t range from 0 to 1 μs. Then the following values are obtained.

$t = 0.1$ 0.15   0.2   0.4   0.6   0.8   0.85   0.9   1.0

$e = 0.2234$ 0.316   0.398   0.64   0.788   0.877   0.893   0.9074   0.9315

At $t = 0.141$, $e = 0.3$, and at $t = 0.87$, $e = 0.9$.

∴   The front time is

$$t_f = 1.67 \,(0.87 - 0.141) = 1.215 \text{ μs}.$$

At $t = 50$ μs, the voltage is $e = 0.501$. Therefore, the given values of E, α and β satisfy the rqeuirements for a standard 1.2/50 μs waveshape. The peak value $V_p = 1$ is obtained at $t = 2.22$ μs.

Example 13.2:   Given $E = 1.13$, $\alpha = 326$, and $\beta = 11 \times 10^3$, examine the waveshape $e = 1.13\,(e^{-326 \times t} - e^{-11 \times 10^3 \, t})$ for a switching impulse.

Solution: At $t = 29$ μs $= 29 \times 10^{-6}$ sec, $e = 0.3$.

At $t = 172$ μs $= 172 \times 10^{-6}$ sec, $e = 0.9$.

∴ $t_f = 1.67 \,(172 - 29) = 238$ μs.

At $t = 2500$ μs, $e = 0.5$.

These fall within the required tolerances. The wave reaches a peack value of 0.985 at 360 μs.

13.3.1   Lightning Impulses: Values of E, α, β

For design purposes, Fig. 13.3 (a), (b), (c) provide information about the three parameters in equation (13.1) for

$$t_f = 1.2 \pm 30\% \text{ and } t_t = 50 \pm 20\% \text{ μs}$$

timings. The method of calculating these is quite lengthy and will be outlined in section 13.4 next. It requires the solution of a set of non-

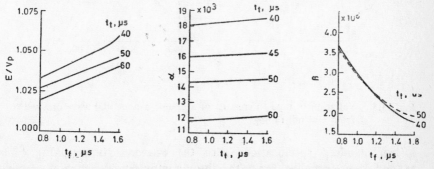

Fig. I3. 3.   Variation of E, α, β of a double-exponential voltage wave of lightning type with front time t_f. The tail time t_t is used as parameter.

linear equations and has been carried out by an iteration procedure using the Digital Computer. Simplified methods yielding quick results for engineering purposes will also be discussed later.

13.3.2   Switching Impulses: Values of E, α, β

For $t_f = 175$ to 325 μs and $t_t = 2000$ to 3000 μs,

the range of variation of E, α and β are shown in Fig. 13.4.

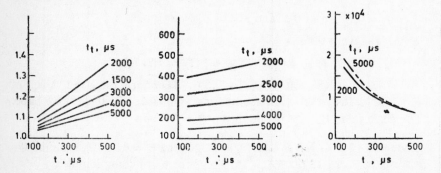

Fig. 13.4.   Variation of E, α, β for switching-impulse type of voltage wave with front time t_f.

In certain testing procedures, laboratories in various parts of the world have not maintained uniformity in reporting insulation properties as related to the waveshape timings. Some report the waveshape used in terms of the standard wavefront/wavetail timings while others use wavecrest/wavetail timings. In order to correlate the wavefront time t_f

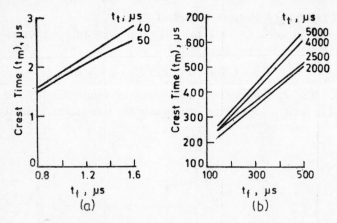

Fig. 13.5.   Variation of time to crest t_m of a double-exponential waveform with front and tail timings.

as per standard specifications with the wavecrest time t_m, Fig. 13.5 (a) and (b) are provided for both lightning impulse and switching surges. Of these, the latter is more important.

For switching-surge testing of insulation structures and to understand breakdown phenomena, one important parameter or time of the voltage is the duration during which the magnitude remains in excess of 90% crest. This depends on both the wavefront time and the time to

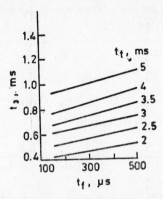

Fig. 13.6. Variation of time duration t_3 in which magnitude is in excess of 90% crest with front and tail timing of a double-exponential wave.

50% on the tail, t_f and t_t. Figure 13.6 provides this information. The following range of values are typical as observed from this figure.

Front Time t_f	$t_3 = t_{1t} - t_{1f}$	(for t_t from 2000 to 5000 μs)
150	420 to 930 μs	(2.8 t_f to 6.2 t_f)
250	470 to 980 μs	(1.88 t_f to 3.92 t_f)
500	590 to 1120 μs	(1.18 t_f to 2.24 t_f).

13.4 Procedures for Calculating α, β, E

The three quantities for a double-exponential wave are α, β, E which will completely specify the wave as given in equation (13.1). These have to be determined from the three known quantities for a surge, namely, (i) the crest value V_p, (ii) the front time t_f as defined by the standard

$$e(t) = E(e^{-\alpha t} - e^{-\beta t})$$

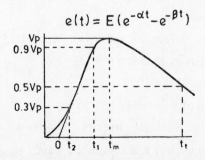

Fig. 13.7. Double-exponential waveshape with important timings and magnitudes.

specifications, and (iii) the time t_t to 50% value on the tail. The values of α, β, and E thus determined from V_p, t_f and t_t will be used to calculate the waveshaping resistors used in a generator as will be discussed later on. The complete waveshape is sketched in Fig. 13.7 along with important voltage magnitudes and timings, from which the following six properties or equations can be written down:

(1) At $t = t_2$, $e = 0.3\ V_p$ $\quad \therefore \quad 0.3\ V_p = E(e^{-\alpha t_2} - e^{-\beta t_2})$ $\qquad$ (13.3)

(2) At $t = t_1$, $e = 0.9\ V_p$ $\quad \therefore \quad 0.9\ V_p = E(e^{-\alpha t_1} - e^{-\beta t_1})$ $\qquad$ (13.4)

(3) At $t = t_t$, $e = 0.5\ V_p$ $\quad \therefore \quad 0.5\ V_p = E(e^{-\alpha t_t} - e^{-\beta t_t})$ $\qquad$ (13.5)

(4) $\quad t_f = (t_1 - t_2)/0.6$, by definition $\qquad$ (13.6)

(5) At $t = t_m$, $e = V_p$ $\quad \therefore \quad V_p = E(e^{-\alpha t_m} - e^{-\beta t_m})$ $\qquad$ (13.7)

(6) At $t = t_m$, $de/dt = 0$ $\quad \therefore \quad \alpha e^{-\alpha t_m} = \beta e^{-\beta t_m}$ $\qquad$ (13.8)

Of these, only V_p, t_f and t_t are known. We define

$$x = \beta/\alpha, \ y = t_2/t_m, \text{ and } z = t_t/t_m \qquad (13.9)$$

Also, let $p = 0.6\ t_f/t_t$, which is a known quantity. $\qquad$ (13.10)

Then, by manipulating the equations properly, the following equations for x, y, z are determined, which are non-linear.

$$x^{(pz+y)} - 1 = 3\ (x^y - 1) \cdot x^{pzx/(x-1)} \qquad (13.11)$$

$$3(x^z - 1) = 5(x^y - 1) \cdot x^{x(z-y)/(x-1)} \qquad (13.12)$$

$$3(x - 1) = 10(x^y - 1) \cdot x^{-x(y-1)/(x-1)} \qquad (13.13)$$

These have to be solved for x, y, z in terms of p, the only known quantity. In order to do so, the following approximations are made to obtain one equation relating x to p, x to y, and x to z so that by solving for x in terms of p, the values of y and z are determined.

(1) In practice, $x = \beta/\alpha$ is much larger than unity and x is also much greater than $y = t_2/t_m$. From examples shown previously, for the standard lightning impulse,

$$x = \beta/\alpha = 169 \gg 1, \text{ and } y = t_2/t_m \approx 1/3 \ll x.$$

For switching impulse,

$$x = \beta/\alpha = 33.75 \gg 1 \text{ and } y = t_2/t_m = 1/25 \ll x.$$

Using these properties in equation (13.13) gives

$$3(x - 1) = 10\ (1 - x^{-y}) \cdot x^{x/(x-1)} \qquad (13.14)$$

This gives the value of y in terms of x as

$$y = [\ln \{10/(10 - 3 \cdot x^{-1/(x-1)})\}]/\ln (x) \qquad (13.15)$$

(2) In a similar manner, y is eliminated from equation (13.12) and (13.13) to yield an equation relating z and x as follows:

$$(x^z - 1)/(x - 1) = 0.5 \cdot x^{x(z-1)/(x-1)} \tag{13.16}$$

Now, x is a quantity usually ranging from 50 to 200 and $z \approx t_t/t_m$ ranges from 10 to 20. Therefore, $(x^z - 1)/(x - 1) \approx x^{(z-1)}$. Using this in (13.16) gives

$$z = 1 + (x - 1) \cdot \ln(2)/\ln(x) \tag{13.17}$$

(3) Equations (13.15) and (13.17) for y and z are substituted in equation (13.11) to finally yield an equation for x in terms of $p = 0.6\ t_f/t_t$, the known quantity.

When the value of x is obtained by an iteration procedure, all other quantities, namely, $(y, z, \alpha, \beta, t_1, t_2, t_m$ and $E)$ can be calculated as follows:

(a) y from equation (13.15); (b) z from equation (13.17); (c) $t_m = t_t/z$; (d) $\alpha = \ln(x)/t_m(x-1)$; (e) $\beta = \alpha \cdot x$; (f) $t_2 = y \cdot t_m$; (g) $t_1 = 0.6\ t_f + t_2$; and finally (h) $E/V_p = (e^{-\alpha t_m} - e^{-\beta t_m})^{-1} = 0.5\,(e^{-\alpha t_t} - e^{-\beta t_t})^{-1}$.

Figures 13.3 aud 13.4 have been plotted from this procedure.

Case II. When the surge waveshape is defined through the time t_m to crest instead of the wavefront time t_f, the evaluation α, β, E becomes simpler. Referring to Fig. 13.7, the following 3 equations can be written.

(1) At $t = t_m$, $e = V_p$ $\therefore$ $V_p = E(e^{-\alpha t_m} - e^{-\beta t_m})$ $\qquad$ (13.18)

(2) At $t = t_t$, $e = 0.5 V_p$ $\therefore$ $0.5 V_p = E(e^{-\alpha t_t} - e^{-\beta t_t})$ $\qquad$ (13.19)

(3) At $t = t_m$, $de/dt = 0.$ $\therefore$ $0 = \beta e^{-\beta t_m} - \alpha \cdot e^{-\alpha t_m}$ $\qquad$ (13.20)

The resulting equation for $x = \beta/\alpha$ in terms of the known quantity $z = t_t/t_m$ becomes

$$(x - 1)/\ln(x) = (z - 1)/\ln(2) \tag{13.21}$$

The value of x is easily determined by trial and error, after which the values of α, β, E are calculated from (i) $\alpha = \mathrm{Ln}\,(2)/(t_t - t_m)$; (ii) $\beta = \alpha x$; and (iii) $E/V_p = (e^{-\alpha t_m} - e^{-\beta t_m})^{-1}$.

13.5 Waveshaping Circuits: Principles and Theory

A double-exponential waveshape of voltage across a capacitive load can be generated from the circuit shown in Fig. 13.8. A capacitor

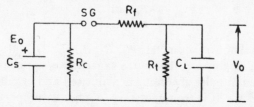

Fig. 13.8. Equivalent circuit of an impulse voltage generator for calculating waveshaping components.

C_s of the source is charged to a direct voltage E_0 through a charging resistor shown as R_c. The actual arrangement of C_s and R_c will be described in a later section, but when the waveshape is generated, they are in parallel. At the required instant of time, the spark gap SG is triggered. The value of $R_t \gg R_f$ in practice so that the source capacitance C_s charges the load capacitance C_L very quickly through the series resistance R_f. This gives the steep wavefront for the output voltage V_0. When C_L is sufficiently charged, both C_s and C_L discharge through R_t giving a slower wavetail portion. The waveshape of output voltage V_0 will be double-exponential as will be shown by an analysis of the circuit using Laplace Transforms. The values of C_s, C_L, R_f and R_t determine (a) the wavefront time, t_f; (b) the wavetail time t_t to 50% peak; and the peak value V_p of the output voltage V_0 in relation to the charging voltage E_0, that is, the voltage efficiency.

The charging resistor R_c is extremely high and is in parallel with C_s so that its effect on the waveshape of output voltage is negligible. How-ever, we will retain R_c in deriving the equations and then let $R_c \to \infty$. The operational form of the relation between the output voltage $V_0(s)$ and the charging voltage (E_0/s) when the spark gap SG is conducting will be

$$V_0(s) = \frac{E_0}{R_f C_L}\left[s^2 + \left(\frac{1}{R_f C_s} + \frac{1}{R_f C_L} + \frac{1}{R_t C_L} + \frac{1}{R_c C_s}\right)s \right.$$
$$\left. + \frac{R_c + R_f + R_t}{R_c R_f R_t C_s C_L}\right] \qquad (13.22)$$

Now, if the output voltage is $V_0(t) = E(e^{-\alpha t} - e^{-\beta t})$, its Laplace Transform is

$$V_0(s) = E(\beta - \alpha)/[s^2 + (\alpha + \beta)s + \alpha\beta] \qquad (13.23)$$

Comparing (13.22) and (13.23), and equating coefficients of like powers of s, there result the required equations between the generator compo-nents and the values of α, β, E determined from required timings as described in the previous section. These are:

$$\frac{1}{R_f C_s} + \frac{1}{R_f C_L} + \frac{1}{R_t C_L} + \frac{1}{R_c C_s} = \alpha + \beta \qquad (13.24)$$

$$(R_c + R_f + R_t)/R_c R_f R_t C_s C_L = \alpha\beta \qquad (13.25)$$

and
$$R_f C_L = \frac{E_0}{E} \cdot \frac{1}{\beta - \alpha} \qquad (13.26)$$

There are 5 unknown parameters R_c, R_f, R_t, C_s, C_L and only 3 equa-tions. However, in practice, the source capacitance and charging resistor are fixed when an impulse generator is purchased and installed in a high-voltage laboratory. It is then necessary to choose proper values of R_f and R_t for a given load capacitance C_L. The values of α, β, E are taken from Figs. 13.3 and 13.4 for the generation of required waveshapes for lightning and switching impulses.

When R_c is very large (> 1 Megohm), we obtain the following equa-tions from (13.24) and (13.25):

$$\frac{1}{R_f C_s} + \frac{1}{R_f C_L} + \frac{1}{R_t C_L} = \alpha + \beta, \quad \text{and} \quad 1/R_f R_t C_s C_L = \alpha\beta \quad (13.27)$$

Solving these for R_f and R_t, there result

$$R_t = \frac{\alpha + \beta}{2\alpha\beta\,(C_s + C_L)}\left[1 \pm \sqrt{1 - \frac{4\alpha\beta\,(C_s + C_L)}{C_L(\alpha + \beta)^2}}\,\right] \quad (13.28)$$

from which

$$R_f = 1/R_t C_s C_L \alpha\beta \quad (13.29)$$

From equation (13.28), we observe that the discriminant must be positive in order to obtain a real value for R_t. This requires

$$(C_L/C_s) \geqslant 4\alpha\beta/(\beta - \alpha)^2 \quad (13.30)$$

Voltage Efficiency

Equation (13.26) can be used for calculating the voltage efficiency defined as $\eta_v = V_p/E_0$. At $t = t_t$, $e = 0.5\,V_p$ so that

$$V_p = 2E\,(\exp\,(-\,\alpha t_t) - \exp\,(-\,\beta t_t))$$

$$= \frac{2E_0}{R_f C_L(\beta - \alpha)}\,(\exp(-\alpha t_t) - \exp\,(-\,\beta t_t)) \quad (13.31)$$

Therefore

$$\eta_v = V_p/E_0 = \frac{2}{R_f C_L(\beta - \alpha)} \cdot (\exp\,(-\,\alpha t_t) - \exp\,(-\,\beta t_t)) \quad (13.32)$$

Now, by imposing the condition that the voltage efficiency must be less than or equal to 1, (since we are not considering overshoot which cannot take place in the absence of inductance), we obtain the limiting value for $(R_f C_L)$ to be

$$R_f C_L \geqslant \frac{2}{\beta - \alpha}\,(\exp(-\,\alpha t_t) - \exp\,(-\,\beta t_t)) \quad (13.33)$$

13.5.1 Limitations on Values of Generator Components

There are certain limiting equations for the circuit components which can generate the required waveshape and efficiency. These are

(1) $C_s/C_L \leqslant (\beta - \alpha)^2/4\alpha\beta$ $\qquad\qquad\qquad\qquad\qquad$ (13.34)

(2) $R_f C_L \geqslant 2(\exp(-\,\alpha t_t) - \exp(-\,\beta t_t))/(\beta - \alpha)$ $\qquad$ (13.35)

(3) Since $R_f C_L = 1/\alpha\beta\,C_s R_t$, a limit on $R_t C_s$ is also placed, namely,

$$R_t C_s \geqslant \frac{1}{R_f C_L \,\alpha\beta} \leqslant \frac{\beta - \alpha}{2\alpha\beta}\,(\exp(-\,\alpha t_t) - \exp(-\,\beta t_t))^{-1} \quad (13.36)$$

Example 13.3:   Find suitable values for generator capacitance C_s and the resistance R_f and R_t for generating a 1.2/50 μs lightning impulse across a load $C_L = 2000$ pF. The values of α and β are $\alpha = 14.5 \times 10^3$ and $\beta = 2.45 \times 10^6$.

Solution: $(\beta - \alpha)^2/4\alpha\beta = 41.74$ and $41.74\,C_L = 83.5$ nF.

Therefore, any value of C_s less than 83.5 nF, according to equation (13.34) will yield a reasonable value for R_t, the resistance in parallel with the load C_L.

Also, $\quad R_f\,C_L \geqslant 2(e^{-14.5\times10^3\times50\times10^{-6}} - e^{-2.45\times50})/(2.45-0.0145)\,10^6$

$$\geqslant 0.398\times10^{-6}, \text{ from equation (13.35)}$$

Therefore $\quad R_f \geqslant 0.2\times10^3 = 200$ ohms for $C_L = 2000$ pF.

Comments. If the generator consists of 6 stages of capacitors connected in series while SG discharges, then each can be 500 nF$=0.5$ μF giving a total capacitance of 83.3 nF. We can select a resistance $R_f = 300$ ohms which gives $R_t = 1/\alpha\beta\,C_sC_LR_f = 563$ ohms. The series resistance R_f can consist of 6×50 ohms and R_t made from 6×94 ohms for waveshaping. According to equation (13.32) the voltage efficiency falls to 66.3%. In order to increase the efficiency, a lower value of R_f can be used which will require a higher value for R_t. For example, with $R_f = 250$ ohms, the efficiency is increased to 83% and $R_t = 680$ ohms.

Example 13.4:　The generator of previous example with $C_s = 83.3$ nF is required to generate a switching impulse of approximately 250/2500 μs waveshape for which $\alpha = 365$ and $\beta = 11\times10^3$. Design the values of C_L, R_f, R_t and find the efficiency.

Solution: $C_L/C_s \geqslant 4\alpha\beta/(\beta - \alpha)^2$ from equation (13.34). With given values, $C_L \geqslant 11.83$ nF.

We observe that the load capacitance is much higher for generating switching surges than for lightning impulse generation.

$$R_f\,C_L \geqslant 2(\exp\{-\alpha t_t\} - \exp\{-\beta t_t\})/(\beta - \alpha) = 75.5\times10^{-6}$$

Therefore $R_f \geqslant 6382$ ohms.

Let values of $C_L = 12$ nF and $R_f = 6.4$ kilohms be chosen.
Voltage efficiency $\eta_v = 75.5\times10^{-6}/(6.4\times10^3\times12\times10^{-9}) = 98.3\%$.

13.5.2　Approximate Equations for Quick Estimation of R_f and R_t
The previons discussion used a rigorous method of determining the values of waveshaping resistors with given generator capacitance and required load capacitance. A quick but approximate set of values for R_f and R_t can be obtained by the following procedure, on the lines of the following ideas.

(1) During the generation of the fast wavefront, the parallel resistance R_t has little influence over charging the load capacitance C_L from the generator capacitance C_s.

(2) Similarly, during the generation of the wavetail portion of the

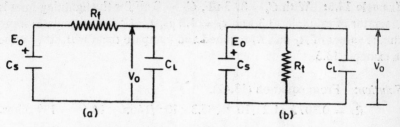

Fig. 13.9. Equivalent circuits during formation of wavefront and wavetail for quick estimation of R_f and R_t.

impulse, when both capacitances discharge through R_t, the series resistor exercises little influence.

These suggest the equivalent circuits, Fig. 13.9 (a) and (b) for the generation of wavefront and wavetail.

For the circuit of 13.9 (a),

$$v_0(t) = \frac{E_0 C_s}{C_s + C_L} (1 - \exp\{- t/T_f\})$$ (13.37)

and for 13.9 (b),

$$v_0(t) = \frac{E_0 C_s}{C_s + C_L} \cdot \exp(-t/T_t)$$ (13.38)

The two time constants T_f and T_t are, respectively,

$$T_f = R_f C_s C_L/(C_s + C_L) = R_f \times (C_s \text{ and } C_L \text{ in series})$$ (13.39)

and

$$T_t = R_t (C_s + C_L) = R_t \times (C_s \text{ and } C_L \text{ in parallel})$$ (13.40)

When the front and tail timings t_f and t_t of the wave are specified, the values of R_f and R_t can be calculated as shown below in terms of C_s and C_L.

At $t = t_2$, $v_0 = 0.3$ giving $1 - \exp(-t_2/T_f) = 0.3$, from Fig. 13.7,

At $t = t_1$, $v_0 = 0.9$ giving $1 - \exp(-t_1/T_f) = 0.9$.

These yield $t_2 = 0.3567 T_f$ and $t_1 = 2.303 T_f$.

But by definition,

$$t_f = 1.67 (t_1 - t_2) = 3.25 T_f = 3.25 R_f C_s/(C_s + C_L)$$ (13.41)

Therefore $\qquad R_f = 0.3077 t_f \cdot (C_s + C_L)/C_s C_L.$ (13.42)

Again, from the definition of wavetail timing, $\exp(-t_t/T_f) = 0.5$ giving $t_t/T_t = \ln 2 = 0.6931$. Therefore $T_t = t_t/0.6931 = 1.4428 t_t$.

But

$$R_t = T_t/(C_s + C_L) = 1.4428 t_t/(C_s + C_L)$$ (13.43)

Therefore, equations (13.42) and (13.43) give values for R_f and R_t very quickly when C_s, C_L, t_f and t_t are known and does not require a knowledge of α and β. However, these resistance values are approximate and in a laboratory some adjustment will be necessary after looking at the actual waveform on an oscilloscope.

Example 13.5: With $C_s = 83.3$ nF, $C_L = 2$ nF for the lightning-impulse generator of example 13.3 and $t_f = 1.2$ μs, $t_t = 50$ μs, calculate approximate values of R_f and R_t required and compare them with the values in example 13.3.

Solution: From equation (13.42),

$$R_f = 0.3077 \times 1.2 \times 10^{-6} \times 85.3 \times 10^{-9}/166.6 \times 10^{-18} = 189 \text{ ohms}.$$

Note that R_f was calculated to have a minimum value of 200 ohms in the previous example.

From equation (13.43)

$$R_t = 1.4428 \times 50 \times 10^{-6}/85.3 \times 10^{-9} = 846 \text{ ohms}.$$

From the exact expression R_t can be calculated for $R_f = 200$ ohms which will be $R_t = 1/\alpha\beta\, C_s\, C_L\, R_f = 846$ ohms. This is exactly the same value as obtained from the approximate equivalent circuit, Fig. 13.9(b).

Example 13.6: Estimate the values of R_f and R_t quickly for the switching-surge generator with $C_s = 83.3$ nF, $C_L = 12$ nF, $t_f = 250$ μs, and $t_t = 2500$ μs.

Solution: $R_f = 0.3077 \times 0.25 \times 10^{-3} \times 95.3 \times 10^{-9}/(83.3 \times 12 \times 10^{-18})$
  $= 7.33$ Kilohms.

The value obtained in example 13.4 was 6.4 Kilohms.

$$R_t = 1.4428 \times 2.5 \times 10^{-3}/95.3 \times 10^{-9}$$
$$= 38 \text{ Kilohms against 39 Kilohms obtained before.}$$

From equation (13.37) and Fig. 13.9(a) we observe that R_f controls the wavefront time and so is called the wavefront resistor. Similarly, R_t may be designated the wavetail resistor, as shown in equation (13.38) and Fig. 13.9(b).

13.6 Impulse Generators with Inductance

In section 13.5, the several inductances inherent in an impulse generator and the test circuit were not considered. There exist inductances in the generator itself, the connecting leads, and measuring system which give rise to a deformation of the ideal double-exponential waveform on the front, peak, and tail. Thus, no waveshape generated in an impulse generator is truly double exponential. These inductances are

(a) inductance in the generator capacitance, resistors, and leads connecting them internally;
(b) inductance in the high-voltage lead connecting the generator output terminal with the voltage divider and the test object:
(c) inductance in the voltage divider column; and
(d) inductance in the ground connection.

13.6.1 Inductance in Generator

Figure 13.10(a) an inductance L is shown in series with C_s. In normal practical impulse generator circuits, the arrangement of R_f and R_t can

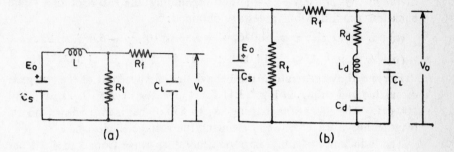

Fig. 13.10. (a) Impulse generator with inductance.
(b) Impulse voltage divider with resistance and inductance connected to impulse generator and load capacitance.

change with the generator connection. In this circuit, R_t is shown on the source side but this does not alter the characteristics to any great extent from the previous section. The operational expression for the output voltage is

$$V_0(s) = E_0 \frac{R_t}{(R_f + R_t) LC_L} \cdot \frac{1}{s^3 + as^2 + bs + c} \tag{13.44}$$

where $\quad a = (L + R_f R_t C_L)/(R_f + R_t) L C_L,$

$$\left. \begin{array}{l} b = (R_f C_L + R_t C_L + R_t C_s)/(R_f + R_t) L C_s C_L \\ \text{and} \qquad c = 1/(R_f + R_t) L C_s C_L \end{array} \right\} \tag{13.45}$$

Two cases arise for the roots of the denominator,

$$s^3 + as^2 + bs + c = 0.$$

These are:

(1) All three roots are real. Let these be $(-\alpha, -\beta, -\gamma)$. Then the time response of output voltage will be

$$v_0(t) = E_0 \cdot \frac{R_t}{(R_f + R_t) LC_L} \cdot \frac{1}{(\alpha-\beta)(\beta-\gamma)(\gamma-\alpha)} [(\gamma-\beta) e^{-\alpha t}$$
$$+ (\alpha-\gamma) e^{-\beta t} + (\beta-\alpha) e^{-\gamma t}] \tag{13.46}$$

(2) One real root $-\alpha$, and a pair of complex conjugate roots $-\beta \pm j\gamma$. For this case, the inverse transform of equation (13.44) is

$$v_0(t) = E_0 \frac{R_t}{(R_f + R_t) LC_L} [Ae^{-\alpha t} + e^{-\beta t} (B \cos \gamma t + D \sin \gamma t)] \tag{13.47}$$

where $A = 1/\{(\beta-\alpha)^2 + \gamma^2\}$, $B = -A$,
and $\quad D = \{1-(\beta^2 + \gamma^2 - \alpha\beta) A\}/(\alpha\gamma)$ $\left. \begin{array}{l} \\ \end{array} \right\}$ $\tag{13.48}$

Example 13.7: For an impulse generator, take $C_s=20$ nF, $C_L=2$nF, $R_f=376$ ohms, $R_t=3120$ ohms. For $L = 40$ μH and 100 μH, obtain the values of α, β and γ.

Solution: Using Newton's method for finding the real root of a cubic equation, the following values are obtained:

(a) $L = 40$ μH: $\alpha = 14{,}046$, $\beta = 1.83 \times 10^6$, $\gamma = 6.688 \times 10^6$.

(b) $L = 100$ μH: $\alpha = 14{,}050$, $\beta = 1.743 \times 10^6$, $\gamma = 1.363 \times 10^6$.

For the above generator, the behaviour of wavefront of the surge is calculated and shown in Fig. 13.11 for the cases of $L = 0$, 40 μH, and 100 μH. The effect of increasing R_f to 550 ohms is also shown on its lengthening of wavefront and reducing the voltage efficiency.

The inductance of e.h.v. impulse generators range from 3 to 4 μH per stage. A 12-stage generator has a total inductance of about 40 μH when all stages are connected in series during discharge.

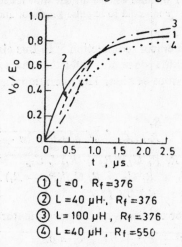

① L = 0, R_f = 376

② L = 40 μH, R_f = 376

③ L = 100 μH, R_f = 376

④ L = 40 μH, R_f = 550

Fig. 13.11. Effect of impulse-generator inductance on wavefront of lightning impulse.

13.6.2 Inductance of Lead Connecting Generator and Load

The physical sizes of extra high voltage equipment, laboratory dimensions and air gap clearances are on the increase so that very long leads are necessary in order to interconnect various components during testing. These give rise to travelling-wave effects when the generator discharges. For example, a lead length of 20 metres has a travel time for the impulse of 1/15th microsecond at light velocity. Repeated reflections from the generator and load give rise to oscillations which will appear on both the load and the measuring system. This affects the buildup of the wavefront. This has been investigated intensively by manufacturers and testing engineers resulting in the invention of the Zaengel Voltage Divider measuring system. It uses a suitable value of resistor at the entrance to the voltage divider high-voltage terminal to damp the oscillations.

13.6.3 Inductance of the Voltage Divider

Voltage dividers are used for stepping down the high voltage to a value suitable for the low-voltage measuring equipment. These dividers may be of the pure resistive type, pure capacitive type, or a combination of resistance and capacitance, the resistor used only to damp any natural oscillation caused by the high-voltage capacitor and the inductance originating from the height of the capacitor itself. For example, a capacitance of 1000 pF with an inductance of 2.5 μH requires a resistance $R_d = 2\sqrt{L_d/C_d} = 100$ ohms for critical damping of the oscillations. The effect of inductance L_d of the divider is to increase the rate of rise of wavefront so that the measured waveform will have a shorter wavefront time than the actual voltage to which the test object will be subjected.

The equivalent circuit of an impulse generator circuit considering the divider capacitance, inductance, and resistance is shown in Fig. 13.10(b). The Laplace-Transform of the output voltage is .

$$V_0(s) = \frac{E_0}{R_f C_L} \cdot \frac{s^2 + as + b}{s^4 + ms^3 + ns^2 + ps + q} \tag{13.49}$$

where $a = R_d/L_d$, $b = 1/L_d C_d$,

$m = R_d/L_d + 1/R_f C_L + 1/R_f C_s + 1/R_t C_s$

$n = 1/L_d C_d + (R_d/L_d)(1/R_f C_L + 1/R_f C_s + 1/R_t C_s)$
$\qquad\qquad\qquad\qquad + 1/R_f R_t C_s C_L$

$p = (1/R_f C_L \quad 1/R_f C_s + 1/R_t C_s)/L_d C_d + (R_d + R_f)/$
$\qquad\qquad\qquad\qquad\qquad (R_f R_t C_L C_d L_d)$

and $\qquad q = \{R_t C_s + C_d(R_f + R_t)\}/(R_f R_t^2 C_s^2 C_L C_d L_d)$

$\qquad\qquad\qquad\qquad\qquad\qquad (13.50)$

The inverse transform can be evaluated when numerical values for all circuit components are given Fig. 13.12 shows waveforms (with an expanded wavefront) for an impulse generator whose details are as follows:

$$C_s = 20.8 \text{ nF}, \ C_L = 1 \text{ nF}, \ R_f = 276 \text{ ohms}, \ R_t = 3120 \text{ ohms}$$

The voltage divider details are

(a) $R_d = 100$ ohms, $C_d = 1$ nF, $L_d = 2.5 \mu$H.

(b) $R_d = 50$ ohms, $C_d = 2$ nF, $L_d = 1.25 \mu$H.

Set (b) consists of two of set (a) connected in parallel. As a comparison, the output voltages when the divider inductance is neglected are also shown. Note that the presence of L_d has given a steeper rise in output voltage.

13.7 Generation of Switching Surges for Transformer Testing

The equivalent circuit of an impulse generator with its waveshaping circuit

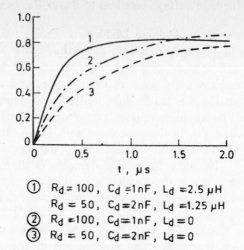

① $R_d = 100$, $C_d = 1nF$, $L_d = 2.5 \mu H$
  $R_d = 50$, $C_d = 2nF$, $L_d = 1.25 \mu H$
② $R_d = 100$, $C_d = 1nF$, $L_d = 0$
③ $R_d = 50$, $C_d = 2nF$, $L_d = 0$

Fig. 13.12. Load voltage waveshapes with R_d and
L_d of the voltage divider.

and transformer to be tested is shown in Fig. 13.13(a). With R_{d1} omitted,
C_s and C_{sh} are in parallel and the circuit is simplified to Fig. 13.13(b).

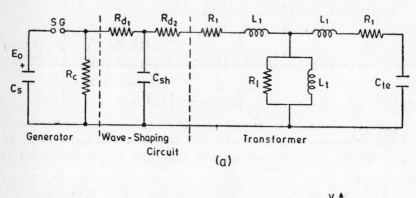

(a)

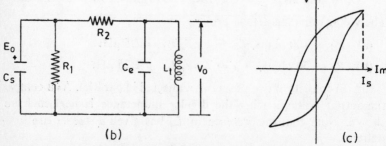

(b)

(c)

Fig. 13.13. (a) Complete equivalent circuit of transformer undergoing
switching impulse test.
(b) Simplified equivalent circuit.
(c) B-H curve of transformer core illustrating saturation current.

The figure also shows the B-H curve of the transformer core. Since the transformer undergoing test is unloaded, the leakage inductance L_1 is omitted.   Also let core loss be ignored.

Normally, negative polarity is used since positive polarity yields erratic results.   When core saturation takes place, the inductance L_t of transformer becomes very low and the voltage collapses when the saturation current I_s is reached.   The standard specifications require that the wavefront should be at least 20 μs, the total length of the switching surge at least 500 μs, and the magnitude should be in excess of 90% of crest value for at least 200 μs. We assume that the length of surge is denoted by the time from first zero to the instant the magnetizing current I_m equals I_s. These form the design constraints.

The output voltage V_0 across the transformer in operational form is

$$V_0(s) = \frac{E_0}{R_2 C_e} \frac{s}{s^3 + as^2 + bs + c} \tag{13.51}$$

where
$$a = L_t(R_1 C_s + R_1 C_e + R_2 C_e)/(R_1 R_2 C_s C_e L_t)$$
$$b = (R_1 R_2 C_s + L_t)/(R_1 R_2 C_s C_e L_t) \tag{13.52}$$
and
$$c = (R_1 + R_2)/(R_1 R_2 C_s C_e L_t)$$

Let the roots of the polynomial $(s^3 + as^2 + bs + c) = 0$ be denoted as $s = -p, -q + jr, -q - jr$.   Then, the time variation of the output voltage will be

$$V_0(t) = \frac{E_0}{R_2 C_e} [Ae^{-pt} + \sqrt{B^2 + C^2} \cdot e^{-qt} \cdot \cos(rt - \varphi)] \tag{13.53}$$

where
$$A = -B = -p/\{(q - p)^2 + r^2\}$$
$$C = (q^2 + r^2 - pq)/[r\{(q - p)^2 + r^2\}] \tag{13.54}$$
and
$$\tan \varphi = C/B.$$

The current through the magnetizing inductance is

$$I_m(s) = V_0(s)/sL_t \tag{13.55}$$

Its time variation will be

$$i_m(t) = \frac{E_0}{R_2 C_e L_t} \frac{1}{(q - p)^2 + r^2} \left[e^{-pt} + \frac{\sqrt{(p + q)^2 + r^2}}{r} e^{-qt} \cdot \cos(rt - \varphi') \right] \tag{13.56}$$

where
$$\tan \varphi' = (q - p)/r \tag{13.57}$$

Saturation will set in at a time t_0 when $i_m(t_0) = I_s$, from which the time of collapse of the surge will be determined and hence the length of the switching surge.

Figure 13.14 shows the output voltage for two sets of impulse-generator component values.

(1) $C_s = 0.125$ μF, $R_1 = 22$ K, $R_2 = 3.7$ K, $C_e = 6$ nF.
(2) $C_s = 20.1$ nF, $R_1 = 132$ K, $R_2 = 22.2$ K, $C_e = 10$ nF.

The second set consists of 6 stages, each stage having the component values given by set (1). The inductance of the transformer, L_t, has been varied from 10 Henry to 100 Henry. On these figures, saturation of core has not been considered.

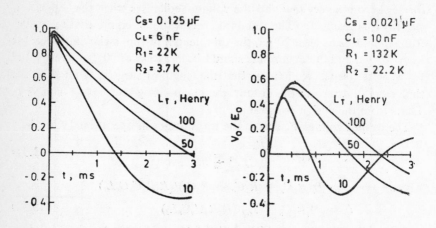

Fig. 13.14. Voltage waveshapes across transformer showing variation of impulse-generator circuit components and transformer inductance.

We observe that with lower transformer inductances the length of wave decreases and oscillations set in. With a smaller value of generator capacitance, the magnitude V_p/E_0, i.e. the voltage efficiency, decreases, and the wave is shorter. In certain cases, the desired time of 200 μs for the voltage to remain in excess of 90% crest values may be difficult to achieve. The wavefront becomes steeper with a higher generator capacitance.

Many problems in laboratory waveshape generation and e.h.v. line design have been worked by Ms. Amruthakala using Computer Graphics. (see Ref. 14, 'Other Journals' in Bibliography).

13.8 Impulse Voltage Generators: Practical Circuits

The general principles of impulse generators for generating desired waveshapes have been discussed in previous sections. The circuits used in practice will be described here. Voltage dividers used for measuring the high voltages as well as voltmeters, oscilloscopes, and other equipment will be described in the next section.

The important components of an impulse generator are basically (a) the impulse capacitor, C_s; (b) the load capacitor C_L; (c) the wavefront-shaping resistor, R_f; and (d) the wavetail-shaping resistor, R_t. In addition, other components are the charging transformer-rectifier, the charging resistor, spark gaps and other auxiliaries. In some generators, the series resistor R_f is inserted between the high-voltage output terminal and the terminal common to the voltage divider and test object. In other designs, it consists of several smaller resistors connected in series with the impulse capacitors in each stage in order to damp out any oscillations occurring due to stage inductance. In principle, the generation of extremely high voltages up to 6000 kV or 6 MV requires a very large number of stages of capacitors, each rated for about 200 kV, to be charged in parallel from the low voltage, and discharged in series.

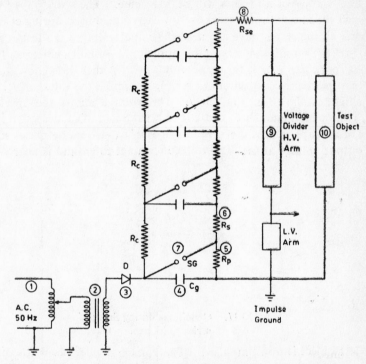

Fig. 13.15. Practical 4-stage impulse generator circuit layout.

Figure 13.15 shows a complete schematic diagram of one type of Marx Circuit. Figure 13.16 shows the overall equivalent circuit when discharging. Only 4 stages are shown for clarity but this could be extended to any number, say 30, built on the same principle.

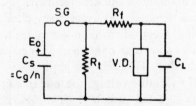

Fig. 13.16. Equivalent circuit of generator when discharging.

Referring to Fig. 13.15, 50 Hz ac is supplied to the regulating transformer (1) through an isolating transformer (not shown). The isolating transformer is usually a 3-phase transformer at power-frequency voltage with only two of the three windings used on the secondary side. Thus, line-to-line voltage is fed to the regulating winding (1). The diode (3) delivers a rectified ac to charge the impulse-generator capacitance C_g. While the first or lowest stage is charged directly, the second and subsequent-stage capacitances are charged through a combination of R_c, R_s, and R_p. The higher stages therefore acquire charges at slower rates but the final voltage of all stages will be the same. The spark gaps SG are set such that they do not sparkover at the voltage used during charging.

After all the stages have acquired their final voltage, the test engineer can 'fire' the impulse generator by applying a suitable potential to the lowest spark gap (7). If the stage voltage is E_s, then breakdown of the first gap places two capacitors in series, assuming the value of R_c to be very high, as shown in Fig. 13.17. The voltage across the spark gap SG_2 is now high enough to cause it to breakdown. The successive sparkover of all gaps places the several stages of capacitors in series so that the output voltage at the high-voltage terminal to ground is nearly $n E_s$.

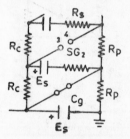

Fig. 13.17.   Conditions during firing of
series spark gap.

The test object is now impulsed. The voltage divider used for supplying a suitable low voltage to the measuring system usually consists of (a) a high-voltage high writing-speed oscilloscope, (b) a peak voltmeter, or (c) a digital reading device of the voltage.

An arrangement for initiating the triggering of the lowest gap is variously known as a 3-electrode arrangement or a trigatron. Figure 13.18 shows a simplified diagram of the arrangement. A low-voltage impulse generator discharges through a coupling capacitor to a third electrode called the Triggering Electrode which will place a voltage of opposite polarity to the sphere shown at right thereby increasing the voltage drop across SG momentarily. This fires the gap.

In order to obtain variable voltage output to suit the test object, the charging voltage is adjusted suitably. This also requires that the gap spacings of all spark gaps such as SG must be changed to prevent anomalous firing or not firing at all. In practical generators one set of spheres, say the ones at left, in all the multitude of spark gaps are fixed to an insulating column as tall as the impulse generator itself. The other companions are attached to another insulating column which is movable

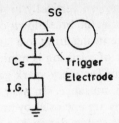

Fig. 13.18.   Three-electrode trigger gap for
firing impulse generator.

through a motor. The operator has to set the gap properly to suit every
test voltage value required.

The net equivalent circuit of Fig. 13.15 when all spark gaps SG have
fired is shown progressively in Fig. 13.19 for the 4-stage Marx Circuit.
For identical stages, currents in leads $a-b$ and $b-c$ are zero. Also, for

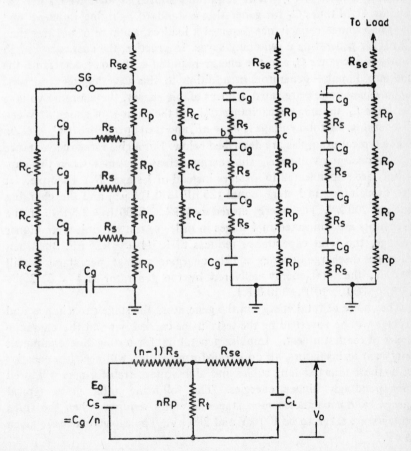

Fig. 13.19.   Equivalent circuit for Fig. 13.15.

very high values of R_c, they can be considered as open. Normally, all triggering gaps SG are arranged vertically in one column which allows the ultra-violet radiation from the breakdown of a lower gap help trigger the one above it. This prevents misfiring to a large extent. As the number of stages increases, the risk of misfiring is minimized still further by providing a triggering generator for every spark gap since collection of dust on the spheres is a great nuisance. In very high voltage generators, above 3 MV, the spark gaps are enclosed by an insulating cylinder. The proper operation of the test system depends so much on these spark gaps that very great care must be given to them. They also act as spark transmitters and disturb high-rate-of-rise phenomena such as the wavefront of a lightning impulse. This problem will be discussed later on.

13.9 Energy of Impulse Generators

We observed in Section 13.5, equation (13.34), that the load capacitance must have a minimum value, $C_L \geqslant C_s \cdot 4\alpha\beta/(\beta-\alpha)^2$, and in Examples 13.3 and 13.4, a value of $C_L = 2$ nF required a source capacitance C_s not exceeding 41.74 times C_L for generating a standard lightning impulse, and the same source capacitance required a load capacitance of not less than 12 nF for generating a switching surge. In practice, the energy stored in source capacitance C_s and the energy required in the load determine the design of impulse generators in addition to the waveshaping resistors. Impulse generators are rated in terms of this energy, the charging voltage per stage E_s, the stage capacitance C_g, and the number of stages n. When discharging the total voltage is $V_0 = nE_s$ and the capacitance is $C_s = C_g/n$. Some typical examples are described below. For a given stage capacitance C_g and charging voltage E_s, the energy increases n times when the capacitors are re-connected in n stages instead of 1 stage. For example, let the capacitance in 1 stage equal 125 nF $= 0.125$ μF, and the charging voltage 200 kV. The energy stored is $\frac{1}{2}cV^2 = 2500$ J $= 2.5$ KJ. If the capacitors are connected in 4 stages in order to test an object at higher voltage, the total capacitance reduces to $125/4 = 31.25$ nF. But when all stages discharge in series and the charging voltage per stage is still 200 kV, the total energy delivered by the generator is $\frac{1}{2} \times 31.25 \times 10^{-9} \times (800 \times 10^3)^2 = 10.0$ KJ.

The choice of total energy in the generator, the stage capacitance and voltage will be governed by the tests to be carried out and the characteristics of the test object. Impulse generators for testing line equipment only such as insulators, air-core transformers, etc., will need low energies while those that test long cables and high-voltage transformers will need correspondingly higher energies. The following table gives typical energies and capacitance in n stages in series required when the stage voltages are taken to be 100 kV and 200 kV. The capacitances are given in nF.

Table: *Total Capacitance of Impulse Generators when Discharging (nF)*

Total Energy, KJ	Stage Voltage 100 kV		Stage Voltage 200 kV	
	$n = 1$	$n = 10$	$n = 1$	$n = 10$
2.5	500	5	125	1.25
5	1000	10	250	2.5
10	2000	20	500	5
20	4000	40	1000	10
30	6000	60	1500	15

The relation between energy, stage voltage, number of stages in series when discharging and total capacitance of generator when discharging is given by

$$KJ = \tfrac{1}{2}C_s \cdot (n \times E_s)^2 \times 10^{-6} \qquad (13.58)$$

where C_s = capacitance of generator in nF and E_s = stage voltage in kV.

Example 13.8: An impulse generator consists of 6 stages with each stage having a capacitance of 0.125 μF (125 nF). The charging voltage is 200 kV. Calculate (a) the energy output of generator when all stages are connected in series, and (b) the energy when the 6 capacitors are reconnected in 3 stages and the charging voltage per stage is kept the same as 200 kV.

Solution: The two circuits are shown in Fig. 13.20.

(a) Total generator capacitance when discharging = $\tfrac{1}{6} \times 125$
$$= 20.83 \text{ nF}.$$
Total voltage when discharging = $6 \times 200 = 1200$ kV.
Therefore, energy = $\tfrac{1}{2} \times 20.83 \times 10^{-9} \times 1200^2 \times 10^6$
$$= 15{,}000 \text{ Joules} = 15 \text{ KJ}.$$

(b) When the generator is reconnected in 3 stages, each stage consists of 2 capacitors in parallel, Fig. 13.20(b), with a total stage capacitance of 250 nF. When the 3 stages discharge in series, the total generator capacitance is $250/3 = 83.3$ nF, and the output voltage is 600 kV.

Therefore, energy = $\tfrac{1}{2} \times 83.3 \times 10^{-9} \times 600^2 \times 10^6$
$$= 15{,}000 \text{ Joules} = 15 \text{ KJ}.$$

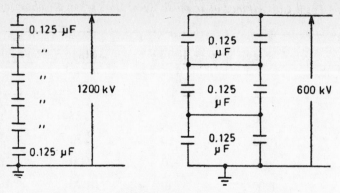

Fig. 13.20.   6-stage impulse generator reconnected for
energy relations.

Equation 13.58 can also be used for the two cases.

The capacitances of typical loads on an impulse generator may be summarized as follows:

Type of Test Object	Capacitance Range, C_L
(1) Line Insulators	50 pF for each disc. Total capacitance of string does not exceed a few pF.
(2) Bushings of Apparatus	100 to 500 pF.
(3) Power Transformers	1000 pF up to 1000 KVA 1000–10,000 pF over 1000 KVA
(4) Cables	150 to 300 pF per metre length.

The suggested minimum capacitance in generator when discharging is $5 \times$ load capacitance ($C_s > 5C_L$).   Therefore, when the flashover voltage is determined for the test object, the voltage output of generator is fixed. The number of stages, the stage capacitance, and charging voltages are then determined. These are normally fixed in a laboratory when once the generator is purchased. The following examples will provide guidelines for selection criteria of impulse generators for a small h.v. testing laboratory and a very large e.h.v. laboratory.

Example 13.9:   A small laboratory for testing apparatus up to 66 kV level (power frequency) requires an impulse generator.   Assume that the largest piece of load to be tested is 10 metres of cable having 200 pF/ metre capacitance. The impulse withstand is to be 325 kV, crest. Assume the breakdown voltage to be 20% higher and that a test voltage up to 450 kV should be available for research and development purposes (Research factor of 450/390 = 1.15). Design a suitable impulse generator for this laboratory.

Solution.   Cable capacitance = $200 \times 10 = 2000$ pF.   Assume a capacitive voltage divider of 1000 pF.

∴ Total load on generator, $C_L = 3000$ pF.

Approximate generator capacitance when discharging $C_s = 5C_L = 15$ nF. Selecting a standard charging voltage of 100 kV per stage will require 5 stages with a voltage efficiency in the range of 90% when the output is 450 kV. The capacitance per stage $C_g = C_s \cdot n = 15 \times 5 = 75$ nF.

Therefore, total energy when discharging $= \frac{1}{2} \times 15 \times (5 \times 100)^2 \times 10^{-6}$

$$= 1.875 \text{ KJ.}$$

Energy in each stage $= \frac{1}{2} \times 75 \times 100^2 \times 10^{-6} = 0.375$ KJ.

Example 13.10: An e.h.v. laboratory for testing equipment up to 765 kV level requires an impulse withstand level of 2400 kV, peak of lightning. voltage.

[*Note*: IEC and ISI have a minimum level of 1950 kV, but this will be increased in future to 2400 kV. This is not yet a standard.]

The flashover voltage is expected to be 20% higher and a R. and D. factor of 1.11 is to be incorporated (10% higher than flashover value). The load capacitance is 5 nF with the voltage divider expected to provide a further 500 pF capacitance. The stage charging voltage is 200 kV.

Select the required stage capacitance, number of stages, and the energy of generator when discharging.

Solution: The highest voltage required across the high voltage terminal of the generator is $2400 \times 1.2 \times 1.11 = 3200$ kV.

A voltage efficiency of 80% requires a total of 4000 kV in charging. Therefore, number of stages $= 4000/200 = 20$. The load capacitance is $C_L = 5 + 0.5 = 5.5$ nF. Hence, the generator capacitance when all 20 stages are discharging in series can be selected as $C_s = 5C_L = 27.5$ nF. Therefore, the capacitance of each stage is $C_g = 20 \times 27.5 = 550$ nF $= 0.55$ μF.

Energy $= \frac{1}{2} \times 27.5 \times (20 \times 200)^2 \times 10^{-6} = 220$ KJ.

Energy per stage $= \frac{1}{2} \times 550 \times 200^2 \times 10^{-6} = 11$ KJ

[*Note*. For such large energies and high voltages, a manufacturer's catalogue must be consulted for nearest standard size of capacitor, energy, and voltage efficiency.]

13.10 Generation of Impulse Currents

Lightning impulse currents up to 200,000 Amperes crest are generated by charging a set of impulse capacitors in parallel and discharging them the same way in a test object, which is usually a surge diverter or high rupturing-capacity fuse. The waveshape has a time to crest of 8 μs, and 20 μs to 50% value on tail, as shown in Fig. 13.2 before. Figure 13.21(a) shows the actual connection diagram and 13.21(b) the electrical equivalent circuit when the capacitors are discharging through the test object.

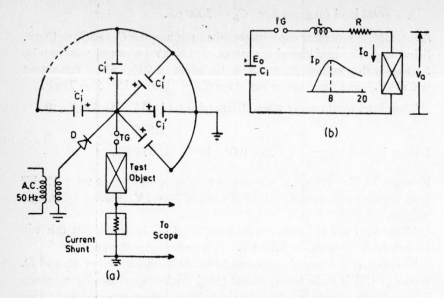

Fig. 13.21. Impulse current generator: (a) Actual Schematic of
Component Layout. (b) Equivalent Circuit.

In some special tests where the impulse current is to be super imposed
on a power-frequency voltage at a desired point on the wave, the circuit
shown in Fig. 13.22 is used. The trigger gap TG is arranged to fire
at a selected point on the 50 Hz wave. Both the impulse current and
the power-follow current are measured by utilizing the voltage drop produced on suitable non-inductive shunts. The power-frequency voltage
can be measured by a potential transformer PT while a resistive voltage
divider is used on the impulse-generator side beyond the diode D to
measure to charging voltage. A capacitive voltage divider is connected
across the test piece to measure the impulse voltage. The charging voltage
of the capacitors is again in the range of 100 kV to 200 kV.

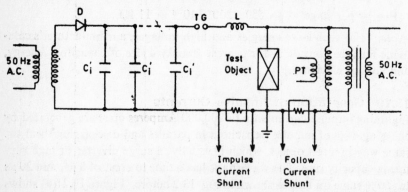

Fig. 13.22. Circuit diagram for simultaneous impulse current test and
power-frequency test on an arrester.

13.10.1 Analysis of Impulse Current Generator Circuit

Equivalent circuit of the impulse current generator is shown in Fig. 13.23. For the sake of analysis, we assume that the voltage across the test object remains constant at all values of current, that is, the test object is a metal oxide surge arrester. In gap-type SiC arresters, the current is a function of the voltage and nonlinear.

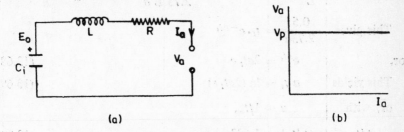

(a)   (b)

Fig. 13.23. Equivalent circuit of impulse-current generator for calculating circuit components, and arrester I-V characteristics (ZnO arrester).

For the gapless ZnO arrester, the following operational expression for current is valid

$$L \frac{di}{dt} + Ri + \frac{1}{C_i} \int_0^t i \, dt + V_a = E_0 \qquad (13.59)$$

$$I(s) = \frac{E_0 - V_a}{L} \cdot \frac{1}{s^2 + (R/L)s + (1/L\,C_i)} \qquad (13.60)$$

The requirement of the generator is to develop a unidirectional current with a peak value I_p occurring at time t_m and falling to 50% peak value at time t_t. For the sake of preliminary estimate, let us assume that the value of circuit resistance R can be adjusted to give critical damping of the circuit. Then,

$$R = 2\sqrt{L/C_t}. \quad \text{For this case,}$$

$$I(s) = \frac{E_0 - V_a}{L} \cdot \frac{1}{(s + a)^2} \qquad (13.61)$$

where $\qquad a = R/2L. \quad \text{Also,} \quad R^2/L^2 = 4/LC_t. \qquad (13.62)$

Then, the time variation of current through arrester is

$$i(t) = \frac{E - V_a}{L} \cdot t \cdot e^{-at} \qquad (13.63)$$

Its peak occurs when $t_m = 1/a = 2L/R$. Therefore, $at_m = 1$ (13.64)

This is obtained by differentiating (13.63) with respect to t and letting $di/dt = 0$ to find t_m. Since $t_m = 8$ μs, there results $R = 2L/t_m$ (13.65) Therefore, the required value of R is obtained in terms of the circuit inductance L, and known value of crest time t_m.

The peak value of current is

$$I_p = \frac{E - V_a}{L} \cdot t_m \cdot e^{-a\,t_m} = \frac{E - V_a}{L} \cdot t_m \cdot e^{-1} = \frac{E - V_a}{2.73} \cdot \sqrt{\frac{C_i}{L}} \quad (13.66)$$

At $t = t_t$, $i(t) = 0.5\,I_p$

Therefore, $\quad \dfrac{E - V_a}{L} \cdot t_t \cdot e^{-at_t} = 0.5 \cdot \dfrac{E - V_a}{2.73\,L\,a}, \quad$ since $t_m = \dfrac{1}{a}$ (13.67)

This gives $\quad \dfrac{0.5}{2.73a} = t_t \cdot e^{-at_t}$

or, $\qquad\qquad\quad e^{at_t} = 2at_t\,e \qquad\qquad\qquad\qquad\qquad (13.68)$

This yields $\qquad a\,t_t = \ln(2at_t\,e) \qquad\qquad\qquad\qquad (13.69)$

or, with $\qquad\quad a = 1/t_m,$

$$t_t/t_m - \ln(t_t/t_m) = 1.693 \qquad\qquad\qquad\qquad (13.70)$$

This equation determines the time t_t to half value when crest time t_m is known and for the circuit critically damped.

Example 13.11: An R-L-C circuit is critically damped. It generates a current of the type $i(t) = J.t.e^{-at}$ with peak occurring at $t_m = 8$ μs. (a) Calculate the time to 50% value on tail. (b) If the circuit inductance is $L = 40$ μH, find the values of circuit resistance R and the capacitance C_i of the impulse capacitance when discharging.

Solution: (a) From equation (13.70), a trial and error solution gives

$$t_t/t_m = 2.68.$$

This is nearly equal to the standard 20 μs. Therefore, the resistance is correct.

(b) $\qquad\qquad R = 2L/t_m = 2 \times 40 \times 10^{-6}/8 \times 10^{-6} = 10$ ohms.

$\qquad\qquad\qquad C_i = 4L/R^2 = 1.6$ μF.

If the damping is insufficient, the inverse transform of equation (13.60) should be worked out but with $R > 2\sqrt{L/C}$ which gives two real roots for the polynomial in the denominator. Let the waveshape be assumed a double exponential. Then,

$$I(s) = \frac{E_0 - V_a}{L} \frac{1}{\beta - \alpha} \left(\frac{1}{s + \alpha} - \frac{1}{s + \beta} \right) \qquad (13.71)$$

where $\quad \alpha = \dfrac{R}{2L}\left[1 - \sqrt{1 - 4L/R^2 C_i}\,\right] = a\left[1 - \sqrt{1 - 1/a^2 L\ \bar{C_i}}\,\right]$

and $\qquad \beta = \dfrac{R}{2L}\left[1 + \sqrt{1 - 4L/R^2\ C_i}\,\right] = a\left[1 + \sqrt{1 - 1/a^2 L\ C_i}\,\right]$ (13.72)

Also, $\beta - \alpha = \dfrac{R}{L} \sqrt{1-4L/R^2 C_i} = 2a\sqrt{1-1/a^2L\,C_i}$ \hfill (13.73)

Therefore

$$i(t) = \frac{E - V_a}{R[1-4L/R^2\,C_i]^{1/2}} \ (e^{-\alpha t}-e^{-\beta t}) = I_0 \cdot (e^{-\alpha t} - e^{-\beta t}) \quad (13.74)$$

Referring to equation (13.21) in Section (13.4) where $x=\beta/\alpha$ and $z=t_t/t_m$,

$$(x - 1)/\ln (x) = \frac{(\beta/\alpha) - 1}{\ln (\beta/\alpha)} = \frac{z - 1}{\ln (2)} = \frac{(t_t/t_m)-1}{\ln 2},$$

$\alpha = \ln (2)/(t_t - t_m)$, $\beta = \alpha \cdot x$, and $I_0/I_p = (e^{-\alpha\,t_m} - e^{-\beta\,t_m})^{-1}$

There are now sufficient equations for calculating R, C_i, and L when one of them is known.

Example 13.12: An impulse current generator consists of 10 impulse capacitors each of $0.125\,\mu F$ connected in parallel. The desired output waveshape has $t_m = 8\mu s$ and $t_t = 20\mu s$. Calculate

(a)   the approximate values of resistance and allowable inductance in the circuit to generate a double-exponential waveshape of current,

(b)   the charging voltage to obtain peak value of $I_p = 10$ KA through an arrester whose residual voltage is 25 kV (3 p.u. on 6 kV, r.m.s. 8.486 kV, peak, base).

Solution: Let $x = \beta/\alpha$. Also, $t_t/t_m = 20/8 = 2.5$. Therefore

$$(x - 1)/\ln (x) = (2.5 - 1)/\ln 2 = 2.1642.$$

A trial and error solution yields $x = 4$.

$\alpha = \ln 2/(t_t - t_m) = 5.776 \times 10^4$, so that $\beta = 4\alpha = 23.1 \times 10^4$.

From equation (13.72),

$$5.776 \times 10^4 = a\,(1-\sqrt{1-1/a^2LC_i})$$

and $\qquad\qquad 23.1 \times 10^4 = a\,(1+\sqrt{1-1/a^2LC_i}).$

This yields, $a = R/2L = 14.44 \times 10^4$, and $LC_t = 0.75 \times 10^{-10}$.

The given value of $C_i = 10 \times 0.125\mu F = 1.25 \times 10^{-6}$ Farad.

Therefore $L = 60\,\mu H$ and hence $R = 17.33$ ohms.

From equation (13.74), $I_0 = \dfrac{E - V_a}{R\sqrt{1-1/a^2\,LC_t}} = \dfrac{E_0 - 8.486 \times 3}{10.4}$ KA.

Therefore

$$I_p = 10 \text{ KA} = \frac{E_0 - 25.47}{10.4} \ (e^{-5.776 \times 10^4 \times 8 \times 10^{-6}} - e^{-23.1 \times 8 \times 10^{-6}})$$

giving $E_0 = 245.5$ KV.

The charging voltage is therefore about 250 kV. By increasing the source capacitance C_t the charging voltage can be decreased. Note that in such high current circuits, the voltage efficiency is not a factor of considerable importance.

13.11   Generation of High Alternating Test Voltage

Power frequency voltages up to 2000 kV, r.m.s., and 1 A may be required for testing insulation structures at e.h.v. levels. The short-duration test voltage required is $1.5 \times$ highest equipment voltage $(l - l)$ so that 1150 kV is required for 765 kV level. The flashover voltage is usually 20% higher and with a research and development factor of 1.1, there arises a need for a voltage of 1520 kV in such a laboratory. For 1150 kV transmission equipment, based on a maximum of 1200 kV, a power-frequency testing voltage of about 2000 kV may be required. Such high voltages at low current levels are generated by cascade-connected units where more than 1 transformer is connected in series. For example, at the IREQ laboratories in Canada, 4 units of 600 kV are connected in series and the lower 2 units have two units in parallel making up a total of 6 transformers. The need for the paralleling of the lower units arises for pollution testing of insulators to give a higher current (up to a voltage of 1200 kV) and lower leakage reactance. Normally, at these high voltages, except for insulator tests carried out under polluted condition, the load on a testing transformer is predominantly if not entirely capacitive. The power output of the transformer in MVA is

$$P = wC. \, V^2 \times 10^{-6}, \text{MVA} \tag{13.75}$$

where $w = 2\pi f$, C = load capacitance in μF, and V = test voltage at the high-voltage terminal of transformer in kV, r.m.s.

If the load capacitance is measured in pF, and the transformer power in KVA, then

$$P = wC \, V^2 \times 10^{-9}, \text{KVA} \, (C \text{ in pF}, V \text{ in kV})$$

When wet tests are carried out, there must be a reserve of power over that given above because of leakage currents on the test piece. The load capacitance includes the capacitance of the test object, that of the transformer itself, and the large high-voltage shields and electrodes that are necessary.

Typical ratings of single units might be as follows:

High Voltage Output	Current Range (continuous)	KVA Range
50 kV	40 mA to 400 mA	2 to 20
100	100 mA to 500 mA	10 to 50
200	0.5A to 2A	100 to 400
350	0.5A to 2A	175 to 700
600	1A to 3A	600 to 1800

The short-time current rating will normally be twice the continuous current rating (for about 12 hours in a day), if testing is carried out for 5 minutes with a cool-off period of 1 hour. For partial discharge measurement, the testing has to be carried out for 30 minutes and would be considered as a continuous test and not a short-time test.

Principles of Construction

The testing transformer differs in construction from a power transformer in some essential details. The core is constructed out of laminated grain-oriented magnetic steel of the same type as conventional transformers but the flux density used is below saturation (knee or bend) value and on the linear portion of the B-H curve. This is necessary to prevent voltage distortion caused by large magnetizing current. Because of lack of radiators or forced cooling, the oil content in the tank has to provide the large thermal time constant required so that the enclosure is more bulky to hold sufficient oil for cooling.

Windings are made from double-enamelled copper and very tightly wound to provide good capacitive coupling. For transformers to be cascaded, there are 3 windings on the core as shown in Fig. 13.24. In addition, for transformers built with a reactor for compensating the large capacitive current drawn from the source, the reactor is wound on the same core.

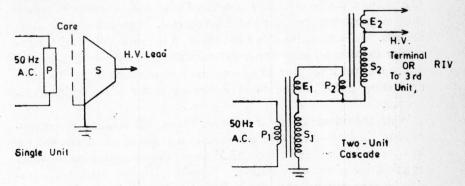

Fig. 13.24. Cascade connection of transformers for generating high power-frequency alternating voltages.

The primary winding is located nearest the core. The secondary turns are so wound that they taper towards the outside giving the same stress-grading effect as used in a condenser bushing. An electrostatic shield is usually used which also acts to prevent any oscillations caused by the inductance and high capacitance, especially during a transient such as breakdown of test object. The excitation winding or the tertiary is also of low voltage just like the primary, and rated for 1000 to 6000 volts. But it is at high potential with one end connected to the high-voltage end of secondary. The tertiary is used only when the transformer forms a lower unit in a cascade-connected multi-unit transformer set. A regulating transformer is provided at the input to the primary of the ground unit to vary the voltage input from zero to full rated value.

The short-circuit reactance of a testing transformer is very important from the point of view of internal voltage drop. It should be as high as possible to prevent damage in the event of high current during flashover of test object, and at the same time low enough to permit drawing high current during pollution tests. For this reason, the lower units can consist of two units in parallel. Typical values of short-circuit voltage at rated current range from 1.5% to 40% of rated value, the higher values applying for units in cascade. Single units may have up to about 10% internal impedance but standard specifications must be consulted for the maximum allowed. When two units are paralleled, it drops to 5% for pollution tests.

Reactive Compensation

When transformer windings, capacitors, and cables are tested, the load on the testing transformer is purely capacitive of a large magnitude. A testing transformer normally works at zero power factor leading in contrast to a power transformer whose power factor is almost always lagging and may be 0.9 or higher. Therefore, compensating reactors are necessary to minimize the total current drawn from the source in laboratories. The reactors can be externally connected with arrangements to switch them at pre-determined levels of input current. This can only be done in steps. An alternative is to build them on the same core as the transformer and connect them in parallel with the primary windings in each unit. It is usual to provide a maximum compensation at rated voltage amounting to 50% of the rated continuous power of the testing transformer.

When the voltage of the transformer varies, the reactive power supplied by the built-in reactors also varies but as the square of the voltage in a parabolic manner. Figure 13.25 shows the compensation provided as a function of voltage variation. Therefore, should extra compensation be necessary, external switched reactors have to be provided across the source to the ground unit. This is always single phase.

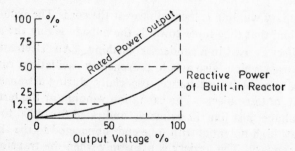

Fig. 13.25. Reactive compensation available with internally-built compensating reactors.

High Voltage Electrodes, Insulating Supports and Enclosures

Because of the extremely high voltage used at the high voltage terminal, very large electrodes are necessary.   The diameter of a nearly hemispherical electrode varies from 3 metres for a 300 kV to 400 kV terminal to 5 metres for 2000 kV units. The minimum clearances required for safety from the transformer range from 1.5 metres for 300 kV to almost 15 m for 2000 kV.   Because of the large size of aluminium electrode required, it can be made in segments using a network of saucer-shaped electrodes each about 25 cm diameter.   Such a construction has a patented name called POLYCON in Sweden.

Testing transformers have insulating cylinders made from reinforced fibre-glass or other insulating materials of high mechanical strength. The entire tank with oil, core, and windings is placed inside this cylinder. This eliminates separate insulating pedestals necessary in cascade-connected transformers where the higher units must be insulated from ground for the voltage of the lower unit. There are several schemes for lowering this insulation requirement such as connecting the high voltage secondary middle to the core, but these are details best discussed with the manufacturers.   However, a transformer using an insulated cylinder is only meant for indoor laboratory with a roof and walls to keep the elements out.   For outdoor use, which eliminates the need for a large building and its attendant costs, conventional steel tanks with porcelain bushings are necessary.   Figure 13.26 shows the two constructions.

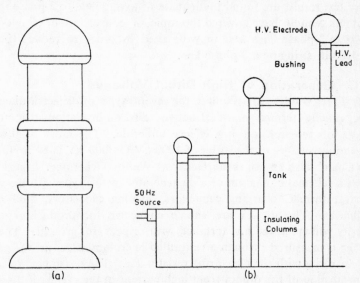

Fig. 13.26.   Overall pictorial representation of cascade-connected transformers with insulating cylinders and steel tanks.

A circuit diagram typical of a 2-unit cascade testing transformer installation is shown in Fig. 13.27, together with source, regulating transformer, voltage divider and peak and r.m.s. voltmeters. One can also observe that with a 3-unit cascade-connected unit, it is possible to use them as 3 separate single-phase units and perform experiments in 3-phase configuration at the rated voltage of each unit. However, a suitable 3-phase source of supply is necessary for such an eventuality. This facility is used when conducting high-voltage measurements on 3-phase test lines to obtain corona loss values and such other properties as radio interference, electrostatic fields, and audible noise in 3-phase configuration, if single-

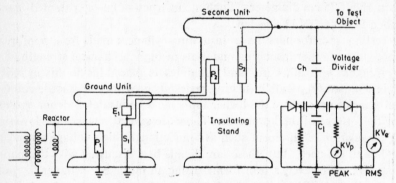

Fig. 13.27.　Internal arrangement of 2-unit cascade-connected transformer in insulating cylinder, with excitation and measuring circuits.

phase test results are found inadequate to extrapolate to 3-phase results. Chapters 5 and 6 have covered the topic of conversion of single-phase results to 3-phase lines and may be used in order to reduce the cost involved in stringing a 3-phase line.

13.12　Generation of High Direct Voltages

High direct voltages are required for several types of dielectric diagnostic tests, usually termed non-destructive tests, on insulation structures, as well as for research on h.v. effects under dc. The most usual way of generating voltages of the order of 1000 kV to 3000 kV is by a 'Greinacher Chain', also known as the 'Cockroft-Walton Generator.' Figure 13.28 shows a schematic diagram of a dc generator of this type with the charging transformer, charging capacitors, loading capacitors, diodes, and auxiliaries. The diodes are shown connected for producing positive polarity voltage at the h.v. terminal with respect to ground. The high voltage is measured through a calibrated micro-ammeter at the ground end in series with the measuring resistor R_m. The resistance R_p is used for protection of the diodes from high current in the event of flashover of test object.

The rectifiers are either selenium or silicon diodes. Their polarity is easily changed by reversing their terminals. For n stages, there are n charging capacitors C_C and an equal number of loading capacitors C_L, but $2n$ diodes D_1 to D_{2n}. The principle of operation consists of diode

D_1 charging C_{C1} when terminal 1 of the high-voltage charging transformer is positive, to the peak value of the input sine wave of voltage. During the next half cycle, terminal 2 is positive with respect to ground so that D_1 is reverse biased. However D_2 is forward biased and the loading capacitor C_{L1} is charged to a voltage $2E_m = \sqrt{2}.2E$, where $E =$ r.m.s. value of output voltage of the transformer. The terminal A is now

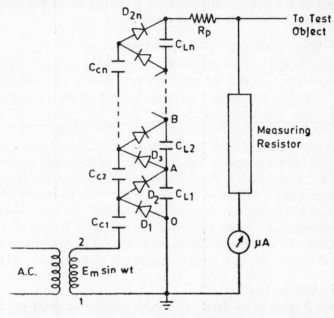

Fig. 13.28. Greinacher Chain or Cockroft-Walton cascade arrangement for generating high direct voltage.

at a voltage of $2\sqrt{2}E$ above ground and positive. During the succeeding half cycle, the diodes D_1 and D_3 help to transfer the charge of C_{L1} to the charging capacitor C_{C2} while C_{C1} receives a charge through D_1 to make up for the charge it lost to C_{L1} during the previous half cycle. The chain of capacitors will get charged in successive half cycles and the full voltage of the generator will be built up rapidly since there are no resistances in the circuit except the leads. When the test object is connected, that is the generator is loaded, some charge is lost which is made up once again from the charging source. The output voltage on no load at the high-voltage terminal is $2n\sqrt{2}E$, while on load there will be a regulation.

The ripple content in the output voltage depends upon the number of stages which is one of the limiting factors for using a very large number of stages. The ripple may be reduced by additional filter units. In one practical scheme, an e.h.v. transformer with a smoothing capacitor in parallel has worked very well at the 400-kV FGH in Germany where for an outdoor research line meant primarily for ac, by merely adding the capacitor, research under high dc voltages was carried out for corona loss, radio interference, etc. Many laboratories in the world also use such simplified arrangements instead of the Greinacher chain although the transformer must be rated for the full value of the high voltage dc instead of only $1/n$ of the voltage required at the high voltage terminal. The current output of such generators is small being about 50 mA which is limited by the diodes. Oil-immersed diodes with proper cooling and connected in parallel can be rated higher.

13.13 Measurement of High Voltages

In a high-voltage laboratory, voltages with a wide range of magnitudes, waveshapes, rates of rise and time durations have to be measured accurately, by meters and other equipment located in the control room at ground level. The distances are quite large and sources of error are numerous. The most versatile equipment is the voltage divider coupled with an oscilloscope, a peak voltmeter, and a digital recorder. Coaxial cables are used everywhere and in modern developments, fibre optic channels are also used. We will discuss the voltage divider in some detail and indicate the errors involved which have to be considered in calibration and accuracy. In some laboratories, the low-voltage input to an impulse generator or transformer is measured instead of using a high-voltage divider, but this lacks accuracy. For routine testing of mass-produced equipment, such quick methods will be found quite adequate.

13.13.1 Voltage Dividers

There are 3 types of dividers: (a) purely resistive divider; (b) a purely capacitive divider; and (c) resistance-capacitance divider. The former two are subject to serious limitations for e.h.v. work if precautions are not taken, which are overcome in the R-C divider.

Resistive Voltage Divider

Figure 13.29(a) shows a purely resistive divider consisting of a high-voltage arm R_h, a low-voltage arm R_1 at the ground end, a cable with surge impedance R_c terminated at the oscilloscope end by a resistance R_c equal to the surge impedance of the cable to prevent reflections from the opened end of the measuring cable. In high-voltage oscilloscopes, the vertical deflection circuit offers almost infinite resistance and there is no resistance or capacitance at the input end as in conventional oscilloscopes. Standard co-axial cables available for measurement have $R_c = 50$ ohms or 75 ohms.

The voltage ratio is

$$r_v = V_1/V_h = \frac{R_1 R_c/(R_1 + R_c)}{R_h + R_1 R_c/(R_1 + R_c)} = \frac{R_1 R_c}{R_h R_1 + R_h R_c + R_1 R_c} \quad (13.76)$$

Example 13.13: Calculate the divider ratio when a 20 Kilohm high-voltage arm is connected to a 100-ohm l.v. arm and the cable has $R_c = 75$ ohms.

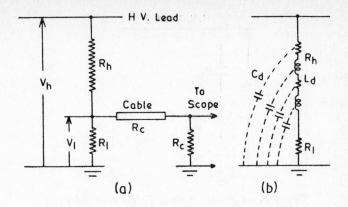

Fig. 13.29. Resistive voltage divider. (a) Equivalent circuit. (b) Stray capacitances of divider.

Solution: The total l.v. arm resistance is $100 \times 75/175 = 42.86$ ohms.

Therefore $r_v = 42.86/(20{,}000 + 42.86) = 1/467.6$

Also, $100 \times 75/(20{,}000 \times 175 + 100 \times 75) = 1/467.6$.

At e.h.v. levels, the height of the high-voltage lead above the grounded floor of the laboratory can be 20 metres so that inductances are present in the resistance. In addition, stray capacitances are present which increase as the ground is neared. A realistic circuit of even a purely resistive divider is depicted in Fig. 13.29 (b). Screening of the resistor is therefore necessary to eliminate these stray capacitances. An advantage of the resistive divider is the damping provided for travelling waves on the high-voltage lead connected from the source to the test object. Serious limitations of heating of the divider limits its use for e.h.v. since the resistance must be low enough to prevent the impedance of the stray capacitances from affecting the measurements. The value of R_h is usually limited to 5 to 20 Kilohms.   Its use is mainly up to 500 kV.

Capacitive Voltage Divider
A simplified diagram of a capacitive voltage divider, Fig. 13.30, consists of the high-voltage arm C_h with a capacitance between 300 and 1000 pF, a low-voltage arm C_1, a measuring cable whose open end is connected to the vertical deflection plates of a CRO or a peak voltmeter or other high-input-impedance measuring apparatus.   It is not possible to terminate the cable at the far end by a resistance R_c since this would offer a very low impedance across the low-voltage arm, thereby making the output voltage nearly zero. Also, the low-voltage capacitor C_1 will discharge rapidly to ground through the low resistance R_c if it is present.   Instead, a series resistance R_c equal to the cable surge impedance is placed at the entrance whose purpose is to place $\frac{1}{2}V_1$ across the cable. At the open end

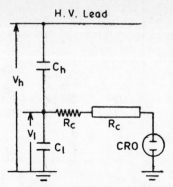

Fig. 13.30   Capacitive voltage divider with resistance
R_c at entrance to measuring cable.

this voltage will double its value by total reflection giving a voltage equal
to V_1 across the measuring instrument. Thus no correction for calibration
is necessary.

This type of divider also suffers from inductance of the high-voltage
arm which is very tall, as well as the effect of stray capacitances. A
typical high-voltage capacitor of 10–20 metre height and 25 cm diameter
has an inductance between 2.5 and 10 μH. The natural frequency for
1000 pF and 5 μH is $f_0 = (2\pi \sqrt{L_d C_d})^{-1} = 2.25$ MHz, and for 500 pF
and 2.5 μH it is 4.5 MHz. A lightning impulse rising to a peak value in
2.5 μs is equivalent to one-quarter cycle of 0.1 MHz waveshape. A steeper rise such as 100 ns to peak value has a frequency of 2.5 MHz. The
oscillations caused by the divider L and C will be superimposed on the
front of the wave giving an impure lightning-impulse voltage which will
be difficult to calibrate for the wavefront timings.

Another source of error with such tall slender equipment is that they
might act as antennas to pick up r-f radiation from the numerous spark
gaps in an impulse generator which form quite powerful spark transmitters. Another effect of these spark gaps in very large halls is cavity
resonance. In most e.h.v. laboratories the radiation from these sparks
is actually required in order to provide a deflection voltage without time
delay to the horizontal deflection plates of the oscilloscope. The length
of measuring cable connected to the vertical deflection plates then acts as
a delay line so that the vertical deflection voltage is impressed with some
time delay to enable the full waveform to be viewed on the screen.

Example 13.14: Select the capacitance required in the l.v. arm of a capacitive divider to provide a peak voltage of 500 volts at the oscilloscope
when the high-voltage arm of the divider has $C_h = 1000$ pF and the voltage to be measured is 1200 kV, peak. Omit the cable surge impedance
and R_c.

Solution: The required voltage ratio is

$$1/r_v = \frac{V_1}{V_h} = \frac{1/C_1}{1/C_1 + 1/C_h} = \frac{500}{1200 \times 10^3} = \frac{1}{2400}, \text{ giving}$$

$$C_1/C_h = 2400 - 1 = 2399.$$

Therefore $C_1 = 2399 \times 10^{-9} \text{F} = 2.399 \ \mu\text{F} = 2400 \ C_h$.

Note the low-voltage arm has the higher capacitance which is almost equal to the high-voltage capacitance times the voltage ratio

$$(C_1 = (r_v - 1) \ C_h).$$

In order to reduce the inductance of the low voltage capacitor and its leads, the total capacitance would be usually made up of several capacitors in parallel. In the above example, 6 capacitors of 0.4 μF each might be soldered to heavy rings at both ends.

In section 13.5.1 and examples 13.3 and 13.4, it was shown how a switching-surge generator requires a higher loading capacitor C_L in comparison to a lightning-impulse generator for the same source capacitance C_s. The values were 12 nF instead of 2nF for the two cases with $C_s = 83.3$ nF. The capacitive voltage divider offers an advantage over a purely resistive divider by acting as an increased capacitive load and the same generator can deliver both a switching surge and a lightning impulse if a resistive divider is used for the latter. These can usually be arranged with a capacitor inside the insulating column surrounded by the high-voltage resistors. The low-voltage arm is provided with a switch which connects either a pure capacitance or a resistance. Proper switching of R_C must also be made for such cases.

Resistive-Capacitive Divider

Figure 13.31 shows the arrangement of resistance and capacitances in the high-voltage arm. The low-voltage arm could consist of a resistance and capacitance to match the high-voltage arm to provide the desired voltage division, but in most cases the very low value of resistance required can be neglected and only a suitable value of low-voltage capacitor C_1 is sufficient.

The resistance serves a dual purpose. It helps to damp reflections in the high-voltage lead to the test object and critically damp the effect of series inductance in the divider column itself. Typical values of total resistance of a divider for 1200 kV is 100 ohms for a capacitance of 1000 pF. This will critically damp any natural oscillation in the high-voltage arm having an inductance of 2.5 μH ($R_d = 2\sqrt{L_d/C_d}$). The effect of R_d-C_d-L_d in the divider on the wavefront of a lightning impulse has been fully analyzed in Section 13.6.3 and shown in Fig. 13.12. The presence of L_d gives a steeper rise of the front of the wave as compared to its absence.

As mentioned earlier, the resistive or pure ohmic voltage divider cannot be normally used for measuring all types of voltage waveshapes

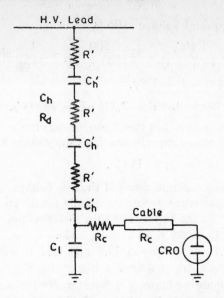

Fig. 13.31. Resistance-Capacitance Voltage divider.

because of the stray earth capacitance whose effect can be minimized if not eliminated by using screening electrodes or using a very low value of resistance of the high-voltage arm. This will, in turn, have a limited thermal rating for the very high voltages to be measured. In order to have a very low electrical time constant to reproduce the voltage faithfully, the product of resistance and effective stray capacitance must be low which again requires a low value of resistance. Furthermore it cannot act as a load for generating composite waves (both lightning and switching) from the same generator.

For the purely capacitive divider, the stray capacitance to earth can be taken into account in the overall response since it adds to the capacitance of the high-voltage arm. The pure capacitance, however, cannot damp oscillations caused by travelling waves or lead inductances. Therefore, a damped capacitive divider of the R-C type is found advantageous and has been applied for measuring voltage variations with as steep a rise to peak value as 100 ns $= 0.1$ μs, and for voltage magnitudes of several megavolts.

There are essentially two types of damped capacitive dividers: (1) the critically-damped divider; (2) the low-damped divider. Both these use series-connected sets of R-C elements. In the critically-damped divider, the total resistance equals $R = 2\sqrt{L_d/C_e}$, where $L_d =$ inductance of h.v. portion of the divider and $C_e =$ capacitance to earth of the voltage divider. This is proved as follows.

(1) Figure 13.32 shows the approximate equivalent circuit where the effective earth capacitance is shown connected to the mid-point of the divider h.v. arm. The impedance Z_d looking between the h.v. lead and earth is

$$Z_d(s) = \frac{\left(s + \dfrac{2R}{L} + \dfrac{1}{LC_s}\right)\left[1 + \dfrac{LC_e}{2}\left(s^2 + \dfrac{2R}{L}s + \dfrac{2C + C_e}{LC\,C_e}\right)\right]}{C_e\left(s^2 + \dfrac{2R}{L}s + \dfrac{2C + C_e}{LC\,C_e}\right)} \qquad (13.77)$$

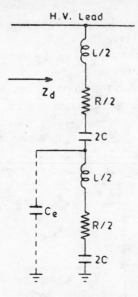

Fig. 13.32. Critically-damped R-C divider with effective stray capacitance connected to middle of divider.

The numerator determines the poles, and for critical damping,

$$s^2 + \frac{2R}{L}s + \frac{4C + C_e}{LC\,C_e} = 0 \text{ with its dicriminant} = 0$$

This gives $\dfrac{4R^2}{L^2} = 4 \times \dfrac{4C + C_e}{LC\,C_e}$ \hfill (13.78)

Now, if the stray capacitance $C_e \ll 4C$, there results

$$R^2 = 4L/C_e \text{ or } R = 2\sqrt{L/C_e} \qquad (13.79)$$

In general the value of resistance required ranges from 400 to 1000 ohms for voltages higher than 1 MV = 1000 kV. C_e is of the order of 50-150 pF and L is 5 to 10 μH.

(2) For a low-damped capacitive divider, the resistance is kept between 0.25 to 1.5 times $\sqrt{L_h/C_h}$, where now $L_h =$ inductance of the entire loop of the high-voltage equipment and measuring system (which may be as much as 20 m $\times$ 20 m), and $C_h =$ capacitance of the h.v. part of the divider only. In this case the h.v. lead has considerable inductance, and the resistance used equals nearly its surge impedance. For a horizontal lead, the surge impedance is 120 ln (2H/r), where $H =$ height of lead above a conducting plane and $r =$ radius of lead. For a vertical

rod, it is nearly 60 ln $(2H/r)$, where H = height of the rod. Based upon these values, the time constant in the divider is kept below 100 ns for reproducing lightning impulses of 1.2/50 μs. This usually requires resistances in the range of 50 to 300 ohms for dividers meant for 1 MV and higher voltages.

Since the critically-damped divider requires a value of resistance higher than the surge impedance of the h.v. lead, it is connected at the h.v. terminal of the impulse generator. The low-damped divider, on the other hand, can be matched to the surge impedance of the lead and provides a good compromise. When the impulse generator with the leads and divider is designed by the same manufacturer, optimum design can be achieved.

We now observe that the best compromise consists of the following scheme, Fig. 13.33. In order to damp the oscillations occurring on the front of the impulse which is caused by the generator inductance, an

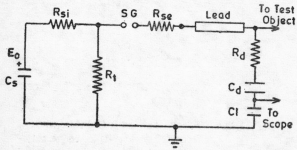

Fig. 13.33. Complete equivalent circuit of impulse generator with internal damping resistor R_{si}, external series resistance R_{se}, wavetail shaping-resistor R_t, damped R-C divider and test object.

internal series resistance in each stage amounting to a total of R_{si} is necessary. The wavetail-shaping resistor R_t is connected inside the generator in each stage. A part of the wavefront-shaping resistor is connected as R_{se} at the head of the h.v. lead to the voltage divider and test object located 15 to 20 m away. The voltage divider can be of low-damped or critically-damped type with an internal resistance R_d.

13.13.2 The High-Speed Oscilloscope

For measuring lightning impulses and chopped waves, very high frequency response is necessary in an oscilloscope. A 50 MHz oscilloscope is usually aimed at. For a full 1.2/50 μs wave, the highest frequency component in a frequency spectrum goes to 1 MHz, and if the wave is chopped at 0.5 μs on the front the frequency required is 20 MHz. When used with voltage dividers, the full-scale deflection of the oscilloscope ranges from 500 volts to 2000 volts so that impulse oscilloscopes must handle much higher voltages than conventional general purpose oscilloscopes. The horizontal deflection should be capable of being expanded to 2 μs full screen in order to view the wavefront properly.

Just as in all oscilloscopes used for viewing very fast phenomena, the vertical deflection plates must get their voltage after a set time delay following the horizontal-axis deflection voltage. A delay line or circuit with very low attenuation is necessary. The delay provided by a coaxial cable having a dielectric of permittivity $\varepsilon = 2.5$ is $T_d = \sqrt{2.5}/3 \times 10^8 = 5.27 \times 10^{-9}$ sec $= 5.27$ ns per metre, since the velocity is proportional to the reciprocal of $\sqrt{\varepsilon}$. Figure 13.34 shows the T-arrangement required to match the cables to the two deflection plates with the incoming-cable

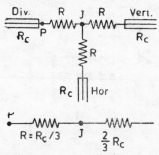

Fig. 13.34.   T-network for matching horizontal and vertical
input cables to oscilloscope with cable from
low-voltage arm of voltage divider.

surge impedance of R_c. When viewed from point P towards the oscilloscope, the impedance must equal R_c so that $R + \frac{1}{2}(R + R_c) = R_c$ giving $R = R_c/3$. For 50-ohm cables, the resistance to be connected in the T-junction is $R = 16\frac{2}{3}$ ohms and for 75-ohm cables, the resistance to be connected in the T-junction is $R = 25$ ohms.

When such matching T-networks are used, the vertical deflection requires a calibration correction because the voltage drop in R is 33.3% of the voltage reaching the end of the lead at P. The deflection observed on the oscilloscope screen will be multiplied by 1.5 to obtain the true voltage of the l.v. arm of the voltage divider.

This can be eliminated if the signal to the horizontal deflection plates is provided by other means. The scheme used is to pick up the radiated h-f energy from the large number of spark gaps, which break down when the impulse generator is fired, by means of a rod antenna. This is connected to the horizontal deflection plates of the oscilloscope through as short a length of coaxial cable as possible. The measuring cable from the l.v. arm of the voltage divider will then act as a delay line or an artificial delay line can be provided for suitable time delay to the vertical deflection plates.

13.13.3   The Peak Voltmeter and the Sphere Gap
The oscilloscope provides a pxcture of the waveshape of an impulse and offers a means of checking the timings and purity of the voltage as required by Standard Specification. Once this is verified and circuit constants

adjusted suitably, it will only be necessary to have an indication of the peak value of the voltage. For this purpose, a simpler instrument can be used. Figure 13.35 shows the schematic circuit diagram of a peak-measuring voltmeter. Here, the diode D will charge the auxiliary capacitor C_m to the peak value of impulse which is obtained across the low-voltage arm of the voltage divider. This is indicated on a voltmeter PVM, or on a micro-ammeter connected in series with the discharge resistor R_d.

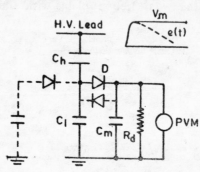

Fig. 13.35. Circuit diagram of Peak Voltmeter.

This resistance is necessary in order that the meter capacitance C_m can be ready for next impulse. A second diode can be used for the opposite polarity of the incoming signal.

Sphere gaps have been standard pieces of measuring equipment in a laboratory, but they are now finding little or no use for e.h.v. needed for testing. The PVM and SG voltmeter cannot indicate the waveshape but only the peak values. In addition, a sphere gap can be adjusted to limit the peak value of voltage across the test piece by suitably adjusting the gap length. The sphere gap breakdown is very consistent and its accuracy is within 3%. Atmospheric correction factors are necessary where the breakdown voltage depends upon the air density factor

$$\delta = \frac{273 + t_0}{273 + t} \cdot \frac{p_{mm}}{760} = \frac{273 + t_0}{273 + t} \cdot \frac{b_m}{1013}, \qquad (13.80)$$

where $t_0 = 20°C$, t = temperature in laboratory °C, p_{mm} = barometric pressure in mm Hg and b_m = barometric pressure in millibars. For lightning impulses, the breakdown voltage varies directly with δ, while for switching surges it varies as $\delta^{0.7}$. One may approximate the barometric pressure to decrease by 7.5 mm or 10 millibars for every 100 metres increase in elevation. Temperature has very little effect on the pressure, but standard specifications must be consulted. According to Indian Standards, the temperature correction is

$$b_m = 1013(1 - 1.8 \times 10^{-4} t) \qquad (13.81)$$

13.13.4 The Digital Recorder

A recent development in storing and displaying an impulse voltage waveform together with all test details is the Digital Recorder, which used the facility of fast sampling of μs-phenomena and storing them. It is believed by those that have developed this instrument that it will eventually replace the high-speed oscilloscope, for impulse and other types of measurement in h.v. testing.

A computer programmed to display all essential properties such as the peak value, time to peak, front time, overshoot, time to 50% on tail, rate of rise in kV/μs, and any other desired property facilitates analysis of the recorded waveform very easily. The waveform can be displayed on a computer graphic screen, stored on magnetic tape for future use, or a hard copy can be obtained for customer distribution or storage.

The waveform is sampled at a very high rate such as 10 ns/sample, and the use of 8-bit word length permits $2^8 = 256$ discrete voltage levels to be recorded and stored. The only limitation for high sampling rate is the band width of the digital recorders, which are however higher than those of conventional oscilloscopes. Some high-frequency oscillation which is lost in an oscilloscope has been handled by the A/D converters in the Digital Recorder. Pre-triggering phenomena, which occur before the waveform commences, can also be stored where necessary.

Because of the use of integrated circuits, which are influenced by the energy radiated by the discharge of spark gaps in an impulse generator, these electronic circuits must be shielded very well, even though they may be installed in a Faraday cage enclosing the control desk and operator. Ideally, the shield should be an entirely enclosed box made from thick sheet metal with an extremely low resistivity. The currents induced in the metal will set up fields to counter the external disturbing e.m. fields. Any obstruction to the flow of this current, such as holes or high-resistance welds and seams or doors with gaps, will reduce the effectiveness of the shielded enclosure. Even the lead-in for cables will affect its performance. Research and development of such Digital Recorders with display and storage is progressing rapidly and this is a very new technique.

13.13.5 Use of Fibre Optics in H.V. Measurements in Laboratories and Stations

Fibre optics and associated transmitter-receivers have now assumed very great importance and wide use in high-frequency communications. Their insulating properties and freedom from e.m. interference make them excellent replacements for conventional coaxial cables for high-voltage measurements. Future measurements, control, and communication can be said to belong to optical fibres. The sizes of e.h.v. switchyards and substations have become so large that the length of control and communication cables are subject to interference due to ground-mat potential rise which could interfere even with proper relay operation. Furthermore, the use of mini-computers and micro-processors are also subject to such potential rises and interference. Optical technology can provide eht electrical isolation required and freedom from injected interfering noise. They also provide faster transmission or transfer of data than conventional cables. Schemes are now under construction whereby carrier-type communication and control can be carried out between generating and receiving stations by fibre optical links, since they can carry a very large number of channels. Microwave communication is impeded by hills and other obstructions which are eliminated with the use of fibre-optical links.

Optical fibres are also used in high-voltage laboratories for measurement of the extremely high voltages. It has always been the practice to locate the low-voltage divider near the ground plane in order to protect the measuring equipment and the operator from experiencing high voltage. By the use of an optical-fibre link, the low-voltage arm can be placed at the high voltage end of the divider thereby eliminating any stray capacitances caused by nearness to the laboratory ground floor. An optical isolator is used to insulate the measuring equipment from the high voltage. This scheme uses a light-emitting diode (LED) which is energized from the voltage developed across the low-voltage arm. The light intensity, modulated by the impulse or other waveform to be measured, can be made to fall on a photodiode or transistor which sends the signal through the optical fibre link to the measuring equipment. Optical isolators provide effectively the insulating clearances.

13.13.6 Measurement of Partial Discharges

Localized electrical discharges inside high-voltage equipment are very important for their effect on insulation deterioration. They are the result of ionization in cavities or along the surfaces of solid dielectrics. Partial discharges give rise to corona-type pulses over a very wide frequency range, which are propagated from the source of partial discharge through circuit capacitances in the windings of transformers and along the conductors. The measurement of these discharges is recommended by IEC Publication No. 270, 1980 and in IS 2026: Power Transformers.

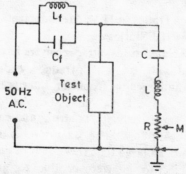

Fig. 13.36. Circuit for measurement of
Partial Discharges.

One example of the many schemes available is given in Fig. 13.36.

The measuring device M consists of an oscilloscope or a radio noise meter for displaying the q-V relationship as a Lissajous figure or the conducted noise in microvolts. In most transformers 1 pico coulomb corresponds to 1 μV on the noise meter, but this depends upon the capacitance of the winding from the source of p.d. to the high voltage terminal where the measuring system is located.

Voltage is applied to the object under test, E. The transformer must be free of partial discharge up to the test voltage. A filter F is interposed between the transformer and test object so that all energy in the partial discharge goes to the measuring system and none into the source transformer which presents a low impedance to the h-f pulses. The p.d. pulses are transmitted to the coupling capacitor and a damped tuned circuit consisting of the voltage-divider inductance and a resistor. The entire measuring system must also be free from partial discharges. The coupling capacitor is filled with nitrogen at high pressure. The entire system must be enclosed in a Faraday cage and screened from external discharges. The design of a properly shielded enclosure which prevents external e.m. disturbances from reaching the p.d. measuring equipment is a highly specialized topic and will not be discussed here. P.D. measurements are diagnostic tests and can give an indication of a fault that might develop progressively inside high-voltage equipment such as transformers, cables and other equipment using very thick insulation.

13.14 General Layout of E.H.V. Laboratories

We end this chapter with some information on the design and layout of e.h.v. laboratories, chiefly of the indoor type. Outdoor laboratories for extremely high voltages are also used in order to save the high cost of the building, but the equipment cost is higher. A combination of indoor and outdoor laboratories also exist since some tests such as pollution and radio interference require closed chambers. Outdoor equipment requires steel tanks and procelain insulator housing to guard against the elements. We shall give some examples of laboratory layout. It is assumed that the highest voltages required in the testing equipment has been decided by the designer based on the type of tests to be performed.

Clearances

The dimensions of laboratories will depend upon:
    (a) the highest voltage for which the lab is designed,
    (b) the number of pieces of equipment required,
    (c) the air gap clearances to be maintained, and
    (d) the types of tests to be conducted and sizes of test objects.

Over and above the requirement of a main hall, annexes have to be provided for research, development and auxiliary equipment, as well as offices. The following recommended clearances can be used as a guide for design.

(1) For Power Frequency Tests

Up to 1000 kV a linearly increasing clearance of at least 4 metres will be necessary. The 50% flashover value of a rod-plane gap for power-frequency is given approximately by $V_{50} = 652.d^{0.576}$ kV peak with d in metres. The following gap lengths are then necessary: 1000 kV r.m.s. — 3.835 metres, 1500 kV — 7.754 metres, 2000 kV — 12.8 metres.

Example 13.15: An airgap clearance is to be designed for a withstand voltage of 1500 kV, r.m.s. The withstand voltage is 80% of 50% flashover voltage. For a rod-plane gap, calculate the minimum clearance required.

Solution: 50% flashover voltage = $1500 \sqrt{2}/0.8 = 2651.65$ kV, peak.

Therefore, $2651.65 = 652.d^{0.576}$ giving $d = 11.42$ metres.

(2) Lightning-Impulse Clearances

For voltages up to the highest value used for lightning-impulse tests (6 to 7 MV) the required clearance rises linearly with voltage, as has been discussed in Chapter 11. A clearance of 2 metres for each 1000 kV is necessary. The impulse flashover formula is $V_{50} = 500\, d$ for a rod-plane gap (V_{50} in kV and d in metres). The withstand voltage can be assumed to be 80% of V_{50}, so that a 2-metre gap will withstand 800 kV, crest. For other types of gap, the same gap length can withstand higher values of voltage.

(3) D.C.

A clearance of 4 metres for every 1000 kV is recommended.

(4) Switching Surges

A given electrode geometry exhibits the weakest electrical strength for positive-polarity switching surge, as described in chapter 11. The recommended clearance is $d = 4(MV)^2$, which gives 4 metres for 1000 kV crest and 16 metres for 2000 kV, crest.

The rod-plane CFO is $V_{50} = 500.d^{0.6}$ kV crest according to Paris' formula. This yields $d = 3.2$ m for 1000 kV and 10.1 metres for 2000 kV, crest. With the withstand voltage being 80% of CFO, the clearances required are 4.6 m for 1000 kV withstand and 14.6 m for 2000 kV withstand.

General Layout

The clearances given above will help to dimension the layout of equipment. In order to reduce the size required for a hall, one might observe that there is no need for using all the sources of high voltage simultaneously. For example, very rarely if ever is a switching impulse test carried out simultaneously with power-frequency tests. This enables the transformer to be moved out of its permanent location during tests if it is mounted on wheels. Similarly for the impulse generator. When this philosophy is adopted, it is sufficient to construct the floor of steel and pave it with smooth cement. An arrangement for supplying a cushion of pressurized air under the base of the heavy equipment will facilitate easy movement. The approximate weight of a 2000-kV power frequency transformer for indoor use of the insulating-cylinder type is 100 Tonnes. An impulse generator for 6 MV and 30 KJ weighs 25 Tonnes approximately.

Large halls have a metallic screen in their walls to act as shield in order to prevent e.m. radiation from sources external to the building, such as ignition spark radiation etc., entering the hall thereby disturbing sensitive measurements such as partial discharges and radio interference (RIV) in the µV range at 0.5 to 1.5 MHz. However, with the very large number of spark gaps discharging the impulse generator, this metallic shield in the walls will reflect the energy back into the hall and disturb not only RIV measurements but also the steep-rising front of the lightning impulse itself. This effect is known as Cavity Resonance. A semiconducting coating on the inside of all walls of the huge hall will prevent the re-radiation. Many laboratories have used chicken-wire mesh on the inside. An outdoor laboratory is free from cavity resonance but it will be subject to e.m. interference from sources outside laboratory area such as passing vehicles (ignition radiation), welding equipment nearby, motors, etc. In every case, therefore, a compromise has to be effected such as constructing specially-sheilded rooms for sensitive measurements and lead the high-voltage connection through a wall bushing from an outdoor generator. Such bushings are very heavy and long, requiring proper support for the cantilever action.

Grounding grids in e.h.v. laboratories and earthing wells need very special care. It is considered beyond the scope of this book to deal with this problem.

Example 13.16: Design a possible layout of a laboratory to be equipped with the following high-voltage equipment. Other specifications are also supplied.

(1) *Lightning and Switching Impulse Generator*
4000 kV output voltage. Charging voltage per stage 200 kV, stage capacitance 1.5 µF energy 30 KJ, number of stages 20. Generator height 15 metres, electrode height 2 m. Required clearances: Lightning 8 m, switching 26 m. Base 3 m on side.

(2) *Power Frequency Transformer*
3-unit cascade, 1500 kV, 1A. Insulated cylinder type. Approximate height to the top of the high-voltage electrode —15 m. Required clearance is 8 m all round. Provide one more ground unit for higher current output connected in parallel with the first or base unit of 3-unit cascade.

Clearance = 3 m. Bases: 1500 kV −5 m on side. 500 kV −3.5 m on side.

(3) *D.C. Generator*

2000 kV, 10 mA. Height 10 m, base 2 m on side.

(4) *Voltage Divider*

Same height as impulse generator. Distance from generator = 20 m.

(5) *Space for Test Object*

10 m along the length of hall.

(6) *Control Room*

As annexe to the main hall. 15 m long, 6 m wide.

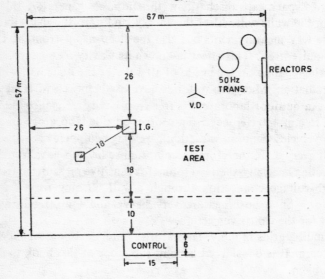

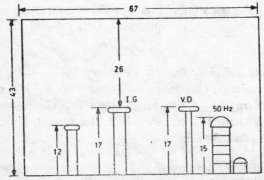

Fig. 13.37. Plan and elevation of general layout of a e.h.v. laboratory for 4000 kV impulse, 1500 kV power frequency and 2000 kV dc voltages.

Solution. Figure 13.37 shows a possible plan and elevation of the laboratory showing dimensions of equipment and required clearances. No attempt has been made to move equipment from their permanent locations. The main points to note are the following: clearance from IG to walls —26 m for switching surge. Distance between IG and Voltage divider —20 m.

Clearance from IG to test object —18 m.

Distance from 50 Hz transformer cascade to wall —8 m; to voltage divider —5 m.

Compensating reactors are placed against walls.

Review Questions and Problems

1. What are the standard waveshapes used for the following types of surges? (a) Lightning voltage, (b) Lightning Current, (c) Switching impulse for testing line material, (d) Switching impulse for transformer tests? Give neat sketches and indicate all magnitudes and timings.

2. A double-exponential voltage rises to peak value in $t_m = 0.1$ μs and falls to 50% value on tail at $t_t = 0.3$ μs. Determine its equation $e(t) = E(e^{-\alpha t} - e^{-\beta t})$ by calculating E/V_P, α and β. What type of voltage is this?

3. An impulse generator with source capacitance $C_s = 125$ nF and load capacitance 5 nF is required to generate a double-exponential voltage of 1.2 μs/50 μs. Calculate approximately the values of front and tail resistances R_f and R_t.

4. What are the three places where inductance is present in an impulse generator? Discuss their effect on measured waveforms of voltage by a voltage divider.

5. What is a critically-damped divider and an under-damped divider? Where are they used?

6. In problem 3, there are 12 capacitors with each stage having 125 nF capacitance and the charging voltage per stage is 200 kV. Calculate its energy when (a) all stages are connected in series, (b) the stages are re-connected with 2 capacitors in parallel in 6 stages, (c) all 12 capacitors are connected in parallel into 1 stage.

7. An impulse current generator has 6 impulse capacitors each of 0.2 μF in parallel. The desired waveshape is $t_m = 8$ μs and $t_t = 20$ μs. Calculate (a) the approximate values of resistance and inductance allowable in the circuit to generate double-exponential current waveform and (b) the charging voltage to obtain 15 KA through an arrester whose residual voltage stays constant at 20 kV, peak.

8. Draw a neat diagram of an insulating cylinder type of cascade-connected tranformer with 3 stages showing power input, the windings with internal compensating reactors and measuring circuits.

9. A pure capacitive divider has a high-voltage arm of 495 pF rated for 2.4 MV. The divider ratio is 990/1 Calculate the value of capacitance of the low-voltage arm and its voltage rating.

10. Draw the circuit diagram of a peak voltmeter and explain the principle of operation.

CHAPTER 14

Design of EHV Lines Based Upon Steady State Limits and Transient Overvoltages

14.1 Introduction

The concluding chapter of this book will be devoted to a discussion of the application of all the material of previous chapters to evolve methods of design of e.h.v. lines. Two important governing factors stressed here are (1) steady-state operating limits, and (2) insulation design based upon transient overvoltages. Other problems usually associated with transmission line are (a) transient and dynamic stability, and (b) short-circuit currents and voltages. These depend on the characteristics of the connected generating stations and system interconnection and cannot be discussed here. But such a study has to be conducted after the line is designed and some factors might have to be modified.

High-voltage insulation problems can be generalized in sequential steps when the system transient studies are conducted and the air-gap insulation characteristics are understood to offer a proper level of safe operation which is determined on a probability basis of allowing a certain level of line outages. But the steady-state compensation requirements depend upon the power transmitted and line length. The different types of reactive-power compensation discussed in Chapter 12 can be divided into the following categories;

(1) Lines for which compensating equipment is provided at only the receiving-end of the line across the load. Such schemes use synchronous condensers, switched capacitors, static Var systems and regulating tranformers with on-load top-changing facility. These lines are short.

(2) Those for which static shunt-reactor compensation is required at no load conditions for voltage control at the two ends. These are permanently connected and switched capacitors across the load take care of compensation required for load variation at the receiving end. These lines can be termed medium length in so far as e.h.v. transmission is concerned.

(3) Conditions for which the load requirement necessitates series-capacitor

compensation along the line in addition to shunt-reactor compensation at the ends for voltage control on no load. Switched capacitors will take care of load conditions for voltage control. These lines can be considered as long.

(4) Lines which necessitate an intermediate-switching station where shunt reactors are provided in addition to series capacitors in the two line sections along with shunt-compensating reactors for no-load conditions across the source and load. These are extra-long lines.

The theory and analysis of these types have been covered in Chapter 12. Air gap insulation characteristics have been dealt with in Chapter 11. We will discuss design procedures through examples of several types of lines based upon the principles set forth in all the earlier chapters. The governing criteria for line design are postulated below and discussed further.

14.2 Design Factors Under Steady State

All designs can be considered as synthesis of analytical procedures that are available and an enumeration of the limits and constraints under which line designs have to be carried out. The steady-state considerations are the following:

(a) Maximum allowable bus voltages and across equipment for a given voltage level. These are specified by Standards and given in Chapters 2 and 11.

(b) Current density in conductors which determine the cross sectional area, the resulting temperature rise, etc. These were discussed in Chapter 3.

(c) Bundling, corona-inception gradient, and energy loss. These factors are important for fixing the conductor diameter and number of conductors in the bundle and have been discussed in Chapters 4 and 5.

(d) Electrostatic field under the line at 50 Hz covered in Chapter 7.

(e) Radio interference and audible noise discussed in Chapters 5 and 6.

(f) Compensation requirements for voltage control as described in chapter 12.

A summary of equations and constraints can be given here.

(a) The limits for maximum operating voltages are as follows:

Nominal System kV	220	345	400	500	735–765	1000	1150
Maximum Equipment kV	245	362	420	525	765	1050	1200

The power-handling capacity of a single circuit can be assumed for commencing the design to be $P = 0.5 \, V^2/Lx$, MW/circuit with V in kV, line-line, L = line length in km, and x = total series reactance per km per phase. The % power loss is $p = 50\,r/x$, where r = line resistance per km per phase. Based on the total power to be transmitted, the number of circuits can be decided for a chosen voltage level. Several voltage levels should be selected and alternative designs worked out, since design requirements vary from location to location. It has been discussed earlier that one 1200-kV circuit can carry as much power as 3-750 kV lines and 12-400 lines for the same distance of transmission.

(b) Current density normally encountered lies between 0.75 A/mm² to 1 A/mm². At the lower values for carrying the same current (determined by power per circuit and chosen voltage level) conductors tend to be larger than at the higher value of current density which reflect on the surface voltage gradient and corona effects.

(c) Charge of bundled conductors govern the electrostatic field in line vicinity, the surface voltage gradient on conductors and the resulting RI, AN, and corona loss. The charge per unit length on the 3-phase conductors is $[q] = 2\pi e_0 [M][V]$, where the voltage matrix is $[V] = V[1\angle 0°, 1\angle -120°, 1\angle 120°]_t$ in the steady state with V = r.m.s. value of line-to-neutral voltage in volts. The matrix $[M] = [P]^{-1}$, where $[P]$ is the Maxwell's potential-coefficient matrix with the elements $P_{ll} = \ln (2H_l/r_{eq})$, $P_{ij} = \ln (I_{ij}/A_{ij})$ $i \neq j$. Here H_i = height of conductor i above ground, I_{ij} = distance between conductor i and the image of conductor j, A_{ij} = aerial distance between conductors i and j, and r_{eq} = geometric mean radius or equivalent radius of bundle = $R(N \cdot r/R)^{1/N}$, N = number of sub-conductors, r = sub-conductor radius, R = bundle radius, and $i, j = 1, 2, 3$.

The resulting inductance matrix is $[L] = 0.2[P]$ mH/km (or µH/m), and the capacitance matrix is $[C] = 2\pi e_0 [M]$, Farad/m (the factor $2\pi e_0 = 10^{-9}/18 = 55.55 \times 10^{-12}$ F/m, where $e_0 = 8.842$ µµF/m).

For the 3 phases,

$$\left. \begin{aligned} q_1/2\pi e_0 &= V[M_{11}\angle 0° + M_{12}\angle -120° + M_{13}\angle 120°] \\ q_2/2\pi e_0 &= V[M_{21}\angle 0° + M_{22}\angle -120° + M_{23}\angle 120°] \\ q_3/2\pi e_0 &= V[M_{31}\angle 0° + M_{32}\angle -120° + M_{33}\angle 120°] \end{aligned} \right\} \quad (14.1)$$

(d) *Electrostatic Field*

The horizontal and vertical components of field are determined at a point $A(x, y)$ where x and y are its coordinates with reference to a chosen origin. A convenient location is at ground under the centre phase. Then;

(i) Vertical component due to phase i will be

$$E_v(i) = (q_i/2\pi e_0) \cdot [(H_i - y)/D_i^2 + (H_i + y)/(D'_i)^2] \quad (14.2)$$

(ii) Horizontal component is

$$E_h(i) = (q_i/2\pi e_0)(x - x_i)[1/D_i^2 - 1/(D_i')^2] \quad (14.3)$$

where x_i = horizontal coordinate of phase i,

$$D_i^2 = (x - x_i)^2 + (y - H_i)^2$$

and $(D_i')^2 = (x - x_i)^2 + (y + H_i)^2$ (14.4)

(iii) The total field is

$$E_t^2 = [\Sigma E_v(i)]^2 + [\Sigma E_h(i)]^2$$ (14.5)

(e) *Conductor Surface Voltage Gradient*
The quantity q_i calculated earlier is the charge of the entire bundle. The maximum surface voltage gradient on any sub-conductor will be

$$E_{max} = (q_i/2\pi e_0) \cdot (1/N) \cdot (1/r)[1 + (N - 1) r/R]$$ (14.6)

For most practical calculations, the Mangoldt Formula (Markt-Mengele Formula) can be used for horizontal configuration of the three phases (Chapter 4).

(f) *Corona-Inception Gradient*
For a cylindrical conductor above a ground plane, Peek's formula for corona-inception gradient is

$$E_0 = 2140 \cdot m \cdot \delta(1 + 0.0301/\sqrt{r\delta}) \text{ kV/m, r.m.s.}$$ (14.7)

where m = conductor surface roughness factor < 1

δ = air-density factor = $\dfrac{b}{1013}\dfrac{273 + t_0}{273 + t}$

r = conductor radius in metre,
b = barometric pressure in millibars,
t = temperature °C
t_0 = reference temperature, 20°C presently used, but will become 27°C when adopted in future.

The barometric pressure varies with elevation and temperature dropping by 10 millibars for every 100 metre increase in elevation. The temperature correction is insignificant. The value of $m = 0.7$ to 0.8 in fair weather and 0.55 to 0.65 in rain.

(g) *Radio Noise Level* (Chapter 6)
For up to 4 conductors in bundle, the C.I.G.R.E. formula is

$$RI(k) = 0.035 E_m(K) + 1200 r - 33 \log (D(K)/20) - 30 \text{ dB}$$ (14.8)

where $E_m(k)$ = maximum surface voltage gradient on sub-conductor in kV/m, r.m.s., $D(K)$ = aerial distance from phase k to the point where RI is evaluated and $k = 1, 2, 3$.

Rules for adding RI due to all phases are given in Chapter 6. For 1 MHz level, deduct 6 dB from equation (14.8).

(h) *Audible Noise* (Chapter 5)

For $N < 3$, $AN(k) = 120 \log E_m(k) + 55 \log (2r) - 11.4 \log D(k)$

$$-245.4, \text{dB (A)} \tag{14.9}$$

For $N \geqslant 3$, $AN(k) = 120 \log E_m(k) + 55 \log (2r) - 11.4 \log D(k)$

$$+26.4 \log N - 258.4, \text{dB (A)} \tag{14.10}$$

The total AN level of the 3 phases is

$$AN = 10 \log \sum_{k=1}^{3} 10^{0.1\, AN(k)}, \text{dB (A)} \tag{14.11}$$

(i) *Corona Loss.* (Ryan and Henline Formula, Chapter 5).

$$W_L = 4fC\, V(V - V_0) \text{ MW/km, 3-phase.} \tag{14.12}$$

(1) *Voltage Control at Power Frequency* (Chapter 12).

(i) *Line only*:
$$\begin{bmatrix} E_s \\ I_s \end{bmatrix} = \begin{bmatrix} A, & B \\ C, & D \end{bmatrix} \begin{bmatrix} E_R \\ I_R \end{bmatrix} \tag{14.13}$$

$$\left.\begin{aligned}
A = D &= \cosh pL = \cosh \sqrt{ZY} \\
B = Z_0 \sinh pL &= \sqrt{Z/Y}.\sinh \sqrt{ZY} \\
C = \sinh pL/Z_0 &= \sqrt{Y/Z}.\sinh \sqrt{ZY} \\
AD - BC &= 1.
\end{aligned}\right\} \tag{14.14}$$

$$Z = (r + jwl)\, L, \quad Y = jwcL. \tag{14.15}$$

r, l, c = distributed resistance, inductance and capacitance per km
L = line length in km
$w = 2\pi f$

(ii) *Shunt Reactor Compensation.* Reactor admittance $= B_L$.

$$\begin{bmatrix} A_T, & B_T \\ C_T, & B_T \end{bmatrix} = \begin{bmatrix} A, & B \\ C, & D \end{bmatrix}_{line} -jB_L \begin{bmatrix} B, & O \\ 2A-jB_L\,B, & B \end{bmatrix} \tag{14.16}$$

(iii) *Series Capacitor Located at Centre and Shunt Reactors at ends.*

$$\begin{bmatrix} A_T, & B_T \\ C_T, & D_T \end{bmatrix} = \begin{bmatrix} A, & B \\ C, & D \end{bmatrix}_{line} - jB_L \begin{bmatrix} B, & O \\ 2A-jB_L\,B, & B \end{bmatrix}$$

$$-\frac{1}{2}X_c B_L \begin{bmatrix} 1 + \cosh pL, & O \\ \dfrac{2}{Z_0} \sinh pL - j\,B_L\,(1 + \cosh pL), & 1 + \cosh pL \end{bmatrix}$$

$$-\frac{jX_c}{2Z_0} \begin{bmatrix} \sinh pL, & (\cosh pL + 1)\, Z_0 \\ (\cosh pL - 1)/Z_0, & \sinh pL \end{bmatrix} \tag{14.17}$$

Limits. The following limiting values will govern the design of line and the width of line corridor required.

(a) *Electrostatic Field.* Ground-level maximum value is 15 kV/m, r.m.s.
(b) *Radio Interference.* 40 dB above 1 μ V/m at 1 MHz in fair weather at the edge of R-O-W.
(c) *Audible Noise.* 52.5 dB (A) at edge of R-O-W.
(d) *Corona-Inception Gradient.* Margin above maximum voltage gradient is at least 10% on centre phase or the phase with maximum voltage gradient.
(e) *Corona Losses.* With corona-inception gradient being above maximum voltage gradient at the operating voltage, no corona loss is anticipated.
(f) *Line Compensation.* This must be designed to hold the bus voltage within limits given by Standard Specification.

14.3 Design Examples Steady State Limits

The above criteria and equations for all quantities under steady state operating conditions of line will be applied for design of the following lines.

(1) 400 kV 200 km line to transmit 1000 MW using synchronous condenser at load end.
(2) 400 kV 400 km line to transmit 1000 MW using shunt-reactor compensation at both ends for no load condition and switched capacitors for load condition for voltage control at buses.
(3) 400 kV 800 km line to transmit 1000 MW using 50% series capacitor compensation at line centre, shunt reactors at both ends for voltage control at no load, and switched capacitors across load.
(4) 750 kV 500 km line for transmitting 2000 MW with only shunt-reactors at ends for voltage control at no load and switched capacitors under load.

14.4 Design Example—I (400 kV 200 km 1000 MW)

(1) *Number of Circuits*

With equal voltage magnitudes at both ends to commence the design and 30° phase difference between them, the power-handling capacity per circuit is with $x = 0.327$ ohm/km,

$$P = 0.5 \times 400^2/0.327 \times 200 = 1223 \text{ MW, 3-phase.}$$

Therefore, one circuit will be sufficient.

(2) *Line Clearance and Phase Spacing*

NESC recommends a minimum clearance of 29 feet = 8.84 metres based on 17' for first 33 kV and 1' extra for each 33 kV above this or part thereof. The sag is expected to be 12 m.

Therefore, average height for calculation of potential coefficients is $H = 9 + 12/3 = 13$ m in horizontal configuration. The phase spacing required from switching surge considerations is expected to be 12 m. At maximum swing of insulators it should be 7.5 m. Allow 4.3 metre height for insulator string and 30° swing. This aspect will be taken up in Section 14.4 under Design Based upon Transient overvoltages.

(3) Conductor Size and Number in Bundle

The twin conductor has become the accepted standard for 400 kV, although one line in Germany (Rheinau) uses 4 conductors in bundle in 4-circuit configuration. We first try the double Morkulla or Moose with $r = 0.0159$ metre radius at bundle spacing of $B = 0.4572$ metre. This will be checked for (a) corona-inception gradient and margin between this and the maximum surface voltage gradient. Using Mangoldt formula,

For the Centre Phase

$$E_{mc} = V\,[1 + (N-1)\,r/R]/[N.r.\,\ln\,\{2HS/r_{eq}\sqrt{4H^2 + S^2}\}] \quad (14.18)$$

Here, $V = 420/\sqrt{3}$ kV at the maximum operating voltage,

$$N = 2, R = 0.2286, r = 0.0159, H = 13, S = 12$$

$$r_{eq} = \sqrt{rB} = 0.08526 \text{ m.}$$

Therefore, $E_{mc} = \dfrac{420}{\sqrt{3}}\,\dfrac{1.07}{2 \times 0.0159 \times 4.85}$

$$= 1681.5 \text{ kV/m} = 16.815 \text{ kV/cm}$$

For the Outer Phases

$$E_{mo} = \frac{V}{N.r}\,[1 + (N-1)\,r/R]/[\,\ln\,(2H/r_{eq})$$
$$-0.5\,\ln\{\sqrt{(4H^2 + S^2)\,(H^2 + S^2)}/S^2\}] \quad (14.19)$$
$$= 1601.5 \text{ kV/m} = 16.015 \text{ kV/cm.}$$

Corona-Inception Gradient   (at $\delta = 1$). Take $m = 0.75$

$$E_{or} = 2140\,(1 + 0.0301/\sqrt{0.0159}) \times 0.75 = 1988 \text{ kV/m}$$

Margin. For the centre phase, $(1988-1681.5)/19.88 = 15.42\%$. For the outer phases, $(1988-1601.5)/19.88 = 19.44\%$. These are considered satisfactory.

(4) RI Level

At 15 m from the outer phase at ground.

$$E_{mc} = 16.82 \text{ kV/cm}, E_{mo} = 16.02 \text{ kV/cm.}$$

$$D(1) = \sqrt{13^2 + 15^2} = 19.85 \text{ m}, \ D(2) = \sqrt{13^2 + 27^2} = 30 \text{ m},$$

$$D(3) = \sqrt{13^2 + 39^2} = 41.1 \text{ m}. \quad (Note: D(1) \approx 20 \text{ m})$$

Therefore,

$$RI(1) = 0.035 \times 1602 + 1200 \times 0.0159 - 30 = 45.15 \text{ dB}$$

$$RI(2) = 0.035 \times 1682 + 19.08 - 30 - 33 \log (30/20) = 42.14 \text{ dB}$$

$$RI(3) = 0.035 \times 1602 + 19.08 - 30 - 33 \log (41.1/20) = 34.83 \text{ dB}.$$

Therefore, RI level of line = 45.15 dB at 0.5 MHz and 39.15 dB at 1 MHz. This is considered satisfactory since the limit is 40 dB at the edge of R-O-W. The line corridor can therefore be made equal to $2(S+15)=54$ metres.

(5) Audible Noise
At 15 m from outer phase at ground.

$$AN(1) = 120 \log 1602 + 55 \log 0.0318 - 11.4 \log 19.85 - 245.4$$

$$= 42 \text{ dB (A)}$$

$$AN(2) = 120 \log 1682 - 82.37 - 245.4 - 11.4 \log 30 = 42.5$$

$$AN(3) = 120 \log 1602 - 82.37 - 245.4 - 11.4 \log 41.1 = 38.4$$

Therefore,

$$AN = 10 \log (10^{4.2} + 10^{4.25} + 10^{3.84}) = 46.08 \text{ dB (A)}$$

This is less than the limiting value of 52.5 dB (A) and the line corridor governed by RI can be considered as adequate.

(6) Electrostatic Field
The Maxwell Potential coefficient has the elements

$$P_{ll} = \ln (26/0.08526) = 5.716, \quad P_{oc} = \ln (\sqrt{26^2 + 12^2}/12) = 0.87 \text{ and }$$

$$P_{oo} = \ln (\sqrt{26^2 + 24^2}/24) = 0.39.$$

Therefore, $[P] = \begin{bmatrix} 5.716 & 0.87 & 0.39 \\ 0.87 & 5.716 & 0.87 \\ 0.39 & 0.87 & 5.716 \end{bmatrix}$, giving $[M] = [P]^{-1}$

$$= \begin{bmatrix} 0.1728 & -0.026 & -0.008 \\ -0.026 & 0.1772 & -0.026 \\ -0.008 & -0.026 & 0.1728 \end{bmatrix}$$

Then, at $V = 420/\sqrt{3}$ kV,

$$\begin{bmatrix} q_1/2\pi e_0 \\ q_2/2\pi e_0 \\ q_3/2\pi e_0 \end{bmatrix} = \begin{bmatrix} 0.1728\angle 0° -0.026 \angle -120° -0.008 \angle 120° \\ -0.026 \angle 0° +0.1772\angle -120° -0.026 \angle 120° \\ -0.008 \angle 0° -0.026 \angle -120° +0.1728\angle 120° \end{bmatrix} \frac{420}{\sqrt{3}}$$

At a distance d along ground from the line centre, the vertical component of e.s. field comes out to be

$$E_v = \frac{420}{\sqrt{3}}\left[\left\{\frac{4.94}{\sqrt{H^2+(d+S)^2}} - \frac{2.6416}{\sqrt{H^2+d^2}} + \frac{2.116}{\sqrt{H^2+(d-S)^2}}\right\}\right.$$
$$\left. + j\left\{\frac{0.398}{\sqrt{H^2+(d+S)^2}} - \frac{4.576}{\sqrt{H^2+d^2}} + \frac{4.472}{\sqrt{H^2+(d-S)^2}}\right\}\right]$$

The maximum value occurs at $d = 1.4\,H = 18.2$ metres from line centre and is 8.6 kV/m. This is below the limit of 15 kV/m for safe let-go currents.

(7) Corona Loss

Since an ample margin is given between the corona-inception gradient and the maximum surface voltage gradient on conductors, no corona loss is anticipated at the design stage in fair weather.

(8) Line Compensation Requirements

From the $[P]$ and $[M]$ matrices, the average values of positive-sequence inductance and capacitances are calculated as follows:

$$L_s = 0.2\,P_{ii} = 1.144 \text{ mH/km},$$

$$L_m = 0.2\,(2 \times 0.87 + 0.39)/3 = 0.142 \text{ mH/km}$$

Therefore,

$$L_+ = L_s - L_m = 1.144 - 0.142 \cong 1 \text{ mH/km.} \quad (\mu H/m)$$

$$X_+ = 2\pi f\,L_+ = 0.314 \text{ ohm/km.}$$

$$C_s = 2\pi e_0\,M_{ii} = \frac{10^{-9}}{18}\,(0.1728 + 0.1772 + 0.1728)/3$$

$$= 9.68 \text{ nF/km} \quad (pF/m).$$

$$C_m = 2\pi e_0\,(-0.026 - 0.008 - 0.026)/3 = -1.11 \text{ nF/km}$$

Therefore,

$$C_+ = C_s - C_m = 9.68 + 1.11 = 10.79 \text{ nF/km}$$

$$y = wc_+ = 3.39 \times 10^{-6} \text{ mho/km.}$$

$$r = 0.023 \text{ ohm/km}$$

For L = 200 km

$$Z = 4.6 + j62.8 = 62.97\angle 85°.8 \text{ ohm.} \quad Y = j0.68 \times 10^{-3} \text{ mho}$$

Therefore,

$$Z_0 = (Z/Y)^{1/2} = 304.3 \angle -2°.1,$$

$$\sqrt{ZY} = 0.207 \angle 87°.9 = 0.0076 + j0.2069$$

$$\cosh \sqrt{ZY} = 0.9787 + j0.00156 = 0.9788 \angle 0.09°$$

$$\sinh \sqrt{ZY} = 0.0074 + j0.2055 = 0.2056 \angle 87°.94.$$

For the line, $A = D = 0.9788 \angle 0.09°$, $B = 62.56 \angle 85°.84$,

$$C = 0.6756 \angle 90° \times 10^{-3}.$$

As a first assumption, let the receiving-end voltage $E_r = 420$ kV. Then, centre of the receiving-end circle diagram is at

$$x_c = - \mid E_r^2 A/B \mid \cos (\theta_b - \theta_a) = - 204.5 \text{ MW}$$

$$y_c = - \mid E_r^2 A/B \mid \sin (\theta_b - \theta_a) = - 2752 \text{ MVAR}$$

Radius $R = \mid E_r E_s/B \mid = m \cdot 2820$ MVA, where $m = E_s/E_r$.

The receiving-end circle diagram is drawn for several values of m ranging from 0.8 to 1.0. The load line is drawn at 1000 MW at unity power factor. At no load, the reactive power required is found as Q_{or} which is inductive. At full load, the capacitive reactive power is also determined as Q_{Lr}. The value of m that gives $Q_{or}/Q_{Lr} = 0.7$ will give the required value of E_s when a synchronous condenser is utilized for compensation. The two equations are from Fig. 14.1,

$$(Q_{or} + \mid y_c \mid)^2 + x_c^2 = 2820^2 \, m^2 \tag{14.20}$$

and

$$(1000 + \mid x_c \mid)^2 + (\mid y_c \mid - Q_{Lr})^2 = 2820^2 \, m^2$$

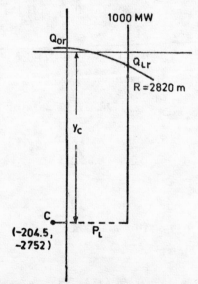

Fig. 14.1. Compensation calculation for Design Example No. 1.

A value of $m = 1.0162$ gives the desired ratio $Q_{or}/Q_{Lr} = 0.7$ with inductive compensation of $Q_{or} = 106.4$ MVAR at no load and capacitive compensation of 151.8 MVAR at full load of 1000 MW at unity power factor. The sending-end voltage will then be $E_s = 1.0162 \times 420 = 426.8$ kV. This is 6.8 kV higher than allowed by Standard Specification. Therefore, the receiving-end voltage should be taken to be less than 420 kV and the compensation requirement re-worked.

A second trial could start at $E_R = 400$ kV.

$$x_c = -185.5 \text{ MW}, \ y_c = -2496.4 \text{ MVAR}, \ R = 2557.5m \text{ MVA}.$$

The equations for Q_{or} and Q_{Lr} are found to be

$$Q_{or} = \sqrt{(2557.5 \ m)^2 - 185.5^2} - 2496.4, \text{ and}$$

$$Q_{Lr} = 2496.4 - \sqrt{(2557.5 \ m)^2 - 1185.5^2}$$

For $m = 1.023$ or $E_s = 409.2$ kV, there result $Q_{or} = 113$ MVAR inductive and $Q_{Lr} = 164$ MVAR capacitive giving a ratio $Q_{or}/Q_{Lr} = 0.69$. This is satisfactory. The synchronous condenser rating will be approximately. 165 MVAR leading for over excitation and 115 MVAR lagging for under-excited operation. The bus voltages are 400 kV at the receiving end and 409.2 kV at the source end. These are within the limits specified by Standards.

The power angle at full-load operation is

$$\delta = \theta_b - \arctan\frac{|y_c| - Q_{Lr}}{|x_c| + 1000} = 85.84 - 63 = 21.84°.$$

The analytical equations for (E_s, I_s) in term of (E_r, I_r) are given below:

$$\begin{bmatrix} E_s \\ I_s \end{bmatrix} = \begin{bmatrix} A, & B \\ C, & D \end{bmatrix} \begin{bmatrix} E_r \\ I_r \end{bmatrix} \tag{14.21}$$

Now, $E_r = 400$ kV $\angle 0°$, line-line.

The total load at the receiving end is

$$W_r = P_r + jQ_r = 1000 - j \ 164 \text{ MVA, 3-phase.}$$

Therefore,

$$I_r = (W_r/E_r)^* = (1000 + j \ 164) \sqrt{3} \times 400$$

$$= 1.443 + j \ 0.237 \text{ Kiloamperes per phase.}$$

$$= 1.462 \ \angle 9.33° \text{ KA.}$$

Therefore,

$$E_s = 0.9788 \ \angle 0°.09 \times \frac{400}{\sqrt{3}} + 62.56 \angle 85°.84 \times 1.462 \angle 9°.33$$

$$= 236.73 \ \angle 22.73°, \text{ kV line-to-ground} = 410 \text{ kV, } 1-1$$

Therefore,

$$\delta = 22.73 \text{ as compared to } 21.84 \text{ from the circle diagram,}$$

$$I_s = j0.6756 \times 10^{-3} \times \frac{400}{\sqrt{3}} + 0.9788\angle0°.09 \times 1.462\angle9.33, \text{ KA}$$

$$= 1.4984\angle28°.33 = 1.319 + j\,0.711 \text{ KA}.$$

The sending-end power is

$$W_s = \sqrt{3}\,E_s\,I_s^* = \sqrt{3} \times 410 \times 1.4984\angle22.73{-}28°.33$$

$$= 1059 - j\,103.84 \text{ MVA}$$

Therefore the line loss is $p = 1059 - 1000 = 59$ MW, 3-phase

The % line loss is $p = 59 \times 100/1059 = 5.57\%$ and the efficiency of transmission is $\eta = 1000 \times 100/1059 = 94.43\%$

14.5 Design Example—II

400 kV, 400 km, 1000 MW with Shunt Compensation

Using $P = 0.5 \times 400^2/400 \times 0.327 = 640$ MW, we find that 2 circuits may be necessary to handle 1000 MW. Design will be based on 500 MW/Circuit.

The high-voltage effects will be the same as the example in Sec. 14.3 for the 200-km line. Only the compensation requirements will be different.

No-Load Compensation

For control of voltage at no load, shunt reactors of 50 MVAR each at both ends will be used. This is a recommended practice in India. But the bus voltages should be checked for other reactive powers. As shown in chapter 12, the total generalized constants are

$$A_T = D_T = A - jB_LB, B_T = B, C_T = C - B_L^2B - j2B_LA \qquad (14.22)$$

where A, B, C, D pertain only to the line.

For 400 km, $Z = 9.2 + j\,124$ ohm, $Y = j\,1.36 \times 10^{-3}$ mho

$$Z_0 = 302.3 \angle- 2°.1, \sqrt{ZY} = 0.411 \angle 88°.$$

Therefore,

$$\cosh\sqrt{ZY} = 0.917 \angle0.36°, \sinh\sqrt{ZY} = 0.4 \angle 88.1°,$$

For the line, $A = D = 0.917\angle0.36°$, $B = 120.77\angle86°$ and

$$C = 1.32 \times 10^{-3} \angle90°.21$$

At 50 MVAR 3-phase and 400 kV, line-to-line, each shunt reactor has an admittance of $B_L = 0.312 \times 10^{-3}$ mho. This give $A_T = D_T = 0.955$.

Selecting a sending-end voltage $E_s = 400$ kV, the load-end voltage will be $E_r = E_s/A_T = 418.85$ kV, which is below the maximum allowable 420 kV.

Now, $B_T = B = 120.77\angle86°$ and $C_T = C - B_L^2B - j\,2B_LA$

$$= j\,0.63 \times 10^{-3} \text{ mho}.$$

Therefore, the percentage compensation afforded by the shunt reactors is $(1.32 - 0.63)100/1.32 = 52.3\%$, which is considered satisfactory. Normal values lie between 40% and 60%.

Receiving-End Power-Circle Diagram

Centre: $x_c = - \mid E_r^2 A_T/B_T \mid \cos (\theta_b - \theta_a) = -96.8$ MW

$\qquad y_c = - \mid E_r^2 A_T/B_T \mid \sin (\theta_b - \theta_a) = -1384$ MVAR

Radius: $R = \mid E_r E_s/B_T \mid = 1387.3$ MVA.

For a load of 500 MW, the compensation required is

$$(500 + 96.8)^2 + (1384 - Q_{Lr})^2 = 1387.3^2$$

giving $Q_{Lr} = 132$ MVAR, capacitive. This will take the form of switched capacitors controlled by an automatic voltage regulator at the receiving end.

The sending-end voltage and current are now calculated. The load at the receiving end is $W_r = 500 - j\,132$ MVA, 3-phase

Therefore, $I_r = \dfrac{500 + j\,132}{\sqrt{3} \times 418.85} = 0.6892 + j\,0.182 = 0.7128\ \angle 14°.8$ KA.

$$\begin{bmatrix} E_s \\ I_s \end{bmatrix} = \begin{bmatrix} 0.955, & 120.77\angle 86° \\ 0.63 \times 10^{-3}\ \angle 90°, & 0.955 \end{bmatrix} \begin{bmatrix} 418.85/\sqrt{3} \\ 0.7128\ \angle 14°.8 \end{bmatrix}$$

$$= \begin{bmatrix} 230.86 & \angle 21°.5 \\ 0.7346 & \angle 26°.37 \end{bmatrix}$$

This gives $E_s = 400$ kV, line-to-line.

Therefore, sending-end power is $W_s = \sqrt{3}\,E_s I_s^* = 507 - j\,43.3$ MVA. This gives a line loss of 7 MW and efficiency $\eta = 500 \times 100/507 = 98.62\%$.

14.6 Design Example—III

400 kV, 800 km, 500 Mw/circuit, 50% series-capacitor compensation, and shunt reactors at both ends.

For 400 km: $Z = 18.4 + j\,248 = 248.6\ \angle 85°.8$,

$\qquad\qquad Y = 2.72 \times 10^{-3}\ \angle 90°,\ Z_0 = 302.3\ \angle -2°.1$,

$\qquad \sqrt{ZY} = 0.822\ \angle 88°,\ \cosh \sqrt{ZY} = 0.6817\ \angle 1.765°$,

$\qquad \sinh \sqrt{ZY} = 0.733\ \angle 88°.5$.

For the line only,

$$A = D = 0.6817\ \angle 1.765°,\ B = 221.5\ \angle 86°.37,$$

$$C = 2.424 \times 10^{-3}\ \angle 90°.6.$$

For the series capacitor, $X_c = 124$ ohms.

The voltages selected are $E_r = 420$ kV and $E_s = 400$ kV.

Therefore, $\qquad |A_T| = |E_s/E_r| = 0.9524.$

Now, from Chapter 12,

$$A_T = A - j B_L B - j(X_c/2Z_0) \sinh \sqrt{ZY} - 0.5\ X_c B_L\ (1 + \cosh \sqrt{ZY})$$

or, $\qquad 0.9524 = (0.83166 + 116.81\ B_L) + j\ (0.021 - 15.33\ B_L).$

This yields $B_L = 1.0333 \times 10^{-3}$ mho and at 400 kV, the shunt reactors have a rating of 165 MVAR at each end. Using this value of B_L gives $A_T = 0.9524\ \angle 0.313°$. Then,

$$B_T = B - j\,0.5\ X_c\ (1 + \cosh \sqrt{ZY}) = 117.8\ \angle 82°.5,$$

and

$$C_T = C - B_L^2\ B - j\,2B_L A - j\,0.5\ (X_c/Z_0^2)\ (\cosh pL - 1)$$

$$- (X_c B_L/Z_0) \sinh \sqrt{ZY} + j\,0.5 X_c\ B_L^2\ (\cosh pL + 1)$$

which yields $C_T = 0.802 \times 10^{-3}\ \angle 89°.7.$

% compensation provided by the shunt reactors is

$$(2.424 - 0.802) \times 100/2.424 = 67\%$$

which is quite a satisfactory figure.

Receiving-end Power-Circle Diagram

Centre: $x_c = -193.5$ MW, $\quad y_c = -1413$ MVAR.

Radius: $R = 1426$ MVA for $E_s = 400$ kV and $E_r = 420$ kV.

The required switched capacitors at load and for 500 MW at unity power factor will be 167 MVAR which is nearly equal to the shunt reactor MVAR provided at load end.

14.7 Design Example—IV

750 kV, 500 km, 2000 MW (with only shunt-reactors)

(Details of calculation are left as an exercise for the reader)

Using an average value of $x = 0.272$ ohm/km, the power/circuit is $P = 0.5 \times 750^2/500 \times 0.272 = 2080$ MW. Therefore, one circuit can handle 2000 MW.

Line clearance and phase spacing from NESC recommendations and switching-surge insulation clearance will be taken as $H_{min} = 12$ m and $S = 15$ m. The sag is expected to 18 m so that the average height for calculations is $H = 12 + 6 = 18$ m.

A 4-conductor bundle is normally used for 750 kV. The conductor size is in the range 1.2″ to 1.4″ diameter ($r = 0.015$ to 0.0175 m). These will be checked for corona-inception voltage and margin between this and the maximum surface voltage gradient. At a roughness factor $m = 0.75$ and $\delta = 1$, Peek's formula gives

(a) $E_{or} = 2140\ (1 + 0.0301/\sqrt{0.015}) \times 0.75 = 2000$ kV/m $= 20$ kV/cm

(b) $E_{or} = 2140\ (1 + 0.0301/\sqrt{0.0175}) \times 0. /5 = 1970\,\mathrm{kV/m} = 19.7\,\mathrm{kV/cm}$

for the two conductor sizes.

By Mangoldt's formulae, the maximum surface-voltage gradient in horizontal configuration on the centre phase turns out to be 19.9 kV/cm. Therefore, there is on margin between this and E_{or}. The conductor size is too small. But at $r = 0.0175$ m, the centre-phase gradient is calculated to be 1740 kV m = 17.4 kV/cm. The margin is $(1970-1740)/19.7$ = 11.7%. For the outer phases, it will be greater than this value. Therefore, the choice of conductor size will fall on $r = 0.0175$ metre. Further design calcuations will be based on this value.

RI Level. The three *RI* levels at 15 metres along ground from the outer phases are

$$RI(1) = 46.5\ \mathrm{dB},\ RI(2) = 43.9\ \mathrm{dB},\ \text{and}\ RI(3) = 36\ \mathrm{dB}.$$

Therefore $RI = \tfrac{1}{2}\,(46.5 + 43.9 + 3)$

$$= 46.7\ \mathrm{dB\ at\ 0.5\ MHz\ and\ 40.7\ dB\ at\ 1\ MHz.}$$

This is considered adequate since there is a dispersion of ± 6 dB in the CIGRE formula. Also, the value of 40 dB at 1 MHz is meant only for preliminary estimates. Other consideration such as S/N ratio will have to be considered in final decision. The width of R-O-W based on a line corridor extending to 15 m from outer phase is $2\ (15 + 15) = 60$ metres.

Audible Noise Level. At 15 m from the outer phase, AN(1) = 48, AN(2) = 49, and AN(3) = 44.3 dB (A). The resulting AN level of line is $AN = 10\ \log_{10}\ (10^{4.8} + 10^{4.9} + 10^{4.43}) = 52.3$ dB (A). For no complaints according to the Perry Criterion of Chapter 5, the limit is 52.5 dB (A).

Electrostatic Field. From a calculation of Maxwell's Potential coefficient Matrix [P] and its inverse [M], and using the results of Chapter 7, the maximum ground-level e.s. field works out to be 10.8 kV/m at $x = 1.2$ H from the line centre. This is below the safe limit of 15 kV/m.

Line Constants. These are, from [P] and [M] matrices,

$$L_+ = L_s - L_m = 0.866\ \mathrm{mH/km},\quad C_+ = 13\ \mathrm{nF/km}$$

The conductor resistance is $r = 0.0136$ ohm/km.

Therefore $z = 0.0136 + j\,0.272$ ohm/km, and $y = j\,3.4 \times 10^{-6}$ mho/km.

Compensation Requirements (*Neglect resistance*)
Several sets of sending-end and receiving-end voltages are selected. At no load, compensation is provided by using shunt reactors at both ends.

Now, $A = D = \cosh \sqrt{ZY}$

$$= \cosh \sqrt{(0.0136 + j\,0.272)\,j\,3.4 \times 10^{-6} \times 500} = 0.866$$

$$B = 129.2 \angle 90°, \text{ and } C = 1.945 \times 10^{-3} \angle 90°$$

Set 1. Try $E_r = E_s = 750$ kV. This gives $B_L = 1.037 \times 10^{-3}$ mho for each reactor and at 750 kV, the MVAR = 585. % compensation comes to be 95%. This is considered too high. A value of 50-70% is aimed at.

Set 2. For 50% shunt-reactor compensation, or about 300 MVAR at each end, for $E_r = 750$ kV, E_s works out to be 700 kV. The voltage can be improved by using some more compensation.

Set 3. For 400 MVAR shunt-reactor capacity at each end, with $E_r = 750$ kV, the sending-end voltage works out to be $E_s = 720$ kV. This is considered satisfactory. By drawing the circle diagram or a geometrical method resulting out of it, for 2000 MW load at unity power factor, capacitive compensation amounting to 500 MVAR in switched capacitors will be required. This completes the design based on steady-state constraints and limits of some examples of 400 kV and 750 kV lines.

14.8 Line Insulation Design Based Upon Transient Overvoltages

The sections that conclude the book will discuss the important topic of selection of long air-gap clearances required between (a) conductor to tower window, (b) conductor to ground, and (c) conductor to conductor to withstand (i) switching surges, (ii) power-frequency voltage, and (iii) lightning. The theory and discussion presented in Chapter 11 will form the basis for air-gap-length selection, while the chapters on Lightning (Chapter 9), Switching Surges (Chapter 10) and Power Frequency (Chapter 12) give indications of magnitudes of overvoltages for which insulation has to be provided. The magnitudes of over-voltages and the probability of their occurrence is an individual characteristic of a system so that no fixed designs can be given in this discussion, but only the guiding principles can be illustrated through examples.

The principles upon which insulation levels are selected are only two: (1) A knowledge of all relevant properties of overvoltages which a system might experience; and (2) A knowledge of insulation characteristics of all types of voltages to which it will be subjected. A design evolved for a given tower-window can only be considered preliminary which will have to be checked and modified suitably in an e.h.v. laboratory by conducting actual tests on a tower mock-up.

14.8.1 Discussion of Rod-Plane Gap Design

The basis for selection of air-gap clearance between any given type of electrode geometry can best be understood by relating it or comparing it with the design of a rod-plane gap, which shows the lowest flashover and withstand voltage of any type of electrode geometry. We illustrate the procedure by

(a) selecting a range of positive switching-surge magnitudes from 1.8 p.u. to 3 p.u. on a 400 kV and a 750 kV system,

(b) then using two representative formulae to calculate the required rod-plane gap length d.

1 P.U. Value of Switching Surge

For 400 kV lines, at the maximum operating voltage of 420 kV, 1 p.u. value of crest line-to-ground voltage is $420\sqrt{2/3} = 343$ kV. At 750 kV, it is $750 \sqrt{2/3} = 612.4$ kV.

Critical Flashover Voltage, Withstand Voltage and Standard Deviation
The assumption that long air-gap breakdown and withstand voltages follow a Gaussian or Normal distribution will be followed. Accordingly, the universally-accepted relation that withstand voltage $=(1-3\sigma)\times 50\%$ flashover voltage will also be considered, with the standard deviation $\sigma = 5\%$ of CFO. As discussed in Chapter 11, this value of σ is very erratic in behaviour, depending upon such factors as waveshape of switching surge, gap length, climatic conditions, etc.

With the above assumptions, as the basis, we will work out the clearances required according to two formulas: (1) Electricité de France (Leroy and Gallet Formula), and (2) Paris' Formula. These are, $V_{50} = 3400/(1 + 8/d)$ and $V_{50} = 500 \cdot d^{0.6}$, with d in metres and V_{50} in kV, crest.

Now, let the switching surge magnitude be K_s p.u. so that for a system voltage V, the magnitude of s.s. is $V_w = K_s \cdot V \sqrt{2/3}$. Then, the 50% flashover value is

$$V_{50} = \frac{V_w}{1 - 3\sigma} = \frac{K_s V \sqrt{2/3}}{1 - 3\sigma} \qquad (14.22)$$

Therefore, according to the above two formulae, the gap length will be

$$\text{(1)} \quad d_1 = 8/(3400/V_{50} - 1) \qquad (14.23)$$

and $\quad$ (2) $\quad d_2 = (V_{50}/500)^{1.667} \qquad (14.24)$

Table 14.1 gives the calculated gap lengths for $V=420$ kV and $V=750$ kV, for $\sigma = 5\% \ V_{50}$.

The values are plotted in Fig. 14.2. In addition to the above formulae, Chapter 11 has listed other formulae to which reference should be made.

For the 400 kV line, an increase in switching surge magnitude from 1.8 to 3 p.u. requires an increase of air gap clearance from 2.1 m to 4.27 m. The increase in s.s. is 66.7% while the air-gap clearance increases by 103.3% according to the Leroy and Gallet Formula. If Paris' Formula is used the increase becomes 133.3%. For the 750 kV line, the increase in air gap is 176% for 66.7% increase in switching surge magnitude.

Table 14.1. Rod-Plane Gap Clearance Based on Switching Surges

P.U. Value K_s	400-kV line Withstand $V_w = 343 K_s$	$V_{50} = V_w/0.85$	d_1	d_2	$V_w = 612.4 K_s$	750-kV line V_{50}	d_1	d_2
1.8	617.4	710	2.1	1.8	1102	1267	4.75	4.71
2.0	686	789	2.42	2.14	1225	1408	5.66	5.62
2.2	754.6	868	2.74	2.51	1347	1549	7	6.59
2.4	823.2	947	3.09	2.9	1470	1690	7.91	7.61
2.6	892	1026	3.455	3.31	1592	1831	9.34	8.7
2.8	960.4	1104	3.85	3.75	1715	1972	11.05	9.85
3.0	1029	1183	4.27	4.2	1837	2113	13.13	11.08

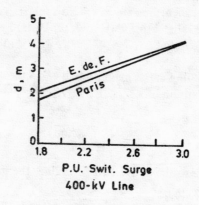

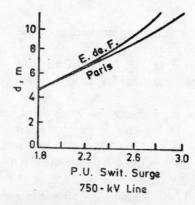

Fig. 14.2. Rod-plane gap clearance for 400 kV and 750 kV systems. Switching surge magnitudes ranging from 1.8 to 3 p.u. (Table 14.1).

14.8.2   Conductor-Tower, Coductor-Ground and Conductor-Conductor Clearances

The other geometries can be handled in the same manner. Paris' Formulae for these cases are used which are as follows:

Conductor-Tower. $V_{50} = 1.3 \times V_{50}$ for rod-plane $= 650 \, d^{0.6}$ (14.25)

For this case, the gap factor is 1.3.

Conductor-Ground.   Having calculated the required conductor-tower clearance d for the anticipated switching-surge magnitudes, the minimum clearance from conductor to ground will be

$$H = 4.3 + 1.4d, \text{ metres} \tag{14.26}$$

Phase-to-Phase Clearance.   This is also described by a 'gap-factor' whose value is 1.8.   Thus,

$$V_{50} = 900 d^{0.6} \tag{14.27}$$

However, in this case, the switching surge is between phases which is not equal to $\sqrt{3} \times$ phase-to-ground magnitude of switching surge. These must be determined by experiments carried out on models or digital computer calculations.

We can now work out the conductor-tower and conductor-ground clearances for the switching surge values ranging from 1.8 p.u. to 3 p.u. for the 400 kV and 750 kV lines.   These are carried out according to the formulae:

Conductor-Tower. $d = (V_{50}/650)^{1.667}, \text{ metres}$ (14.28)

where $V_{50} = V_w/(1 - 3\sigma) = V_w/0.85, \text{ kV}$ (14.29)

and $V_w = $ withstand voltage

$= K_s \times$ (crest line-to-ground voltage of system)   (14.30)

These are plotted in Fig. 14.3.   We observe from the table and the recommended clearance by the N.E.S.C., that a minimum clearance of 9 metres for the 400 kV line can sustain switching-surge magnitudes of over 3 p.u.   The minimum clearance for 750 kV line is 12 metres which can sustain a switching surge magnitude of about 2.6 p.u.

The last clearance is the phase-to-phase gap.   Table 14.3 shows the values for s.s. magnitudes ranging from 2.5 to 3.5 p.u. where 1 p.u. voltage is 343 kV crest for the 400 kV line, and 612.4 kV crest for the 750 kV line.   The required gap length is $d = (V_{50}/900)^{1.667}$, where $V_{50} = V_w/0.85$ and $V_w = K_s \times 1$ p.u. voltage,

These are plotted in Fig. 14.4.

The phase to phase clearance is gaining great importance since the advent of chainette type of construction for e.h.v. lines, Fig. 14.5, where

Table 14.2. Conductor-Tower and Conductor-Ground Clearance

P.U.S.S. K_s	400 kV Line				750 kV Line			
	V_w	V_{50}	$d = (V_{50}/650)^{1.667}$	H	V_w	V_{50}	d	$H = 4.3 + 1.4d$
1.8	617.4	710	1.164	5.924	1102	1267	3.04	8.56
2.0	686	789	1.38	6.232	1225	1408	3.63	9.38
2.2	754.6	868	1.62	6.57	1347	1549	4.25	10.25
2.4	823	947	1.87	6.92	1470	1690	4.92	11.2
2.6	892	1026	2.14	7.3	1592	1831	5.62	12.17
2.8	960	1104	2.42	7.69	1715	1972	6.36	13.2
3.0	1029	1183	2.715	8.1	1837	2113	7.133	14.29

Table 14.3.   Phase-Phase Clearance

P.U.S.S. K_s	2.5	2.7	2.9	3.1	3.3	3.5
400 kV Line						
$V_w = 343 K_s$	857.5	926	995	1063	1132	1200
$V_{50} = V_w/0.85$	986	1065	1144	1223	1302	1381
$d = (V_{50}/900)^{1.667}$	1.164	1.324	1.492	1.67	1.85	2.04
750 kV Line						
$V_w = 612.4 K_s$	1531	1653.5	1776	1898	2021	2143
V_{50}	1761	1902	2042	2183	2324	2465
d	3.06	3.48	3.92	4.38	4.86	5.36

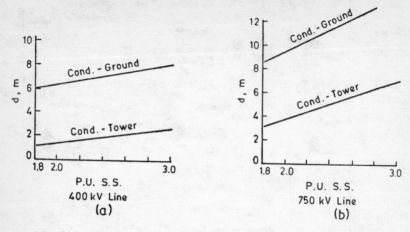

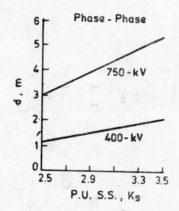

Fig. 14.3.   Conductor-Tower and Conductor-Ground clearances required
for Fig. 14.2 (Table 14.2).

Fig. 14.4.   Conductor-conductor clearance
(Table 14.3).

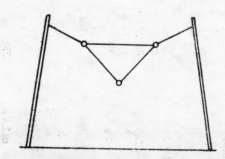

Fig. 14.5.   Chainette construction of e.h.v. and
u.h.v. 3-phase lines with inverted
delta configuration of conductor.

the conductors are supported in an inverted Δ configuration by means of strings of insulators. The tower structure is removed to the sides and except for the outer phases, the line-to-ground switching surge does not determine the insulation clearance between phases.

14.8.3 Air Gap Clearance for Power Frequency and Lightning

The equations for the strength of a long rod-plane air gap for power frequency and lightning are as follows:

Power Frequency: $\quad V_{50} = 652.d^{0.576}$, kV, crest $\qquad$ (14.31)

Lightning: $\qquad V_{50} = 500.d$, kV, crest $\qquad$ (14.32)

We must note that the minimum clearance to tower occurs at maximum swing of the insulator from vertical if an I-string is used whereas with a double − 90° V-string, the swing may not be present at the tower. The swing depends upon wind velocity and in violent storms could be as much as 60°. During thunderstorms, clearances for insulation for lightning can be based on a maximum swing of 30° as compared to power frequency which is on continuously.

Example 14.1: For 400 kV and 750 kV lines, calculate the required conductor-to-tower clearances given the following data: Maximum p.u. value of lightning impulse is 2.8 p.u. for both lines. The standard deviations are 5% for both power frequency and lightning. The gap factor for conductor-tower is 1.3.

(a) *For Power Frequency*

400 kV line. $V_w = 343$ kV, crest. $V_{50} = 343/0.85 = 403.5$ kV.

The 50% flashover voltage is $1.3 \times 652\ d^{0.576}$

Therefore $\qquad d = (403.5/1.3 \times 652)^{1.7361} = 0.276$ metre.

750 kV Line. $V_w = 612.4$ kV, crest. $V_{50} = 612.4/0.85 = 720.5$ kV.

Therefore $\qquad d = (720.5/1.3 \times 652)^{1.7361} = 0.754$ metre.

(b) *For Lightning*

400 kV Line. $V_w = 2.8 \times 343 = 960.4$ kV, crest.

$\qquad V_{50} = 960.4/0.85 = 1130$ kV.

Therefore $\qquad d = (1130/1.3 \times 500) = 1.74$ metres.

750 kV Line. $V_w = 2.8 \times 612.4 = 1714.7$ kV, crest.

Therefore $\quad V_{50} = 1714.7/0.85 = 2017.3$ kV.

$\qquad d = 2017.3/1.3 \times 500 = 3.1$ metres.

Review Questions and Problems

1. In design Examples I–III for the 400 kV line, the margin between corona-inception gradient and maximum surface voltage gradient on the outer conductors at 420 kV is to be maintained at 30%. Select a proper conductor size with roughness factor $m = 0.75$.

2. Then re-work the three examples using this conductor size.

3. Check the Design Example IV for the 750 kV line.

Answers to Problems

Chapter 2

1. 66, 132, 220, 400 kV.

3.

V		400 kV				750 kV		
L, km	200	400	600	800	200	400	600	800
P, MW	1222	611	407	306	5170	2585	1723	1292.5
I, Amp	1767	884	589	442	2984	1992	1328	996

4. *1150 kV–280 km:* $P = 0.5 \times 1200^2/280 \times 0.231$

   $= 11{,}131$ MW/circuit

   800 km: $P = 3896$ MW/circuit without series capacitor

   $= 6500$ MW/circuit with 50% compensation

   750 kV–280 km: $P = 0.5 \times 750^2/280 \times .327 = 3071$ MW/cct.

   3 circuits are required.

   Power loss $= 50\ r/X = 2.5\%$

   Total loss $= 250$ MW

5.

Power, MW	2500	3000	4000	5000	12000
Length, km	250	300	400	300	250 450 1000
Alternatives	1 of$\pm$600 kV	2 of	2 of	3 of$\pm$400 kV	5 of$\pm$600 kV
	2 of$\pm$400 kV	$\pm$400 kV	$\pm$500 kV	2 of$\pm$600 kV	

Chapter 3

1. $R = 0.0617$ ohm

2. (a) $R = 0.0153$ ohm

   (b) 3.75 percent increase

3. Equation; $273 + t = 10^2(771.46 - 12.69t)^{0.25}$; $t = 52°C$

4. (a)
$$[P] = \begin{bmatrix} 4.98, & 0.85, & 0.38 \\ & 4.98, & 0.85 \\ & & 4.98 \end{bmatrix}$$

   (b) $[L]_{ut} = 0.2[P]$ mH/km, $L_s = 0.996$ mH/km,

   $L_m = 0.139$ mH/km

$$[C]_{ut} = \begin{bmatrix} 10.7, & -1.738, & -0.52 \\ & 10.95, & -1.738 \\ & & 10.7 \end{bmatrix} \text{nF/km} \qquad \begin{array}{l} C_s = 10.783 \text{nF/km} \\ C_m = -1.332 \text{ nF/km} \end{array}$$

(c) $L_0 = 1.274$ mH/km, $L_1 = L_2 = 0.857$ mH/km

$C_0 = 8.12$ nF/km, $C_1 = C_2 = 12.115$ nF/km.

(d) $[R_g] 0.75 [D]$ ohm/km, $[L_g] \approx 0.4 [D]$ mH/km

5. (a)

$$[P] = \begin{bmatrix} 5.03, & 0.94, & 0.94 \\ 0.94, & 4.339, & 0.59 \\ 0.94, & 0.59, & 4.339 \end{bmatrix} \begin{array}{l} \text{Top} \\ \text{Bottom} \\ \text{Bottom} \end{array}$$

(b) $[L]_{ut} = 0.2 [P]$ mH/km

$$[C]_{ut} = \begin{bmatrix} 11.9, & -2.27, & -2.27 \\ -2.27, & 13.49, & -1.33 \\ -2.27, & -1.33, & 13.49 \end{bmatrix} \text{nF/km}$$

$L_s = 0.914$ mH/km, $L_m = 0.165$ mH/km.

$C_s = 12.96$ nF/km, $C_m = -1.96$ nF/km.

(c) $L_0 = 1.244$ mH/km, $L_1 = L_2 = 0.7492$ mH/km.

$C_0 = 9.046$ nF/km, $C_1 = C_2 = 14.917$ nF/km.

(d) $[R_g] \approx 1[D]$ ohm/km.

$$[L_g] \approx \begin{bmatrix} 0.14, & 0.1, & 0.1 \\ 0.1, & 0.28, & 0.1 \\ 0.1, & 0.1, & 0.28 \end{bmatrix} \text{mH/km}$$

6. See equation (3.85)

8. (a) 12.157, 10.6, 8.043

(b)

$$[T]^{-1} = \begin{bmatrix} 0.4842, & -0.7587, & 0.4358 \\ 0.7083, & -0.0251, & -0.7055 \\ 0.5286, & 0.6563, & 0.5383 \end{bmatrix} \cdot [T] = [T]_t^{-1}$$

9. *750 kV*: $I_c = 659$ A/ph; *1150 kV*: $I_c = 1244$ A/ph.

Chapter 4

1. (a) $Q_1 = 5.55 \times 10^{-9}$ coul $Q_1' = 0.125 Q_1$, $Q_1'' = 0.031 Q_1$,

$$Q_1''' = 0.0045 Q_1,$$

$S_1 = 0$ $\qquad S_1' = 0.1875$, $S_1'' = 0.1909$,

$$S_1''' = 0.191, \text{ etc.}$$

$$Q_2 = -0.333Q_1, \qquad Q_2' = -0.0476Q_1, \; Q_2'' = -0.0118Q_1,$$
$$Q_2''' = -.00172Q_1,$$

$$S_2 = 0.1667 \text{ m}, \qquad S_2' = 0.1905, \; S_2'' = 0.191, \; S_2''' = 0.191$$

- (b) 311 V/m.
- (c) 964.3 KV.

2. $E_{out} = 16.17$ KV/cm, r.m.s., $E_{cen} = 17.075$ KV/cm, r.m.s.

3. $E_{or} = 18.593$ KV.

   Margins.   outer = 13 per cent; centre = 8.16 per cent.

4. *0,25°*   *1000,15°*   *2000,5°*   *3000,0°*

   $\delta = 0.9832$ 0.917   0.8459   0.7554

   $E_{os} = 25.88$ 24.3   22.6   20.43 KV/cm, r.m.s.

5. $\dfrac{E}{E_{ref}} = \sqrt{\dfrac{\delta}{\delta_{ref}}} \dfrac{0.0301 + \sqrt{r.\delta}}{0.0301 + \sqrt{r.\delta_{ref}}}$, $\delta_{ref} = 0.8862$ at 1000 m and 25°C.

   (a) $\delta = 0.9513 \cdot \dfrac{E_a}{E_{ref}} = 1.036 \dfrac{0.0301 + 0.9753\sqrt{r}}{0.0301 + 0.9414\sqrt{r}}$

   (b) $\delta = 0.8165 \cdot \dfrac{E_b}{E_{ref}} = 0.96 \dfrac{0.0301 + 0.9036\sqrt{r}}{0.0301 + 0.9414\sqrt{r}}$

   (*r* in metre).

6. (a) Gradient factor $= E_r/V = 12.78$ Volt/metre per volt.

   (b) $C = 0.426$ nF;   (c) $Z_0 = 234.7$ ohms.

7. (a) $g = 13.75$ volt/metre per volt.

   (b) $C = 0.834$ nF;   (c) $Z_0 = 120$ ohms.

Chapter 5

4. (a) 105.12×10^6 KW–h,   (b) 11.68×10^6 KW–h,   (c) 11.1 per cent.

6. 51.45 dB (A).

7. 56.2 dB (A).

Chapter 6

2. Ratio = 20/1

3.
f, MHz	0.5	0.75	0.9	1.0	1.1	1.2	1.3	1.4	1.5
Station, dB	50	60	75	70	65	52	80	75	60
$f^{-1.5}$, dB	+9	+3.75	+1.37	0	−1.23	−2.38	−3.4	−4.37	−5.28
Allowed Noise	57	51.75	49.4	48	46.8	45.6	44.6	43.6	42.7
S/N ratio	−7	8.25	25.6	22	18.2	6.4	35.4	31.4	17.3

Stations at 0.5, 0.75, 1.1, 1.2, 1.5 MHz have S/N ratio below 22 dB.

4.

$d/H =$	0	0.1	0.2	0.4	0.6	0.8	1.0	1.5	2.0	2.5	3.0
F_{1a}		2.22	2.21	2.2	2.13	2.04	1.89	1.7	1.14	.72	.48	.33
F_{2a}		0	.12	.24	.45	.62	.72	.72	.51	.3	.18	.11
F_{3a}		.78	.76	.68	.45	.18	.06	.2	.21	.12	.06	.03

5.

F_{1d}	1.8	1.79	1.76	1.65	1.47	1.26	1.05	.65	.42	.285	.2
F_{2d}	0	.11	.2	.34	.4	.38	.33	.19	.11	.065	.04

6. $RI_1 = 50.66$ dB, $RI_2 = 48.9$ dB, $RI_3 = 40.2$ dB

$RI = 51.28$ dB at 0.5 MHz $= 45.28$ at 1 MHz.

With ± 6 dB dispersion, $RI = 51.28$ to 39.28 at 1 MHz.

This is just sufficient if limit is set at 40 dB at 1 MHz.

7. (a) $L_f = 4.532$ mH, $R_f = 14.84$ Kilohms

    (b) 46.5 dB at 1.5 MHz; 36.5 dB at 0.8 MHz.

Chapter 7

3. 47.86 kV

4. 8.565, 14.96, 1.87 kV/m

Chapter 8

1. 1800 kV, 585 kV

2. 994.5 kV

3. (a) light velocity; (b) 300 ohms; (c) 0.2667N, 2.317 dB;

    (d) 1.532 p.u.

4. $K_t = 1.25$, $K_r = + 0.25$

Chapter 9

1. 140 to 188 for 400 km per year.

3. 280 Kiloamps crest.

6. (a) $I_w = 9.09$ KA, crest. (b) $I_a = 12.73$ KA. (c) $R_a = 117.9$ ohms.

Chapter 10

1. $I_{sc} = 7.7$ KA, r.m.s. $= 2.5$ p.u.

5. (a) $\alpha = 15$. (b) $W_0 = 1000$. $f_0 = 159$ Hz. $T = 6.28$ ms.

    (c) 2.82 p.u. (d) 0.91.

6. (a) 1005. (c) 1 p.u.

Chapter 11

1. (1) 6.16 m. (2) 6.1 m. (3) 5.96 m.

5. $V_{50} = 2439$ KV. $V_{83\cdot3} = 2585$ KV.

Chapter 12

2. (a) $A = D = 0.81 \angle 1°.13$,

   $B = 159 \angle 86°.6$, $C = 2.1665 \times 10^{-3} \angle 89°.1$.

   (b) $I_c = CV_R = 938$ Amps/phase

   $Q_o = 987$ MVAR, 3-phase

   (c) $X_c = - 182$ MW, $Y_c = - 2314$ MVAR,

   (d) SIL $= 2076.4$ MW

3. $E_r = 838$ KV

4. 838 KV

6   38.7 Hz $(0.775f)$

Chapter 13

2. $E/V_P = 1.68$, $\alpha = 3.466 \times 10^6$, $\beta = 21.9 \times 10^6$

   Voltage developed across 1 ohm by positive corona pulse.

3. $R_f = 77$ ohms, $R_t = 555$ ohms.

6. (a) 30 KJ, (b) 30 KJ, (c) 30 KJ.

7. (a) $R = 9$ ohms, $L = 62.5$ μH.

   (b) 350 KV.

9. $C_1 = 0.49$ μF. voltage rating ≈ 2.5 KV

Bibliography

Books

1. Alston, L.L.: High Voltage Technology, Oxford University Press, 1968. (Harwell Post-Graduate Series).
2. Aseltine, J.A.: Transform Methods in Linear System Analysis. McGraw-Hill, 1980.
3. Bewley, L.V.: Travelling Waves in Transmission Systems. John Wiley. (1933, 1951). Dover (1963).
4. Bewley, L.V.: Two-Dimensional Fields in Electrical Engineering. (McMillan). Dover 1963.
5. Bowdler, G.W.: Measurements in High Voltage Test Circuits, Oxford Pergamon Press. 1973.
6. Craggs, J.D. and Meek, J.M.: High Voltage Laboratory Technique. Butterworth Scientific Publications. 1954.
7. Craggs, J.D. and Meek, J.M.: Electrical Breakdown of Gases. Oxford Clarendon Press. 1953.
8. Craggs, J.D. and Meek, J.M.: Electrical Breakdown of Gases. (Ed.)., Chichester, John Wiley. 1978.
9. Edison Electric Institute: E.H.V. Transmission Reference Book. (General Electric Co.) 1968.
10. Electrical Power Research Institute, Palo Alto, Calif. Transmission line Reference Book Above 345 kV. 1975.
11. Greenwood, A: Electrical Transients in Power Systems. Wiley-Interscience. 1971.
12. Hayashi, Shigenori: Surges on Transmission Systems. Denki-Shoin, Kyoto, Japan, 1955.
13. Kind, Dieter: An Introduction to High-Voltage Experimental Technique. (Translated by Y. Narayana Rao). Wiley Eastern, 1979.
14. Kuffel, E. and Abdullah, M.: High Voltage Engineering. Oxford, Pergamon Press. 1970.
15. Kuffel, E. and Zaengel, W.S.: High Voltage Engineering. Oxford, Pergamon Press. 1984.
16. Lee, T.H.: Physics and Engineering of High Power Switching Devices. M.I.T. Press, Cambridge, Mass. 1975.
17. Lewis, W.W.: Protection of Transmission Systems Against Lightning. John Wiley and Sons. 1950.
18. Loeb, L.B.: Basic Processes of Gaseous Electronics. U. of Calif. Press, Berkeley, 1960.
19. Loeb, L.B.: Electrical Coronas. U. of Calif. Press, 1965.
20. Naidu, M.S. and Kamaraju, V.: High Voltage Engineering. Tata McGraw-Hill Publishing Co., 1982.
21. Radzevig, D.K.: High Voltage Engineering. (Translated by M. Chaurasia) Khanna Publishers, New Delhi.

22. Rudenberg, R.: Transient Performance of Electric Power Systems. McGraw-Hill Book Co., 1950.
23. Rudenberg, R.: Electric Shock Waves in Power Systems. Harvard U. Press, Cambridge, Mass. 1968.
24. Scott, E.J.: Transform Calculus. Harper and Row, New York, 1955.
25. Stevenson, W.D. Jr.: Elements of Power System Analysis, 4th Ed., McGraw-Hill International Book Co., 1982.
26. Westinghouse Transmission and Distribution Reference Book. IBH-Oxford. 1950.
27. Zahn, M.: Electromagnetic Field Theory—A Problem Solving Approach. John Wiley, 1979.

Monographs and Catalogues

1. High Voltage Test Systems—ASEA, Haefely, Micafil. High Voltage Laboratory Equipment.
2. Int'ntl Symposium on Controlled Reactive Compensation IREQ-EPRI. 1979.
3. Morched, A.S.: Transmission Line Energization Transients. Division of High Voltage Systems. The University of Trondheim, Norway. 1972.
4. I.E.E. Monograph Series 18: Computation of Power System Transients. J.P. Bickford et. al., Peter Pergrinus, London.
5. I.E.E.E. Brown Book. Industrial and Commercial Power System Analysis. IEEE Std.—399;1980.
6. Electrical Manufacturing Co. Calcutta. Catalogue on Line Material and Hardware.
7. UPSEB Statistics at a Glance. Planning Wing, Uttar Pradesh State Electricity Board, Shakti Bhavan, Ashok Marg, Lucknow, U.P., 1984.
8. ALCOA: Conductor Motion Control. Feb. 1976.

Begamudre, Rakosh Das

— Successive Refraction of Travelling Waves on Multi-conductor Transmission Lines. Journal, Institution of Engineers, India, April 1957.
— Investigation of Corona Loss and RI from H.V. Overhead DC Transmission Lines. With R.M. Morris. Transactions I.E.E.E., Jan., 1964, pp. 10-17.
— Corona Pulses and RI under H.V. D.C., ibid, May 1964, pp. 483-91.
— Radio Noise Levels of D.C. Transmission Lines, Bulletin of the Radio and EE Div., National Research Council of Canada, Ottawa, Sept. 1964.
— Corona Pulses and RN Levels from a Water Drop under Several Types of Excitation. Letter to Editor, Proceedings I.E.E.E., 1967, p-2164.

— RI Levels of Transmission Lines and Corona Pulse Characteristics. Trans. Canadian Eng. Inst. Oct., 1970, p. 1-7.

— The Absolute Method of Pre-determination of RN Levels of EHV AC and DC Transmission Lines. Symposium of Selection of Next EHV Transmission Voltage for India. Central Board of Irrigation and Power, New Delhi, Publication No. 142, May 1980.

— Voltage Gradient Calculation of 345 kV to 1500 kV Transmission Line Conductors. Ibid.

— Next Higher Transmission Voltage for India. C.B.I.P. Publication No. 162, Dec. 1980.

— Switching-Surge Testing of Transformers. Indian Transformer Manufacturers' Association Symposium. December 1980, New Delhi.

— Design of Coils for Representing Long Lines on a TNA. Part I: Theoretical Considerations, Part II: Practical Considerations. C.B.I.P. 49th Research Meeting Proceedings September 1981.

— Effect of Line Coil Representation in a TNA on Switching Transients of Transmission Lines. (With S. Subhash, K.S. Meera and M. Kanya Kumari). C.B.I.P. Publication No. 170, 51st Research Meeting.

— Impulse Generators for Lightning and Switching Surges. Design and Performance. Technical Report No. 106, Central Power Research Institute, Bangalore.

— Design of Cages for Corona Studies. Technical Report No. 109, C.P.R.I. Bangalore.

— Limits for Interference Fields from EHV and UHV Power Transmission Lines: RI, AN and E.S. Fields. C.B.I.P. Publication No. 170.

— Formulas for Calculation of Excitation Function of Corona from RN fields of Long EHV Lines. Conference on EHV Technology, IEEE-IISc-IE(I), Bangalore, August 1984.

— Effect of conductor Size on EHV Transmission-Line Problems for 750 kV, 1000 kV, and 1150 kV, Proceedings, 2nd National Power Systems Planning Conference, Hyderabad, India. September 1983. (with S. Ganga).

— Procedures for EHV Line Designs Based upon Steady-State Operating Limits. C.B.I.P. Publication No. 177, Aurangabad, 52nd Research Meeting, 1985.

— Procedures for EHV Line Insulation Designs Based Upon Transient Overvoltages. Ibid.

— A Method for Solving Surge Phenomena on EHV Lines Including Attenuation Caused by Corona Loss and Resistance. Ibid.

— Energy and Electrical Power Equipment Development. State of the Art Speech. Ist National Power Systems Planning Conference, Annamalai University, India, September 1982.

— Energy Development and Human Progress. Keynote address, Institution of Engineers (India), Kanpur. 15 September 1983.

Brown Boveri Review

1. J. Glavitsch: Power Frequency Overvoltages in EHV Systems. 1964, pp. 21–32.
2. W. Frey: Considerations Governing Choice of Voltage Level for Long-Distance High-Capacity Transmission Systems. 1964, pp. 5–9.
3. P. Althammer and R. Petitpierre: Switching Operations and Switching Surges in Systems Employing E.H.V. 1964, pp. 33–46.
4. A Goldstein: Coordination of Insulation and Choice of Lightning Arresters, 1964, pp. 47–55.
5. E. Ruoss, and P. Djurdjevic: Network Analyzer for Studying Transient Phenomena in H.V. Networks. 1968, pp. 734–739.
6. A. Braun, A. Eidinger and E. Ruoss: Interruption of Short-Circuit Currents in H.V.A.C. Networks, 1979, pp. 240–254.
7. J. Kopainsky and E. Ruoss: Interruption of Low Inductive and Capacitive Currents in H.V. Systems, 1979, pp. 255–261.
8. E. Ruoss: Overvoltages on Energizing HV Lines. 1979, pp. 262–270.
9. U. Burger: Insulation Coordination and Selection of Surge Arrester. 1979, pp. 271–280.

Emile Haefely Publications

1. K. Feser: Transient Behaviour of Damped Capacitive Dividers of Some MV. IEEE Transactions, PAS-93, 1974, pp. 116–121. (Reprint by E. Haefely et cie).
2. K. Feser: A New Type of Voltage Divider for the Measurement of High Impulse and A.C. Voltages. (Translated from Bulletin ASE, Nr. 19, pp. 929–935.
3. K. Feser: Problems Related to Switching Impulse Generation of HV in Test Plants. Bulletin SAE, Vol. 65, 1974, pp. 496–508.
4. A. Rodewald, Transient Phenomena in the Marx Multiplying Circuit After Firing the First Coupling Spark Gap. Bull. ASE, Vol. 60, 1969, pp. 37–44.
5. K. Feser, Influence of Corona Discharges on the Breakdown Voltage of Airgaps-Proc. IEE, Vol. 118, No. 9, Sept. 1971, pp. 1309–1313.
6. K. Feser: Mechanism to Explain the Switching Impulse Phenomena. Schweiz Tech. Zeit., 1971, pp. 937–946.
7. P. Mathiessen: La Production de Hautes Tensions d' Essais. 'Le Monteur-Electricien,' No. 4, 1972, pp. 11–22.
8. W. Müller, A. Rodewald, and H. Steinbigler: Stoβstrom labor.
9. G. Reinhold: U.H.V. D.C. Power Supplies for Large Currents.
10. Th. Praehauser: Measurement of Partial Discharges in H.V. Apparatus with Balanced Circuit, Bull. ASE, 1973, pp. 1183–1189.
11. Th. Praehauser: Locating Partial Discharges in H.V. Equipment. Bull. ASE, 1972, pp. 893–905.

I.E.E.E. Transactions—Power Apparatus and Systems

1. C.F. Wagner and A.R. Hileman: The Lightning Stroke-II. 1961, pp. 622–642.

2. C.F. Wagner and A.R. Hileman: Mechanism of Breakdown of Laboratory Gaps. 1961, pp. 604-622.

3. E.C. Sakshaug, J.S. Kresge, and S.A. Miske, Jr.: A New Concept in Station Arrester Design. 1977, pp. 647-656.

4. E.C. Sakshaug: Current Limiting Gap Arresters—Some Fundamental Considerations. 1971, pp. 1563-1573.

5. J.R. Hamann, S.A. Miske, Jr. I.B. Johnson, and A.R. Courts. A Zinc Oxide Varistor Protective System for Series Capacitors. 1981, pp. 929-937.

6. M. Kobayashi, et. al.: Development of ZnO Non-Linear Resistors and Their Application to Gapless Surge Arresters, 1978, pp. 1149-1158.

7. IEEE Working Group on Insulation Switching Surges: Guide for Application of Insulators to withstand Switching Surges. 1974, pp. 58-67.

8. P. Sarma Maruvada, Helene Menemenlis and R. Malewski: Corona Characteristics of Conductor Bundles under Impulse Voltages. 1977, pp. 102-115.

9. J.R. Stewart and D.D. Wilson: High Phase Order Transmission—A Feasibility Analysis. Part I—Steady State Considerations. Part-II —Overvoltages and Insulation Requirements. 1978, pp. 2300-2317.

10. R. Malewski, D. Train, and A. Dechamplain: Cavity Resonance Effect in Large HV Laboratories Equipped with E.M. Shield. 1977, pp. 1863-1871.

11. R. Malewski and A. Dechamplain: Digital Impulse Recorder for HV Laboratories. 1980, pp. 636-649.

12. D.C. Erickson: The Use of Fibre Optics for Communications, Measurement and Control Within HV Substations. 1980, pp. 1057-1065.

13. S.E. Klersztyn: Numerical Correction of HV Impulse Deformed by the Measuring System. 1980, pp. 1980-1995.

14. S. Goldberg and W.R. Schmus: SSR and Torsional Stresses in T–G Shafts. 1979, pp. 1233-237.

15. R.G. Farmer, A.L. Schwalb and E. Katz: Navajo Project Report on SSR Analysis and Solutions. 1977, pp. 1226-1232.

16. IEEE SSR Working Group: Countermeasures to SSR Problems. 1980, pp. 1810-1818.

17. IEEE SSR Task Force: First Benchmark Model for Computer Simulation of SSR. 1977, pp. 1565-1572.

18. IEEE RN Subcommittee—AN Task Force: A Guide for the Measurement of Audible Noise from Transmission Lines. 1972, pp. 853-856.

19. D.E. Perry: An Analysis of AN Levels Based upon Field and 3-Phase Test Line Measurements. 1972, p. 857-865.

20. H. Kirkhan and W.J. Gajda, Jr.: A Mathematical Model of Transmission Line AN, 1983, pp. 710-728.

21. V.L. Chartier and R.D. Stearns: Formulas for Predicting AN from Overhead HV AC and DC Lines. 1981, pp. 121–128.

22. IEEE Task Force on Corona: A Comparison of Methods for Calculating Audible Noise of HV Transmission Lines. 1982, pp. 4090–4099.

23. N. Kalcio, J. Diplacido and F.M. Dietrich: Apple Grove 750 kV Project—2-Year Statistical Analysis of AN. 1977, pp. 560–570.

24. C.H. Gary: The Theory of the Excitation Function, 1972, pp. 305–310.

25. M.R. Moreau and C.H. Gary: Pre-determination of RI Level of HV Transmission Lines. I—Predetermination of the Excitation Function. II—Field Calculating Methods. 1972, pp. 284–304.

26. IEEE RN Subcommittee: A Field Comparison of RI and TVI Instrumentation, 1977, pp. 863–875.

27. N. Giao Trinh and P. Sarma Maruvada: A Method of Predicting the Corona Performance of Conductor Bundles Based upon Cage Test Results. 1977, 312–325.

28. W.C. Guyker, J.E. O'Neil and A.R. Hileman: Right of Way and Conductor Selection for the Alleghany Power System 500-kV Transmission System. 1966, 624–632.

29. P. Sarma Maruvada and N. Giao Trinh: A Basis for Setting Limits to RI from HV Transmission Lines. 1975, pp. 1714–1724.

30. IEEE RN Noise and Corona Subcommittee: Review of Technical Considerations on Limits to Interference from Power Lines and Stations. 1980, pp. 365–384.

31. B. Bozoki: Effect of Noise on Transfer-Trip Carrier Relaying. 1968, 173–179.

32. W.R. Lauber: Amplitude Probability Distribution Measurements at the Apple Grove 775 kV Project. 1976, 1254–1266.

33. CIGRE/IEEE Survey on EHV Transmission Line Radio Noise, and Comparison of RN Prediction Methods with CIGRE/IEEE Survey Results. 1972, pp. 1019–1042.

34. G.W. Juette and L.E. Zaffanella: RN Currents and AN on Short Sections of U.H.V. Bundled Conductors. 1970, pp. 902–913.

35. G.W. Juette and L.E. Zaffanella: RN, AN, and Corona Loss of EHV and UHV Lines under Rain: Predetermination based on Cage Tests. ibid, pp. 1168–78.

36. G.E. Adams: Voltage Gradients on HV Transmission Lines. 1955 (A.I.E.E. Trans., PAS), pp. 5–11.

37. A.S. Timascheff: Fast Calculation of Gradients of a 3-phase Bundle-conductor Line with any Number of Sub-conductors. 1971, pp. 157–164. Also, 1975, pp. 104–107, and 1961 (AIEE Trans.) pp. 590–597.

38. S.P. Maruvada and W. Janischewskyj: Electrostatic Field Calculation of Parallel Cylindrical Conductors. 1969, pp. 1069–1079.

39. M.S. Abou-Saeda and E. Nasser: Digital Computer Calculation of

Potential and its Gradient of a Twin Cylindrical Conductor. 1969. pp. 1802–1814.

40. R.M. Morris and P.S. Maruvada: Conductor Surface Voltage Gradients on Bipolar HV DC Transmission Lines. 1976, pp. 1934–1945.

41. C.H. Gary: Corona Inception Field Calculation: Generalization of Peek's Law. 1972. pp. 2262–2263.

42. IEEE Corona and RN Working Group: A Survey of Methods for calculating Voltage Gradients. 1979. pp. 1966–2014.

43. G. Gallet and G. Leroy: IEEE Conference Paper No. C–73–408–2, 1973.

44. L. Paris: Influence of Airgap Characteristics on Line-to-ground Switching-Surge Strengths. 1967, pp. 936–947.

45. H.S.H. Goff, D.G. McFarlane, and F.J. Turner: Switching Surge Tests on Peace River Transmission Line Insulation. 1966, pp. 601–613.

46. J.K. Dillard and A.R. Hileman: U.H.V. Transmission Tower Insulation Tests. 1970. pp. 1772–1784.

47. IEEE Working Group on Insulator Switching Surges: Guide for Application of Insulators to withstand Switching Surges. 1975. pp. 58–67.

48. Y. Watanabe: Switching Surge Flashover Characteristics of Extremely Long Air Gaps. 1967, pp. 933–936.

49. F.S. Young, E.M. Schneider, Y.M. Gutman, and N.N. Tikhodeyev: USA-USSR Investigation of 1200-kV Tower Insulation. 1980. pp. 462–470.

50. J. Sabath, H.M. Smith and R.C. Johnson: Analog Computer Study of Switching Surge Transients for a 500-kV System. 1966. pp. 1–9.

51. D.D. Wilson: Phase-Phase and Phase-Neutral Switching Surges on 500 kV Open-Ended Lines. 1969. pp. 660–665.

52. S. Annestrand, E.F. Bossuyt and N. Dag Reppen: Insulation Performance Analysis of a 500-kV Transmission Line Design. 1970. pp. 429–437.

53. IEEE Working Group on Switching Surges, Part IV: Control and Reduction on AC Transmission Lines. 1982, pp. 2694–2702.

54. R. Cortina, M. Sforzini and A. Taschini: Strength Caracteristics of Air Gaps Subjected to Interphase Switching Surges. 1970. pp. 448–452.

55. A.R. Hileman, P.R. Leblane and G.W. Brown: Estimating the Switching Surge Performance of Transmission Lines. 1970. pp. 1455–1469.

56. T. Udo: Sparkover Characteristics of Long Air Gaps and Long Insulator Strings. 1964. pp. 471–483.

57. C.L. Wagner and J.W. Bankoske: Evaluation of Surge-Suppression Resistors in HV Circuit Breakers. 1967. pp. 698–707.

58. IEEE Substation Committee Working Groupe 78.1; Safe Substation Grounding—Part II. 1982 pp. 4006–4023.

59. IEEE Committee Report: Report on Industry Survey of Protective Gap Applications in HV Systems. 1967. pp. 1432–1437.

60. N. Giao Trinh and Claire Vincent: Statistical Significance of Test Methods for Low Probability Breakdown and Withstand Voltages. 1980 pp. 711–719.

61. J.G. Kassakian and D.M. Otten: On the Suitability of a Gaussian Stress Distribution for a Statistical Approach to Line Insulation Design. 1975. pp. 1624–1628.

62. L.O. Barthold, I.B. Johnson and A.J. Schultz: Switching Surges and Arrester Performance on HV Stations. AIEE Trans., 1956. pp. 481–491.

63. E.W. Kimbark: Bibliography on Single-Pole Switching. 1975 pp. 1072–1077.

64. IEEE Working Group on E.S. Effects: Electrostatic Effects of Overhead Transmission Lines. Part I: Hazards and Effects. Part II: Methods of Calculation. 1972. pp. 422–444.

65. J.W. Kalb: How Switching Surge Affects Line Insulation. 1963. pp. 1024–1033.

Other Technical and Scientific Journals

1. J.R. Carson: Wave Propagation in Overhead Wires with Ground Return. Bell System Technical Journal. Vol. 5, October 1926, pp. 539–554.

2. G.D. Friedlander: UHV—Onward and Upward. IEEE Spectrum. Feb. 1977, pp. 56–65

3. R.L. Retallack; The Chainette Structure Concept. American Power Conference Proceedings, Chicago. 1982. pp. 660–664.

4. J.E. Dulius and R.Y. Shuping: Design of 500 kV Transmission Line Cross-Rope Suspension Structures. Same. pp. 665–670.

5. S.H. Sarkinen and D.A. Bradley: Impulse Tests on UHV Airgap Configurations. Same. pp. 678–682.

6. S.Y. King: The Electric Field Near Bundled Conductors and Their Electrostatic Properties. Proceedings I.E.E. (London), Part C: Monograph 338S, June 1959, pp. 200–206. Also, 1963. pp. 1044–1050.

7. H. Parekh, M.S. Selim, and E. Nasser: Computation of Electric Field and Potential Around Stranded Conductor. Same. 1975. pp. 547–550.

8. G. Quilico: The Electric Field of Twin Conductors. CIGRE Proceedings 1956, Paper No. 214.

9. N.N. Tichodeev: On the calculation of Initial Voltages of the Total Corona on DC Transmission Lines. Electrichestvo. No. 10, 1957.

10. G. Veena: Voltage Gradient Calculation—Selection of Conductor Size and Configuration for EHV Transmission Lines. Technical Report No. 91, Central Power Research Institute, Bangalore, India. 1979.

11. Sujatha Subhash: Computation of Radio Noise Levels of Existing and Proposed EHV Lines in India. Same. Tech. Rep. No. 101, 1980.

12. S. Subhash, K.S. Meera and K.S. Jyothi: Digital Computation and Model TNA Investigations of Switching Surges of 3-Phase Transmission Systems. C.B.I.P. Publication No. 162, pp. 63–71.

13. S. Subhash, K.S. Meera and M. Kanya Kumari: Switching Overvoltage Study on Transmission Lines with Complex Terminations by TNA Models and Digital Calculation. Same. pp. 73–81.

14. A. Amruthakala: EHV and UHV Transmission Line Design Using Computer Graphics. Same. pp. 95–100.

15. S. Ganga: Surface Voltage Distribution on Bundled Conductors. C.B.I.P. 52nd Research Meeting, Aurangabad, February 1985. (Also, C.P.R.I. Technical Report No. 118).

16. The Swedish 800 kV System, CIGRE Proceedings, 1974. Paper No. 31–11. Also, 1978, Paper No. 31–03.

17. A Study of the Design Parameters of Transmission Lines Above 1000 kV. Same. 1972. Paper No. 31–15.

18. Bonneville Power Administration's 1200 kV Transmission Line Project. Same. 1978. Paper 31–09.

19. Design of EHV 1150 kV AC Transmission Line of the USSR. Same. 1976. Paper No. 31–03.

20. Switching of Transformers, Reactors and Long Transmission Lines, Field Tests in German 400 kV Networks. Same 1980. Paper No. 13–08.

21. British Investigations on the Switching of Long EHV Transmission Lines. Same. 1970. Paper No. 13–02.

22. B.J. Trager, et. al.: Control of Switching-Surge Overvoltages: Metal Oxide Surge Arresters versus Circuit Breaker Resistors. Proceedings, American Power Conference. 1982. pp. 769–775.

23. D.T. Poznaniak, E.R. Taylor, Jr., and R.T. Byerly: Static Reactive Compensation for Power Transmission Systems. Same. pp. 671–677.

24. C.B. Rawlins: Power Imparted by Wind to a Model of a Vibrating Conductor. ALCOA, Massena, N.Y. Report No. 93-82-1, 1982 April.

25. P.R. Emtage: The Physics of Zinc Oxide Varistors. Journal of Applied Physics. Vol. 48, No. 10, Oct. 1977., pp. 4372–4384.

26. L.M. Levinson and H.R. Philipp: The Physics of Metal Oxide Varistors. Ibid, March 1975, pp. 1332–1341.

27. M. Matsuoka: Nonohmic Properties of Zinc Oxide Ceramics, Japanese Journal of Applied Physics. Vol. 10, No. 6, June 1971. pp. 736-746.

Index

ERRATA

Page	Line	In place of	Read
3	2	1200)	12,000)
26	3	$=\dfrac{\pi}{4}$ sq. cm.	delete
27	25	$I = P/\sqrt{3V}$	$I = P/\sqrt{3}\,V$
67	14	$[U]\tfrac{1}{2}/g^2$	$[U]/g^2$
68	8	TV, F)	TVF)
82	6	$\theta_m = \arctan\left[\sqrt{3}(M_{13}-M_{12})/\right.$ $(2M_{11}-M_{12}-M_{13})]$	$\theta_m = \arctan\left[\sqrt{3}(M_{13}-M_{12})\right.$ $(2M_{11}-M_{12}-M_{13})]^{-1}$
107	20	$=\dfrac{q}{2\pi e_0}\ln\dfrac{(X-S)^2+y^2}{(X+S)^2+y^2}$	$=\dfrac{q}{4\pi e_0}\ln\dfrac{(X-S)^2+y^2}{(X+S)^2+y^2}$
142	3rd line from bottom	4.365 lines	4.365 times
146	Fig. 6.1	$i = k - i_p t^{-1.5},\, e^{(-r/t-8t)}$	$i_- = k_- i_p \cdot t^{-1.5} \cdot e^{(-r/t-8t)}$
147	25	strength a	strength as
151	Eq. 6.13	$\dfrac{4A}{2}\dfrac{f}{f_0}$	$\dfrac{4A}{2\pi}\dfrac{\Delta f}{f_0}$
159	Eq. 6.18	$\dfrac{1}{1(d+S)^2/H^2}$	$\dfrac{1}{1+(d+S)^2/H^2}$
170	Eq. 6.51	$\dfrac{1}{3v\sqrt{2a}}$	$\dfrac{I}{3v\sqrt{2a}}$
171	20	$m\sqrt{}$	$\sqrt{m}$
177	22	and 5.222 kilohms	and $R_f = 5.222$ kilohms
183	Eq. 7.12	$(q_i/2\pi e_0$	$(q_i/2\pi e_0)$
	Eq. 7.14	$\sum\limits_{i=1}^{n} E_{nl}$	$\sum\limits_{i=1}^{n} E_{hl}$
184	14	$J_{h1}\,J_{h1}$	$J_{h1}\,J_{h2}$
192	Last line	$c = e_0 A/D$	$c = e_0 A/d$
210	Eq. 8.39	$\left.\begin{array}{c} s=jWn \\ \\ pL=j(2n+1)\,\pi/2 \end{array}\right|$	$\left.\begin{array}{c} s=jwn \\ pL=j(2n+1)\,\pi/2 \end{array}\right.$
213	15	$s = \alpha,\,-\beta,$ and	$s = -\alpha,\,-\beta,$ and
216	Eq. 8.71	(l/z)	$(1/z)$
217	2nd line from bottom	Lighting	Lightning
218	Eq. 8.81	$\displaystyle\int_0^\infty (e^{-\alpha t} - e^{-\beta t})^2$	$\displaystyle\int_0^\infty (e^{-\alpha t} - e^{-\beta t})^2 \cdot dt$
	8	lighting	lightning
	19	μv/km	μ mho/km
	20	μv/km	μ mho/km
220	Eq. 8.97	$\cosh(m+jn)=\cos m\cdot\cos n$ $\cosh m\cdot\cosh n$	$\cosh(m+jn)=\cosh m\cdot\cos n$ $\cosh m\cdot\cos n$

Page	Line	In place of	Read
231	12	in	is
236	vertical driven rod	$\ln \dfrac{4L}{a} - 1$	$\ln \dfrac{4l}{a} - 1$
247	2	10000	100,000
270	23	Time for l travel	Time for 1 travel
276	21	$L_s \quad L_m$	$l_s \quad l_m$
	22	L_g	l_g
280	7	$\measuredangle [\cos (wt + \phi)]$	$L[\cos (wt + \phi)]$
283	Eq. 10.88	$\dfrac{R_s}{z_0} \dfrac{1}{s}$	$\dfrac{R_s}{z_0} + \dfrac{L_s s}{z_0}$
291	2nd line from bottom	problem 4	problem 5
292	Eq. 11.20	$3.775 \, pd/T$	$3.775 \sqrt{pd/T}$
311	3	can pe shown	can be shown
312	21	Fur	For
313	35	air gps	air gaps
314	2	relation	relative
	17	$l/v_0 c$	$1/v_0 c$
326	26	$A \cosh pL$	$A = \cosh pL$
332	Eq. 12.53	$\cot 2\pi L/\lambda$.	$\cot 2\pi L/\lambda$.
333	Eq. 12.56	$(A - E_s/E_r) \, jB$	$(A - E_s/E_r)/jB$
335	2nd line from bottom	$2B_{sh}^2 B$	$2B_{sh}^2 B^2$
362	12	ganerated	generated
363	20	peack	peak
364	5	3000	5000
367	15	aud	and
399	3	2440	2400